Electronic Circuit Analysis

PASSIVE NETWORKS

McGraw-Hill
Book Company, Inc.
NEW YORK TORONTO LONDON

Electronic Circuit Analysis

PASSIVE NETWORKS

Phillip Cutler

Autonetics Division, North American Aviation, and Orange Coast College, Costa Mesa, California

With the assistance of

Hardy Hoover, Ph.D.

Atomics International Division, North American Aviation; Pierce College and University of California at Los Angeles

ELECTRONIC CIRCUIT ANALYSIS
Passive Networks

Library of Congress Catalog Card Number 60-6967

14985

IV

To my dear friend and instructor
Charles G. Chamberlain

PREFACE

This text is the first of several volumes which bridge the gap between engineering and technician-level electronics. The need for this type of intermediate-level training has been felt for years within the industry.

Although an increasing number of intermediate courses are being offered in schools and industrial training programs, there is no suitable text available. Most texts in basic electronics explore the subject on what might be termed a familiarization level; and this they do well. However, technicians with this level of training generally require close supervision and lack flexibility when shifted to new projects. Further training is clearly indicated. Specifically, it should be directed towards developing a real insight into the basic principles of circuit analysis, since the problems of analysis, design, and maintenance are intimately involved. A further justification for training in this area is that a solid foundation in circuit analysis is necessary for further training in vacuum tube, semiconductor, and other types of circuitry.

Essentially this volume is a slightly expanded version of a 90-hour course taught in collegiate classes. The material has also been tested over a period of years within the industry as part of a highly successful industrial training program. It has been most gratifying to see many former students promoted to engineering and supervisory positions. Also, I have been pleased to note the relatively large percentage of engineers who attend these courses in order to review and correlate their diversified backgrounds with the problems of practical application.

This text has been written in lecture style and with extensive detail in order to be self-teaching. This feature plus numerous solved and unsolved problems should considerably lessen the instructor's work load. In order to keep the student motivated, the theoretical concepts are continuously interwoven with the problems of practical application. Whenever pertinent, the mathematical derivations are done in great detail so that the techniques of algebraic manipulation are evident.

The scope of this volume is limited to passive networks, although Chap. 5, in some aspects, penetrates the realm of active networks. Future volumes will consider active networks in detail and expand upon the analysis of passive networks.

Specifically, Chap. 1 introduces the concept of notation for voltages and currents. Network theorems and numerous practical illustrations are investigated. A clear insight into the loop and nodal methods of analysis is developed and a brief review of determinants is presented for the reader unfamiliar with this powerful method of solving simultaneous equations.

Chapter 2 reviews a-c fundamentals and applies the analytic tools of Chap. 1 to a-c circuits. A pronounced effort is made to develop the reader's ability to manipulate and interpret equations involving complex numbers.

Chapter 3 is a detailed discussion of concepts pertinent to transient analysis. The reader is shown how to analyze numerous and complicated *RC*, *RL*, and *RLC* circuits while, at all times, maintaining a physical insight into their behavior. Practical applications are stressed throughout.

The analysis of inductively coupled circuits is developed in Chap. 4. Various transformer equivalent circuits and their interpretation and transient response are investigated.

Chapter 5 introduces the reader to the concepts of graphic analysis applicable to nonlinear elements such as vacuum tubes and transistors. Graphic solutions, obtained by load-line techniques, are carefully explained rather than treated as mere mechanical drawings.

To master this text the reader requires a working knowledge of basic electronics and high-school algebra. While calculus is not a prerequisite, it can easily be applied to the text material if the reader possesses this knowledge.

I wish to thank my students and colleagues for their many helpful suggestions and criticisms. In particular I am grateful to Ralph Bahnsen of International Business Machines for many illuminating discussions, Dr. Hardy Hoover for editing this manuscript, and Mrs. Bertha K. Rapaport, my attorney, for organizational and administrative help.

Phillip Cutler

CONTENTS

Chapter 1

NETWORK ANALYSIS AND THEOREMS

IN THIS CHAPTER we shall see how Ohm's law and some useful network theorems help us to understand complex electronic circuits.

Consider, for example, Fig. 1-1. Predicting what a voltmeter should read when connected between terminal A and ground might prove difficult to many of us with our present assortment of analytical tools. But soon, at the end of the section pertaining to Thévenin's theorem, we shall have mastered the principles needed to make circuit problems of this nature quite simple. This exact circuit will be analyzed in that section.

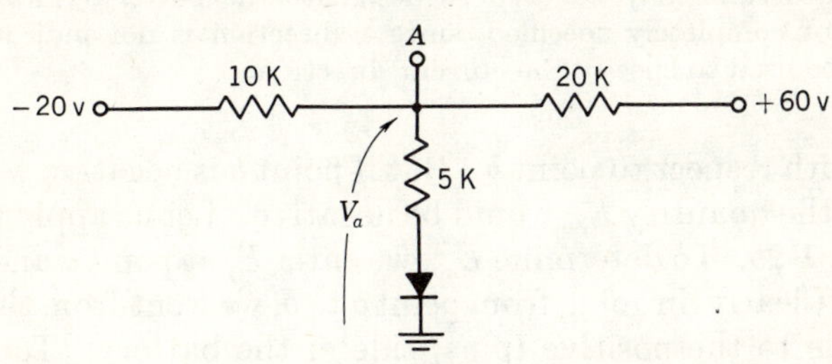

FIG. 1-1. A diode circuit we shall soon learn to analyze.

1-1. Notation. The very first thing we must do is to agree upon some method of designating the voltages and currents with which we shall be dealing. Consider Fig. 1-2a. How would one describe the voltage of the indicated battery? To say that E_1 is a 10-volt battery is not sufficient. It tells us nothing about which terminal is positive or negative. Neither would we know how to connect a voltmeter to the battery if it were enclosed in a box with just the two terminals sticking out. Without any other information, the voltmeter could just as readily have been connected to read backwards, or downscale, as upscale.

To make the description of a voltage complete, let us add two subscripts, say a and b, as shown in Fig. 1-2b. Let E_{ab} mean the voltage at point b with respect to point a. Conversely, E_{ba} would mean the voltage at point a with respect to point b. The first subscript may be considered as the reference point to which the voltage at the second subscript is compared. Furthermore, let us agree that E_{ab} is a positive quantity if point b is positive with respect to point a or, in other words, if point a is

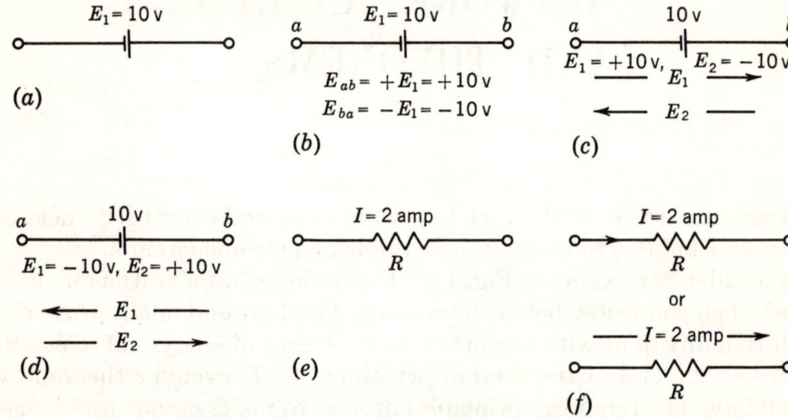

FIG. 1-2. (a) The terminals should be labeled to specify E_1 completely. (b) An example of the double subscript notation we have adopted. (c and d) Arrows or double subscripts may be used to designate the battery voltage. (e) The current is not completely specified, since a direction is not indicated. (f) An arrow may be used to specify the current direction.

negative with respect to point b. But if point b is negative with respect to point a, the quantity E_{ab} would be negative. Let us apply this definition to Fig. 1-2b. To determine E_{ab}, we enter E_1 at point a and come out at point b. Clearly, in going from point a to b, we went from the negative (minus) side to the positive (plus) side of the battery. Terminal b is therefore positive with respect to terminal a, which by our definition means that E_{ab} is a positive quantity, or a plus E_1. Conversely, E_{ba} would be $-E_1$, since in going from b to a we emerge at point a, which is negative with respect to point b. Plainly, too, $E_{ab} = -E_{ba}$, since $E_{ba} = -10$ volts and $-(-10 \text{ volts}) = +10$ volts $= E_{ab}$. *Reversing the order of the subscripts changes the sign of the quantity.*

An alternative means of designating a voltage is to use an arrow. In order to make the arrow notation in the figures compatible with the double-subscript notation, let us agree that the *tail* of the arrow corresponds to the first subscript, and the *head* to the second subscript. The arrow in Fig. 1-2c is then read as "E_1 is the voltage at the head of the arrow with respect to the tail." As in double-subscript notation, if the arrowhead lies at a point which is positive with respect to the tail, E_1 will be a positive quantity equal to $+10$ volts, just as E_{ab} is. The voltage E_2, on the other hand, equals -10 volts, since the head of E_2 is at a point which is negative with respect to the tail. In Fig. 1-2*d*, $E_1 = -10$ volts $= E_{ba} = -E_{ab}$, while $E_2 = E_{ab} = +10$ volts $= -E_{ba}$.

Now let us define current. First of all, when we speak of current in the text, we mean *electron* flow unless otherwise specified. To draw a resistor and say that a current of 2 amp flows through it, as in Fig. 1-2*e*, does not tell us enough. We have no idea which way the current is flowing through R. Without further information, we are unable to tell definitely how an ammeter should be connected so as to measure the current through R.

In order to designate the direction of current flow, we shall use either one of two schemes. The first is to draw an arrow beside or through R to designate the direction in which the current is flowing, as shown in Fig. 1-2*f*. The other alternative is to use double-subscript notation and define I_{ab} as the current flowing from point a to point b. If the current really flows from b to a, then I_{ab} will be a negative quantity equal to $-I_{ba}$. Generally, the current direction will be specified with an arrow rather than with double subscripts.

Let us now try a few problems involving notation:

Problem 1. Satisfy yourself that the results of each case in Fig. 1-3*a* are correct.

Problem 2. In Fig. 1-3*b* a current I flows between terminals a and b in the direction shown. Express E_{ab} and E_{ba} in terms of I and R.

Solution. Since the current (electron flow) enters terminal a first, terminal a will be negative with respect to terminal b. The polarity of the voltage developed across R because of the current through it is indicated by the $-$ and $+$ marks. We see then that in traversing the path from a to b, point b is positive with respect to point a. Accordingly, E_{ab} is a positive quantity which by Ohm's law is equal to IR. On the other hand, E_{ba} will be a negative quantity, since point a is negative with respect to point b. Therefore, $E_{ba} = -IR$.

Problem 3. Express E_{ca} in Fig. 1-3*c*, *d*, *e*, and *f*.

Solution. It should be understood that the current (I_1) finds a complete path through the rest of the network, which is not shown. At the moment we are concerned only with that part of the network existing between terminals a and c, which is illustrated. In each of the figures the path from point c to point a is completed by first going from c to b and noting E_{cb}. Then we proceed from b to a and note E_{ba}. Plainly, $E_{ca} = E_{cb} + E_{ba}$. The quantities E_{cb} and E_{ba} may, however, be negative, depending upon which figure we are discussing.

In Fig. 1-3*c*,

$$E_{ca} = E_{cb} + E_{ba} = +E_1 + I_1R_1$$

In Fig. 1-3*d*,

$$E_{ca} = E_{cb} + E_{ba} = +E_1 - I_1R_1$$

In Fig. 1-3*e*,

$$E_{ca} = E_{cb} + E_{ba} = -E_1 + I_1R_1$$

In Fig. 1-3*f*,

$$E_{ca} = E_{cb} + E_{ba} = -E_1 - I_1R_1$$

Problem 4. Suppose you were told that $E_{ab} = -12$ volts. Construct an equivalent circuit which will correspond to the information given.

Solution. The first thing we know is that we have two terminals, *a* and *b*. According to our definition of E_{ab}, the *b* terminal must be negative with respect to the *a* terminal if E_{ab} is to be a negative quantity. Therefore, $E_{ab} = -12$ volts must appear as in Fig. 1-3*g*.

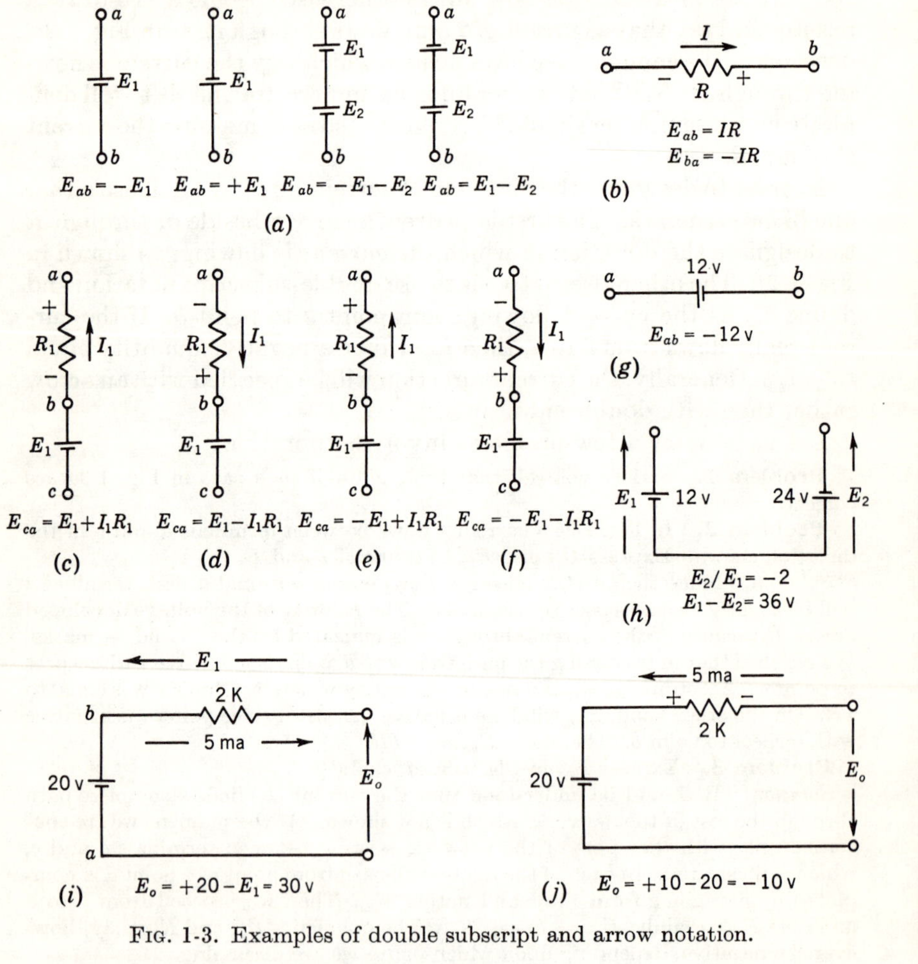

FIG. 1-3. Examples of double subscript and arrow notation.

Problem 5. In Fig. 1-3*h*, what is the ratio of E_2 to E_1? To what is $E_1 - E_2$ equal?

Solution. According to the arrow directions, we have

$$E_1 = +12 \text{ volts}$$

$$E_2 = -24 \text{ volts}$$

Therefore,
$$\frac{E_2}{E_1} = \frac{-24}{+12} = -2$$
$$E_1 - E_2 = 12 - (-24) = 36 \text{ volts}$$

Problem 6. What is E_o in Fig. 1-3*i*?

Solution

$$E_o = E_{ab} - E_1$$
$$E_1 = -5 \text{ ma} \times 2 \text{ kilohms} = -10 \text{ volts}$$
$$E_o = 20 - (-10) = +30 \text{ volts}$$

Problem 7. What is E_o in Fig. 1-3*j*?

Solution. To get from the tail of the arrow to the head, we go from − to +, or up in potential by 10 volts through the 2-kilohm resistor. In going through the battery, we go down (+ to −) 20 volts. Therefore,

$$E_o = +10 - 20 = -10 \text{ volts}$$

1-2. Voltage Dividers. A point which frequently disturbs the inexperienced technician is the use of voltage dividers connected between power supplies which are positive and negative with respect to ground.

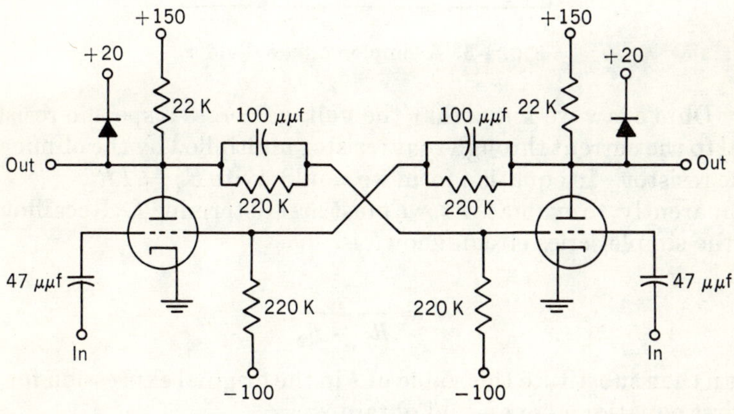

FIG. 1-4. A bistable multivibrator which utilizes power supplies that are positive and negative relative to ground potential.

Figure 1-4 is a typical bistable multivibrator circuit used in digital computers. Note that three different supply voltages with respect to ground are shown. Later we shall analyze this type of circuit in detail, but for now let us master the more elementary concepts.

The first application of Ohm's law that we shall consider is the voltage divider. Precisely what do we mean by a voltage divider? In general, we might say that if we have some network with E_{in} volts applied to it and we get E_2 volts out, we have divided or dropped E_{in} down until E_2 is left at the output terminals.

Voltage dividers have an endless variety of forms and may involve reactive as well as resistive elements.

Consider the circuit of Fig. 1-5, which represents a simple voltage divider. Physically we know that if E_{in} volts is applied to the input, the output E_2 will be less than E_{in} because some of E_{in} is used up in forcing current through resistor R_1. The amount of voltage used up in R_1 is E_1.

Likewise, the voltage appearing across R_2 is E_2, the output voltage in this circuit. Now we should like to find a ready means of determining the voltage across R_2, which we are calling E_2. (E_{in}, R_1, and R_2 are assumed known.)

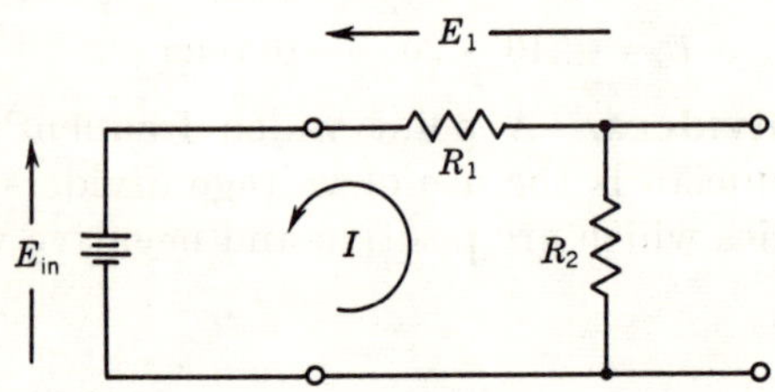

FIG. 1-5. A simple voltage divider.

By Ohm's law we know that the voltage across a specific resistor is equal to the current through that resistor multiplied by the ohmic value of the resistor. In equation form we would write $E_2 = IR_2$.

Apparently, to evaluate E_2, we must first determine I. Recalling that I in the simple series circuit shown is

$$I = \frac{E_{in}}{R_1 + R_2}$$

we can then substitute this value of I in the original expression for E_2—the first equation above—and obtain

$$E_2 = E_{in} \frac{R_2}{R_1 + R_2}$$

The quantity $R_2/(R_1 + R_2)$ is then seen to be the ratio between the output and the input voltages.

If we were interested in the voltage across R_1, it would be equal to

$$E_1 = E_{in} \frac{R_1}{R_1 + R_2}$$

At this point let us apply what we have learned to a few sample problems:

Problem 1. What is (*a*) E_o in the circuit of Fig. 1-6*a*, (*b*) the voltage across the 30-kilohm resistor?

Solution

(*a*) $$E_o = \frac{E_{in}R_2}{R_1 + R_2} = \frac{120 \text{ volts} \times 60 \text{ kilohms}}{30 + 60 \text{ kilohms}} = 80 \text{ volts} \qquad \textit{Ans.}$$

(*b*) $$E_1 = \frac{E_{in}R_1}{R_1 + R_2} = \frac{120 \text{ volts} \times 30 \text{ kilohms}}{30 + 60 \text{ kilohms}} = 40 \text{ volts} \qquad \textit{Ans.}$$

Note that E_1 could have been found by subtracting E_o from E_{in}:

$$E_1 = E_{in} - E_o = 120 - 80 = 40 \text{ volts}$$

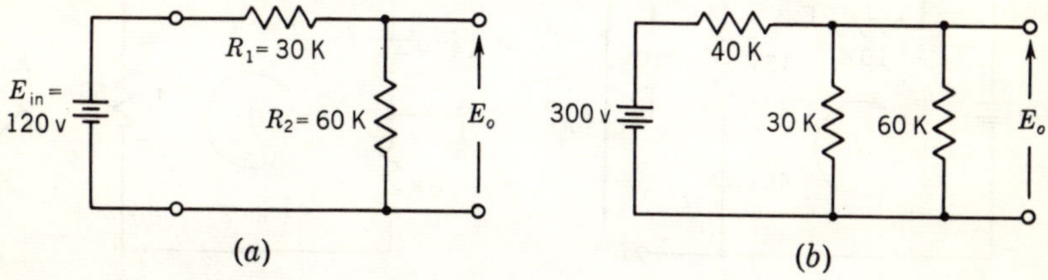

FIG. 1-6. Voltage-divider problems.

Problem 2. What is E_o in the circuit of Fig. 1-6*b*?

Solution. Here we have to decide just what the equivalent of R_2 is. Is it the 60- or the 30-kilohm resistor, or the combined effect of both? We can see that what was R_2 in Prob. 1 is really the 30- and 60-kilohm resistors in parallel here. We can then say the equivalent of R_2 is

$$R_{eq} = \frac{(30)(60)}{30 + 60} = 20 \text{ kilohms}$$

Therefore, $$E_o = \frac{(300 \text{ volts})(20 \text{ kilohms})}{40 + 20 \text{ kilohms}} = 100 \text{ volts} \qquad \textit{Ans.}$$

Problem 3. What is E_{BA} in the circuit of Fig. 1-7?

Solution. First combine the two 10-kilohm resistors to give 5 kilohms. Next determine the total voltage which forces current to flow in this circuit. It is 240 + 40 volts, since the two supplies are series-aiding. So we have 280 volts total. The voltage across the 15-kilohm resistor is then

$$\frac{(280 \text{ volts})(15 \text{ kilohms})}{5 + 15 \text{ kilohms}} = 210 \text{ volts}$$

This is not E_{BA}, however. E_{BA} is between terminals B and A, which includes the effect of the 40-volt battery. If we are interested in the potential of A with respect to B, let us start at point B and go up to A. In going through the 40-volt battery, we have *dropped* in potential by 40 volts with respect to B. This can be written as a minus 40 volts. In going through the 15-kilohm resistor, we note we are going up in potential by plus 210 volts. This is obvious, since the electron current through the 15-kilohm resistor polarizes it as shown. The net voltage of A with respect to B is then

$$E_{BA} = -40 + 210 = +170 \text{ volts} \qquad \textit{Ans.}$$

Problem 4. Find the voltage of point A with respect to ground in Fig. 1-8.

Solution. First, determine the direction of electron flow and note, with the aid of + and − signs, how the electron flow polarizes each resistor. Of course, electrons are leaving the negative terminal of the battery.

The next thing we do is to start at ground and arrive at point A, while keeping track of how far up or down we have gone in potential. There are two paths to take, either of which will bring us to point A from ground. We can either (1) go through the battery and the 40-kilohm resistor or (2) go through the 20- and 60-kilohm resistors.

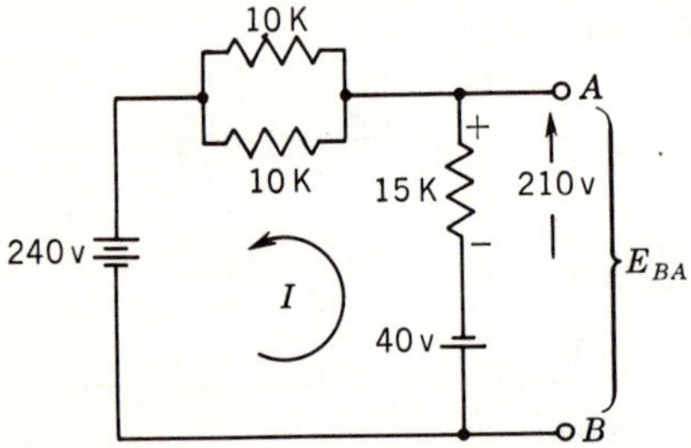

FIG. 1-7. The voltage E_{BA} is to be determined.

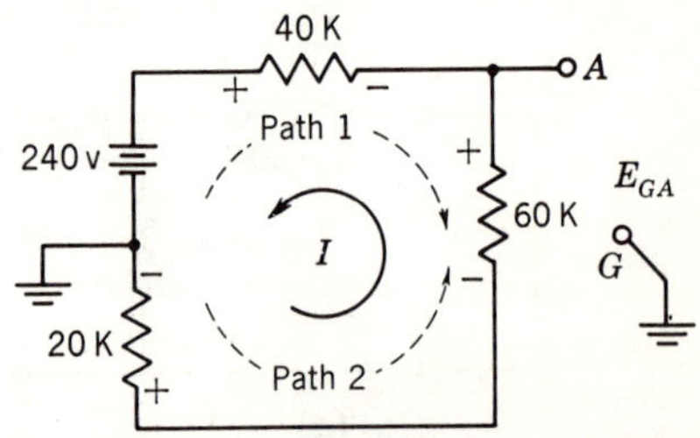

FIG. 1-8. The circuit may be analyzed by following either path.

1. If we take the former path, we must find the voltage developed across the 40-kilohm resistor.

The current in the entire circuit is clearly

$$I = \frac{240}{40 + 60 + 20} = \frac{240 \text{ volts}}{120 \text{ kilohms}} = 2 \text{ ma}$$

The drop across the 40-kilohm resistor is then

$$E = R \times I = (40 \text{ kilohms})(2 \text{ ma}) = 80 \text{ volts}$$

We could also have found the drop across this 40-kilohm resistor by using the voltage-divider relationship as follows:

$$\text{Drop across 40 kilohms} = 240 \text{ volts} \times \frac{40 \text{ kilohms}}{40 + 60 + 20 \text{ kilohms}} = 80 \text{ volts}$$

The voltage E_{GA} around this path is then

$$E_{GA} = +240 - 80 = 160 \text{ volts} \qquad \textit{Ans.}$$

2. If we choose to arrive at point A by going the other way around, through the 20- and 60-kilohm resistors, then with ground, which is zero volts, as our starting point, we would see a rise in potential of 40 volts in going through the 20-kilohm resistor. Going through the 60-kilohm resistor, we see a rise of 120 volts. Therefore,

$$E_{GA} = +40 + 120 = 160 \text{ volts}$$

the same result that we obtained the other way around path 1.

Problem 5. What is the voltage on the grid of the tube with respect to the cathode (E_{KG}) in Fig. 1-9?

Solution. We can start at the cathode (K) and note that we go from the plus end of the resistor to the minus end, which means we have dropped (5 ma)(1 kilohm) = 5 volts in potential.

This tells us that the cathode-to-ground voltage is -5 volts. We still have to go through the 1-meg resistor to get to the grid. In going from ground up to the grid, we encounter no voltage developed across the grid resistor, since $I_g = 0$.

Therefore, the grid is at ground potential or 5 volts negative with respect to the cathode. *Ans.*

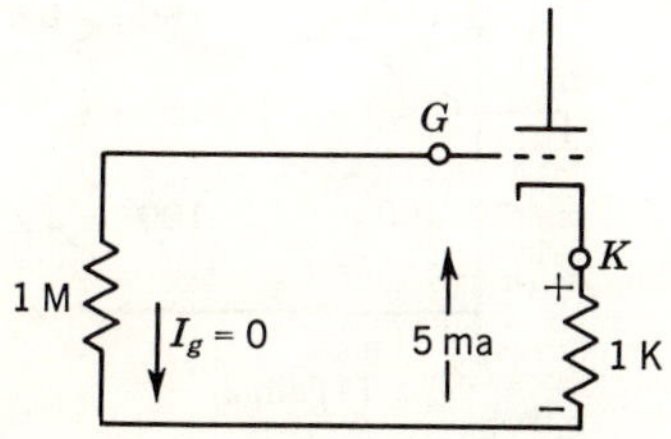

FIG. 1-9. Determine the bias on the tube.

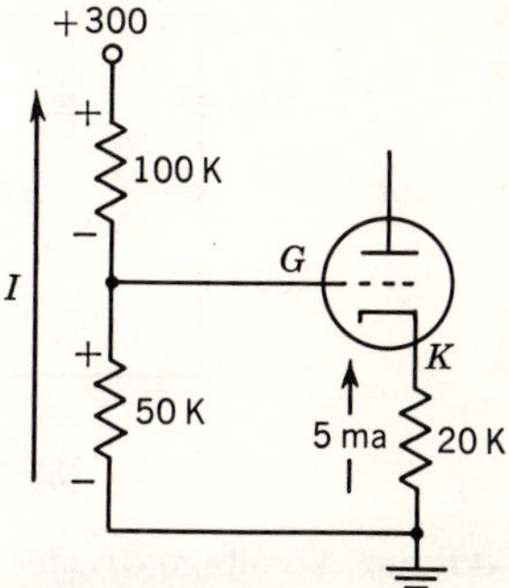

FIG. 1-10. A complex bias circuit we can analyze.

We could just as well have started at G and traveled around to K, in which case we would have evaluated $E_{GK} = +5$ volts. Since $E_{GK} = -E_{KG}$, $+5$ volts for E_{GK} means that $E_{KG} = -5$ volts.

Problem 6. Find (a) the voltage of the grid with respect to ground in Fig. 1-10, (b) the voltage of ground with respect to the cathode, (c) the cathode voltage with respect to the grid. Assume that the grid current equals zero in all cases.

Solution. (a) The voltage at the grid with respect to ground is the voltage across the 50-kilohm resistor, with the grid end being at the higher potential because of the divider current:

$$E_{\text{ground-grid}} = \frac{300 \text{ volts } (50 \text{ kilohms})}{50 + 100 \text{ kilohms}} = +100 \text{ volts} \qquad \textit{Ans.}$$

(b) $$E_{\text{cathode-ground}} = -(5 \text{ ma})(20 \text{ kilohms}) = -100 \text{ volts} \qquad \textit{Ans.}$$

The answer is negative because of our agreed method of designating a voltage.

(c) The grid-to-cathode voltage is found by starting at the grid, going down through the 50-kilohm resistor, where we drop 100 volts in potential (-100 volts), and then going up through the 20-kilohm resistor, where we go up 100 volts. Therefore,

$$E_{\text{grid-cathode}} = -100 + 100 = 0 \text{ volts} \qquad \textit{Ans.}$$

Problem 7. What is the voltage at point A in Fig. 1-11a? (Perhaps some readers are wondering what the other point is, since voltage must be measured between two points. The other point is always understood to be ground, unless some other reference is specified.)

Where is ground in Fig. 1-11*a*? (The experienced technician knows that Fig. 1-11*a* and 1-11*b* are identical.)

Solution. We see that the voltages in Fig. 1-11*a* are specified with respect to a common ground, which is between the two power supplies.

Without performing any calculations, we can make one rather obvious deduction concerning the potential of point A. If we were to reduce gradually the value of the 6-kilohm resistor, point A would be pulled up in potential by the +150-volt supply until, with the 6-kilohm resistor reduced to 0 ohms, point A would be at +150 volts.

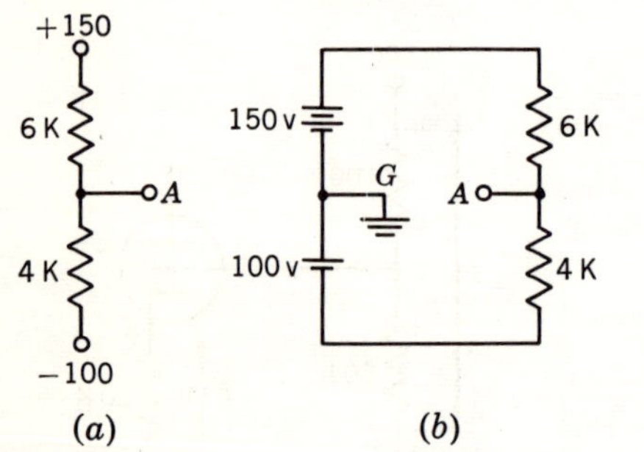

FIG. 1-11. (*a*) A voltage divider with positive and negative power supplies relative to ground potential. (*b*) The complete equivalent circuit of Fig. 1-11*a*.

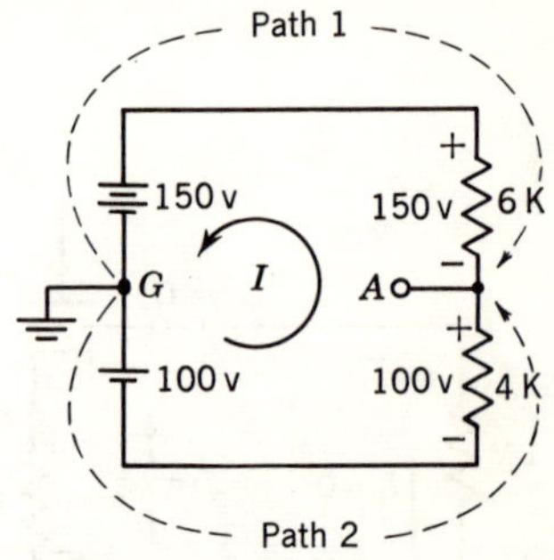

FIG. 1-12. The circuit may be analyzed by tracing either path.

On the other hand, if we were to hold the 6-kilohm resistor constant and keep reducing the 4-kilohm resistor, point A would gradually be pulled down towards −100 volts.

Conversely, if we increased the 4-kilohm resistor, we should be tending to isolate point A from the −100-volt supply, and so point A would go up in potential.

From this discussion we can see that point A is at a potential somewhere between −100 and +150 volts.

For an exact evaluation of the voltage of point A, say E_{GA}, we must first determine the direction of electron flow and note how it polarizes the resistors, as shown in Fig. 1-12.

Starting at ground, we then have our usual choice of two paths, either of which will bring us to point A. In each of these paths we must first determine the voltage across a resistor. To do this, we find the current I:

$$I = \frac{150 + 100}{6 + 4} = \frac{250 \text{ volts}}{10 \text{ kilohms}} = 25 \text{ ma}$$

The two batteries were added together because they are series-aiding. The voltage across the 6-kilohm resistor is 6 kilohms × 25 ma = 150 volts. The voltage across the 4-kilohm resistor is 4 kilohms × 25 ma = 100 volts. Starting at ground and going around *path 1*, we have

$$E_A = +150 - 150 = 0 \qquad \textit{Ans.}$$

Starting at ground and going around *path 2*, we have

$$E_A = -100 + 100 = 0$$

Either way, then, the potential of point A with respect to ground is zero volts. Remember our unfailing check—if traversing both paths does not give the same E_A, then an error has been made.

Following are a number of undiscussed problems, with and without given solutions. The reader should be sure to work every one for himself and then check his results.

Problem 8. Find E_o in Fig. 1-13.
Solution

$$E_o = 240 \text{ volts} \times \frac{4 \text{ kilohms}}{4 + 8 \text{ kilohms}} = 80 \text{ volts} \qquad \textit{Ans.}$$

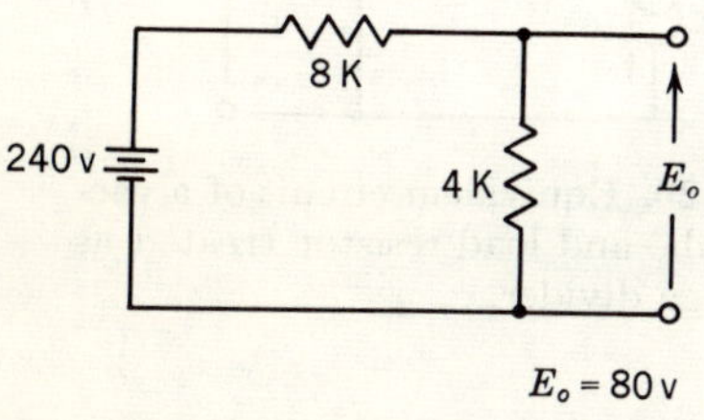

FIG. 1-13

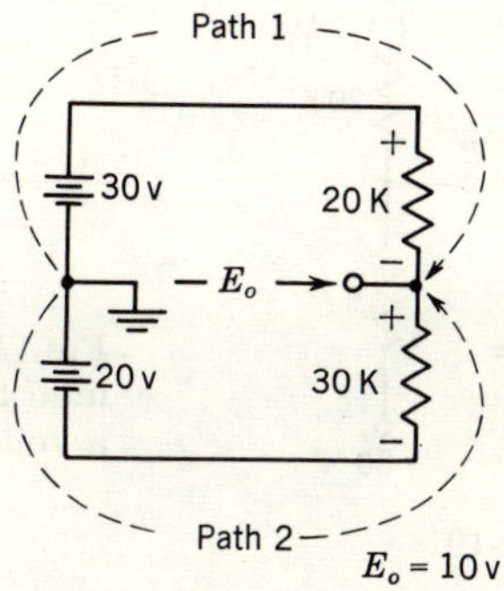

FIG. 1-14

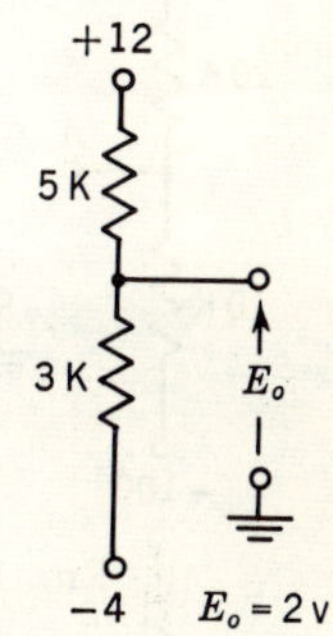

FIG. 1-15

Problem 9. Find E_o in the circuit of Fig. 1-14.
Solution. Using path 1,

$$E_o = +30 \text{ volts} - (30 + 20 \text{ volts}) \times \frac{20 \text{ kilohms}}{20 + 30 \text{ kilohms}} = +10 \text{ volts} \qquad \textit{Ans.}$$

Alternative solution, using path 2,

$$E_o = -20 \text{ volts} + (30 + 20) \text{ volts} \times \frac{30 \text{ kilohms}}{20 + 30 \text{ kilohms}} = +10 \text{ volts} \qquad \textit{Ans.}$$

Problem 10. Find E_o in Fig. 1-15.
Solution

$$E_o = 2 \text{ volts} \qquad \textit{Ans.}$$

How did you do it? What steps did you adopt?

Problem 11. Find E_o in Fig. 1-16.
Solution

$$E_o = 40 \text{ volts} \qquad \textit{Ans.}$$

Problem 12. Find E_o in Fig. 1-17. Solve and check.
Problem 13. Find E_o in Fig. 1-18. Solve and check.

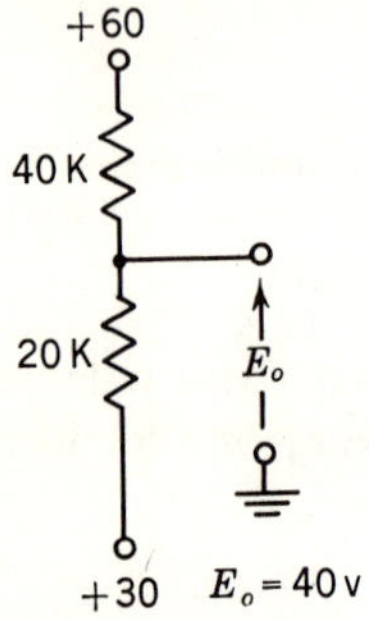

FIG. 1-16

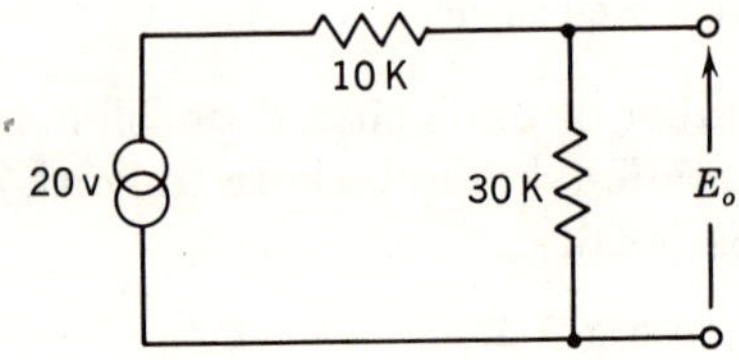

FIG. 1-17

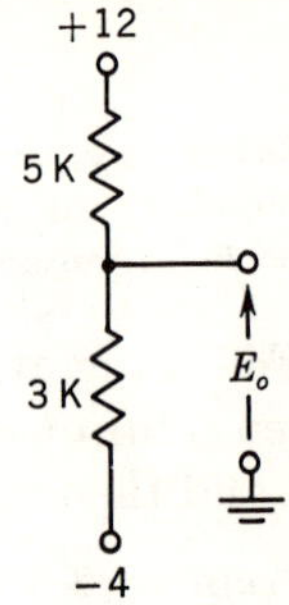

FIG. 1-18

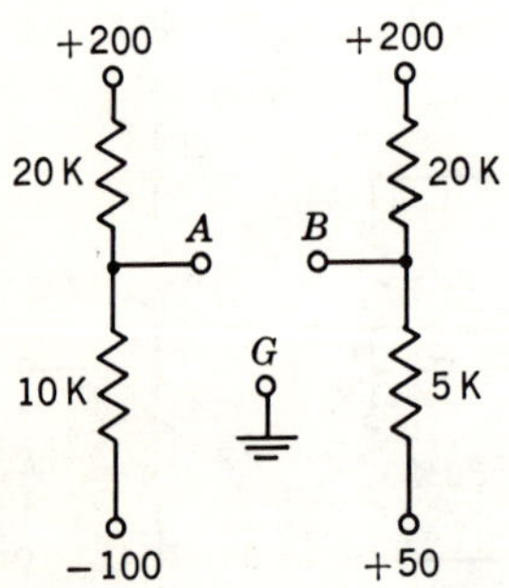

FIG. 1-19

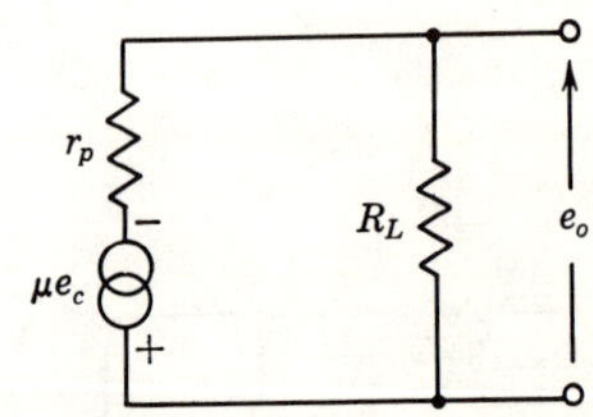

FIG. 1-20. Equivalent circuit of a vacuum tube and load resistor treated as a voltage divider.

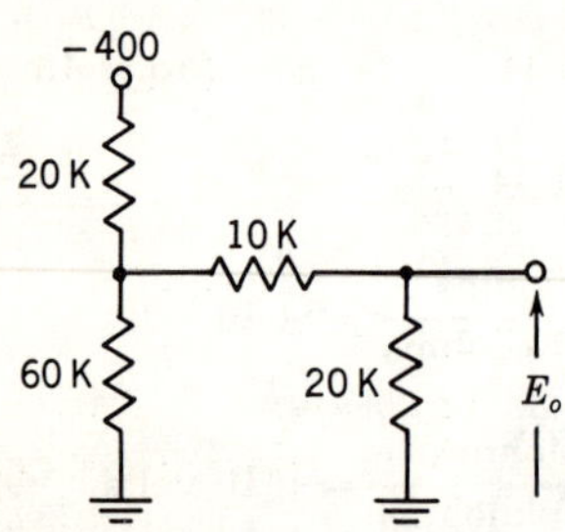

FIG. 1-21. Determine E_o in this illustrative problem.

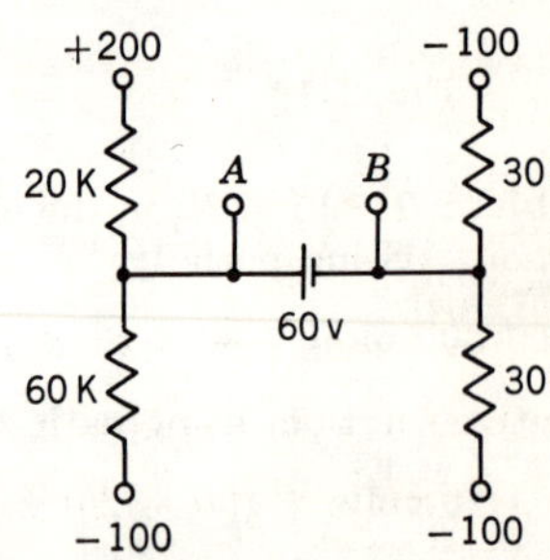

FIG. 1-22. E_{AB} will always be held (clamped) at −60 volts in this problem.

Problem 14. Find E_{AB} in Fig. 1-19.

Solution

$$E_{AB} = E_{AG} + E_{GB}$$

$$E_{GA} = 200 \text{ volts} - \frac{300 \text{ volts} \times 20 \text{ kilohms}}{10 + 20 \text{ kilohms}} = 200 - 200$$

$$= 0 \text{ volts} \qquad \text{or} \qquad E_{AG} = 0 \text{ volts}$$

$$E_{GB} = 200 \text{ volts} - \frac{150 \times 20}{20 + 5} = 200 - 120 = +80 \text{ volts}$$

$$E_{AB} = 0 + 80 = +80 \text{ volts}$$

Problem 15. Find e_o in Fig. 1-20 if $\mu = 100$, $r_p = 20$ kilohms, $R_L = 60$ kilohms, and $e_c = 0.24$ volt.

Solution

$$e_o = -18 \text{ volts} \qquad \textit{Ans.}$$

Problem 16. Find E_o in Fig. 1-21.

Solution

$$E_o = -133.4 \text{ volts} \qquad \textit{Ans.}$$

Problem 17. Find E_{AB} in Fig. 1-22.

Solution

$$E_{AB} = -60 \text{ volts} \qquad \textit{Ans.}$$

1-3. Current Dividers.

A current divider may be considered basically as some combination of impedances that will cause the output current of a network to be less than the input current. We shall restrict our discussion of current dividers to their simplest forms.

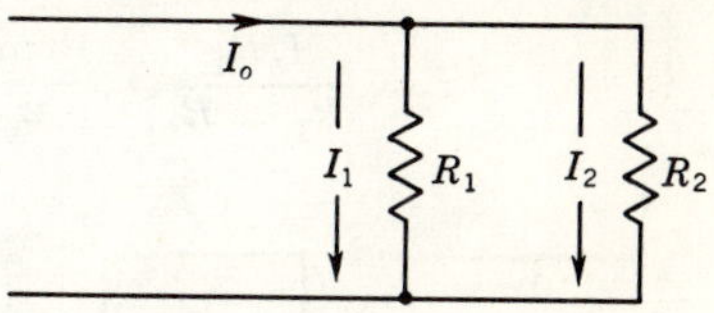

FIG. 1-23. A current divider.

Consider Fig. 1-23, which shows a line current I_o feeding two branches so that I_o splits into I_1 and I_2. We should like to find a simple relationship between I_2 and I_o in terms of the branch resistances R_1 and R_2.

Since R_1 and R_2 are in parallel, the voltage across them must be the same. Therefore, $I_1R_1 = I_2R_2$. We want I_o and not I_1 in our final expression. Therefore, let us express I_1 in terms of I_o. Obviously, $I_o = I_1 + I_2$. Therefore, $I_1 = I_o - I_2$. By substitution, $(I_o - I_2)R_1 = I_2R_2$. Solving for I_2, we obtain

$$I_2 = \frac{I_oR_1}{R_1 + R_2}$$

At this point, we want to show what a simple phrase such as "solving for I_2" implies. The following algebraic manipulations were involved:

$$I_oR_1 - I_2R_1 = I_2R_2$$

Therefore,

$$I_oR_1 = I_2R_2 + I_2R_1 = I_2(R_2 + R_1)$$

which as stated above yields

$$I_2 = \frac{I_oR_1}{R_1 + R_2}$$

This example shows why a background of high school algebra is assumed when reading this book. Where the reader lacks background, he should make it up by studying from the bibliography, and then continue with this book. For the solution of equations, for example, we recommend Chap. 12 of "Basic Mathematics for Electronics."[1] This chapter is specifically on equations.

To continue with our problem, it can be shown similarly that

$$I_1 = \frac{I_oR_2}{R_1 + R_2}$$

Notice here that, in *this* type of formula, as compared with that for the voltage divider, the numerator contains the shunt impedance, whose current we are *not* interested in. Let us apply the formula to a few problems.

Problem 1. Find I_2 in the accompanying circuit of Fig. 1-24.
Solution. Filling in the values for the symbols of Fig. 1-23, we have

$$I_2 = \frac{I_oR_1}{R_1 + R_2} = \frac{10 \text{ ma} \times 2 \text{ kilohms}}{2 + 1 \text{ kilohms}} = 6.67 \text{ ma} \qquad \textit{Ans.}$$

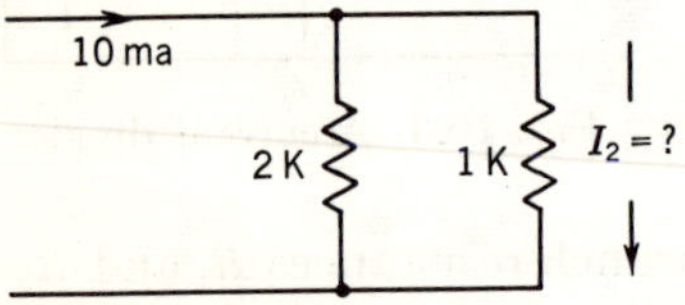

FIG. 1-24. An illustrative current-divider problem.

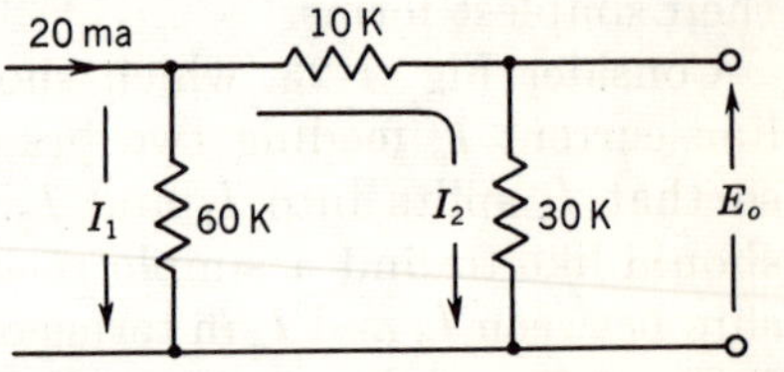

FIG. 1-25. Determining an output voltage using the current-divider relationship.

Problem 2. Find E_o in the circuit of Fig. 1-25.
Solution

$$I_2 = \frac{I_oR_1}{R_1 + R_2} = 20 \text{ ma} \times \frac{60 \text{ kilohms}}{60 + 40 \text{ kilohms}} = 12 \text{ ma}$$

$$E_o = -12 \text{ ma} \times 30 \text{ kilohms} = -360 \text{ volts} \qquad \textit{Ans.}$$

[1] Nelson M. Cooke, "Basic Mathematics for Electronics," 2d ed., McGraw-Hill Book Company, Inc., New York, 1960.

Problem 3. Estimate I_2 in the circuit of Fig. 1-26.

Solution. Since the combination of R_1 and R_2 in parallel is slight as compared with R_o, we can safely assume that

$$I_o \simeq \frac{100 \text{ volts}}{10 \text{ kilohms}} = 10 \text{ ma}$$

$$\text{Then } I_2 = \frac{I_o R_1}{R_1 + R_2}$$

$$= 10 \text{ ma} \times \frac{30 \text{ ohms}}{30 + 60 \text{ ohms}}$$

$$= 3.33 \text{ ma} \quad \textit{Ans.}$$

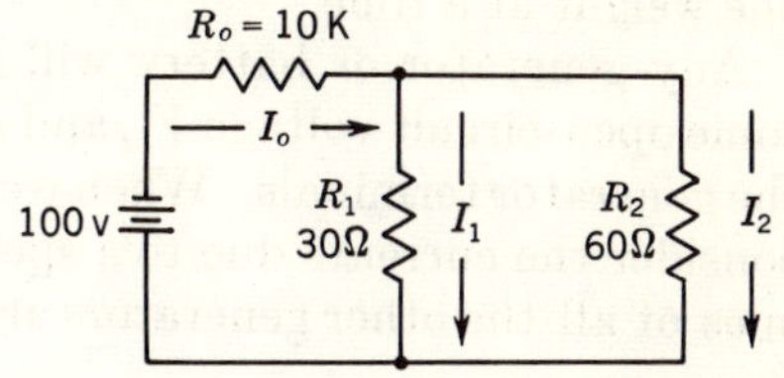

FIG. 1-26. The current I_o is essentially determined by R_o.

1-4. Superposition Theorem. The superposition theorem states that in any linear bilateral network containing one or more generators, the current at any point within the network is the algebraic sum of the currents which would be due to each generator considered individually, and with all other generators being replaced by their internal impedances only.

The above statement is quite a mouthful, and a few words of explanation are in order. First of all, what do we mean by a linear network? A linear network is one in which the current is at all times proportional to the voltage. For example, an air-core coil is a linear element, and doubling the alternating voltage across the coil would double the alternating current. But if the coil has an iron core, it may be a nonlinear network, since doubling the voltage might cause sufficient current to flow through the coil so that the core is saturated over a fraction of a cycle.

When saturated, the coil impedance may drop to a low value which permits a much greater current to flow than would be expected. In fact, doubling the voltage might cause three times rather than twice as much current to flow.

Any device exhibiting a curved volt-ampere characteristic is nonlinear. It may, however, be linear over a specific range of operation.

What do we mean by *bilateral*? A bilateral element is one which carries currents through it equally well for any applied polarity of voltage. For example, an ordinary resistor does not care which end is positive with respect to the other. The same current will flow as long as the voltage stays constant. A diode vacuum tube, on the other hand, is not bilateral. It will readily pass current if the plate is positive with respect to the cathode, but negligible current will flow if the plate is negative with respect to the cathode.

As a physical example that illustrates the idea behind the superposition theorem let us consider a spring-balance type of scale. Hanging a 5-lb weight on the spring might stretch it by 1 in. If another 5 lb were added, you would expect the spring to stretch by 2 in. The total spring deflection is equal to the sum of the individual deflections, considering one weight at a time.

Any generator or battery will have an equivalent circuit—that is, some open-circuit voltage V_{oc} and some impedance Z_{th} seen looking into the generator terminals. When we apply the superposition theorem and consider the currents due to a specific generator, the open-circuit voltages of all the other generators are considered zero, but the equivalent

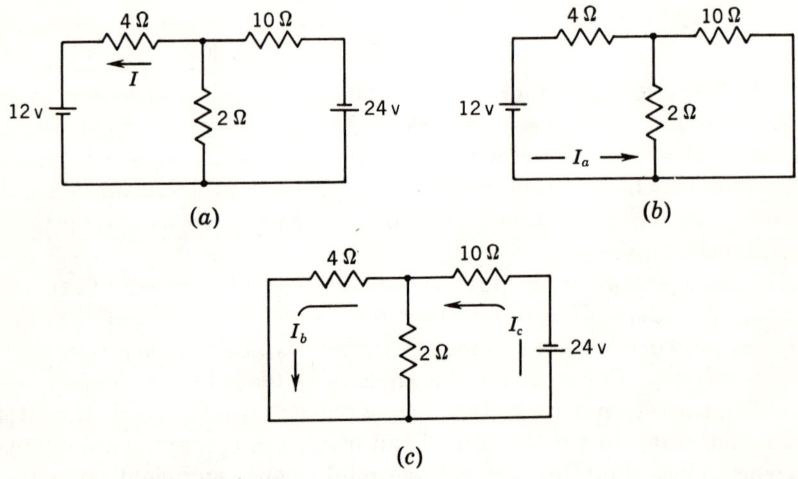

FIG. 1-27. (a) A network to be analyzed by the superposition theorem. The current through the 4-ohm resistor due to (b) the 12-volt source only, and (c) the 24-volt source only.

impedance is not dropped, since it will have an effect on the resultant current. In most of our problems, the source impedance of any generator will be considered zero or lumped with the external impedances.

As an example of the superposition theorem, we seek the current I in Fig. 1-27a. If we consider just one generator at a time, Fig. 1-27a can be separated into Figs. 1-27b and c.

$$I_a = \frac{12 \text{ volts}}{4 + 1.67 \text{ ohms}} = 2.11 \text{ amp}$$

$$I_c = \frac{24 \text{ volts}}{10 + 1.33 \text{ ohms}} = 2.12 \text{ amp}$$

$$I_b = \frac{24 - (2.12)10}{4} = \frac{2.8 \text{ volts}}{4 \text{ ohms}} = 0.7 \text{ amp}$$

Therefore, $I_1 = I_a + I_b = 2.11 + 0.7 = 2.81$ amp *Ans.*

Problem 1. Solve for the current in R_3 by the superposition method. We shall use Figs. 1-28a, b, c, and d.

Solution. Let us first determine the current through R_3 due to the 240-volt battery. We replace the 120-volt battery by its internal impedance, which is zero, as shown in Fig. 1-28*c*.

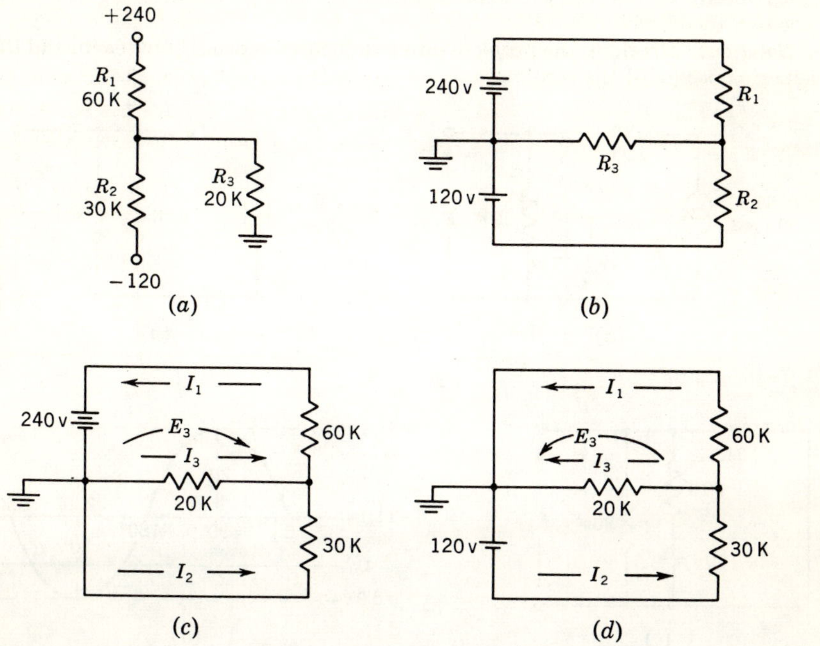

Fig. 1-28. (*a*) A network to be analyzed by the superposition theorem. (*b*) Same network redrawn to facilitate analysis. (*c*) Currents due to the 240-volt source only. (*d*) Currents due to the 120-volt source only.

There are a number of ways to find I_3. We can first find I_1, then the voltage across R_3, and then the current through R_3.

$$I_1 = \frac{240}{60 + \dfrac{(20)(30)}{20 + 30}} = \frac{240 \text{ volts}}{72 \text{ kilohms}} = 3.33 \text{ ma}$$

$$E_3 = 240 - 60I_1 = 240 - 199.8 = 40.2 \text{ volts}$$

$$I_3 = \frac{E_3}{R_3} = \frac{40.2 \text{ volts}}{20 \text{ kilohms}} = 2.01 \text{ ma} \qquad \textit{Ans.}$$

This time, let us replace the 240-volt supply by its internal impedance, which is zero, and determine the effect of the 120-volt supply acting alone on R_3. In Fig. 1-28*d*,

$$I_2 = \frac{120}{30 + \dfrac{(20)(60)}{20 + 60}} = \frac{120 \text{ volts}}{45 \text{ kilohms}} = 2.67 \text{ ma}$$

$$E_3 = 120 - (2.67 \text{ ma})(30 \text{ kilohms}) = 40.1 \text{ volts}$$

$$I_3 = \frac{40.1 \text{ volts}}{20 \text{ kilohms}} = 2.01 \text{ ma}$$

The value of I_3 will be the difference between I_3 as found in Fig. 1-28*c* and I_3 in Fig. 1-28*d*. Since I_3 of Fig. 1-28*d* is equal and opposite to I_3 of Fig. 1-28*c*, the net current through the 20-kilohm resistor is zero.

Problem 2. Derive the expression for E_o in Fig. 1-29*a*, using the superposition theorem.

Solution. Break up the problem into two separate parts. Solve each, and then superimpose (add) the results.

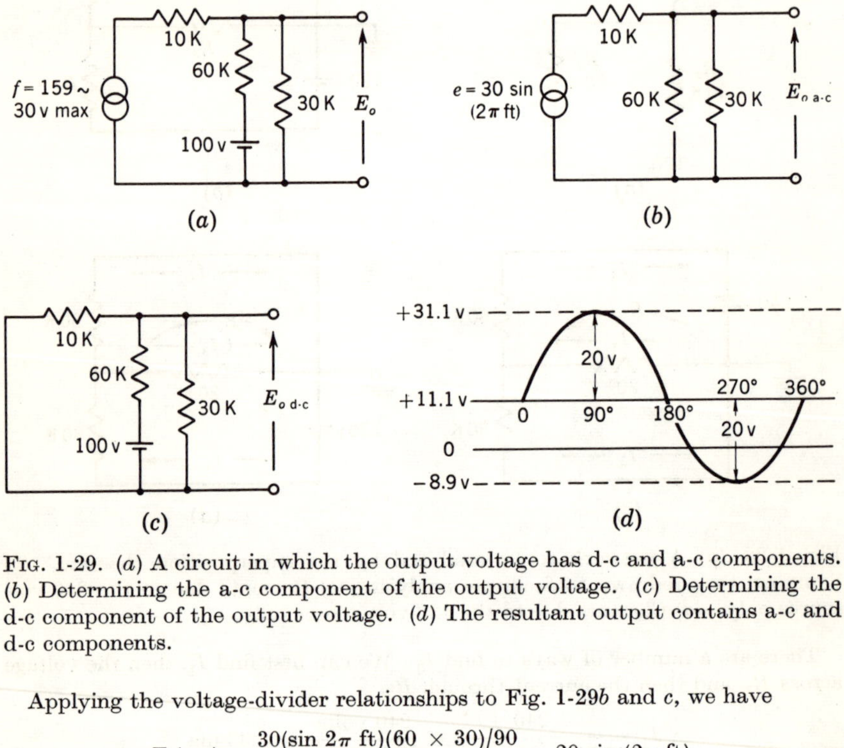

FIG. 1-29. (*a*) A circuit in which the output voltage has d-c and a-c components. (*b*) Determining the a-c component of the output voltage. (*c*) Determining the d-c component of the output voltage. (*d*) The resultant output contains a-c and d-c components.

Applying the voltage-divider relationships to Fig. 1-29*b* and *c*, we have

$$E_o(a\text{-}c) = \frac{30(\sin 2\pi\ \text{ft})(60 \times 30)/90}{10 + (60 \times 30)/90} = 20 \sin (2\pi\ \text{ft})$$

$$E_o(d\text{-}c) = \frac{100 \times (10 \times 30)/40}{60 + (10 \times 30)/40} = 11.1 \text{ volts}$$

Continuing, $E_o = E_o(a\text{-}c) + E_o(d\text{-}c) = 20 \sin (2\pi\ \text{ft}) + 11.1$ volts. Let $2\pi\ \text{ft} = \theta$. Then, when $\theta = 90°$, $E_o = 20 + 11.1 = 31.1$ volts. And when $\theta = 270°$, $E_o = -20 + 11.1 = -8.9$ volts. *Ans.* (See Fig. 1-29*d*.)

The superposition principle is frequently used when dealing with vacuum-tube and transistor circuits. Assuming that we are concerned only with a-c signal variations, we can neglect the d-c power supplies and consider just the a-c sources. The a-c equivalent circuit can then be analyzed for the a-c output due to an a-c input signal.

1-5. Thévenin's Theorem. Thévenin's theorem states that any linear network of impedances and generators, if viewed from any two points in the network, can be replaced by an equivalent voltage source V_{oc} and by an equivalent impedance Z_{th} in series.

This theorem means we can take for our two points any two terminals, *a* and *b*, of the generalized network contained in the box of Fig.

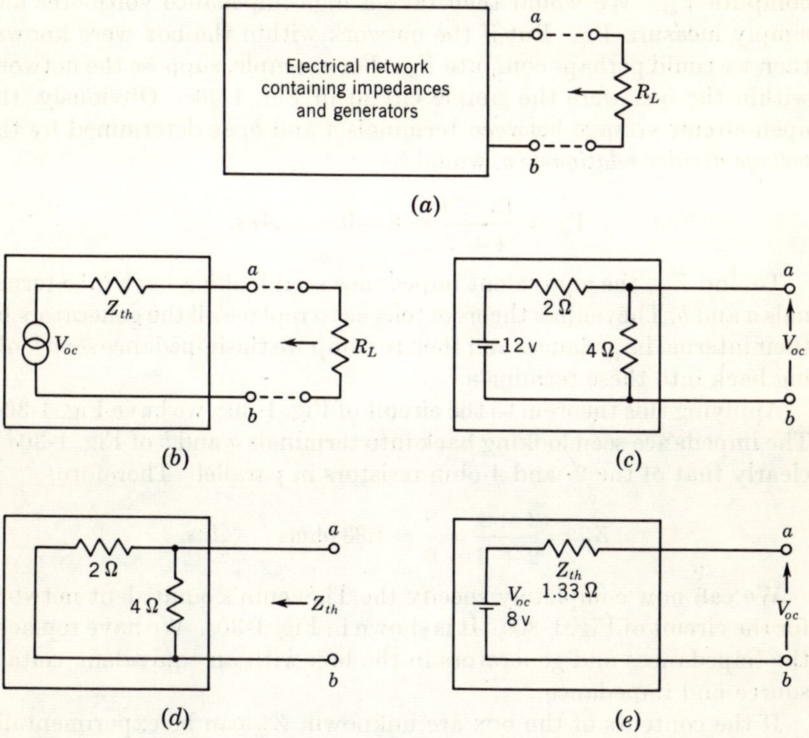

FIG. 1-30. (*a*) The network to the left of the *a-b* terminals may frequently be simplified by Thévenin's theorem. (*b*) The Thévenin's equivalent circuit will consist of a voltage source in series with an impedance. (*c*) A simple circuit whose Thévenin's equivalent is to be determined. (*d*) Determining the equivalent source impedance. (*e*) The complete Thévenin's equivalent circuit for *c*.

1-30*a* and, looking back into the box from the same two terminals, replace the network in the box with an impedance Z_{th} and with a voltage source V_{oc} in series with it. Figure 1-30*b* shows Z_{th} and V_{oc}, the two equivalents of the impedances and generators in the box.

If we were to connect a load R_L to the two terminals *a* and *b* in Fig. 1-30*a*, some current would flow through R_L. If the network of Fig. 1-30*b* is to be equivalent, then exactly the same current must flow through R_L in this simplified network.

Our problem then is to determine specifically a value for V_{oc} and Z_{th}.

The equivalent voltage source V_{oc} can be determined from the original network of Fig. 1-30*a* by measuring the voltage between terminals *a* and *b* with no load connected to them. This voltage is referred to as the *open-circuit* voltage.

If we did not know what was in the box of Fig. 1-30*a*, we could not compute V_{oc}. We would then take a high-impedance voltmeter and simply measure V_{oc}. But if the network within the box were known, then we could perhaps compute V_{oc}. For example, suppose the network within the box were the simple circuit of Fig. 1-30*c*. Obviously, the open-circuit voltage between terminals *a* and *b*, as determined by the *voltage-divider relationship*, would be

$$V_{oc} = \frac{12 \times 4}{4 + 2} = 8 \text{ volts} \qquad \textit{Ans.}$$

To find Z_{th}, the equivalent impedance seen looking back into terminals *a* and *b*, Thévenin's theorem tells us to replace all the generators by their internal impedances and then to compute the impedance seen looking back into these terminals.

Applying this theorem to the circuit of Fig. 1-30*c*, we have Fig. 1-30*d*. The impedance seen looking back into terminals *a* and *b* of Fig. 1-30*d* is clearly that of the 2- and 4-ohm resistors in parallel. Therefore,

$$Z_{th} = \frac{2 \times 4}{2 + 4} = \frac{8}{6} = 1.33 \text{ ohms} \qquad \textit{Ans.}$$

We can now completely specify the Thévenin's equivalent network for the circuit of Fig. 1-30*c*. It is shown in Fig. 1-30*e*. We have replaced the impedances and generators in the box with an equivalent voltage source and impedance.

If the contents of the box are unknown, Z_{th} can be experimentally determined by the following method. First determine V_{oc} with a voltmeter. Then short-circuit terminals *a* and *b* through an ammeter to read the short-circuit current I_{sc} between these terminals. Z_{th} can then be evaluated from $Z_{th} = V_{oc}/I_{sc}$.

To show that this experimental method enables us to evaluate Z_{th}, let us look again at Fig. 1-30*c*. If we were to short-circuit terminals *a* and *b*, the 4-ohm resistor would be shorted; and the short-circuit current through terminals *a* and *b* would be

$$I_{sc} = \frac{12 \text{ volts}}{2 \text{ ohms}} = 6 \text{ amp}$$

We previously determined that $V_{oc} = 8$ volts. Therefore,

$$Z_{th} = \frac{V_{oc}}{I_{sc}} = \frac{8 \text{ volts}}{6 \text{ amp}} = 1.33 \text{ ohms} \qquad \textit{Ans.}$$

Problem 1. Find I_1, using Thévenin's theorem, in Fig. 1-31*a* and *b*.

Solution. Let us conveniently break the circuit at *xx* and then look back—this time, to the right:

$$V_{oc} = -24 \text{ volts} \times \frac{2}{2 + 10} = -4 \text{ volts} \qquad \textit{Ans.}$$

$$Z_{th} = \frac{2 \times 10}{2 + 10} = 1.67 \text{ ohms} \qquad \textit{Ans.}$$

$$I_1 = \frac{12 + 4 \text{ volts}}{4 + 1.67 \text{ ohms}} = 2.82 \text{ amp} \qquad \textit{Ans.}$$

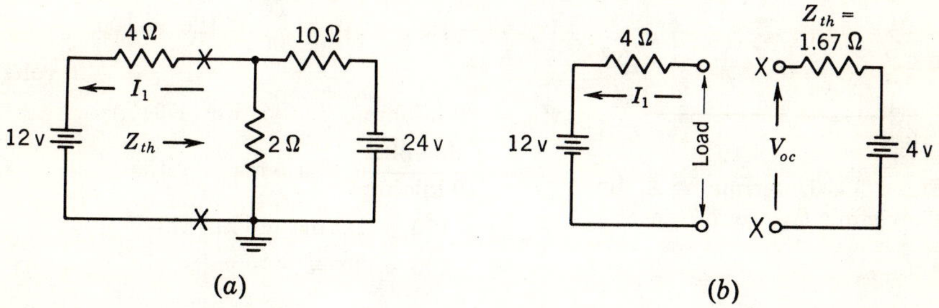

FIG. 1-31. (*a*) Determine I_1 by applying Thévenin's theorem. (*b*) The 12-volt battery and 4-ohm resistor are considered the load.

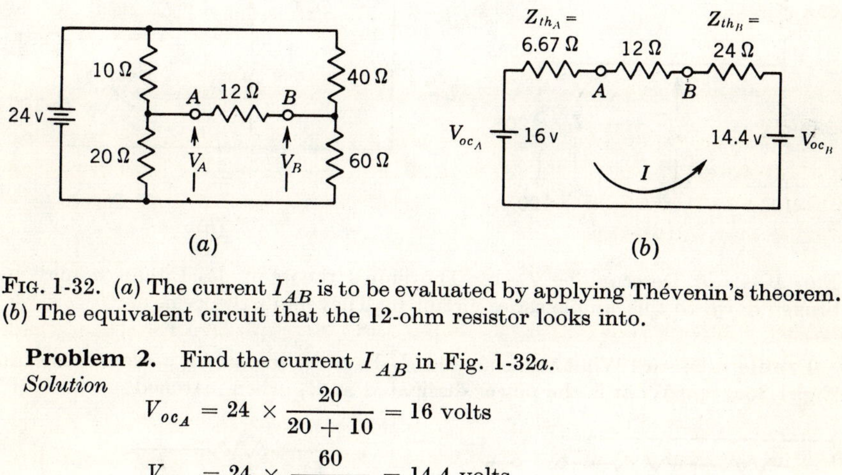

FIG. 1-32. (*a*) The current I_{AB} is to be evaluated by applying Thévenin's theorem. (*b*) The equivalent circuit that the 12-ohm resistor looks into.

Problem 2. Find the current I_{AB} in Fig. 1-32*a*.

Solution

$$V_{oc_A} = 24 \times \frac{20}{20 + 10} = 16 \text{ volts}$$

$$V_{oc_B} = 24 \times \frac{60}{60 + 40} = 14.4 \text{ volts}$$

$$Z_{th_A} = \frac{10 \times 20}{10 + 20} = 6.67 \text{ ohms}$$

$$Z_{th_B} = \frac{40 \times 60}{40 + 60} = 24 \text{ ohms} \qquad \text{(See Fig. 1-32}b\text{.)}$$

$$I = \frac{16 - 14.4}{6.67 + 12 + 24} = \frac{1.6 \text{ volts}}{42.67 \text{ ohms}} = 0.0375 \text{ amp}$$

$$I_{AB} = -I = -0.0375 \text{ amp} \qquad \textit{Ans.}$$

Problem 3. Find I_L in Fig. 1-33.
Solution

$$V_{oc} = 120 \times \frac{40}{100} = 48 \text{ volts}$$

$$Z_{th} = 24 \text{ kilohms}$$

$$I_L = \frac{48 \text{ volts}}{30 \text{ kilohms}} = 1.6 \text{ ma} \qquad \textit{Ans.}$$

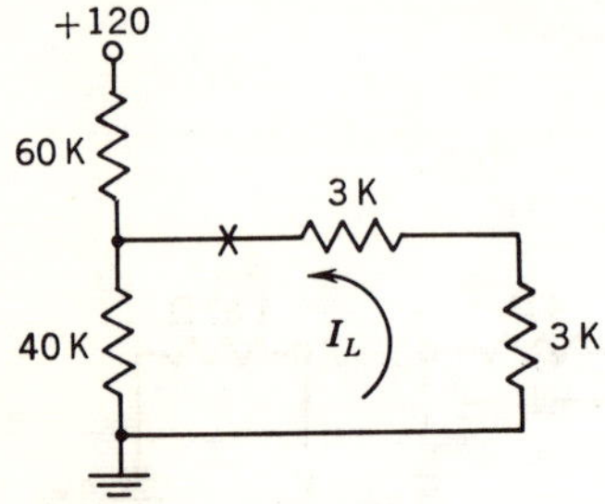

FIG. 1-33. Determining I_L by Thévenin's theorem.

Problem 4. Find I_L in Fig. 1-34*a*.
Solution

$$V_{oc} = +120 - 240 \times \frac{60}{90} = 120 - 160 = -40 \text{ volts}$$

$$Z_{th} = 20 \text{ kilohms} \qquad \text{(See Fig. 1-34}b\text{.)}$$

$$I_L = \frac{60 \text{ volts}}{40 \text{ kilohms}} = 1.5 \text{ ma} \qquad \textit{Ans.}$$

$$V_L = -100 + 1.5 \text{ ma (20 kilohms)} = -100 + 30 = -70 \text{ volts}$$

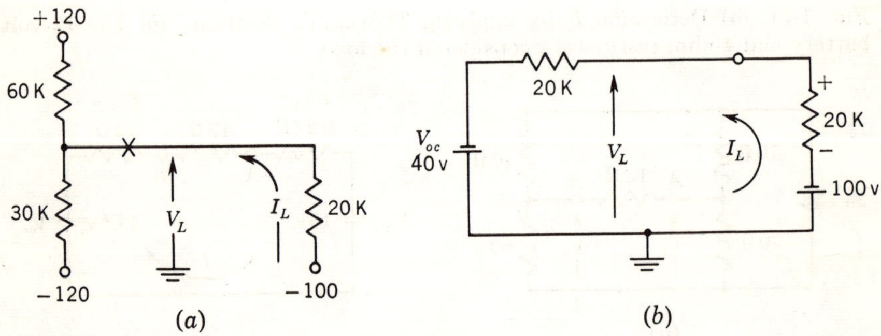

FIG. 1-34. (*a*) Determining I_L by Thévenin's theorem. (*b*) Circuit *a* may be broken down to this simple series circuit by Thévenin's theorem.

Problem 5. (*a*) What value should R_L have for maximum power transfer in Fig. 1-35*a*? (*b*) What is the power dissipated in R_L when matched?

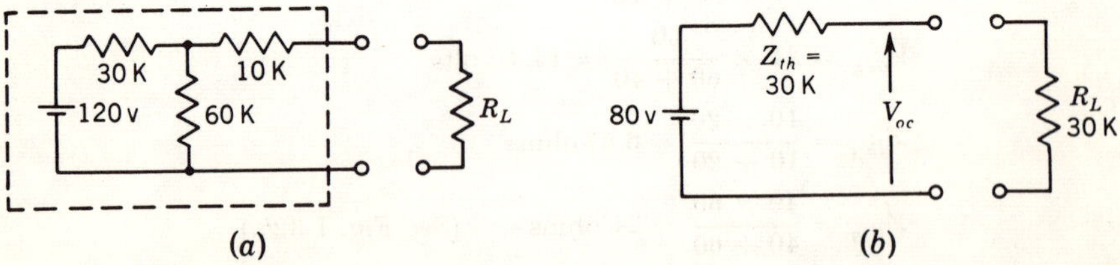

FIG. 1-35. (*a*) Determining R_L for maximum power transfer. (*b*) The load (R_L) should equal the apparent source resistance (Z_{th}) for maximum power transfer.

Solution. (*a*) For maximum power transfer, R_L should equal the impedance seen looking back into the network. Thevenizing, we see, looking back,

$$Z_{th} = 10 + \frac{30 \times 60}{30 + 60} = 30 \text{ kilohms}$$

Therefore, $$R_L = Z_{th} = 30 \text{ kilohms} \qquad \textit{Ans.}$$

(*b*) To find the power in R_L, let us first completely thevenize the network. This is done with the load disconnected. The open-circuit voltage V_{oc} is simply the voltage across the 60-kilohm resistor:

$$V_{oc} = \frac{120 \text{ volts} \times 60 \text{ kilohms}}{60 + 30 \text{ kilohms}} = 80 \text{ volts} \qquad \text{(See Fig. 1-35}b\text{.)}$$

With the load connected, the load power is

$$P_L = \frac{E_L^2}{R_L} = \frac{(40)^2}{30} = 53.3 \text{ mw} \qquad \textit{Ans.}$$

Problem 6. Find the current and voltage between terminals A and B in Fig. 1-36*a*.

Solution. Remove the load and, looking into terminals A and B, replace the generator with its internal impedance (see Fig. 1-36*b*), which is zero ohms in this case. From Fig. 1-36*c*, we can plainly see that

$$Z_{th} = 24 + 20 = 44 \text{ kilohms}$$

$$V_{oc_A} = 90 \text{ volts} \times \frac{60 \text{ kilohms}}{60 + 40 \text{ kilohms}} = 54 \text{ volts}$$

$$V_{oc_B} = 90 \text{ volts} \times \frac{30 \text{ kilohms}}{60 + 30 \text{ kilohms}} = 30 \text{ volts}$$

$$V_{oc_{AB}} = -54 + 30 = -24 \text{ volts}$$

The complete Thévenin's equivalent is shown in Fig. 1-36*d*. With the R_L connected, the load voltage E_{AB} is

$$E_{AB} = -24 \text{ volts} \times \frac{20 \text{ kilohms}}{64 \text{ kilohms}} = -7.5 \text{ volts}$$

$$I = \frac{7.5 \text{ volts}}{20 \text{ kilohms}} = 0.38 \text{ ma}$$

but, by our agreed definition, I_{AB} must equal *a* minus I, or -0.38 ma.

Problem 7. The circuit shown in Fig. 1-37 is the small-signal high-frequency equivalent circuit for a triode amplifier. C_o is made up of stray capacitances, the output capacitance of the stage shown, and the net input capacitance of the following stage.

Derive an expression for that frequency at which the output will be 0.707, or $1/\sqrt{2}$, of its mid-frequency value, $E_{o,mid}$. This frequency is sometimes referred to as the *upper half-power* frequency.

Solution. The mid-frequency gain can be determined from Fig. 1-37 for this problem, except that C_o is omitted, since its effect is negligible at mid-frequencies. Therefore,

$$E_{o,mid} = \frac{\mu e_c R'_g}{r_p + R'_g}$$

where $$R'_g = \frac{R_L R_g}{R_L + R_g}$$

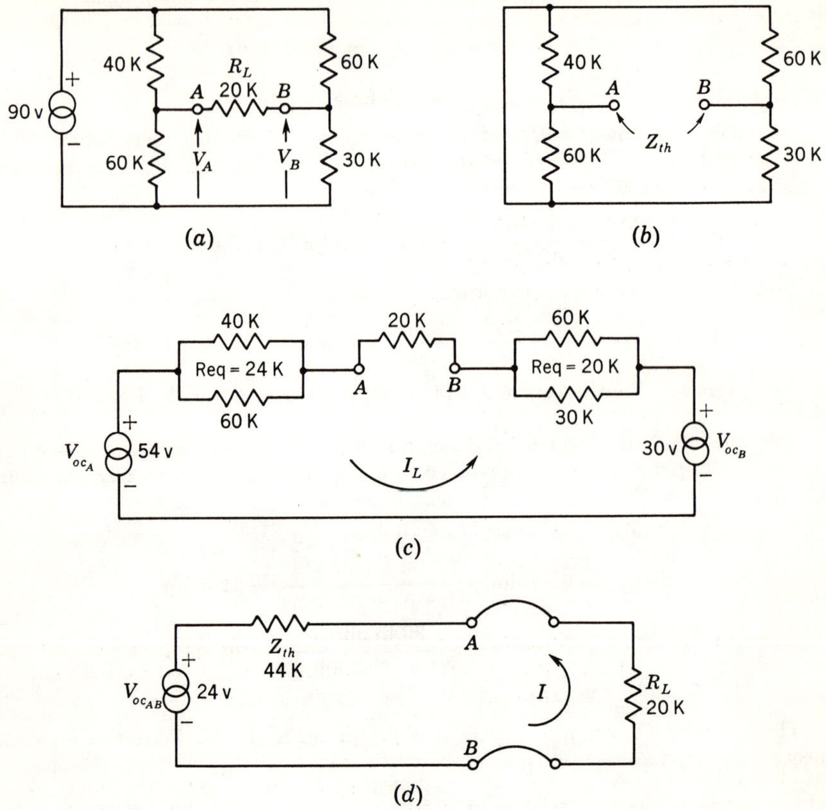

FIG. 1-36. (*a*) Determine the current through R_L. (*b*) The equivalent circuit for determining the source impedance that R_L looks into. (*c*) A preliminary equivalent circuit. (*d*) The final Thévenin's equivalent circuit.

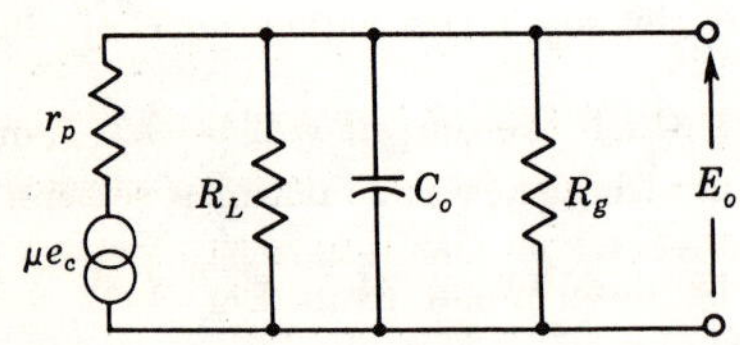

FIG. 1-37. The high-frequency equivalent circuit of a vacuum-tube amplifier.

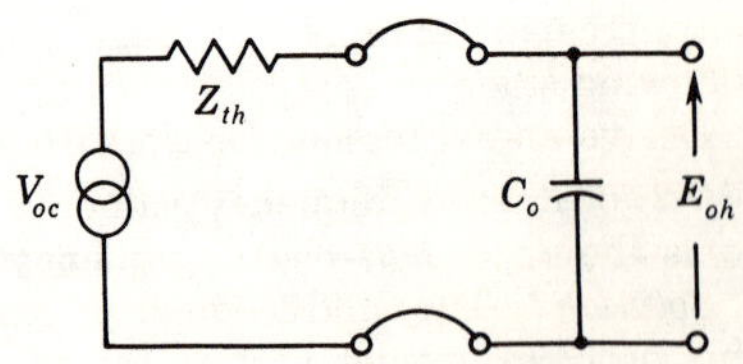

FIG. 1-38. The Thévenin's equivalent circuit that the output capacitance looks back into.

Let us first obtain a general expression for the output at any high frequency, E_{oh}, where C_o must be considered; and then compare it with the mid-frequency output. Since the output appears across the capacitor, as well as across R'_g, let us thevenize, looking back from the capacitor terminals:

$$V_{oc} = \frac{\mu e_c R'_g}{r_p + R'_g}$$

(Note that V_{oc} just happens to equal $E_{o,mid}$.)

$$Z_{th} = \frac{r_p R'_g}{r_p + R'_g}$$

(The resultant circuit is shown in Fig. 1-38.)

$$E_{oh} = \frac{V_{oc}(-jX_C)}{Z_{th} - jX_C} = \frac{E_{o,mid}(-jX_C)}{Z_{th} - jX_C}$$

If our concern is solely with the magnitude of the output, and not with the phase angle, then

$$E_{oh} = \frac{E_{o,mid}\,(X_C)}{\sqrt{Z_{th}^2 + X_C^2}}$$

If
$$E_{oh} = 0.707E_{o,mid} = \frac{1}{\sqrt{2}}E_{o,mid}$$

Then
$$\frac{E_{oh}}{E_{o,mid}} = \frac{1}{\sqrt{2}} = \frac{X_C}{\sqrt{Z_{th}^2 + X_C^2}}$$

Since our interest is at the frequency where this relationship is true, let us first solve for X_C, which is the frequency-containing term. Squaring both sides,

$$\frac{1}{2} = \frac{X_C^2}{Z_{th}^2 + X_C^2}$$

Solving for X_C, $2X_C^2 = Z_{th}^2 + X_C^2$, or $X_C^2 = Z_{th}^2$. Therefore, $X_C = Z_{th}$. At the upper half-power frequency,

$$\frac{1}{2\pi f_C C_o} = Z_{th} \quad \text{and} \quad f = \frac{1}{2\pi Z_{th} C_o}$$

This shows that for higher frequency response, not only must C_o be small, but Z_{th} also. This is why R_L is made small in video amplifiers, even though the mid-frequency output suffers.

Remember the problem at the beginning of this chapter? (See Fig. 1-39*a*.) Let us now use Thévenin's theorem to tell us the voltage V_{GA}. The arrow-bar symbol represents a diode. In this book, the arrowhead is analogous to the plate of a rectifier tube, while the bar is analogous to the cathode (Fig. 1-39*b*).

Now if the anode of a diode is positive in potential with respect to the cathode, it will conduct and can for most purposes be considered as possessing negligible resistance. When conducting, the conducting resistance is referred to as the forward resistance R_f.

If the anode is negative with respect to the cathode, the diode will not conduct, and in most cases may be considered as having an infinite non-conducting, or back, resistance R_b. At times, if the back resistance cannot be neglected, then a specific value will be called out.

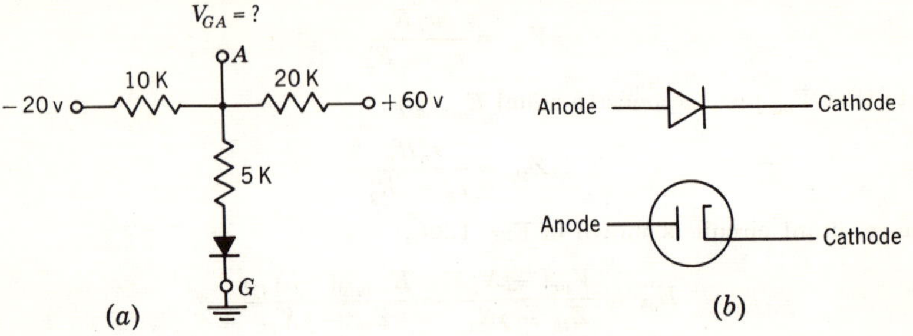

FIG. 1-39. (a) A diode network readily analyzed by Thévenin's theorem. (b) Symbols for a semiconductor and a vacuum-tube diode.

To get back to our problem, the voltage at point A with respect to ground will certainly depend upon whether the diode is or is not conducting. Therefore, our first objective is to establish the state of the diode. We can readily do this by considering the diode and 5-kilohm

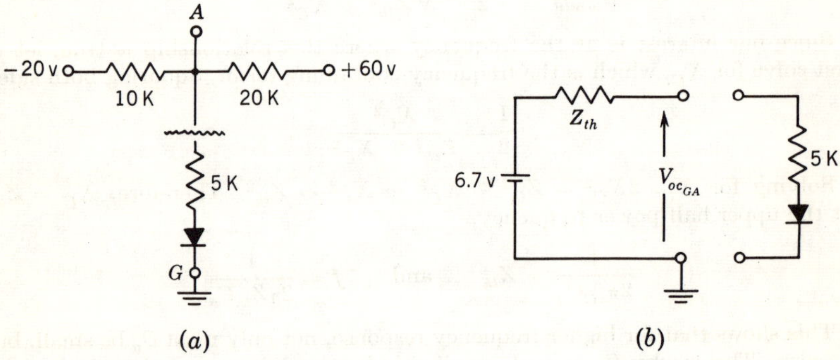

FIG. 1-40. (a) Determining the open circuit voltage that the diode looks into. (b) The simplified equivalent circuit.

resistor as our load and then thevenizing the rest of the network. If the anode side of the diode sees an open-circuit voltage which is positive with respect to the cathode side, the diode conducts. Otherwise, it does not. (See Figs. 1-40a and b.)

$$V_{oc,GA} = -20 + (60 + 20)\,\frac{10 \text{ kilohms}}{10 + 20 \text{ kilohms}}$$

$$= -20 + 80\left(\frac{10}{30}\right) = -20 + 26.7 = 6.7 \text{ volts}$$

Therefore, we see that the open-circuit voltage at point A is of such polarity that the diode will conduct.

To find V_{GA}, we need consider only the voltage divider made up of Z_{th}, the 5-kilohm resistor, and the diode forward resistance, which we can neglect:

$$Z_{th} = \frac{(10)(20)}{10 + 20} = 6.7 \text{ kilohms}$$

$$V_{GA} = 6.7 \text{ volts} \times \frac{5 \text{ kilohms}}{6.7 + 5 \text{ kilohms}}$$

$$= \frac{(6.7)(5)}{11.7} = 2.86 \text{ volts}$$

1-6. Norton's Theorem. Norton's theorem states that any linear network of impedances and generators, if viewed from any two points in the network, can be replaced by an equivalent current source

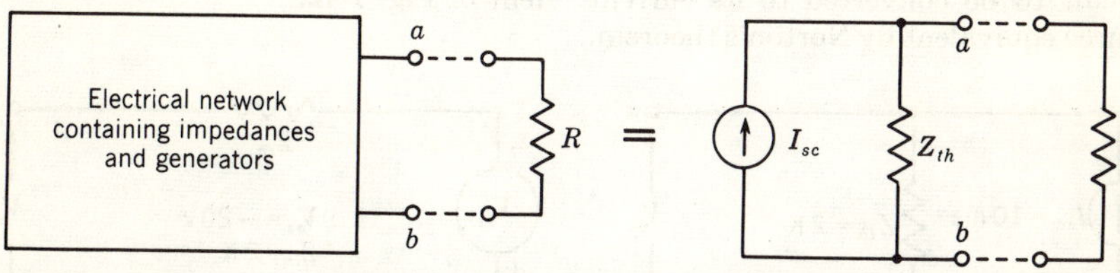

FIG. 1-41. The network to the left of the *a*-*b* terminals may be simplified by Norton's theorem.

I_{sc} and an equivalent impedance Z_{th} in shunt with the current source. This source supplies a current I_{sc} equal to the current that would flow through the two terminals in question if they were shorted. The shunt impedance Z_{th} is the impedance seen looking into the two terminals and is equal to the Thévenin's equivalent impedance. Figure 1-41 illustrates the generalized equivalent current source. The arrowhead in the current source symbol represents the direction of electron flow.

The reader may be puzzled by the concept of a current source. If so, let him imagine a Thévenin's equivalent circuit with an enormously high internal impedance. The Z_{th} is so large that any external load impedance connected in series with Z_{th} will cause the total circuit impedance to *increase* by a negligible amount. The current through the load impedance will then be determined by Z_{th} and V_{oc}, irrespective of the load. Under these circumstances, we say the load "sees" a constant current source. For example, consider a case in which $V_{oc} = 100$ volts and $Z_{th} = 1$ meg. If the output terminals of this Thévenin's equivalent circuit are shorted, the short-circuit current I_{sc} would be 100 μa. Suppose a 1-ohm load were connected between the output terminals. The total circuit resistance is 1,000,001 ohms. For all practical purposes, the current is still 100 μa. Even if the load were 10,000 ohms, the current would still be considered 100 μa for most purposes. An ideal current source would have an infinitely high internal impedance, as compared with zero internal impedance for an ideal voltage source.

Problem 1. Convert the Thévenin's equivalent circuit of Fig. 1-42 to the Norton's in Fig. 1-43, and show that they are the same by solving for the load current in each.

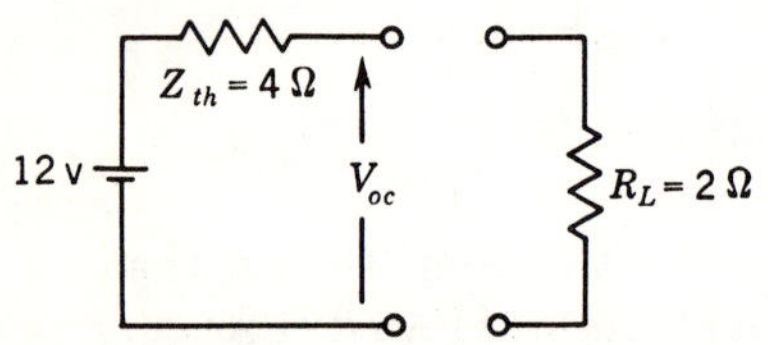

FIG. 1-42. A voltage source equivalent circuit to be converted to its current source equivalent by Norton's theorem.

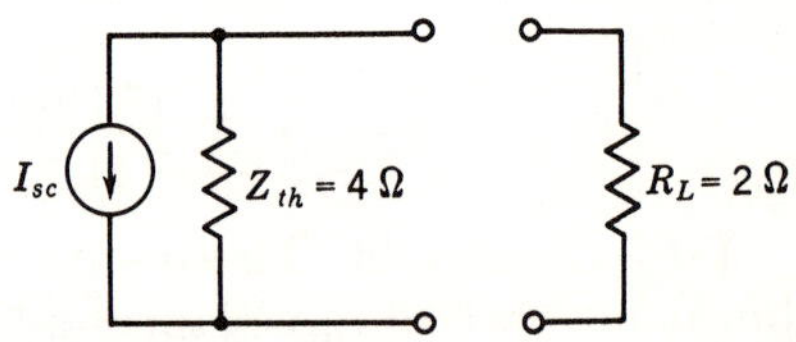

FIG. 1-43. The current source equivalent of Fig. 1-42.

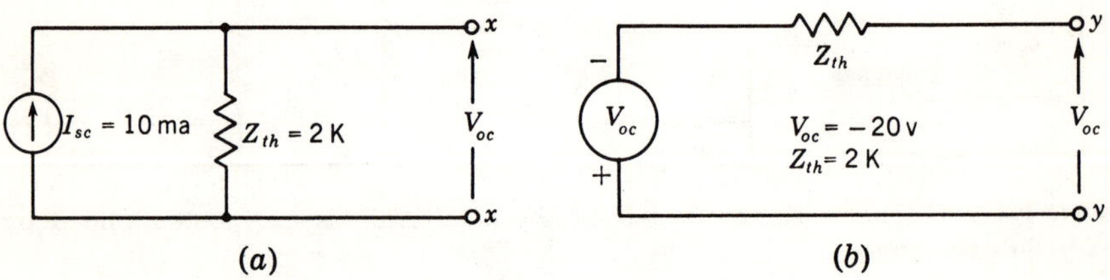

FIG. 1-44. (*a*) A current source equivalent circuit to be converted to its voltage source equivalent. (*b*) The voltage source equivalent of *a*.

Solution. First, find the short-circuit current I_{sc} between the two output terminals:

$$I_{sc} = \frac{V_{oc}}{Z_{th}} = \frac{12 \text{ volts}}{4 \text{ ohms}} = 3 \text{ amp}$$

Now draw the Norton's equivalent. Solve for the load current in Fig. 1-43 by using the current-divider relationship:

$$I_L = I_{sc} \frac{Z_{th}}{Z_{th} + R_1} = 3 \text{ amp} \times \frac{4 \text{ ohms}}{4 + 2 \text{ ohms}} = 2 \text{ amp}$$

Solve for I_L in the Thévenin's circuit of Fig. 1-42:

$$I_L = \frac{V_{oc}}{Z_{th} + R_1} = \frac{12 \text{ volts}}{4 + 2 \text{ ohms}} = 2 \text{ amp}$$

The results are identical.

Problem 2. Convert the Norton's equivalent of Fig. 1-44*a* to the Thévenin's equivalent as shown in Fig. 1-44*b*.

Solution. If the circuits are to be equivalent, then the voltage measured between terminals *xx* must equal the voltage between the terminals *yy*. Also

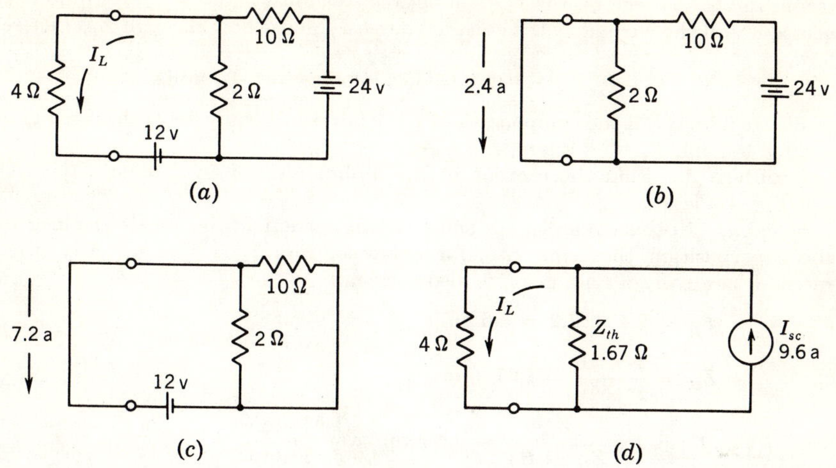

FIG. 1-45. (a) The current source equivalent circuit that the 4-ohm load looks into is to be determined. (b) The short-circuit current may be determined by the superposition theorem. Only the 24-volt source is considered in this figure. (c) The short-circuit current due to the 12-volt source. (d) The Norton's equivalent circuit for 1-45a.

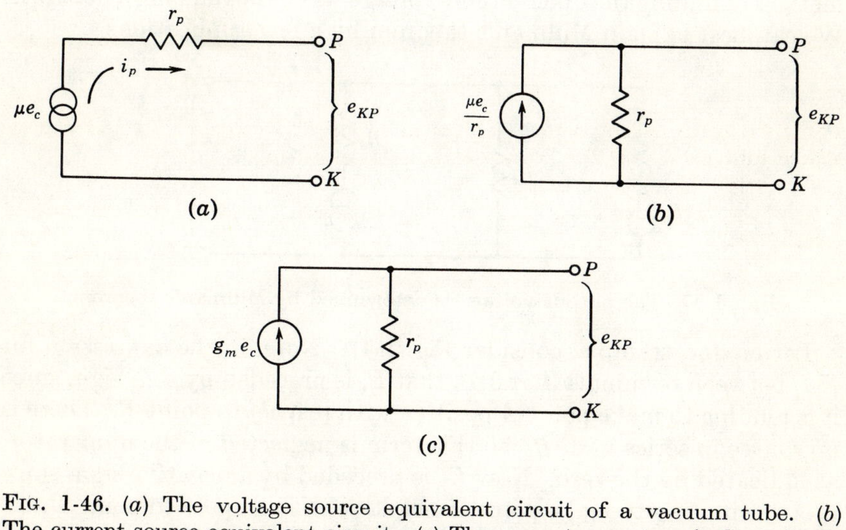

FIG. 1-46. (a) The voltage source equivalent circuit of a vacuum tube. (b) The current source equivalent circuit. (c) The current source equivalent circuit in terms of the tube transconductance and plate resistance.

the voltage between terminals yy with no load must be the V_{oc} seen looking into terminals xx. Therefore, the voltage between xx must equal V_{oc}. With no load across xx, all the current of the current source I_{sc} must flow through Z_{th}. Therefore,

$$V_{oc} = -I_{sc}Z_{th} = -10 \text{ ma} \times 2 \text{ kilohms} = -20 \text{ volts}$$

We then know all the components of the Thévenin's equivalent, that is, $V_{oc} = -20$ volts, and $Z_{th} = 2$ kilohms. *Ans.*

Problem 3. Find the current in the 4-ohm resistor by Norton's theorem (see Fig. 1-45*a* and *b*).

Solution. In this example, the short-circuit current I_{sc} is first determined by the superposition theorem. See, for example, Fig. 1-45*c* and *d*. Any other method, however, of finding I_{sc} is also correct

$$I_{sc} = 2.4 + 7.2 = 9.6 \text{ amp}$$

$$Z_{th} = \frac{2 \times 10}{2 + 10} = 1.67 \text{ ohms}$$

$$I_L = I_{sc}\frac{Z_{th}}{Z_{th} + R_L} = 9.6 \text{ amp} \times \frac{1.67 \text{ ohms}}{5.67 \text{ ohms}} = 2.82 \text{ amp}$$

Problem 4. Convert the circuit of Fig. 1-46*a* from the voltage source form to an equivalent current source.

Solution. Figure 1-46*c* is derived from Fig. 1-46*b* by using the relationship $\mu = g_m r_p$, to which the reader may have been exposed in basic electronics.

1-7. Millman's Theorem.

Millman's theorem is an excellent method of finding the open-circuit voltage when thevenizing a network. We can best explain Millman's theorem by a few applications.

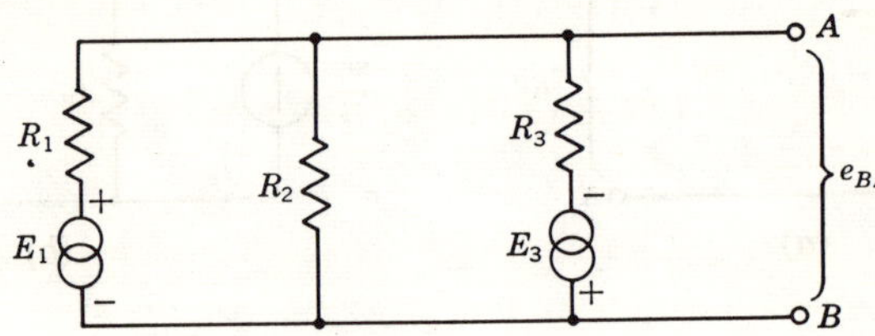

FIG. 1-47. The output voltage is determined by Millman's theorem.

Let us, for example, consider Fig. 1-47. Note, in the expression for E_{BA} between terminals A and B, that E_1 is preceded by a + sign, since it is tending to make point A positive with respect to point B. There is no source in series with R_2, so this term is neglected in the numerator, as indicated by the zero. Now E_3 is preceded by a negative sign, since it is attempting to make point A negative with respect to point B.

$$E_{BA} = \frac{\dfrac{E_1}{R_1} + \dfrac{0}{R_2} + \dfrac{(-E_3)}{R_3}}{\dfrac{1}{R_1} + \dfrac{1}{R_2} + \dfrac{1}{R_3}}$$

Problem 1. Find the currents in Fig. 1-48*a* by Millman's theorem.

Solution

Let $E_1 = 12$ volts $\quad E_2 = 0 \quad E_3 = -24$ volts

$R_1 = 4$ ohms $\quad R_2 = 2$ ohms $\quad R_3 = 10$ ohms

$I_1 = ? \quad I_2 = ? \quad I_3 = ?$

$$E_{BA} = \frac{\dfrac{12\text{ volts}}{4\text{ ohms}} + \dfrac{0}{2\text{ ohms}} + \dfrac{-24\text{ volts}}{10\text{ ohms}}}{\dfrac{1}{4\text{ ohms}} + \dfrac{1}{2\text{ ohms}} + \dfrac{1}{10\text{ ohms}}} = \frac{3 - 2.4}{0.25 + 0.5 + 0.1}$$

$$= \frac{0.6}{0.85} = 0.706 \text{ volts}$$

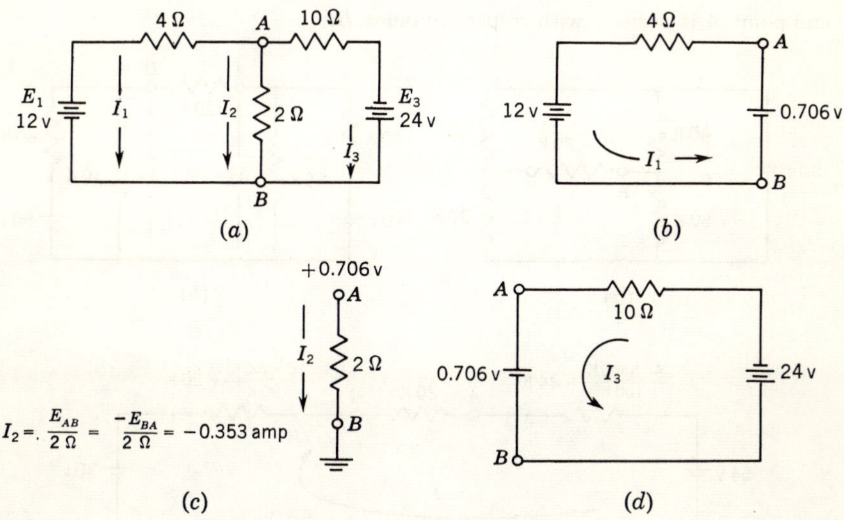

FIG. 1-48. (*a*) Using Millman's theorem, we find $E_{BA} = 0.706$ volt. (*b*) With E_{BA} known, I_1 may be evaluated from this equivalent circuit. (*c*) I_2 must flow opposite to the assumed positive direction, since point A is positive with respect to point B. Remember that we are using electron flow. (*d*) The current I_3 is readily evaluated, once E_{BA} is known.

To find the currents (see Fig. 1-48*b*, *c*, and *d*),

$$I_1 = \frac{12 - 0.706\text{ volts}}{4\text{ ohms}} = 2.82\text{ amp}$$

$$I_2 = \frac{0.706\text{ volts}}{2\text{ ohms}} = -0.353\text{ amp}$$

$$I_3 = \frac{24 + 0.706\text{ volts}}{10\text{ ohms}} = 2.47\text{ amp}$$

Let us work further with Millman's theorem.

Problem 2. Suppose we try to solve the bridge network of Fig. 1-49*a*, which was previously discussed in the section on Thévenin's theorem, by using Millman's theorem.

Solution. Redraw the problem (Fig. 1-49*a*) as shown in Fig. 1-49*b*. With R_L out, the open-circuit voltages are

$$V_{A_{oc}} = \frac{90/40 + 0/60}{1/40 + 1/60} = \frac{90/40}{100/2{,}400} = 54 \text{ volts}$$

$$V_{B_{oc}} = \frac{90/60 + 0/30}{1/60 + 1/30} = \frac{90/60}{90/1{,}800} = 30 \text{ volts}$$

We cannot at this point assume that $V_{A_{oc}} - V_{B_{oc}} = V_{AB}$, with R_L in the circuit. We must thevenize, looking into A and B. The final equivalent circuit is then as shown in Fig. 1-49*c*. The voltage across the 20-kilohm load is then

$$V_{BA} = \frac{(54 - 30) \text{ volts} \times 20 \text{ kilohms}}{64 \text{ kilohms}} = +7.5 \text{ volts}$$

and point A is positive with respect to point B.

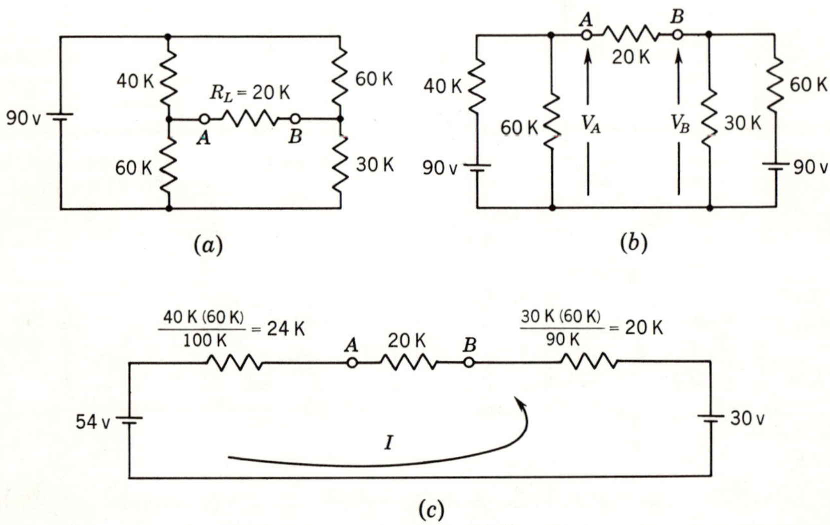

FIG. 1-49. (*a*) Millman's theorem is used in this problem to evaluate the open-circuit voltages at points A and B. (*b*) The circuit of *a* redrawn. (*c*) Millman's theorem in conjunction with Thévenin's theorem yields this equivalent circuit for *a*.

Problem 3. In Fig. 1-50*a*, K_1 and K_2 are relays that pull in with 2 ma. Which relay, if any, is closed? We can consider the forward or conducting diode resistance, R_f, as zero; and the back resistance of the nonconducting diode as infinite.

Solution. Break the circuit at X, and by Millman's theorem find V_{oc} with respect to ground.

$$V_{oc,GX} = \frac{(40/10 + 200/60 - 100/30)10^{-3}}{(1/10 + 1/60 + 1/30)10^{-3}}$$

$$\frac{4 + 3.33 - 3.33}{0.1 + 0.0167 + 0.033} = \frac{4}{0.1500} = \frac{40}{1.5} = 26.6 \text{ volts}$$

The equivalent source impedance looking into point X and ground is

$$Z_{th} = \frac{1}{(1/10 + 1/60 + 1/30)10^{-3}} = \frac{10^3}{0.150} = 6.8 \text{ kilohms}$$

Since point X is positive with respect to ground, D_2 will conduct, and K_2 will try to close. If the current through $K_2 \geqq 2$ ma, it *will* close.

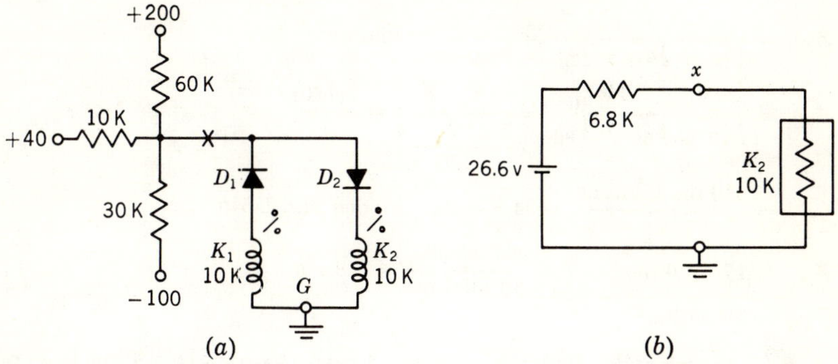

FIG. 1-50. (*a*) Millman's theorem is used to determine which, if any, relay is closed. (*b*) The equivalent circuit that K_2 looks into.

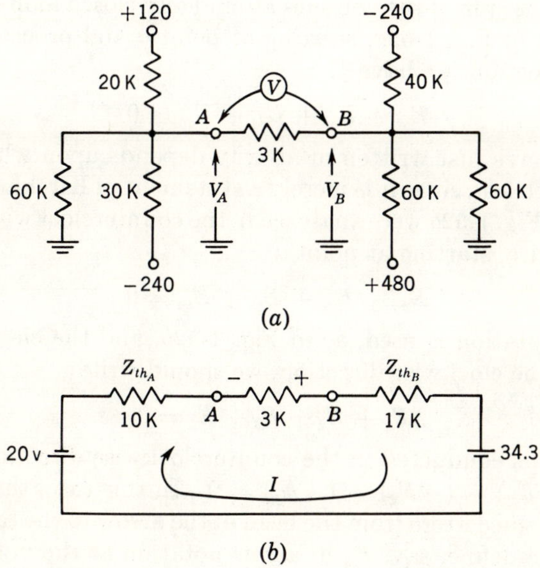

FIG. 1-51. (*a*) A complex circuit in which the voltmeter reading is to be determined. (*b*) An equivalent circuit.

The equivalent circuit is then as in Fig. 1-50*b*; that is,

$$I = \frac{26.6 \text{ volts}}{16.8 \text{ kilohms}} = 1.6 \text{ ma}$$

Apparently K_2 cannot close.

Problem 4. What voltage would a meter read when connected between terminals A and B in Fig. 1-51a?

Solution

$$V_{a_{oc}} = \frac{120/20 - 240/30}{1/20 + 1/60 + 1/30} = \frac{6 - 8}{\dfrac{3 + 1 + 2}{60}} = \frac{-2(60) \text{ ma kilohms}}{6} = -20 \text{ volts}$$

$$Z_{th_A} = \frac{1}{1/20 + 1/60 + 1/30} = \frac{1}{6/60} = 10 \text{ kilohms}$$

$$V_{B_{oc}} = \frac{480/60 - 240/40}{1/40 + 1/60 + 1/60} = \frac{8 - 6}{\dfrac{3 + 2 + 2}{120}} = \frac{2(120)}{7}$$

$$= \frac{240 \text{ ma kilohms}}{7} = 34.3 \text{ volts} \qquad \text{(See Fig. 1-51}b\text{.)}$$

$$Z_{th_B} = 17 \text{ kilohms} \qquad I = \frac{54.3 \text{ volts}}{30 \text{ kilohms}} = 1.8 \text{ ma}$$

$$V_{AB} = 5.4 \text{ volts}$$

1-8. Kirchhoff's Voltage Law—Loop Analysis. This law provides a systematic method of treating circuits so that the currents in complex networks can be evaluated. Simply stated, the voltage law tells us that the sum of the voltages around any closed loop equals zero. For example, in Fig. 1-52a, starting at point a and proceeding in the clockwise direction, we have

$$E_{ab} + E_{bc} + E_{cd} + E_{da} = 0$$

What we have just written in no way depends upon what is in the boxes of Fig. 1-52a, since it is merely a statement of Kirchhoff's law. If the circuit of Fig. 1-52a were analyzed in the counterclockwise direction, we should write, starting at point a,

$$E_{ad} + E_{dc} + E_{cb} + E_{ba} = 0$$

If arrow notation is used, as in Fig. 1-52b, and the circuit is again analyzed in the clockwise direction, we should write

$$E_1 + E_2 + E_3 + E_4 = 0$$

If analysis is conducted in the counterclockwise direction, we have $(-E_4) + (-E_3) + (-E_2) + (-E_1) = 0$. In this case, the quantities are negative, since we go from the head of the arrow to the tail. You will recall that we define, say, E_4 in arrow notation as the voltage at the head of the arrow with respect to the tail. We have also agreed that E_4 will be a positive quantity when the head of the arrow lies on a point more positive than the tail. If the reason for the sign inversion is not clear, especially because nothing is indicated in the boxes, consider Fig. 1-52c. Plainly, $E_A = +10$ volts, while $E_B = -10$ volts, by our definition. A minus E_A must therefore equal -10 volts, but this is identically equal to E_B. We see, then, that $-E_A$ equals $+E_B$, or that going from the head to the tail of E_A is the same as the path from tail to head of E_B (see Fig. 1-52*d*). Perhaps this is explaining the obvious, but understanding the notation can be very important. Let us try a few problems.

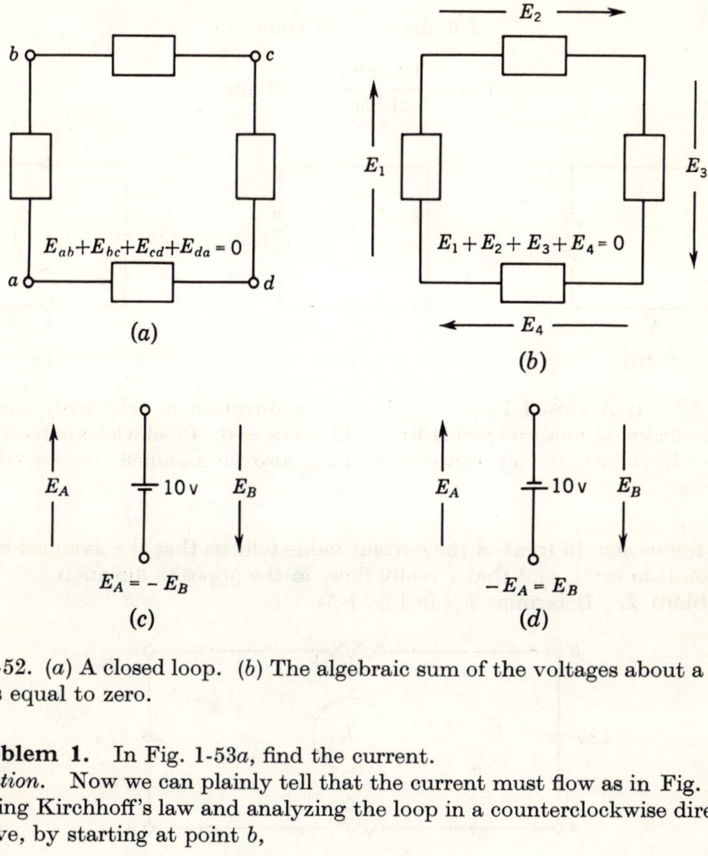

Fig. 1-52. (*a*) A closed loop. (*b*) The algebraic sum of the voltages about a closed loop is equal to zero.

Problem 1. In Fig. 1-53*a*, find the current.

Solution. Now we can plainly tell that the current must flow as in Fig. 1-53*b*. Applying Kirchhoff's law and analyzing the loop in a counterclockwise direction, we have, by starting at point *b*,

$$E_{ba} + E_{ab} = 0$$

$$I(6 \text{ ohms}) + (-12 \text{ volts}) = 0$$

$$I = \frac{12 \text{ volts}}{6 \text{ ohms}} = 2 \text{ amp}$$

If we chose to analyze the circuit in the clockwise direction, and again started at *b*, we would have

$$E_{ba} + E_{ab} = 0$$

$$12 \text{ volts} - I(6 \text{ ohms}) = 0$$

$$I = \frac{12 \text{ volts}}{6 \text{ ohms}} = 2 \text{ amp}$$

In both cases, I is a positive quantity, which indicates that the direction assumed for I in Fig. 1-53*b* is correct.

Suppose, however, that we were not really sure in which direction I flows in Fig. 1-53*a* and arbitrarily assumed that it flows as in Fig. 1-53*c*. If we start at b and proceed in the counterclockwise direction, we have

$$E_{ba} + E_{ab} = 0$$

$$-I(6 \text{ ohms}) - 12 \text{ volts} = 0$$

$$I = \frac{-12 \text{ volts}}{6 \text{ ohms}} = -2 \text{ amp}$$

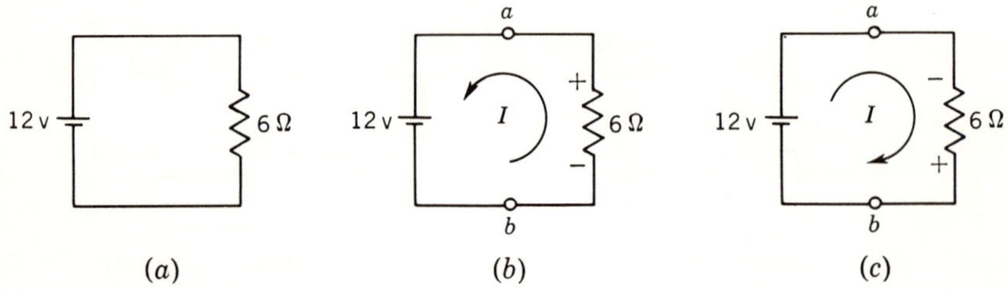

FIG. 1-53. (*a*) A closed loop. (*b*) A current direction is arbitrarily assumed. Counterclockwise analysis yields $6I_1 - 12$ volts $= 0$. Clockwise analysis yields $-6I_1 + 12$ volts $= 0$. (*c*) The current may also be assumed to flow counterclockwise.

The minus sign in front of the current value tells us that the assumed current direction is in error, and that I really flows in the opposite direction.

Problem 2. Determine E_{cb} in Fig. 1-54.

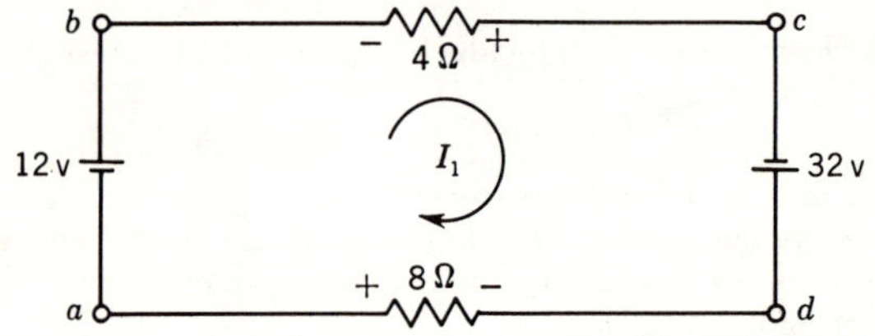

FIG. 1-54. The current is assumed to flow in a clockwise direction about the loop. Summing voltages about the loop in a clockwise direction yields 12 volts + $4I_1$ + 32 volts + $8I_1 = 0$.

Solution. Assume a clockwise current direction, and analyze the circuit in a clockwise direction also. Under these conditions, we should write, starting at point a,

$$E_{ab} + E_{bc} + E_{cd} + E_{da} = 0$$

$$+12 \text{ volts} + 4I_1 + 32 \text{ volts} + 8I_1 = 0$$

$$12I_1 = -44$$

$$I_1 = -3.7 \text{ amp}$$

If the circuit has more than one loop, a current direction is assumed for each loop. The choice of direction is arbitrary and need not be the same in each loop. For example, consider Fig. 1-55*a*. Assuming the directions shown, and starting at point *a*, then analyzing loop one (the current I_1 flows in loop 1) in the clockwise direction, we have

$$E_{ab} + E_{bc} + E_{ce} + E_{ea} = 0$$

$$+360 + (20 \times 10^3)I_1 + 60 \times 10^3(I_1 - I_2) - 240 = 0$$

At this point, the reader may feel unsure about the $60 \times 10^3(I_1 - I_2)$ term. We have $I_1 - I_2$ because the assumed current directions through R_3 are opposite and the net current is the difference between the two. A question might arise as to

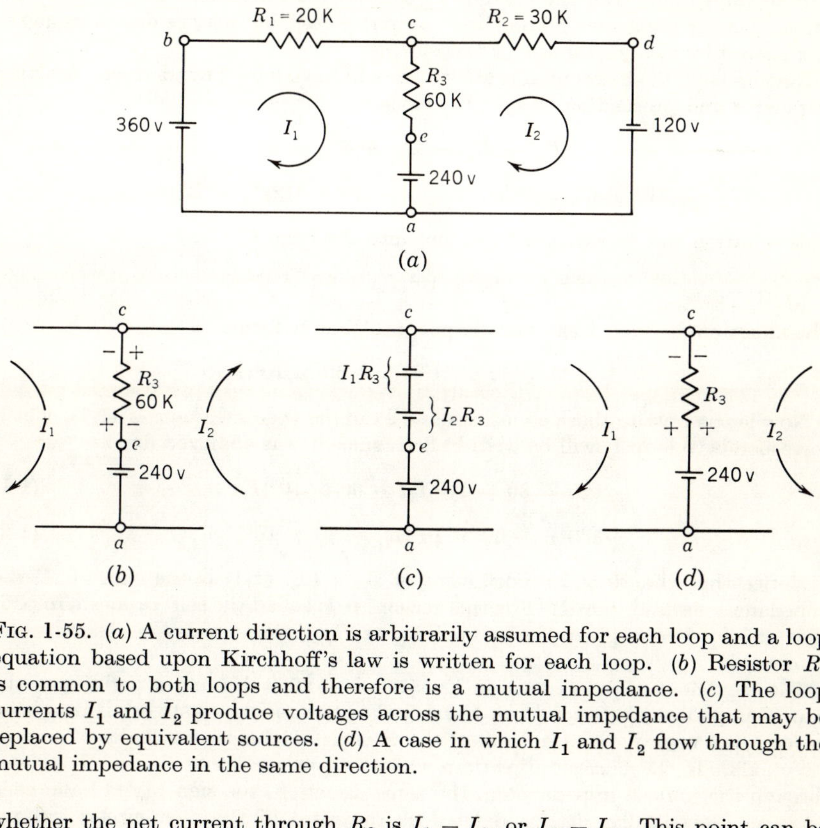

FIG. 1-55. (*a*) A current direction is arbitrarily assumed for each loop and a loop equation based upon Kirchhoff's law is written for each loop. (*b*) Resistor R_3 is common to both loops and therefore is a mutual impedance. (*c*) The loop currents I_1 and I_2 produce voltages across the mutual impedance that may be replaced by equivalent sources. (*d*) A case in which I_1 and I_2 flow through the mutual impedance in the same direction.

whether the net current through R_3 is $I_1 - I_2$, or $I_2 - I_1$. This point can be resolved by examining the branch between terminals *c* and *a*, as shown in Fig. 1-55*b*. The current I_1 polarizes the top end of R_3 negative with respect to the bottom end, as indicated by the left-hand polarity marks on R_3. The current I_2 polarizes R_3, as shown by the right-hand polarity marks. The voltage developed between terminals *c* and *e*, which is due to I_1, will be I_1R_3 volts. The voltage developed between terminals *c* and *e*, which is due to I_2, will be I_2R_3 volts. We could then draw an equivalent circuit based upon these considerations (Fig.

1-55c). Proceeding now in the same clockwise direction from point c to point a, we have

$$E_{ca} = E_{ce} + E_{ea}$$

where $E_{ea} = -240$ volts

$$E_{ce} = +I_1R_3 - I_2R_3 = R_3(I_1 - I_2)$$

$$E_{ca} = R_3(I_1 - I_2) - 240$$

If loop 1 were analyzed in the counterclockwise direction, and with I_1 still assumed in the clockwise direction, we would have written $R_3(I_2 - I_1)$ for E_{ec}. While it is not necessary, we will usually analyze a loop in the same direction in which current has been assumed. The impedance R_3 is common to both loops and for this reason can be considered as a mutual impedance. It is because of this mutual impedance that the current I_2 couples a voltage into loop 1, and vice versa. The coupling need not be direct as in this case, but may be due to magnetic flux shared by two windings in a transformer.

Getting back to the original problem, we still have loop 2 to traverse. Starting at point a and proceeding, say, in the clockwise direction, we have

$$E_{ae} + E_{ec} + E_{cd} + E_{da} = 0$$

$$+240 + 60 \times 10^3(I_2 - I_1) + (30 \times 10^3)I_2 + 120 = 0$$

This equation can be expanded and put into the form

$$-360 = -60 \times 10^3(I_1) + 90 \times 10^3(I_2)$$

The equation for loop 1 can also be put into similar form:

$$-120 = 80 \times 10^3(I_1) - 60 \times 10^3(I_2)$$

Now let us rewrite these equations and examine them. Equation (1-1), which corresponds to loop 1, will be written first, since it was analyzed first:

$$-120 = 80 \times 10^3(I_1) - 60 \times 10^3(I_2) \tag{1-1}$$

$$-360 = -60 \times 10^3(I_1) + 90 \times 10^3(I_2) \tag{1-2}$$

Notice that the 80×10^3 coefficient of I_1 in Eq. (1-1) is the sum of all the impedances around loop 1. For this reason, it is called the self- or mesh, impedance of loop 1, and given the symbol Z_{11} or R_{11} if pure resistive elements are involved.

The coefficient of the I_2 term in Eq. (1-1) is -60×10^3 and is defined as the mutual impedance between loops 1 and 2. For this resistive network it is written as R_{12}. The negative sign in front of the 60×10^3 is due to our choice of current directions. If the current directions chosen were such that I_1 and I_2 flowed through the mutual impedance in the same direction, the sign would have been positive. Specifically, R_{12} is defined as the ratio of the open-circuit voltage induced in loop 1 by a current in loop 2, with all loops except loop 2 open-circuited. In this particular problem, the voltage induced in loop 1 by I_2 is $-I_2R_3$ volts. The sign is minus because this coupled voltage is a negative quantity if loop 1 is analyzed in the same direction in which I_1 is assumed to flow. Therefore,

$$R_{12} = \frac{\text{voltage induced in loop 1}}{\text{current in loop 2}} = \frac{-I_2R_3}{I_2} = -R_3$$

If we examine the equation for loop 2, we see that the coefficient of the I_1 term again is the impedance mutual to loops 1 and 2. Since we are going around loop 2, this coefficient is given the symbol R_{21}:

$$R_{21} = \frac{\text{voltage induced in loop 2}}{\text{current in loop 1}} = \frac{-I_1R_3}{I_1} = -R_3$$

In simple circuits which do not contain amplifiers, R_{12} will always equal R_{21}. The coefficient (90×10^3) of the I_2 term in loop 2 is the sum of all the impedances about loop 2 and is given the symbol R_{22}.

Equations (1-1) and (1-2) constitute a pair of simultaneous linear equations that can be solved by the method of substitution:

Loop (1) $\quad -120 = (80 \times 10^3)I_1 - (60 \times 10^3)I_2$

Loop (2) $\quad -360 = -(60 \times 10^3)I_1 + (90 \times 10^3)I_2$

Solving for I_1 in (1-1), we have

$$I_1 = \frac{-120 + (60 \times 10^3)I_2}{80 \times 10^3} \tag{1-1a}$$

Substituting (1-1*a*) in (1-2),

$$-360 = -(60 \times 10^3)\frac{-120 + (60 \times 10^3)I_2}{80 \times 10^3} + (90 \times 10^3)I_2 \tag{1-2a}$$

and solving for I_2 in (1-2*a*), we obtain

$$I_2 = -10 \text{ ma} \tag{1-2b}$$

Substituting (1-2*b*) in (1-1*a*), and solving for I_1, we have

$$I_1 = \frac{-120 + 60 \times 10^3(-10 \text{ ma})}{80 \times 10^3} = -9 \text{ ma}$$

Kirchhoff's voltage law may be given a physical interpretation. If the sum of the voltages about any closed loop is to equal zero, some of the elements within the loop must be acting as sources, or generators, of voltage and others as sinks, or users, of voltage. In any closed loop, the amount of voltage generated must equal that used up. We can therefore write

$$\Sigma \text{ source voltages} = \Sigma \text{ sink voltages}$$

where Σ is read as "the sum of."

We shall consider batteries, generators, and induced, or coupled, voltages as sources. The voltages developed by the loop current under consideration will be taken as sinks. Clearly, any source can be treated as a sink, and vice versa, if we transpose from one side of the equation to the other with an appropriate sign change. The details of this method of loop analysis are best illustrated with a few examples.

Consider Fig. 1-53*b*, in which an assumed current direction is indicated. Starting at point *a* and following the current around in the assumed direction, we encounter the 12-volt battery, which is a source. Specifically, it is a positive source, since the polarity of the battery is such that it is trying to pump electrons in the direction we have assumed them to flow. We would therefore write +12 volts on the source side. If you go in at the plus side of a source and come out the minus side, it is a positive source on the source side of the equation. We now follow the current through the 6-ohm resistor, where it develops a sink voltage of $6I$ volts, which is written as $+6I$ volts on the sink side of the equation. This brings us back to our starting point. We have

$$\Sigma \text{ sources} = \Sigma \text{ sinks}$$

$$+12 \text{ volts} = 6 \text{ ohms}(I)$$

$$I = \frac{12 \text{ volts}}{6 \text{ ohms}} = 2 \text{ amp}$$

Notice that I is a positive quantity, which indicates that the assumed current direction is correct.

Now consider Fig. 1-53*c*. Starting at *a* and following I around in the assumed direction, we encounter a positive sink voltage equal to $6I$ volts, which is written with a plus sign on the sink side. Coming out of the resistor, we encounter a 12-volt source which is trying to pump electrons out of its negative terminal, that is, opposite to our assumed direction of electron flow. It therefore looks like a negative, or bucking, source and is written with a minus sign on the source side of our equation. We then have

$$\Sigma \text{ sources} = \Sigma \text{ sinks}$$

$$-12 \text{ volts} = 6I$$

$$I = -\frac{12 \text{ volts}}{6 \text{ ohms}} = -2 \text{ amp}$$

Notice that the current now has a negative sign before it, which indicates that the assumed direction was in error.

If the reader will apply the same techniques to Fig. 1-54*a*, the results will be

$$\Sigma \text{ sources} = \Sigma \text{ sinks}$$

$$-12 - 32 \text{ volts} = 4I_1 + 8I_1$$

$$-44 \text{ volts} = 12I_1$$

$$I_1 = \frac{-44 \text{ volts}}{12 \text{ ohms}} = -3.7 \text{ amp}$$

As a final illustration, let us try Fig. 1-55*a*. If we follow I_1 around its loop in the assumed direction, the 360-volt supply appears as a negative source, while the 240-volt supply looks like a positive source to I_1. The positive sink voltages around loop 1 are $I_1(20 \times 10^3) + I_1(60 \times 10^3)$ or

$I_1(80 \times 10^3)$. At first glance, this might seem to take care of everything in loop 1. But what about the effect of the voltages that I_2 couples into loop 1? The current I_2, in flowing through the mutual impedance R_3, develops a voltage across R_3 having the polarity indicated by the plus and minus signs on the right-hand side of R_3 in Fig. 1-55*b*. If we replace this coupled voltage with an equivalent battery having the same relative polarity, we should draw the I_2R_3 battery of Fig. 1-55*c*. Notice that this coupled battery is of such polarity that it looks like a positive or aiding source to the current I_1. Therefore, we should write a plus I_2R_3 on the source side. However, since it is desirable to keep the unknown current terms on the same side of the equation, we shall consider the I_2R_3 term as a *negative* sink. Our circuit equation for loop 1 is, then,

$$-360 + 240 = (20 \times 10^3 + 60 \times 10^3)I_1 - (60 \times 10^3)I_2$$

$$-120 = (80 \times 10^3)I_1 - (60 \times 10^3)I_2 \qquad (a)$$

Now we must pursue I_2 around loop 2 in its assumed direction. In so far as I_2 is concerned, both the 240- and 120-volt batteries look like negative sources. The positive sink voltages developed by I_2 are $I_2R_3 + I_2R_2$, or $(90 \times 10^3)I_2$. We must still allow for the voltage coupled into loop 2 by the current I_1. The polarity of the voltage developed across R_3 by I_1 is indicated by the plus and minus signs on the left-hand side of R_3 in Fig. 1-55*b*. This voltage across R_3 due to I_1 is coupled into loop 2, and is equivalent to the I_1R_3 battery of Fig. 1-55*c*. Note that for the assumed direction of I_2, this coupled voltage looks like a positive, or aiding, source. We should then add a plus I_2R_3 term on the source side, or a minus I_2R_3 term on the sink side. The loop equation for loop 2 is, then,

$$-240 - 120 = -(60 \times 10^3)I_1 + (90 \times 10^3)I_2$$

$$-360 = -(60 \times 10^3)I_1 + (90 \times 10^3)I_2 \qquad (b)$$

Note that the two equations (*a*) and (*b*) developed by equating sources and sinks are identical with equations previously obtained by equating all loop voltages to zero. Both approaches should be mastered, since the source-sink approach sometimes gives a better feeling for what is going on in a circuit. Both types of analysis are used in this text.

If the assumed direction of I_2 were reversed in Fig. 1-55*a*, the loop equations would be

$$-120 = (80 \times 10^3)I_1 + (60 \times 10^3)I_2 \qquad (a)$$

$$360 = (60 \times 10^3)I_1 + (90 \times 10^3)I_2 \qquad (b)$$

With this assumed direction for I_2, the voltages developed by I_1 and I_2 will polarize R_3, as shown in Fig. 1-55*d*. The I_2R_3 voltage coupled into loop 1 is now a negative source, or positive sink. Similarly, the voltage coupled into loop 2 because of I_1 is also a negative source, or positive sink. You will recall that we stated that the mutual impedance is a positive quantity if the assumed loop currents flow through it in the same direction and is a negative quantity if the current directions are opposite through it. From a source-sink point of view, if two loop currents flow through a mutual impedance in the same direction, the coupled voltages will appear as negative sources in the adjacent loop. For example, consider Fig. 1-55*d*. We saw that I_2 coupled a voltage into loop 1 which bucked I_1. Therefore, I_2R_3 was a negative source, or positive sink, and there was a plus sign in front of the mutual term on the sink side.

When I_1 and I_2 flow through the mutual impedance in opposite directions, as in Fig. 1-55*b*, the coupled voltages appear as positive sources, or negative sinks. In this case, the sign before the mutual term is negative on the sink side. Just remember that when the voltage coupled into loop k, due to a current in loop j, looks like an opposing source, or positive sink, to the assumed direction of I_k, the sign of the mutual impedance is positive; otherwise, it is negative. Mutual impedance is also discussed in Chap. 4.

1-9. Determinants. The majority of simultaneous equations encountered in this text will be of the form

$$v_1 = i_1r_{11} + i_2r_{12} \tag{1-3}$$

$$v_2 = i_1r_{21} + i_2r_{22} \tag{1-4}$$

Usually, v_1, v_2, r_{11}, r_{12}, r_{21}, and r_{22} will be known. The problem will be to evaluate the currents i_1 and i_2. A solution can always be obtained by substitution. That is, we can solve for i_1 in terms of i_2 in Eq. (1-3) and obtain

$$i_1 = \frac{v_1 - i_2r_{12}}{r_{11}}$$

Substituting this value of i_1 in Eq. (1-4) gives us

$$v_2 = \left(\frac{v_1 - i_2r_{12}}{r_{11}}\right)r_{21} + i_2r_{22}$$

The only unknown is now i_2, which can be solved for, to yield

$$i_2 = \frac{v_2r_{11} - v_1r_{21}}{r_{22}r_{11} - r_{12}r_{21}}$$

Substituting this value of i_2 in Eq. (1-3) gives us

$$v_1 = i_1r_{11} + \left(\frac{v_2r_{11} - v_1r_{21}}{r_{22}r_{11} - r_{12}r_{21}}\right)r_{12}$$

which can be solved for i_1, to yield

$$i_1 = \frac{v_1 r_{22} - v_2 r_{12}}{r_{22} r_{11} - r_{12} r_{21}}$$

If you will go through the labor of the intermediate steps, I am sure you will welcome an easier method of finding i_1 and i_2. Furthermore, consider the fact that even though you may want to solve only for either i_1 or i_2, you still must solve for one in terms of the other by the method of substitution.

We shall now discuss a method known as solution by determinants for solving simultaneous equations. The main advantages of this method are speed and decreased chance of error as the method becomes more mechanical.

A determinant is simply an orderly array of rows and columns involving the known coefficients of the variable (r values) and the known dependent variables (v terms). At this point, I must ask the reader to accept what follows without any particular reason except that it works. Just perform the indicated operations, which will serve solely as a mathematical tool.

Given

$$v_1 = i_1 r_{11} + i_2 r_{12} \tag{1-3}$$

$$v_2 = i_1 r_{21} + i_2 r_{22} \tag{1-4}$$

If it is i_1 for which you wish to solve, write the following array of rows and columns.

$$i_1 = \frac{\begin{vmatrix} v_1 & r_{12} \\ v_2 & r_{22} \end{vmatrix}}{\begin{vmatrix} r_{11} & r_{12} \\ r_{21} & r_{22} \end{vmatrix}}$$

The vertical lines merely represent boundaries of the array, and the horizontal line represents the ratio between the array in the numerator and the array in the denominator. We shall perform certain operations on the numerator and denominator, and wind up with some number in the numerator and some number in the denominator. The ratio of these numbers is i_1.

Notice that the array in the denominator is simply the coefficients of the i terms in the same order as they appear in the original equations:

$$v_1 = \textcircled{r_{11}}\, i_1 + \textcircled{r_{12}}\, i_2$$

$$v_2 = \textcircled{r_{21}}\, i_1 + \textcircled{r_{22}}\, i_2$$

Denominator determinant: $\begin{vmatrix} r_{11} & r_{12} \\ r_{21} & r_{22} \end{vmatrix}$

This is how the denominator array for i_1 is obtained. It so happens that the denominator array for i_2 is identical with that of i_1. The differences are in the numerators. You will notice that in the numerator array for i_1, the coefficients of i_1 are replaced by the v terms in the same order. That is, v_1 is substituted for r_{11}, and v_2 for r_{21}.

Similarly, we should expect to find the numerator array for i_2 to contain the i_1 coefficients, with v_1 replacing r_{12}, and v_2 replacing r_{22}. The denominator array, as we said, is the same for both i_1 and i_2. Therefore, we can write

$$i_2 = \frac{\begin{vmatrix} r_{11} & v_1 \\ r_{21} & v_2 \end{vmatrix}}{\begin{vmatrix} r_{11} & r_{12} \\ r_{21} & r_{22} \end{vmatrix}}$$

Now, what do we do to these arrays to get a numerical solution for either or both i_1 and i_2? Let us evaluate the i_2 determinant:

$$i_2 = \frac{\begin{vmatrix} r_{11} & v_1 \\ r_{21} & v_2 \end{vmatrix}}{\begin{vmatrix} r_{11} & r_{12} \\ r_{21} & r_{22} \end{vmatrix}}$$

In the numerator array, we multiply r_{11} by v_2 and from it subtract $r_{21} \times v_1$. In the denominator, we multiply r_{11} by r_{22} and from it subtract $r_{21} \times r_{12}$. The procedure can be remembered readily by drawing diagonals through the array.

$$i_2 = \frac{\begin{vmatrix} r_{11} & v_1 \\ r_{21} & v_2 \end{vmatrix}}{\begin{vmatrix} r_{11} & r_{12} \\ r_{21} & r_{22} \end{vmatrix}}$$

The diagonal product pointing down has a + sign in front of it, and the other diagonal has a − sign before it. Therefore, we have

$$i_2 = \frac{r_{11}v_2 - r_{21}v_1}{r_{11}r_{22} - r_{21}r_{12}}$$

Let us evaluate the i_1 array:

$$i_1 = \frac{\begin{vmatrix} v_1 & r_{12} \\ v_2 & r_{22} \end{vmatrix}}{\begin{vmatrix} r_{11} & r_{12} \\ r_{21} & r_{22} \end{vmatrix}} = \frac{v_1r_{22} - v_2r_{12}}{r_{11}r_{22} - r_{21}r_{12}}$$

Notice that the expressions for i_1 and i_2 are identical with those obtained by substitution.

Let us take an actual case. Repeating Fig. 1-27*a*, now shown as Fig. 1-56, find I_1 and I_2. Using Kirchhoff's laws, assume loop currents flowing in a clockwise direction. If we decide to analyze loop 1 in the counterclockwise direction, we have

$$0 = -12 - 4I_1 - 2I_1 + 2I_2$$

or

$$-12 = 6I_1 - 2I_2$$

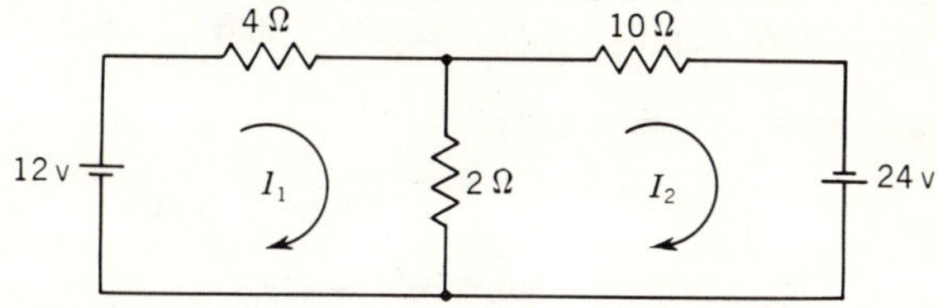

FIG. 1-56. Determining I_1 and I_2 by writing loop equations.

Going around loop 2, we can write

$$0 = -24 - 2I_2 - 10I_2 + 2I_1$$

or

$$-24 = -2I_1 + 12I_2$$

Grouping the equations,

$$-12 = 6I_1 - 2I_2$$

$$-24 = -2I_1 + 12I_2$$

Evaluating I_1 by determinants,

$$I_1 = \frac{\begin{vmatrix} -12 & -2 \\ -24 & 12 \end{vmatrix}}{\begin{vmatrix} 6 & -2 \\ -2 & 12 \end{vmatrix}} = \frac{(-12)(12) - (-24)(-2)}{(6)(12) - (-2)(-2)} = \frac{-192}{68} = -2.82 \text{ amp},$$

which checks our previous result.

Now solve the network of Fig. 1-57*a* by setting up Kirchhoff's loop equations and then evaluating the currents by the method of determinants.

Arbitrarily we can choose loop currents I_1 and I_2 in a clockwise direction, as in Fig. 1-57*b*. Writing the loop equations and solving for I_1, we have

$$-200 - 100 = 90 \times 10^3 I_1 - 30 \times 10^3 I_2 = -300$$

$$100 - 200 = -30 \times 10^3 I_1 + 50 \times 10^3 I_2 = -100$$

$$I_1 = \frac{\begin{vmatrix} -300 & -30 \times 10^3 \\ -100 & 50 \times 10^3 \end{vmatrix}}{\begin{vmatrix} 90 \times 10^3 & -30 \times 10^3 \\ -30 \times 10^3 & 50 \times 10^3 \end{vmatrix}} = \frac{-15 - 3}{4{,}500 - 900} = \frac{-18}{3{,}600} = -5 \text{ ma}$$

Solving for I_2,

$$I_2 = \frac{\begin{vmatrix} 90 \times 10^3 & -300 \\ -30 \times 10^3 & -100 \end{vmatrix}}{(3{,}600 \times 10^3)^2}$$

$$= \frac{-18}{3{,}600} = -5 \text{ ma}$$

Since both I_1 and I_2 have minus signs in front of them, they really flow oppositely to the directions assumed in Fig. 1-57*b*. Let us now redraw Fig. 1-57*b* with the currents correctly represented, as in Fig. 1-57*c*. The current I_1 is identical with that of I_a. Therefore, $I_a = 5$ ma. The current I_2 is identical with that of I_c. Therefore, $I_c = 5$ ma. The current I_b must be zero, since I_1 and I_2 are equal and opposite.

As an added exercise, let us verify the result by Millman's theorem. We choose Millman's theorem because it can be used to find the potential V_{GJ} at the junction point of the resistors:

$$V_{GJ} = \frac{\left(\dfrac{200}{60} - \dfrac{100}{30} - \dfrac{200}{20}\right)10^{-3}}{\left(\dfrac{1}{60} + \dfrac{1}{30} + \dfrac{1}{20}\right)10^{-3}} = \frac{\dfrac{200 - 200 - 600}{60}}{\dfrac{1 + 2 + 3}{60}}$$

$$= \frac{\dfrac{-600 \text{ volts}}{60 \text{ kilohms}}}{\dfrac{6}{60 \text{ kilohms}}} = -100 \text{ volts}$$

Therefore,

$$I_A = \frac{200 - (-100)}{60} = \frac{300 \text{ volts}}{60 \text{ kilohms}} = 5 \text{ ma}$$

$$I_B = \frac{-100 - (-100)}{30} = \frac{0}{30} = 0$$

$$I_C = \frac{-100 - (-200)}{20} = \frac{100 \text{ volts}}{20 \text{ kilohms}} = 5 \text{ ma}$$

which checks.

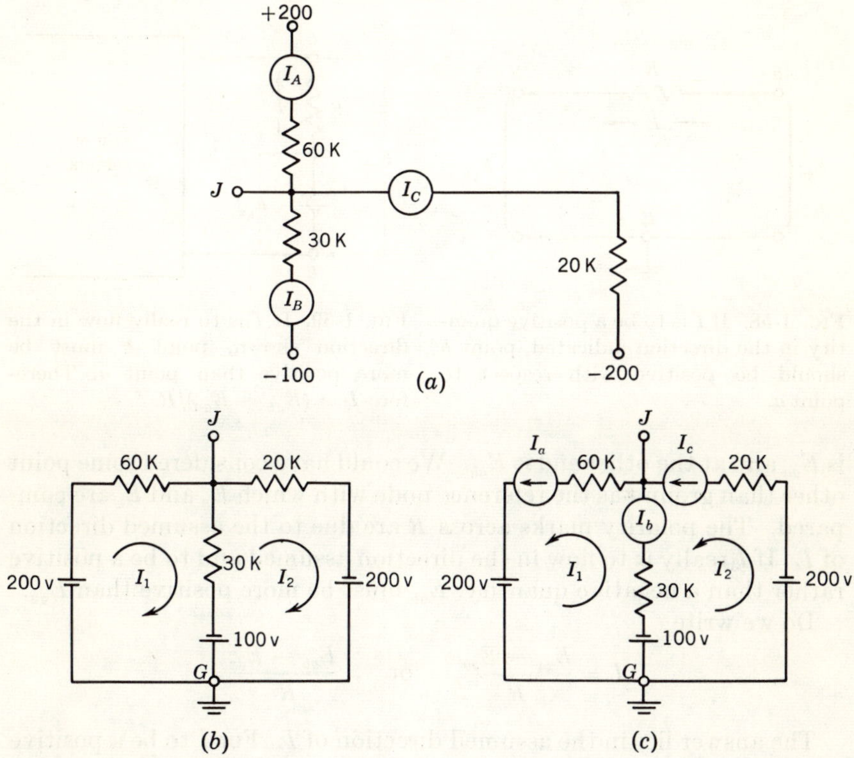

FIG. 1-57. (*a*) An exercise in loop equations. (*b*) Redrawing *a* to clarify the loops. (*c*) The branch currents, I_A, I_B, and I_C may be expressed in terms of the loop currents I_1 and I_2.

1-10. Kirchhoff's Current Law—Nodal Analysis. We call a junction point where two or more currents merge a *node*. For example, in applying Millman's theorem, the point J in Fig. 1-57*c* is a node. The voltage at a node depends on the currents entering or leaving the node. For this reason an analysis based upon node currents is termed a *nodal* *analysis*. In nodal analysis, we solve for node voltages instead of for loop currents.

Nodal equations stem directly from Kirchhoff's current law, which states that the algebraic sum of the currents at a junction point is zero. This law simply means that the currents entering a junction must equal those leaving it. We shall now develop a systematic method of writing nodal equations.

Consider a resistor with both ends at potentials other than ground potential and an arbitrarily assumed direction of electron flow through it, as shown in Fig. 1-58. The voltage at one end with respect to ground

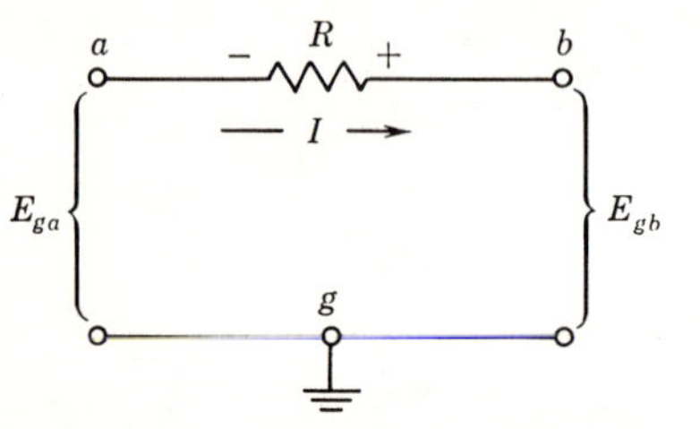

FIG. 1-58. If I is to be a positive quantity in the direction indicated, point b should be positive with respect to point a.

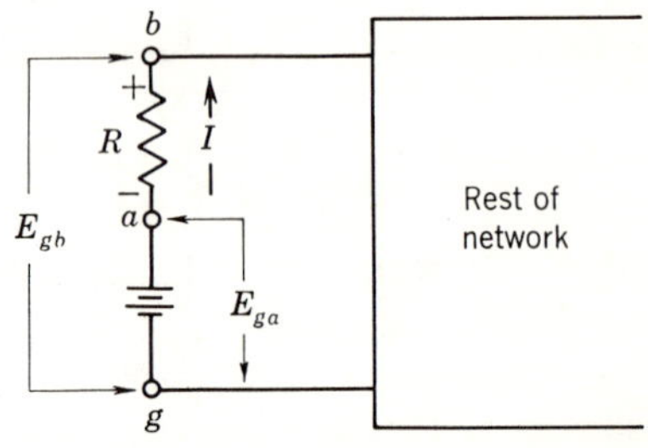

FIG. 1-59. If I is to really flow in the direction shown, point b must be more positive than point a. Therefore $I = (E_{gb} - E_{ga})/R$.

is E_{ga} and at the other end is E_{gb}. We could have considered some point other than ground as the reference node with which E_a and E_b are compared. The polarity marks across R are due to the assumed direction of I. If I really is to flow in the direction assumed and to be a positive rather than a negative quantity, E_{gb} must be more positive than E_{ga}.

Do we write

$$I = \frac{E_{ga} - E_{gb}}{R} \qquad \text{or} \qquad \frac{E_{gb} - E_{ga}}{R}$$

The answer lies in the assumed direction of I. For I to be a positive quantity, E_{gb} must be greater than E_{ga}, so we write E_{gb} first. Therefore,

$$I = \frac{E_{gb} - E_{ga}}{R}$$

We can clarify this point further by considering the example shown in Fig. 1-59. We are assuming that I flows from E_a to E_b; therefore, R is polarized as shown. Let us determine the potential of point b with respect to our ground reference. Starting at ground and going to point a, we go up in voltage by $+E_{ga}$ volts.

In going through the resistor, we go up again by an amount IR. This brings us to point b. Therefore, the potential of point b with respect to ground is $E_{gb} = E_{ga} + IR$. Therefore,

$$I = \frac{E_{gb} - E_{ga}}{R}$$

Suppose I were assumed to be flowing in the opposite direction. Then a positive I would be written

$$I = \frac{E_{ga} - E_{gb}}{R}$$

and a negative I would be written

$$-I = -\frac{E_{ga} - E_{gb}}{R} = \frac{E_{gb} - E_{ga}}{R}$$

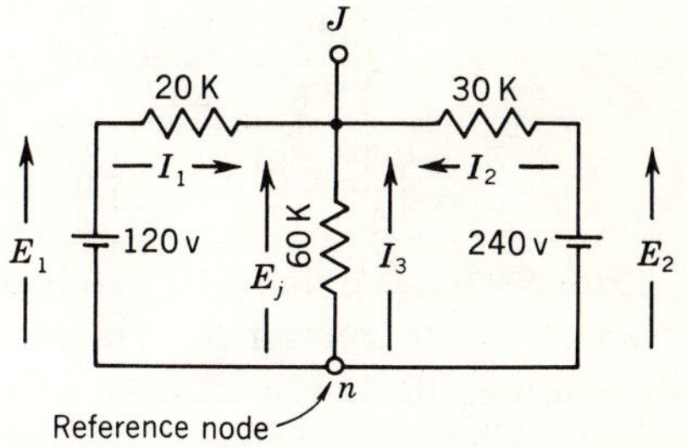

FIG. 1-60. The voltage at node J with respect to the reference node is determined by Kirchhoff's current law (nodal analysis).

Let us apply these principles to some networks. In assuming currents as shown in Fig. 1-60, we can then write the equation, based on Kirchhoff's current law, that the algebraic sum of the currents entering a junction (node) is zero. We arbitrarily consider all currents entering the junction as positive, and leaving the junction as negative: $I_1 + I_2 + I_3 = 0$. We know that at least one of the currents must really flow out of the junction and have a minus sign in front of it, but let us keep to the assumed directions.

All the currents being positively directed or flowing into the junction mean that the junction potential E_{nj} is assumed to be more positive than E_1 or E_2. Therefore,

$$I_1 = \frac{E_j - E_1}{20 \times 10^3}$$

$$I_2 = \frac{E_j - E_2}{30 \times 10^3}$$

$$I_3 = \frac{E_j - 0}{60 \times 10^3}$$

Substituting in $I_1 + I_2 + I_3 = 0$ and noting that $E_1 = +120$ and $E_2 = +240$ with respect to point n, which is our reference node, we have

$$\frac{E_j - 120}{20 \times 10^3} + \frac{E_j - 240}{30 \times 10^3} + \frac{E_j}{60 \times 10^3} = 0$$

Simplifying,

$$\frac{E_j}{20 \times 10^3} + \frac{E_j}{30 \times 10^3} + \frac{E_j}{60 \times 10^3} = \frac{240}{30 \times 10^3} + \frac{120}{20 \times 10^3}$$

$$3E_j + 2E_j + E_j = +2(240) + 3(120)$$

from which

$$E_j = 140 \text{ volts}$$

We can verify this result by Millman's theorem:

$$E_j = \frac{\left(\frac{120}{20} + \frac{240}{30}\right)10^{-3}}{\left(\frac{1}{20} + \frac{1}{60} + \frac{1}{30}\right)10^{-3}} = \frac{\frac{3(120) + 2(240)}{60}}{\frac{3 + 1 + 2}{60}} = 140 \text{ volts}$$

Since we have definitely determined that E_j is at +140 volts with respect to our reference, which in this case is point n, we proceed to determine the true direction of electron flow. Now I_1 was correctly

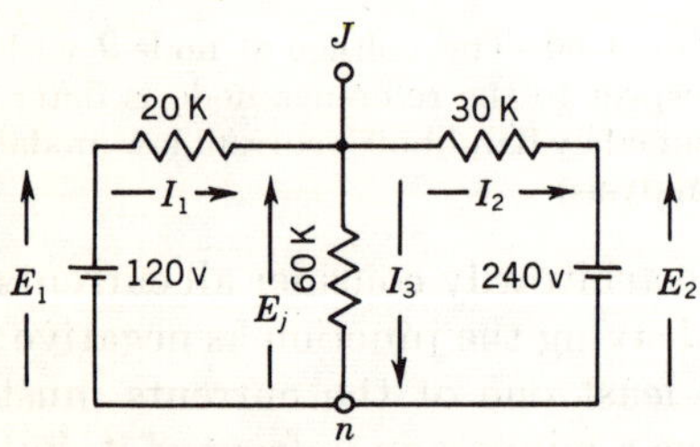

FIG. 1-61. The network of Fig. 1-60 is again analyzed but with some of the assumed current directions changed.

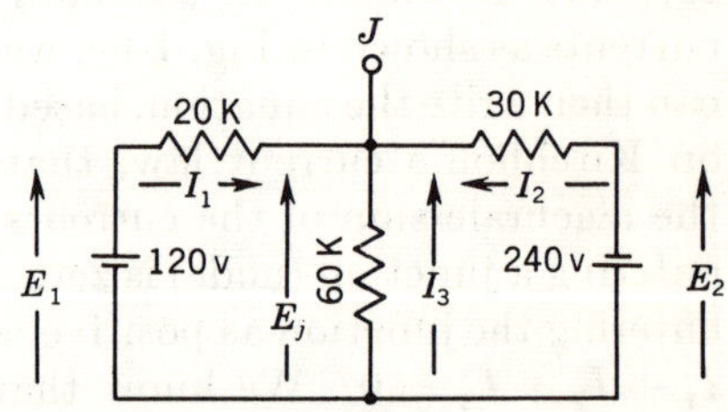

FIG. 1-62. All currents flowing into the junction are assumed positive.

chosen, since it is shown as flowing from E_1 to E_j, E_1 being negative with respect to E_j.

But I_2 is really flowing oppositely to the indicated direction, since E_j is more negative than E_2. The current I_3 was correctly chosen. Substituting +140 volts for E_j; +120 volts for E_1; and +240 volts for E_2 in the expressions for I_1, I_2, and I_3 would result in positive values for I_1 and I_3, which show they flow as indicated, while I_2 would be negative and therefore would flow oppositely to the assumed direction.

Let us do the same problem, but assume that the currents flow as in Fig. 1-61. Again, assuming that currents entering the junction are positively directed and those leaving are negative, we can write

$$I_1 - I_2 - I_3 = 0 \tag{1-5}$$

Now
$$I_1 = \frac{E_j - E_1}{20 \text{ kilohms}}$$

We write E_j first, since it is at a positive potential with respect to E_1 if I_1 is to flow as indicated:

$$I_2 = \frac{E_2 - E_j}{30 \text{ kilohms}} \quad \text{and} \quad I_3 = \frac{0 - E_j}{60 \text{ kilohms}}$$

In the case of I_3, the reference node, or zero potential n must be considered positive with respect to point J if I_3 is to flow in the direction assumed. Substituting in (1-5), we have

$$\frac{E_j - E_1}{20 \times 10^3} - \frac{E_2 - E_j}{30 \times 10^3} - \frac{-E_j}{60 \times 10^3} = 0$$

Collecting terms and simplifying,

$$\frac{E_j}{20 \times 10^3} + \frac{E_j}{30 \times 10^3} + \frac{E_j}{60 \times 10^3} = + \frac{E_1}{20 \times 10^3} + \frac{E_2}{30 \times 10^3}$$

$$= \frac{120}{20 \times 10^3} + \frac{240}{30 \times 10^3}$$

We can see at this point that the form of the equation is exactly the same as in the previous solution. Therefore, we again would get $E_j = +140$ volts. This time, I_1 and I_2 were correctly chosen, but I_3 really flows oppositely to the indicated direction, since

$$I_3 = \frac{-(+140) \text{ volts}}{60 \text{ kilohms}} = -2.3 \text{ ma}$$

Now let us once more try the problem and assume that all currents flow into the junction, but with E_2 reversed as shown in Fig. 1-62. Again assuming that currents flowing into the junction are positive and away negative, we can write

$$I_1 + I_2 + I_3 = 0 \tag{1-6}$$

$$I_1 = \frac{E_j - E_1}{20 \times 10^3} = \frac{E_j - 120}{20 \times 10^3}$$

$$I_2 = \frac{E_j - E_2}{30 \times 10^3} = \frac{E_j - (-240)}{30 \times 10^3} = \frac{E_j + 240}{30 \times 10^3}$$

$$I_3 = \frac{E_j - 0}{60 \times 10^3} = \frac{E_j}{60 \times 10^3}$$

The positive directions indicated by the E_1, E_2, and E_j arrows should be drawn so that they indicate voltages with respect to the reference node. Note that E_2 will be a negative quantity, since the head of the

E_2 arrow lies at a point which is negative with respect to the tail:

$$\frac{E_j - 120}{20 \times 10^3} + \frac{E_j + 240}{30 \times 10^3} + \frac{E_j}{60 \times 10^3} = 0$$

$$E_j\left(\frac{1}{20} + \frac{1}{30} + \frac{1}{60}\right) \times 10^{-3} = \frac{120}{20 \times 10^3} - \frac{240}{30 \times 10^3}$$

$$E_j = \frac{\dfrac{120}{20} - \dfrac{240}{30}}{\dfrac{1}{20} + \dfrac{1}{30} + \dfrac{1}{60}} = \frac{6 - 8 \text{ ma}}{\dfrac{6}{60 \text{ kilohms}}} = -20 \text{ volts}$$

Note that the expression for E_j is identical with that obtained by Millman's theorem.

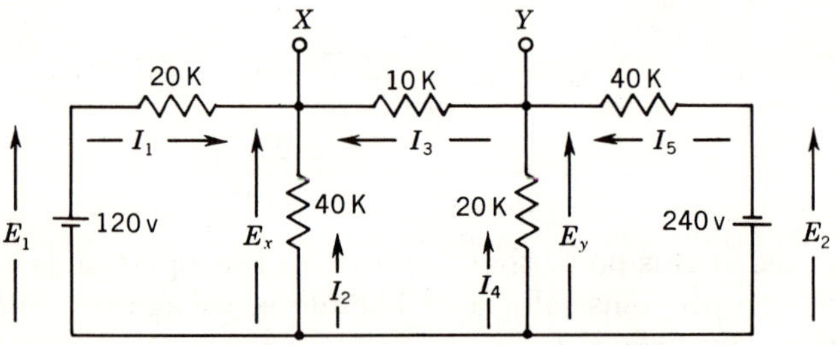

FIG. 1-63. Only two nodal equations (as compared with three loop equations) are required to solve this complex network.

Let us try a more difficult problem. Find the current in the 10-kilohm resistor of Fig. 1-63. To solve this problem, we must first assume current directions, say, as in Fig. 1-63. Then let us call the currents entering positive, and those leaving negative. At node x we then have

$$I_1 + I_2 + I_3 = 0 \qquad (1\text{-}6)$$

and at node y

$$-I_3 + I_4 + I_5 = 0 \qquad (1\text{-}7)$$

$$I_1 = \frac{E_x - E_1}{20 \text{ kilohms}} = \frac{E_x - 120}{20 \text{ kilohms}}$$

$$I_2 = \frac{E_x - 0}{40 \text{ kilohms}} = \frac{E_x}{40 \text{ kilohms}}$$

$$I_3 = \frac{E_x - E_y}{10 \text{ kilohms}}$$

$$I_4 = \frac{E_y - 0}{20 \text{ kilohms}} = \frac{E_y}{20 \text{ kilohms}}$$

$$I_5 = \frac{E_y - E_2}{40 \text{ kilohms}} = \frac{E_y - (-240)}{40 \text{ kilohms}} = \frac{E_y + 240}{40 \text{ kilohms}}$$

Substituting in (1-6), we have

$$\frac{E_x - 120}{20 \text{ kilohms}} + \frac{E_x}{40 \text{ kilohms}} + \frac{E_x - E_y}{10 \text{ kilohms}} = 0 \tag{1-6a}$$

Substituting in (1-7), we have

$$-\frac{E_x - E_y}{10 \text{ kilohms}} + \frac{E_y}{20 \text{ kilohms}} + \frac{E_y + 240}{40 \text{ kilohms}} = 0 \tag{1-7a}$$

Simplifying (1-6a) and (1-7a) for solution by determinants gives

$$E_x\left(\frac{1}{20 \text{ kilohms}} + \frac{1}{40 \text{ kilohms}} + \frac{1}{10 \text{ kilohms}}\right) - E_y\left(\frac{1}{10 \text{ kilohms}}\right) = \frac{120 \text{ volts}}{20 \text{ kilohms}}$$

$$-E_x\left(\frac{1}{10 \text{ kilohms}}\right) + E_y\left(\frac{1}{10 \text{ kilohms}} + \frac{1}{20 \text{ kilohms}} + \frac{1}{40 \text{ kilohms}}\right) = \frac{-240 \text{ volts}}{40 \text{ kilohms}}$$

$$E_x(0.05 + 0.025 + 0.1)10^{-3} - E_y(0.1 \times 10^{-3}) = 6 \times 10^{-3}$$

$$-E_x(0.1 \times 10^{-3}) + E_y(0.1 + 0.05 + 0.025)10^{-3} = -6 \times 10^{-3}$$

Dividing both sides of each equation by 10^{-3} and simplifying,

$$E_x(0.175) - E_y(0.1) = 6$$

and

$$-E_x(0.1) + E_y(0.175) = -6$$

$$E_x = \frac{\begin{vmatrix} 6 & -0.1 \\ -6 & 0.175 \end{vmatrix}}{\begin{vmatrix} 0.175 & -0.1 \\ -0.1 & 0.175 \end{vmatrix}} = \frac{1.050 - 0.6}{0.0306 - 0.01} = \frac{-0.45}{0.0201} = \frac{45}{2.01} = 22.5 \text{ volts}$$

$$E_y = \frac{\begin{vmatrix} 0.175 & 6 \\ -0.1 & -6 \end{vmatrix}}{0.0201} = \frac{-1.050 + 0.6}{0.0201} = \frac{-0.45}{0.0201} = \frac{-45}{2} = -22.5 \text{ volts}$$

Therefore, the current through the 10-kilohm resistor is

$$I_3 = \frac{E_x - E_y}{10 \text{ kilohms}} = \frac{22.5 - (-22.5)}{10} = \frac{45 \text{ volts}}{10 \text{ kilohms}} = 4.5 \text{ ma}$$

Checking by Thévenin's theorem,

$$Z_{th_x} = \frac{20 \times 40}{60} = 13.3 \text{ kilohms} \qquad Z_{th_y} = \frac{40 \times 20}{60} = 13.3 \text{ kilohms}$$

$$V_{ocg_x} = \frac{120 \times 40}{40 + 20} = 80 \text{ volts} \qquad V_{ocg_y} = \frac{-240 \times 20}{20 + 40} = -80 \text{ volts}$$

$$I_3 = \frac{160}{36.6 \text{ kilohms}} = 4.4 \text{ ma}$$

As can be seen by Fig. 1-64, the results are in close agreement.

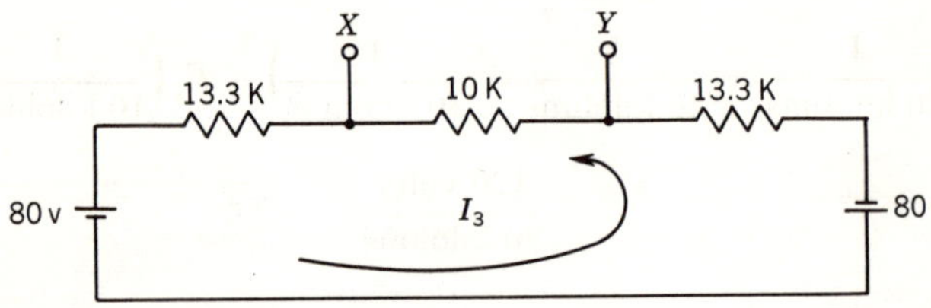

FIG. 1-64. Checking the solution for Fig. 1-63 by Thévenin's theorem.

SUMMARY

In this chapter we have developed basic tools for analyzing networks. The analysis is conducted by finding voltages and currents in various parts of a network. In simple circuits, we can frequently use elementary voltage- and current-divider relationships as extensions of Ohm's law.

As the networks increase in complexity, we can usually apply the superposition, Thévenin's, Norton's, or Millman's theorem. Thévenin's theorem is a very powerful tool for simplifying networks and conveying a feeling for impedance levels. The superposition theorem can be used for solving most networks, but its greatest utility lies in considering the response of a network to mixed voltages such as combined d-c and a-c sources.

We have also discussed Kirchhoff's laws, which enable us to write loop or nodal equations for complex networks. The loop equations allow us to determine the current at any point within a network. The nodal equations permit us to determine the voltages at junction points (nodes) within the network.

As an aid in evaluating the simultaneous equations generated in writing Kirchhoff's laws, we have discussed the method of determinants. The systematic solution of simultaneous equations by determinants hastens the solution, and decreases chances for error.

This chapter is perhaps the most important in the book, as it is the cornerstone of our foundation in circuit analysis.

PROBLEMS

1-1. What is E_o in Fig. 1-65?
Ans. 18 volts.

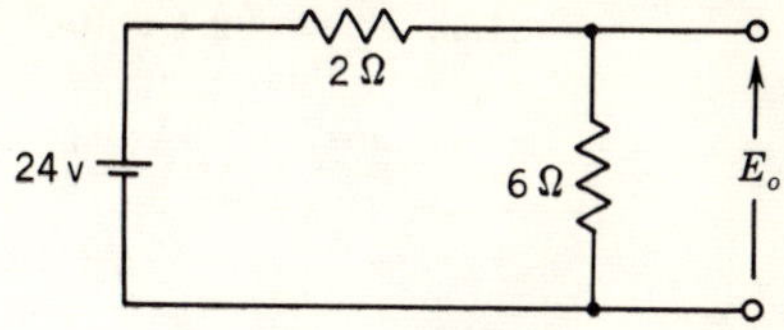

FIG. 1-65. Problem 1-1.

1-2. What is E_o in Fig. 1-66?
Ans. 0 volts.

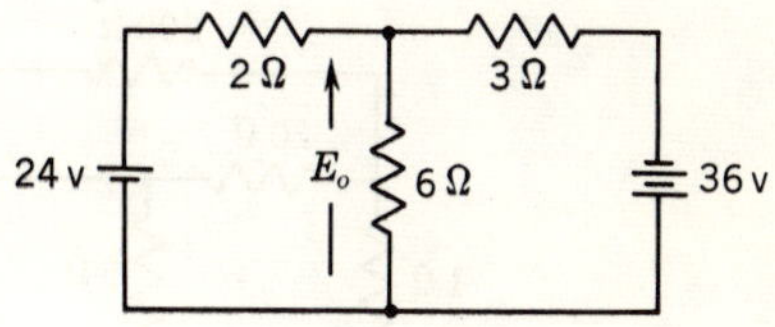

FIG. 1-66. Problem 1-2.

1-3. What is E_o in Fig. 1-67?
Ans. 0 volts.

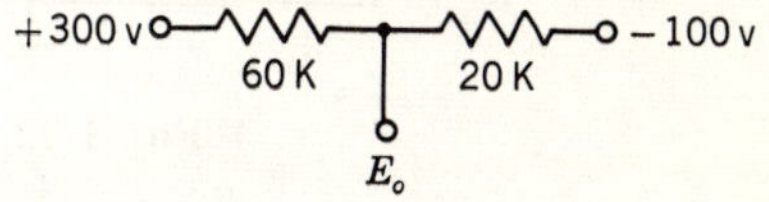

FIG. 1-67. Problem 1-3.

1-4. What is the voltage at the output terminal in Fig. 1-68?
Ans. $E_o = 133$ volts.

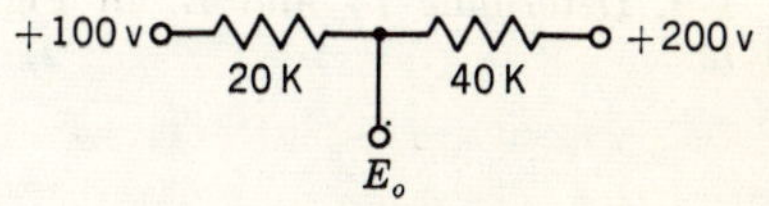

FIG. 1-68. Problem 1-4.

1-5. Approximate E_o and I_L in Fig. 1-69. HINT: The 10-kilohm resistor essentially determines the line current.
Ans. $E_o \approx -0.12$ volts and $I_L \approx 12$ ma.

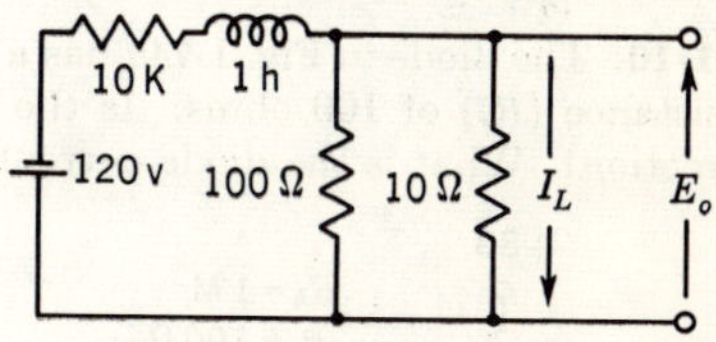

FIG. 1-69. Problem 1-5.

1-6. Determine I_L in Fig. 1-70. HINT: The 100-kilohm resistor essentially determines the line current and the 1-meg resistor negligibly loads the bridge network.

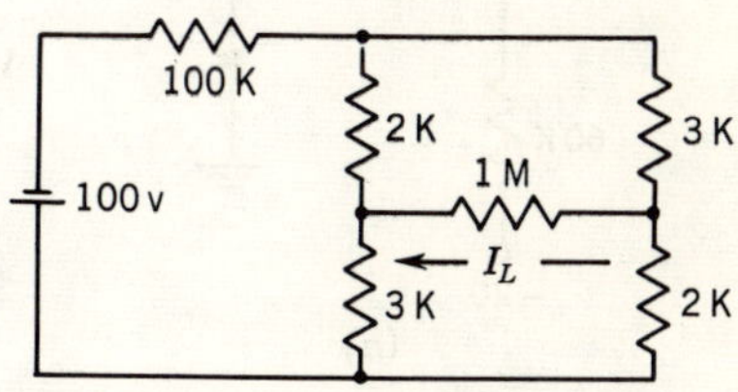

FIG. 1-70. Problem 1-6.

1-7. Determine E_o in Fig. 1-71.
Ans. $E_o = -2.4$ volts.

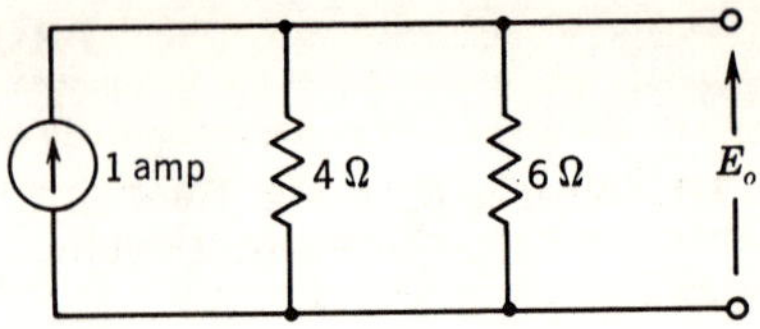

FIG. 1-71. Problem 1-7.

1-8. Estimate E_o in Fig. 1-72. Look for useful approximations.
Ans. $E_o \approx 6.7$ volts.

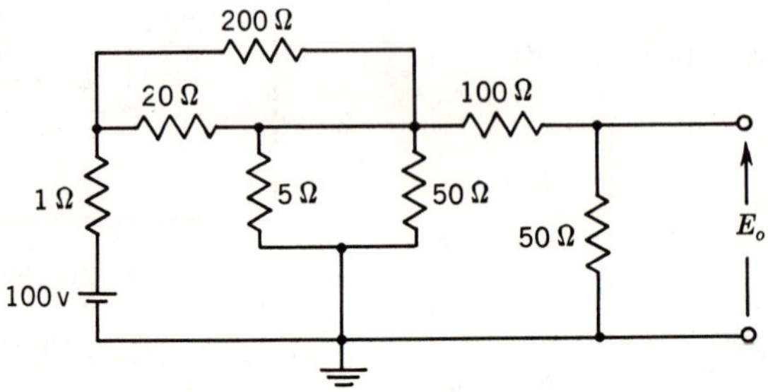

FIG. 1-72. Problem 1-8.

1-9. Determine I_L and E_o in Fig. 1-73.

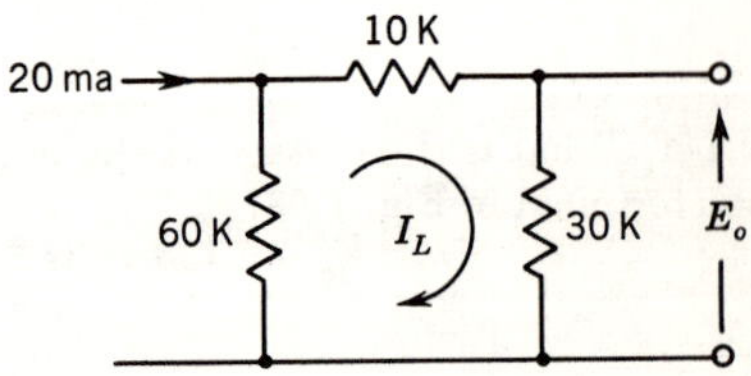

FIG. 1-73. Problem 1-9.

1-10. The diode in Fig. 1-74*a* has a back resistance (R_b) of 1 meg and a forward resistance (R_f) of 100 ohms. Is the diode conducting in the forward or reverse direction? What is the diode current?

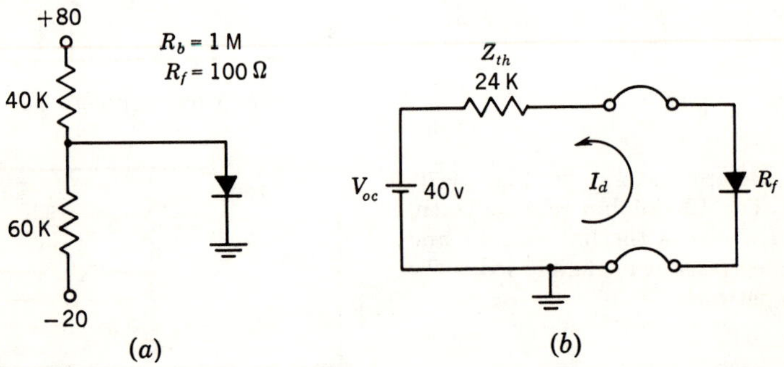

FIG. 1-74. Problem 1-10.

Solution. To determine whether or not the diode is forward-biased, find the polarity of the open circuit voltage that it looks back into. In Fig. 1-74*a*, the V_{oc} the diode sees is 40 volts with the anode end positive with respect to the cathode. The diode is therefore forward-biased and will look like a 100-ohm resistor. The Thévenin's equivalent impedance the diode looks back into is 24 kilohms, as shown in Fig. 1-74*b*. Since the diode forward resistance is so much less than Z_{th}, we can say that $I_d \approx 40$ volts/24 kilohms $= 1.7$ ma. Do not make the error of assuming there are 40 volts across the diode when it is conducting. The actual diode voltage is (1.7 ma)(100 ohms) $= 0.17$ volt. If V_{oc} had been of such polarity that the diode were reverse-biased, the apparent diode resistance would be R_b. In this case, R_b would essentially determine the diode current, since R_b is so much larger than Z_{th}.

1-11. Find I_L, E_o, and that value of load resistance that should be hung across the output terminals for maximum power transfer (Fig. 1-75).

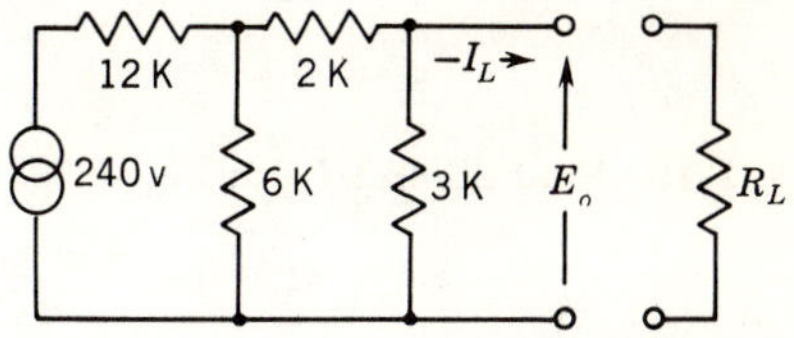

FIG. 1-75. Problem 1-11.

1-12. What is the bias on the tube in Fig. 1-76 with the switch open and closed?

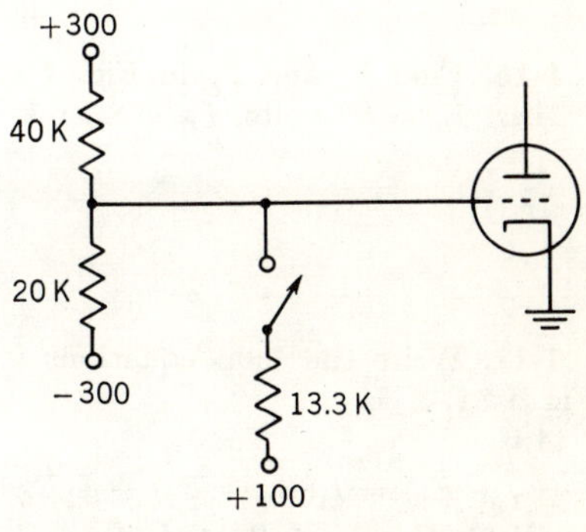

FIG. 1-76. Problem 1-12.

1-13. Find E_o in Fig. 1-77.

Ans. 2 volts.

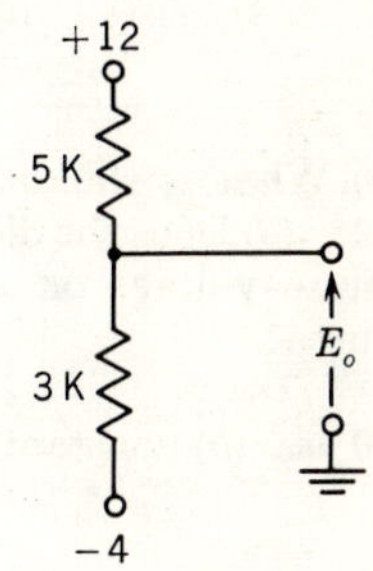

FIG. 1-77. Problem 1-13.

1-14. Find E_o in Fig. 1-78.
Ans. 40 volts.

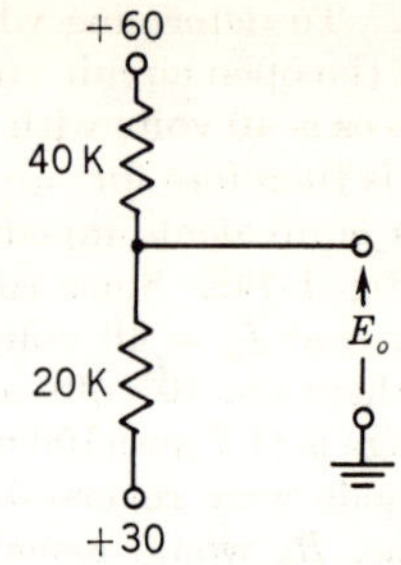

FIG. 1-78. Problem 1-14.

1-15. Find E_o and I_L in Fig. 1-79.

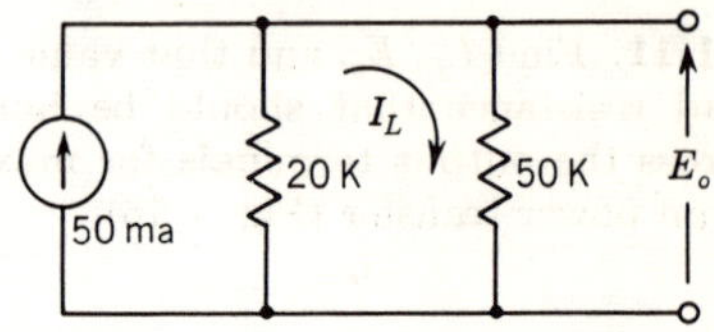

FIG. 1-79. Problem 1-15.

1-16. Find E_o and I_L in Fig. 1-80.
Ans. $E_o = 97$ volts, $I_L = 8.1$ ma.

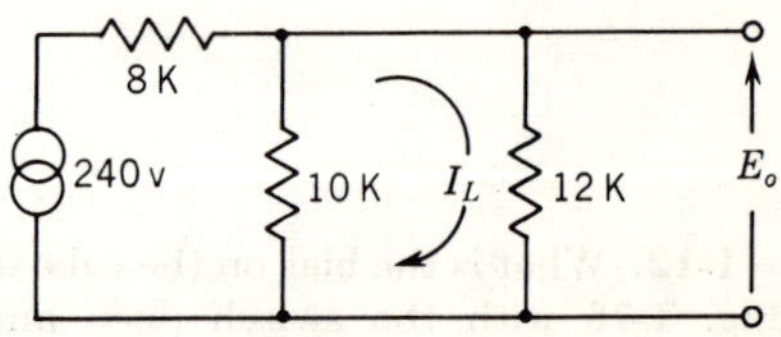

FIG. 1-80. Problem 1-16.

1-17. Write the loop equations for Fig. 1-81.
Ans.

$$-(E_1 + E_2) = I_1(R_1 + R_2) - I_2R_2$$
$$E_2 + E_4 = -I_1R_2 + I_2(R_2 + R_3)$$

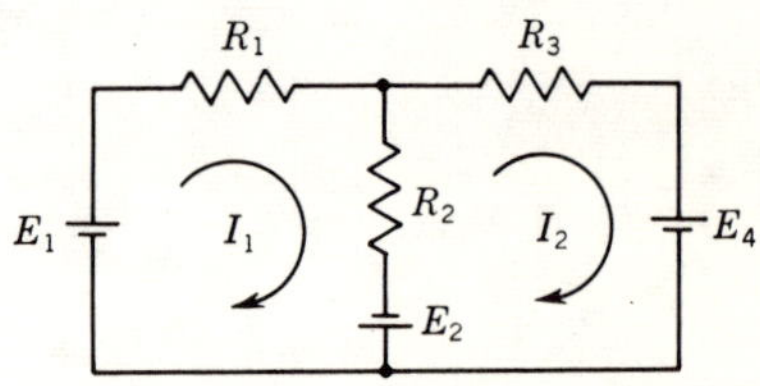

FIG. 1-81. Problem 1-17.

1-18. (*a*) What is the diode current in Fig. 1-82? (*b*) Does the diode tend to see a constant-voltage or a constant-current source?
Ans.
(*a*) $I_d = 60\ \mu$a; (*b*) constant voltage.

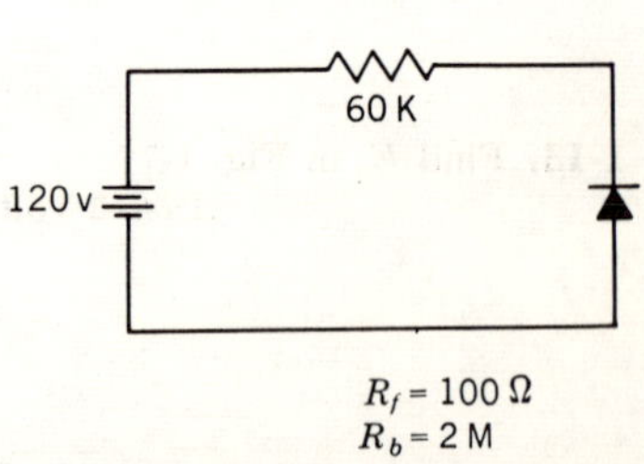

FIG. 1-82. Problem 1-18.

1-19. Write the loop equations for Fig. 1-83.

Ans.

$$E = I_1(R_g + R_1 + R_2) - I_2R_1 - I_3R_2$$
$$0 = -I_1R_1 + I_2(R_1 + R_3 + R_5) - I_3R_5$$
$$0 = -I_1R_2 - I_2R_5 + I_3(R_2 + R_4 + R_5)$$

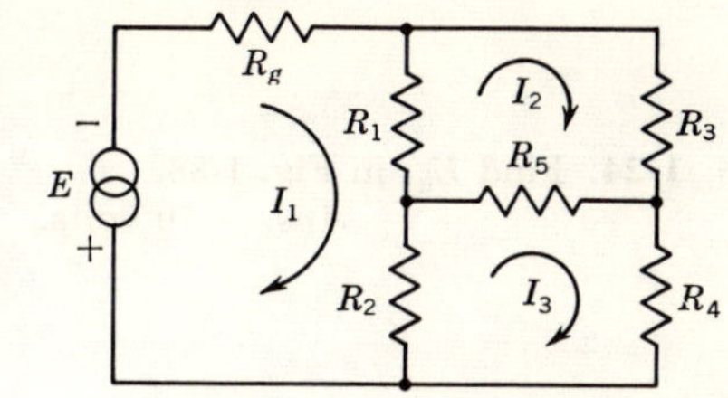

FIG. 1-83. Problem 1-19.

1-20. (*a*) What charging resistance does the capacitor in Fig. 1-84 see when S is closed? (*b*) What voltage will the capacitor ultimately charge up to?

Ans. (*a*) 7.5 kilohms; (*b*) 50 volts.

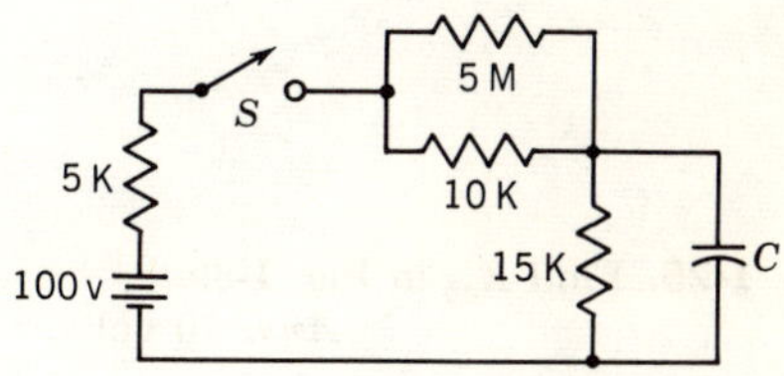

FIG. 1-84. Problem 1-20.

1-21. Find I_L in Fig. 1-85.

Ans. $I_L = 7.5$ ma.

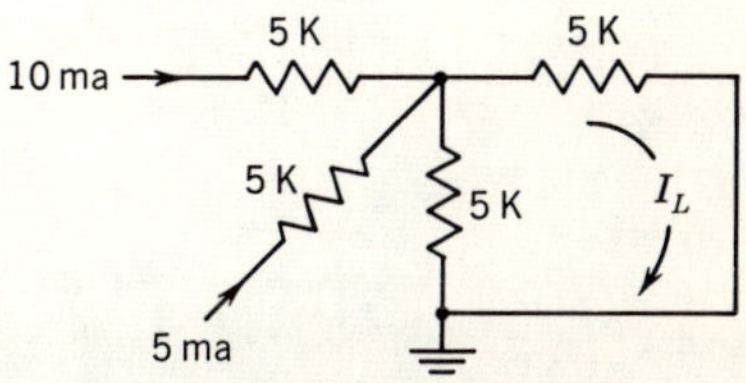

FIG. 1-85. Problem 1-21.

1-22. Find I_L in Fig. 1-86.

Ans. $I_L = 3.33$ ma.

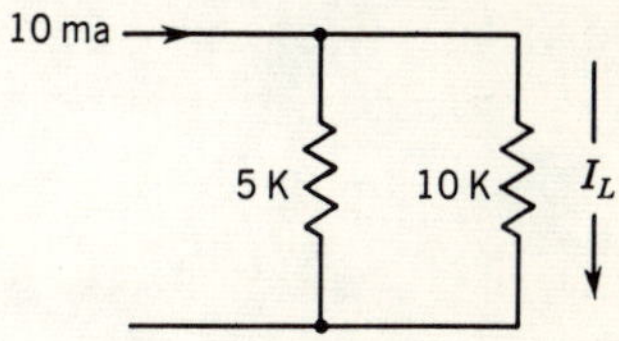

FIG. 1-86. Problem 1-22.

1-23. Find I_{ab} in Fig. 1-87.

Ans. 5 ma.

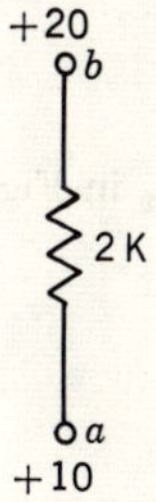

FIG. 1-87. Problem 1-23.

1-24. Find E_{ab} in Fig. 1-88.
Ans. -70 volts.

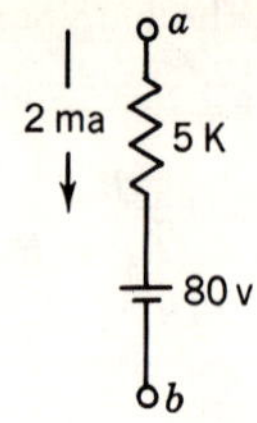

FIG. 1-88. Problem 1-24.

1-25. Find E_{ab} in Fig. 1-89.
Ans. 70 volts.

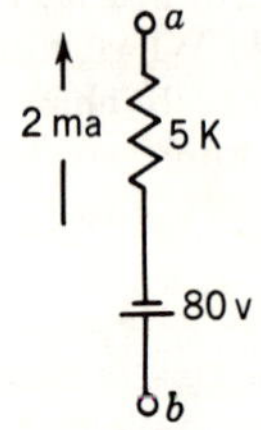

FIG. 1-89. Problem 1-25.

1-26. Find I_{ab} in Fig. 1-90.
Ans. -1 ma.

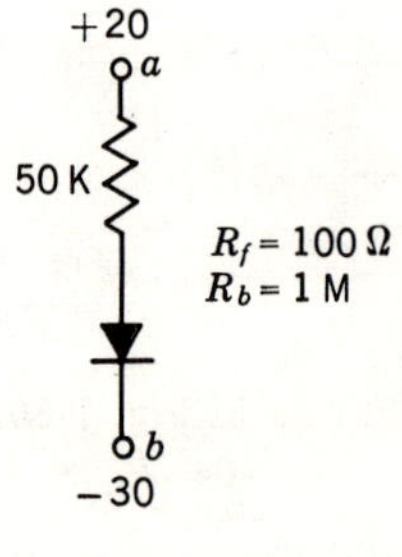

FIG. 1-90. Problem 1-26.

1-27. Find I_{ab} in Fig. 1-91.
Ans. 50 μa.

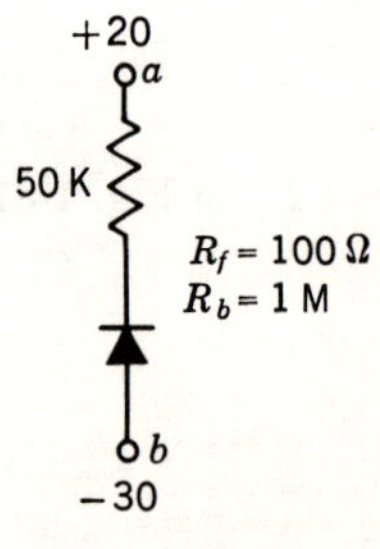

FIG. 1-91. Problem 1-27.

1-28. Sketch the output waveform for Fig. 1-92*a*. *Ans.* See Fig. 1-92*b*.

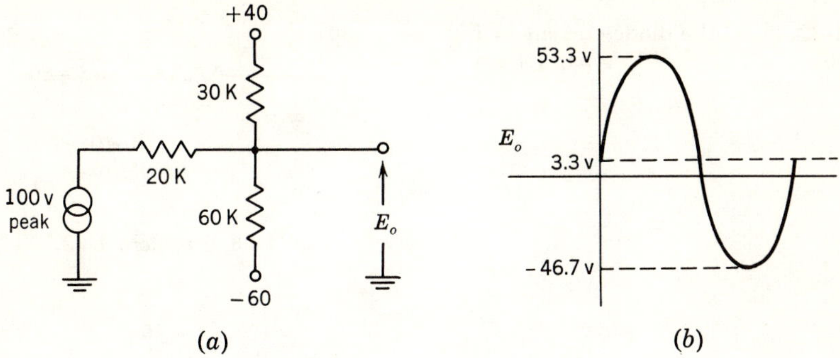

FIG. 1-92. Problem 1-28.

1-29. Find E_o in Fig. 1-93.
Ans.

$$E_o = \frac{(100/20 - 200/50 + 50/10 - 100/50)10^{-3}}{(1/20 + 1/50 + 1/10 + 1/50 + 1/20)10^{-3}} = 16.7 \text{ volts}$$

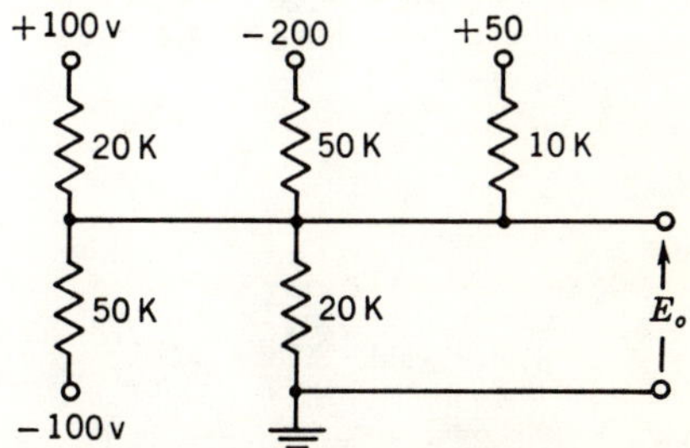

FIG. 1-93. Problem 1-29.

1-30. Find E_o in Fig. 1-94.
Ans. 50 volts.

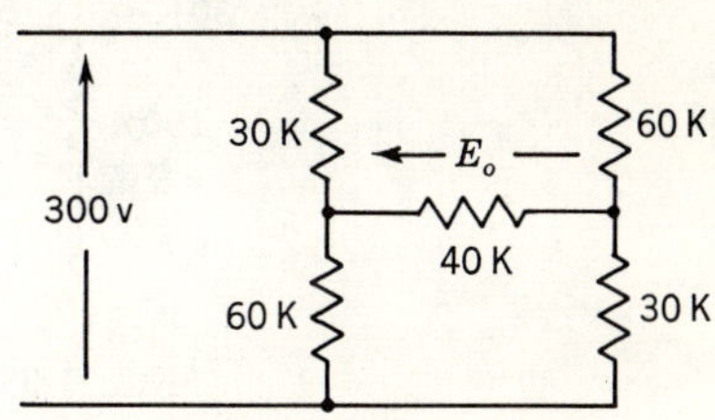

FIG. 1-94. Problem 1-30.

1-31. Find I_L in Fig. 1-95.
Ans. −1 ma.

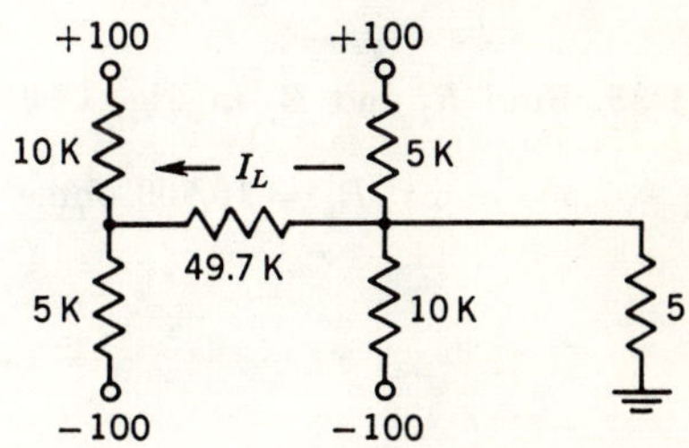

FIG. 1-95. Problem 1-31.

1-32. Find the diode current in Fig. 1-96. *Ans.* 1.4 ma.

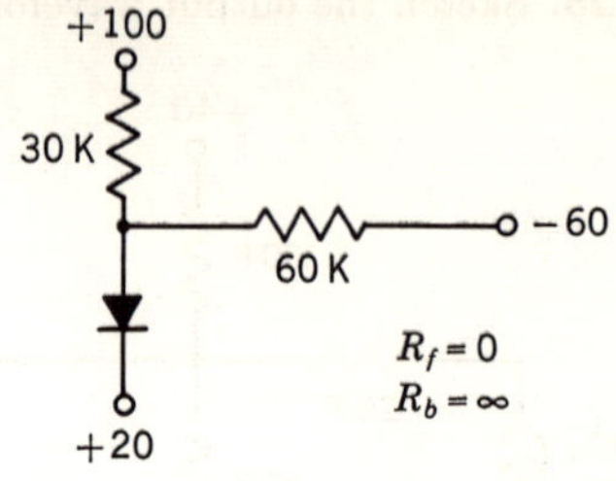

FIG. 1-96. Problem 1-32.

1-33. Find E_o in Fig. 1-97. *Ans.* −4.4 volts.

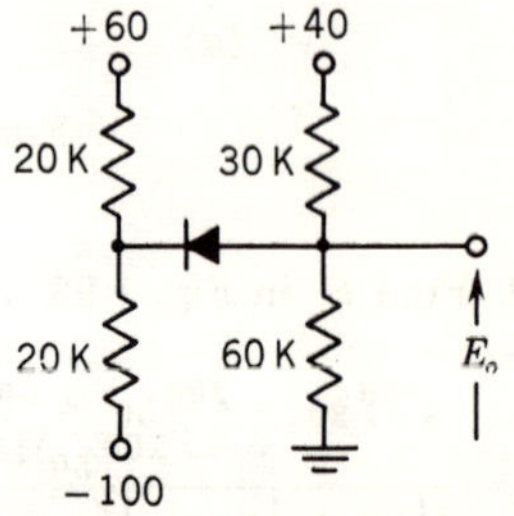

FIG. 1-97. Problem 1-33.

1-34. Find E_o in Fig. 1-98.

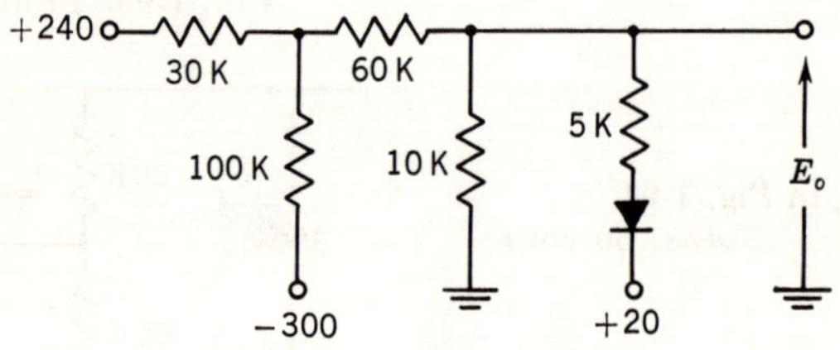

FIG. 1-98. Problem 1-34.

1-35. Find R_1 and R_2 in Fig. 1-99. *Ans.* R_1 = 7,500 ohms, R_2 = 16,700 ohms.

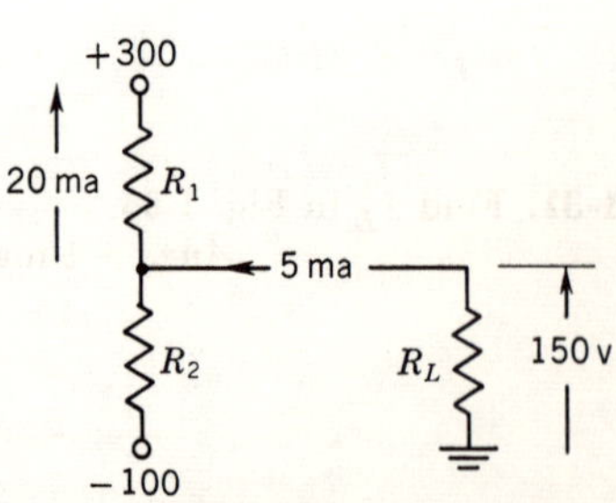

FIG. 1-99. Problem 1-35.

1-36. What polarity and amplitude of trigger pulse *must* be injected at the capacitor in Fig. 1-100 to fire the thyratron if −5 volts is the minimum holdoff voltage with a 300-volt anode supply? *Ans.* Positive 35-volt pulse.

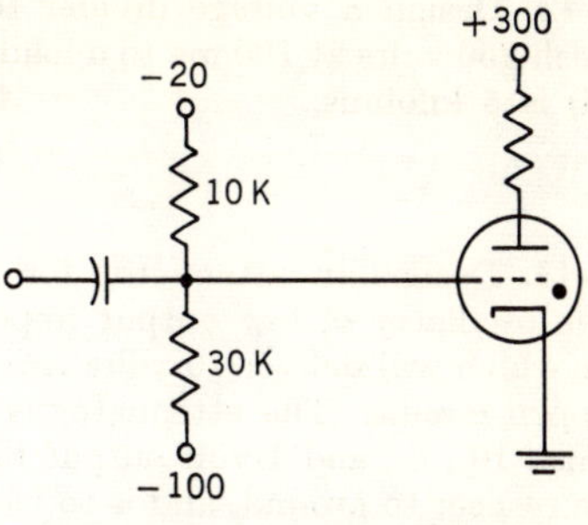

FIG. 1-100. Problem 1-36.

1-37. Design a voltage divider to deliver 0.3 volt at no load. The supply voltage is 28 volts, and the divider current is to be 100 ma.

Ans. $R_1 = 277$ ohms, $R_2 = 3$ ohms.

1-38. Design a voltage divider to deliver 10 ma at 120 volts from a 300-volt supply. The supply drain is to be 40 ma.

1-39. Estimate E_o in Fig. 1-101. *Ans.* 120 volts.

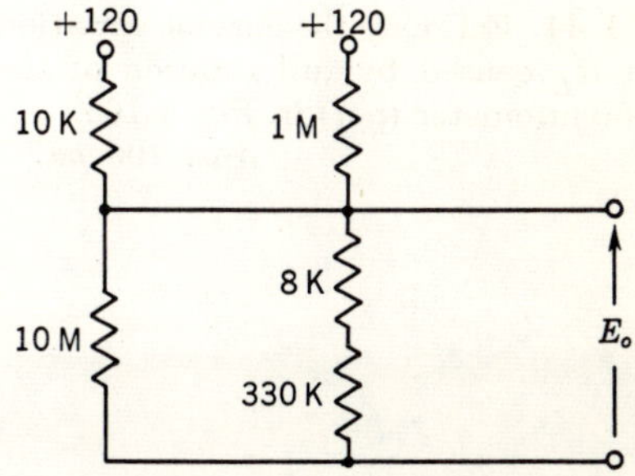

FIG. 1-101. Problem 1-39.

1-40. Estimate E_o in Fig. 1-102. *Ans.* 10 volts.

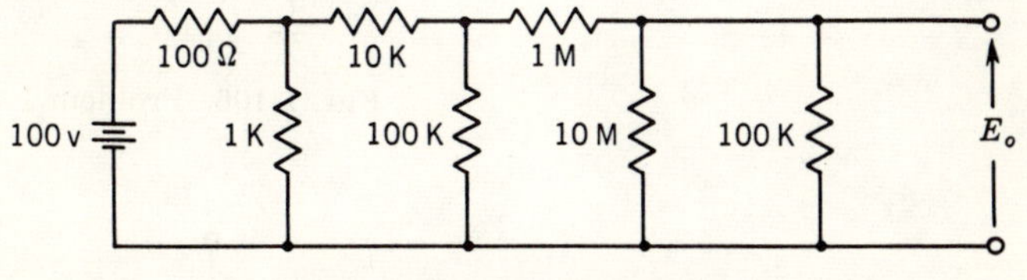

FIG. 1-102. Problem 1-40.

1-41. Estimate E_o in Fig. 1-103. *Ans.* −60 volts.

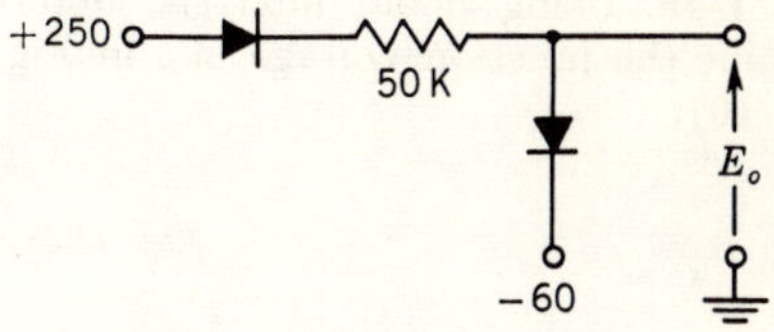

FIG. 1-103. Problem 1-41.

1-42. Design a voltage divider to work from a 500-volt supply which will furnish 250 volts at 100 ma to a load. The total divider resistance (excluding the load) is 5 kilohms. *Ans.* $R_1 = 1{,}450$ ohms, $R_2 = 3{,}550$ ohms.

1-43. Design an attenuator for an audio oscillator of low output impedance which will deliver 10 volts rms to an open circuit. The attenuator is to furnish 10-, 5-, and 1-volt output taps with respect to ground, and is to bleed 20 ma. *Ans.* See Fig. 1-104.

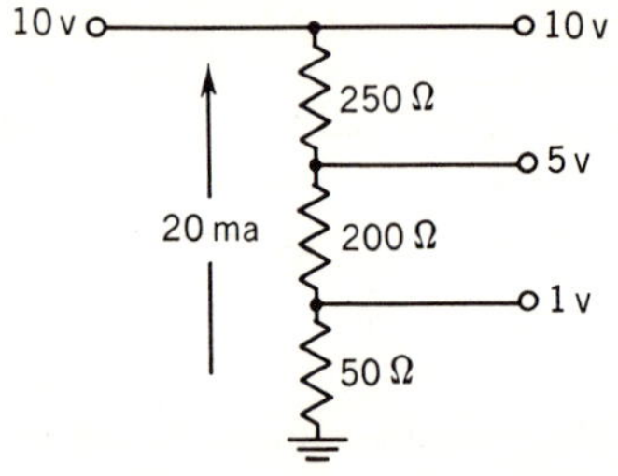

FIG. 1-104. Problem 1-43.

1-44. Estimate the current variation in R_L caused by full rotation of the potentiometer (pot) in Fig. 1-105. *Ans.* 100 μa.

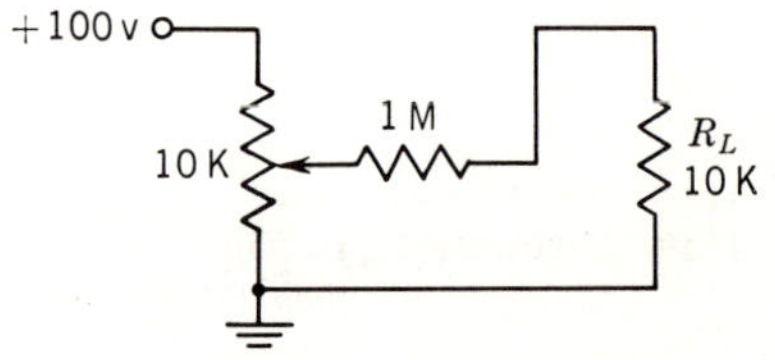

FIG. 1-105. Problem 1-44.

1-45. Estimate the voltage variation across R_L caused by full rotation of the pot in Fig. 1-106. *Ans.* 100 volts.

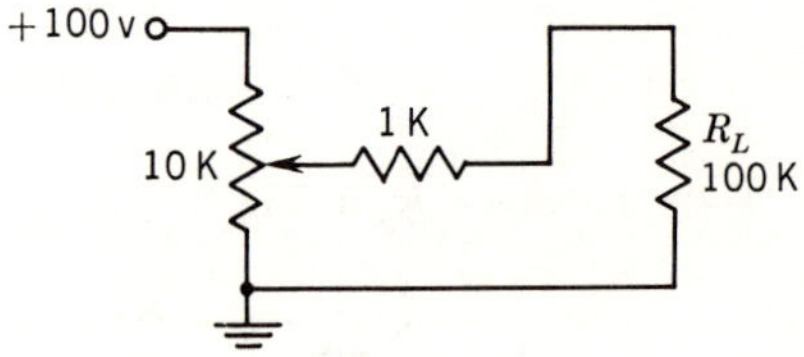

FIG. 1-106. Problem 1-45.

1-46. Using nodal analysis, determine the junction voltage, E_j, in Fig. 1-107.

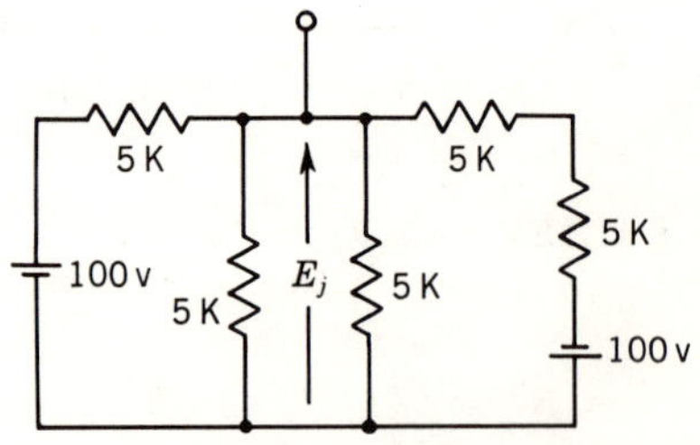

FIG. 1-107. Problem 1-46.

1-47. If $R_1 = R_2 = R_3 \gg R_4$, show that E_o is approximately proportionate to the sum of $E_1 + E_2 + E_3$ in Fig. 1-108. The network of Fig. 1-108 is known as a summing network, and is frequently used in analogue computers and feedback amplifiers. HINT: Use Millman's theorem or nodal analysis or consider each branch a constant-current source.

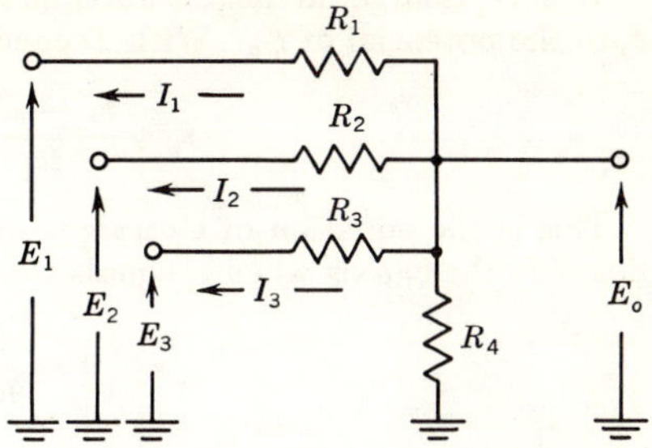

FIG. 1-108. Problem 1-47.

1-48. What is E_o in Fig. 1-109.
Ans. 1.6 volts.

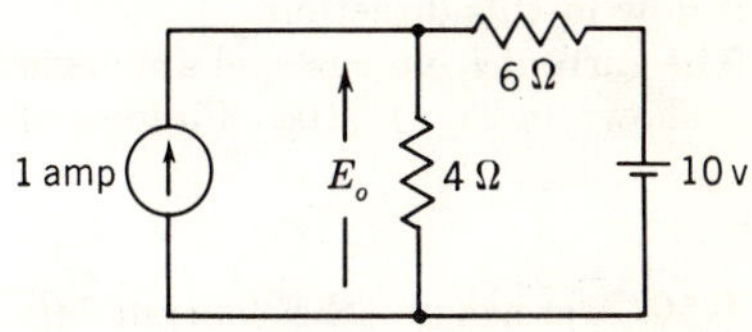

FIG. 1-109. Problem 1-48.

1-49. Assuming the diode in Fig. 1-110*a* is ideal, determine the manner in which e_o varies with e_1 and sketch the response.

Solution. Inspection of the circuit indicates that e_1 forward-biases the diode while E_2 reverse-biases it. When e_1 is less in magnitude than E_2 the open-circuit voltage the diode looks into will reverse-bias it and i_1 will equal zero. The output e_o will therefore equal e_1, since $e_o = e_1 + i_1R_1$. (e_1 is in itself a positive or negative quantity.)

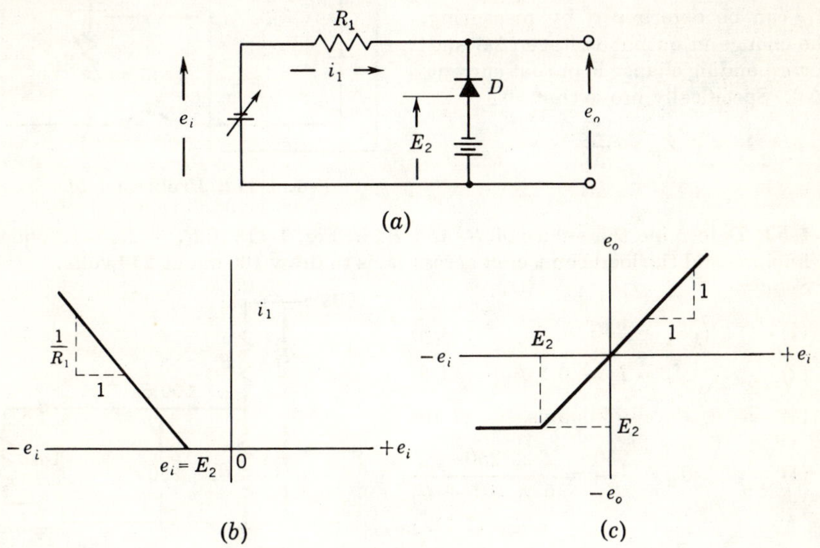

FIG. 1-110. Problem 1-49.

When e_1 is larger in magnitude than E_2, the diode will conduct and force (clamp) e_o to the potential of E_2. With D conducting, i_1 will be expressed by

$$i_1 = \frac{e_1 - E_2}{R_1} = \frac{1}{R_1} e_1 - \frac{E_2}{R_1}$$

This is the equation of a straight line having a slope of $\Delta i_1/\Delta e_1 = 1/R_1$. The line cuts the e_1 axis when i_1 equals zero, which means that

$$0 = \frac{1}{R_1} e_1 - \frac{E_2}{R_1}$$

Therefore, e_1 equals E_2 at the point where i_1 equals zero. We could find what i_1 is when e_1 equals zero as another point on this line by setting e_1 equal to zero. Negative values of i_1 cannot exist in the physical circuit, however, since D will not permit flow in this direction.

The current i_1-versus-e_1 characteristic is plotted in Fig. 1-110*b* and the e_o-versus-e_1 is shown in Fig. 1-110*c*. Circuits of this nature are known as *clipping* circuits.

1-50. Determine the output-input voltage characteristic of the circuit in Fig. 1-111.

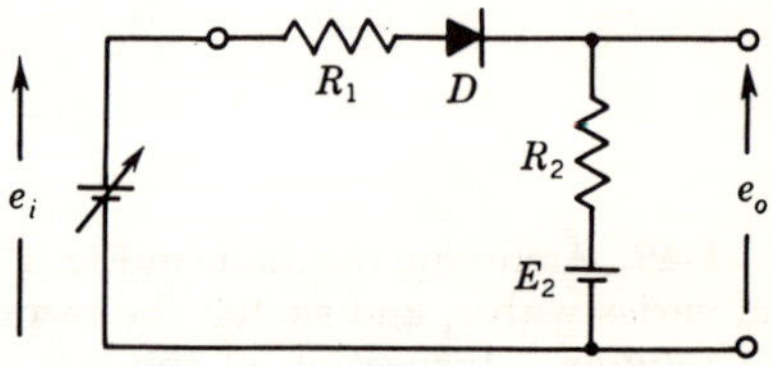

FIG. 1-111. Problem 1-50.

1-51. Referring to Fig. 1-112 show that the internal impedance of the box Z_{th} can be determined by measuring the change in output voltage (Δe) and corresponding change in output current (Δi). Specifically prove that

$$Z_{th} = \frac{\Delta e}{\Delta i}$$

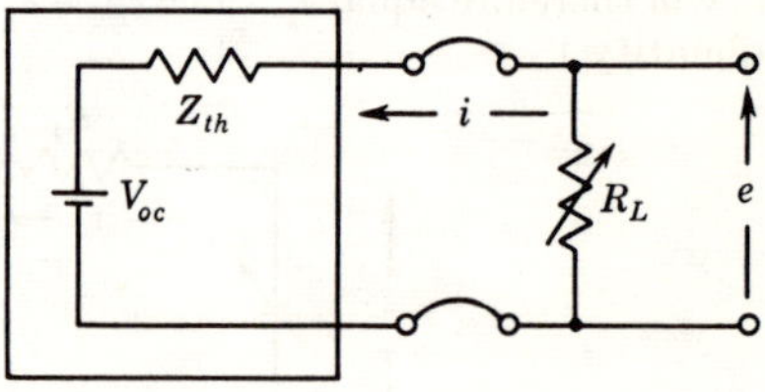

FIG. 1-112. Problem 1-51.

1-52. Determine the values of R_1 and R_2 in Fig. 1-113 if $R_1 + R_2$ is to equal 5 kilohms and the load connected across R_2 is to draw 100 ma at 250 volts.

Solution

(1) $I_1R_1 = 250$ (1-8)

(2) $I_1 = I_2 + 0.1$ amp (1-9)

(3) $R_1 + R_2 = 5$ kilohms (1-10)

(4) $$I_2 = \frac{250}{R_2} = \frac{250}{5 \times 10^3 - R_1}$$ (1-11a)

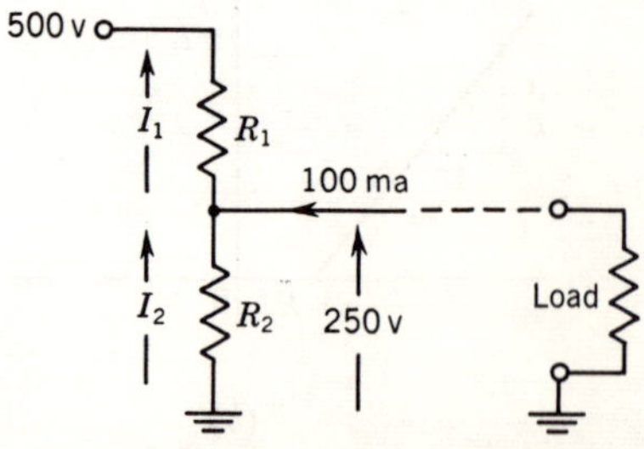

FIG. 1-113. Problem 1-52.

and $$I_1 = \frac{250}{5 \times 10^3 - R_1} + 0.1 \text{ amp}$$

$$= \frac{250 + 500 - 0.1R_1}{5 \times 10^3 - R_1} = \frac{750 - 0.1R_1}{5 \times 10^3 - R_1} \quad (1\text{-}9a)$$

$$\left[\frac{750 - 0.1R_1}{5 \times 10^3 - R_1}\right] R_1 = 250 \quad (1\text{-}8a)$$

$$750R_1 - 0.1R_1^2 - 250(5 \times 10^3) + 250R_1 = 0$$

$$R_1^2 - 10^4R_1 + 0.25(5 \times 10^3)10^4 = 0$$

$$R_1 = \frac{10^4 \pm \sqrt{10^8 - 0.5(10^8)}}{2} = \frac{10^4 \pm 0.71(10^4)}{2}$$

= 8.55 kilohms for this positive radical,
or 1.45 kilohms for negative radical

Plainly, $R_1 = 8.55$ kilohms is extraneous, since R_1 plus R_2 must total 5 kilohms. Then if $R_1 = 1.45$ kilohms, $R_2 = 3.55$ kilohms.

Chapter 2

A-C CIRCUIT ANALYSIS

This chapter assumes that the reader has a working knowledge of basic a-c theory, including the j operator. With this background and the methods of network analysis developed in Chap. 1, specific topics in a-c theory are undertaken. The topics are chosen with two primary objectives in mind.

First of all, we wish to develop manipulative skill in complex algebra. This skill will enable us to write and manipulate equations containing complex numbers into useful forms. The equations may then be examined to determine those frequencies at which the network appears resistive, or has maximum impedance, and so on.

In addition, the topics discussed will aid in the simplification and analysis of a-c networks encountered in future chapters. For example, the section pertaining to Bode diagrams acquaints the reader with this subject at a low level. Familiarity with Bode plots should prove especially helpful to the technician engaged in servo work.

Resonance phenomena and some methods of impedance matching are carefully analyzed. These topics are essential to the analysis of tuned amplifiers and oscillators.

The reader should make every effort to master the material in this section, since alternating currents are primarily used in power and communications work.

2-1. Brief Review. In basic electronics, the reader was undoubtedly exposed to the concepts of reactance, impedance, phase angles, and rotating vectors. Let us briefly summarize the principal points of elementary a-c theory. If the material in this review is unfamiliar, the reader should refer to some texts in basic electronics and mathematics. The author strongly recommends "Basic Mathematics for Electronics."[1] It contains an excellent introduction to the j operator.

[1] Nelson M. Cooke, "Basic Mathematics for Electronics," 2d ed., McGraw-Hill Book Company, Inc., New York, 1960.

1. A sine wave may be considered as the vertical projection y of the rotating vector r in Fig. 2-1. Plainly, $y = r \sin \theta$.

2. The angle θ may be expressed in degrees or radians. There are 2π radians in 360°. An angle in degrees may be expressed in radians by multiplying the angle by $2\pi/360°$. An angle in radians is converted to degrees by multiplying by $360°/2\pi$.

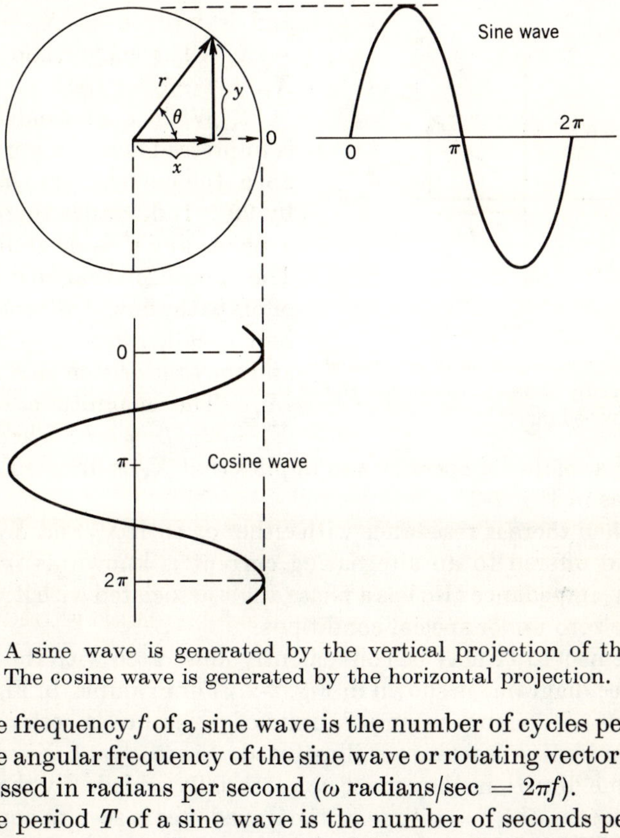

FIG. 2-1. A sine wave is generated by the vertical projection of the rotating vector r. The cosine wave is generated by the horizontal projection.

3. The frequency f of a sine wave is the number of cycles per second.

4. The angular frequency of the sine wave or rotating vector may also be expressed in radians per second (ω radians/sec $= 2\pi f$).

5. The period T of a sine wave is the number of seconds per cycle.

6. The frequency is the reciprocal of the period; that is, $f = 1/T$.

7. The angle θ, in radians, that the rotating vector has swept through is given by $\theta = \omega t$. If there is an initial angular displacement ϕ, at time zero, $\theta = \omega t + \phi$.

8. A cosine wave may be considered as the horizontal projection x of the rotating vector in Fig. 2-1. Clearly, $x = r \cos \theta$. At time zero, the cosine wave is a maximum and the sine wave a minimum. The cosine wave is apparently a sine wave advanced 90°, or $\pi/2$ radians. That is, $x = r \sin (\theta + \pi/2)$.

9. When a sinusoidal voltage is impressed across a pure resistance,

the resultant current is in phase with the applied voltage. Resistance therefore has no phase angle associated with it and is simply written as $R\underline{/0^\circ}$, or just R.

10. When a sinusoidal voltage is impressed across a pure capacitor, the resultant current leads the voltage by 90°. Capacitance therefore has a phase angle associated with it. The opposition that a capacitor offers to the flow of alternating current is called *capacitive reactance* and is written as $X_C\underline{/-90^\circ}$, or $-jX_C$. The magnitude of X_C is $X_C = 1/2\pi fC = 1/\omega C$.

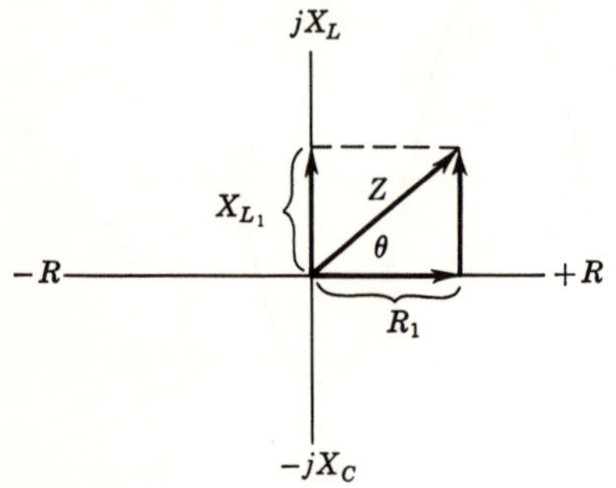

FIG. 2-2. A resistance and reactance are vectorially added to yield the resultant impedance.

11. When a sinusoidal voltage is impressed across a pure inductance, the current lags the voltage by 90°. Inductance therefore has a phase angle associated with it. The opposition that an inductance offers to the flow of alternating current is called inductive reactance and may be written as $X_L\underline{/90^\circ}$, or jX_L. The magnitude of X_L is $X_L = 2\pi fL = \omega L$. The quantity j is called a *rotational operator* and implies that X_L is directed 90° away from R, as in Fig. 2-2.

12. When there is resistance with either or both X_C and X_L, the net opposition offered to an alternating current is known as *impedance*. Generally, impedance also has a phase angle associated with it, although it may be zero under special conditions.

13. Items 9 to 12 may be conveniently illustrated with the aid of an impedance diagram, as shown in Fig. 2-2. For example, in Fig. 2-2 the impedance of a circuit having R_1 units of resistance in series with X_{L_1} units of inductive reactance is illustrated. Since R_1 and X_{L_1} do not lie in the same direction, they must be vectorially added. By the Pythagorean theorem, the magnitude of the impedance is seen to equal $|Z_1| = \sqrt{R_1^2 + X_{L_1}^2}$. The angle θ is the angle whose tangent is X_{L_1}/R_1. This is written in mathematical shorthand as $\theta = \tan^{-1} X_L/R_1$. If an impedance is written without the bracket which designates magnitude only, then the angle is implied.

14. The polar form of denoting a vector quantity is to write the magnitude and associated phase angle. For example, an impedance of 80 ohms at an angle of 30° would be written $Z = 80\underline{/30^\circ}$ ohms. The polar form is a simplified way of writing the exponential form. That is $Z\underline{/\theta} = |Z|\epsilon^{j\theta}$.

15. The rectangular form of denoting a vector is to write its resistive

(real) and reactive (imaginary) components. The reactive component is preceded by a plus or minus j to indicate that it does not lie in the same general direction as does a pure resistance. The rectangular or complex form of the vector in Fig. 2-2 would be written $Z_1 = R_1 + jX_{L_1}$. The angle associated with Z_1 is implied.

16. Conversion between the rectangular and polar forms may be performed with the following relationships:

$$|Z| = \sqrt{R^2 + (X_L - X_C)^2} \qquad \theta = \tan^{-1}\frac{X_L - X_C}{R}$$

$$R = |Z| \cos\theta \qquad X = X_L - X_C = |Z| \sin\theta$$

17. Multiplying a vector by j rotates it 90° counterclockwise. Multiplying a vector by $-j$ rotates it 90° clockwise. Multiplying a vector by j^2 rotates it 180°, which is the same as multiplying by minus 1. That is,

$$j^1 = j \qquad j^2 = -1 \qquad j^3 = -j \qquad \text{and} \qquad j^4 = 1$$

18. Addition and subtraction of vectors are conveniently performed in the rectangular rather than in the polar form. All the resistive components are algebraically added, and then all the reactive components are algebraically added. The result is a vector in rectangular form which is equal to the vector sum of all the individual vectors.

19. Multiplication and division of vectors are conveniently performed in the polar form. In multiplication, the magnitudes are multiplied and the angles added algebraically. In division, the magnitudes are divided, and the angles subtracted algebraically. For example,

$$(8\angle 40°)(4\angle 20°) = 32\angle 60°$$

$$(8\angle 40°)(4\angle -20°) = 32\angle 20°$$

$$\frac{8\angle 40°}{4\angle -20°} = 2\angle 40° - (-20°) = 2\angle 60°$$

$$\frac{8\angle 40°}{4\angle 20°} = 2\angle 40° - 20° = 2\angle 20°$$

The angles are treated as if they were exponents. A vector in exponential form is written $|Z|\epsilon^{j\theta}$, or $Z\epsilon^{j\theta}$, since magnitude is implied when an angle is separately indicated. The product of two vectors is, then,

$$Z_1Z_2 = (Z_1\epsilon^{j\theta_1})(Z_2\epsilon^{j\theta_2}) = Z_1Z_2\epsilon^{j(\theta_1+\theta_2)}$$

20. The reciprocal of an impedance vector, Z, is an admittance vector, y. The resistive part of the admittance vector is called the conductance G, and the reactive component is the susceptance B. The conductance is not necessarily the reciprocal of the resistive component in the impedance vector, nor is the admittance the reciprocal of the reactance.

For example, find the admittance of a circuit consisting of a 3-ohm resistor in series with an inductor having a reactance of 4 ohms.

$$Z = 3 + j4 \text{ ohms}$$

$$Y = \frac{1}{Z} = \frac{1}{3 + j4} = \frac{1}{5\angle 53.2^\circ} = 0.2\angle -53.2^\circ \text{ mhos}$$

$$= (0.12 - j0.16) \text{ mhos}$$

Therefore, $Y = G + jB = (0.12 - j0.16)$ mhos. Notice that the conductance, the resistive part of Y, is not the reciprocal of 3, the resistive part of Z. Nor is the susceptance equal to the reciprocal of the 4-ohm reactance.

Let us refresh our memories by trying a few problems. The calculations shown can readily be done on a trigonometric slide rule. The reader should master this tool, which minimizes the drudgery in a-c calculations.

Problem 1. Express $3 + j4$ in polar form.

Solution

$$3 + j4 = \sqrt{3^2 + 4^2} \angle \tan^{-1} 4/3 = 5\angle 53.2^\circ$$

Problem 2. Express $4\angle 60^\circ$ in rectangular form.

Solution

$$4\angle 60^\circ = 4(\cos 60^\circ + j \sin 60^\circ) = 2 + j3.46$$

Problem 3. Add $2 - j3 + 5\angle 45^\circ$.

Solution

$$2 - j3 + 5(\cos 45^\circ + j \sin 45^\circ) = 2 - j3 + 3.55 + j3.55 = 5.55 + j0.55$$

Problem 4. Evaluate $3 + j4 - \dfrac{2 + j3}{1 - j1}$.

Solution

$$3 + j4 - \frac{3.6\angle 56.3^\circ}{1.4\angle -45^\circ} = 3 + j4 - 2.56\angle 101.3^\circ$$

$$= 3 + j4 - (-0.5 + j2.5) = 3.5 + j1.5$$

NOTE: 101.3° is equivalent to an angle in the second quadrant of 78.7°. The horizontal component is therefore negative.

Problem 5. Evaluate $(2 - j3)(5\angle 45^\circ)$.

Solution

$$(3.6\angle -56.3^\circ)(5\angle 45^\circ) = 18\angle -11.3^\circ$$

Problem 6. Evaluate $(3 + j4)^2$.

Solution

$$(5\angle 53.2°)^2 = 25\angle 106.4°$$

Problem 7. Determine the current in the circuit of Fig. 2-3.

Solution

$$\omega = 2\pi f = 6.28 \times 10^4 \text{ radians/sec}$$

$$X_L = \omega L = 6.28 \times 10^4 \text{ ohms}$$

$$X_C = \frac{1}{\omega C} = 15.9 \times 10^3 \text{ ohms}$$

$$Z = R + jX_L - jX_C = R + j(X_L - X_C)$$

$$Z = 50 + j46.9 \text{ kilohms}$$

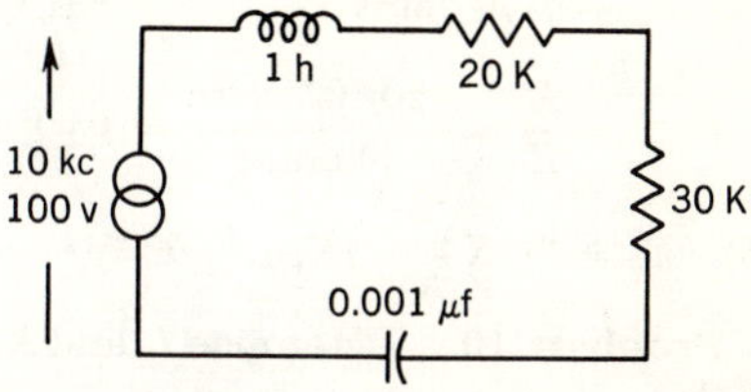

FIG. 2-3. The current is $1.5\angle -43.1°$ ma in this circuit.

$$|Z| = \sqrt{(50)^2 + (46.9)^2} = 68.6 \text{ kilohms}$$

$$\theta = \tan^{-1}(46.9/50) = \tan^{-1} 0.94 = 43.1°$$

$$Z = 68.6 \text{ kilohms } \angle 43.1°$$

Taking the applied voltage at a reference angle of zero degrees,

$$I = 100\angle 0° \text{ volts}/68.6\angle 43.1° \text{ kilohms} = 1.5\angle -43.1 \text{ ma}$$

Since the net reactance of the circuit is inductive, the current lags the applied voltage.

Problem 8. Thevenize the circuit of Fig. 2-4*a*.

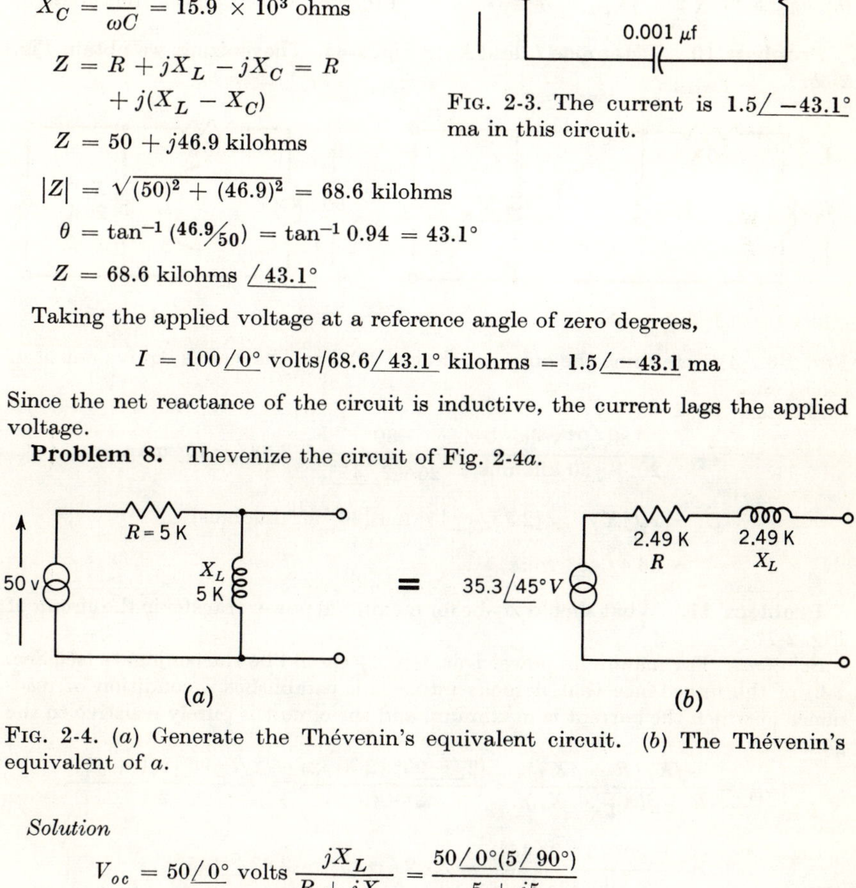

(*a*) (*b*)

FIG. 2-4. (*a*) Generate the Thévenin's equivalent circuit. (*b*) The Thévenin's equivalent of *a*.

Solution

$$V_{oc} = 50\angle 0° \text{ volts } \frac{jX_L}{R + jX_L} = \frac{50\angle 0°(5\angle 90°)}{5 + j5}$$

$$= \frac{50\angle 0°(5\angle 90°)}{7.08\angle 45°} = 35.3\angle 45°$$

$$Z_{th} = \frac{R(jX_L)}{R + jX_L} = \frac{5(5\angle 90°)}{7.08\angle 45°} = 3.53\angle 45° = 2.5 + j2.5 \text{ kilohms}$$

All phase angles in Fig. 2-4*b* are with reference to the applied voltage in Fig. 2-4*a*.

Problem 9. (*a*) Determine the current in Fig. 2-5. (*b*) What is the output voltage, E_o?

Solution

(*a*) $Z = R + j(X_L - X_C) = 30 + j0$ ohms

$$I = \frac{E}{Z} = \frac{120\underline{/0^\circ} \text{ volts}}{30 \text{ ohms}} = 4\underline{/0^\circ} \text{ amp}$$

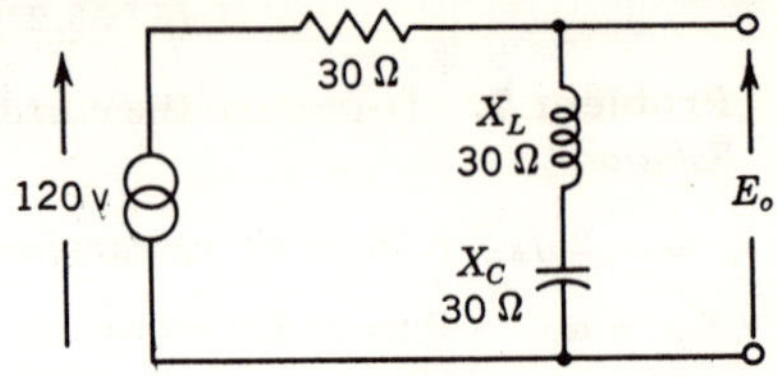

FIG. 2-5. What is the output voltage?

(*b*) $E_o = I(jX_L - jX_C) = 0$

Problem 10. Determine I_2 and E_o in Fig. 2-6*a*. Thevenizing, we obtain Fig. 2-6*b*.

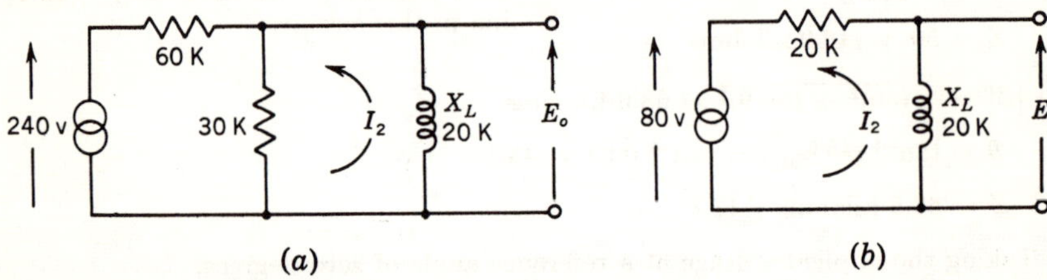

FIG. 2-6. (*a*) Determine the output voltage. (*b*) The Thévenin's equivalent of *a*.

Solution

$$I_2 = \frac{80\underline{/0^\circ} \text{ volts}}{20 + j20 \text{ kilohms}} = \frac{80}{20\sqrt{2}\underline{/45^\circ}} = 2.83\underline{/-45^\circ} \text{ ma}$$

$$E_o = I_2(jX_L) = (2.83\underline{/-45^\circ} \text{ ma})(20\underline{/90^\circ} \text{ kilohms})$$

$$= 56.4\underline{/45^\circ} \text{ volts}$$

Problem 11. What should Z_L be for maximum power transfer in the circuit of Fig. 2-7?

Solution. For maximum power transfer, Z_L should be the conjugate (see Sec. 2-2) of the impedance that it looks into. This establishes a condition of resonance in which the current is maximum, and the circuit is purely resistive to the generator.

$$Z_{th} = \frac{-jX_C(R + jX_L)}{R + j(X_L - X_C)} = \frac{(2\underline{/-90^\circ})(2 + j2)}{2 + j0} = \frac{2\underline{/-90^\circ}[2(1 + j1)]}{2}$$

$$= \frac{(2\underline{/-90^\circ})(2\sqrt{2}\underline{/45^\circ})}{2 \text{ kilohms}} = 2\sqrt{2}\underline{/-45^\circ} = 2.83\underline{/-45^\circ}$$

$$= 2 - j2 \text{ kilohms}$$

Therefore, Z_L should equal $2 + j2$ kilohms, which implies a 2-kilohm resistor in series with an inductance whose reactance is 2 kilohms at the source frequency.

Problem 12. (*a*) Derive an expression which indicates what input frequency is required for the magnitude of E_o in Fig. 2-8 to equal $E_{in}/\sqrt{2}$. (*b*) What is the output phase angle relative to the input at this frequency?

Solution. (*a*) In general

$$\frac{E_o}{E_{in}} = \frac{R}{R - j1/\omega C}$$

Specifically, we are interested in the case where

$$\left|\frac{R}{R - j1/\omega C}\right| = \frac{1}{\sqrt{2}}$$

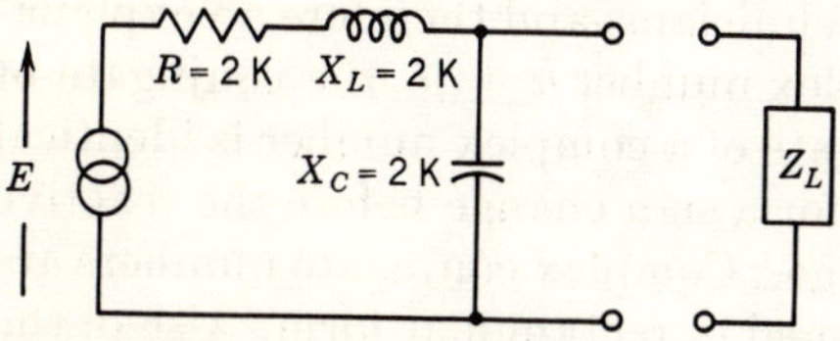

FIG. 2-7. What should the load be for maximum power transfer?

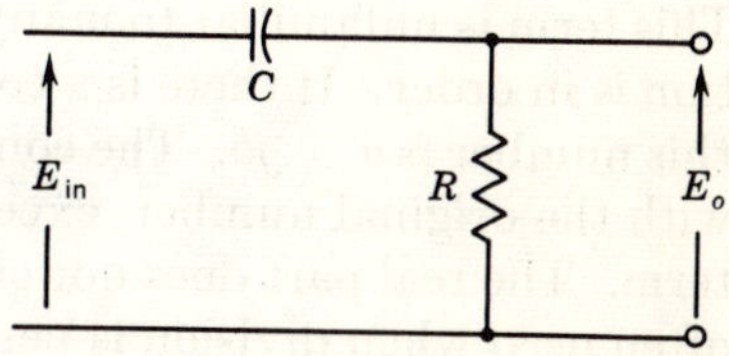

FIG. 2-8. At what frequency is $E_o = E_{in}/\sqrt{2}$?

The magnitude of the numerator is R, and the denominator is $\sqrt{R^2 + 1/(\omega C)^2}$. Therefore,

$$\frac{R}{\sqrt{R^2 + 1/(\omega C)^2}} = \frac{1}{\sqrt{2}}$$

Solving for ω by squaring both sides,

$$\frac{R^2}{R^2 + 1/(\omega C)^2} = \frac{1}{2}$$

$$R^2 + \frac{1}{(\omega C)^2} = 2R^2$$

$$(\omega CR)^2 + 1 = 2R^2(\omega C)^2$$

$$(\omega CR)^2 + 1 = 2(\omega CR)^2$$

$$1 = (\omega CR)^2$$

$$\sqrt{1} = \sqrt{(\omega CR)^2}$$

$$1 = \omega CR$$

$$\omega = \frac{1}{RC}$$

$$f = \frac{1}{2\pi RC}$$

(*b*) In general, if $\omega = 1/RC$,

$$\frac{E_o}{E_{in}} = \frac{R}{R - j1/\omega C} = \frac{R}{R - j1/(1C/RC)} = \frac{R}{R - jR} = \frac{R}{R(1 - j1)}$$

$$= \frac{1}{1 - j1} = \frac{1}{\sqrt{2}\angle -45°} = \frac{1\angle 45°}{\sqrt{2}}$$

Therefore, at $\omega = 1/RC$, the output will lead the input by 45°.

2-2. The Complex Conjugate. In this chapter we shall combine and expand the analytic tools of Chap. 1 with the a-c theory of basic electronics. The reader should have a working knowledge of the j operator and some basic trigonometry. If he is deficient in these subjects, recommended texts are Cooke's "Basic Mathematics for Electronics"[1] and Page's "Algebra of Electronics."[2]

Later we shall frequently refer to the conjugate of a complex number. This term is unfamiliar to many technicians, and therefore an explanation is in order. If there is a complex number $a \pm jb$, the conjugate of this number is $a \mp jb$. The conjugate of a complex number is identical with the original number, except for a sign change before the reactive term. The real part does not change. Complex conjugate numbers are often used when division is performed in rectangular form. Use of the conjugate enables us to express the ratio of two complex numbers as a third complex number. For example, suppose you had to simplify

$$\frac{a + jb}{c - jd} \tag{2-1}$$

At first glance, this expression seems to be in simplest form. But we can multiply both numerator and denominator of the fraction by the conjugate of the denominator. That is,

$$\frac{a + jb}{c - jd}\frac{c + jd}{c + jd} = \frac{(ac - bd) + j(bc + ad)}{c^2 + d^2} = \frac{ac - bd}{c^2 + d^2} + j\frac{bc + ad}{c^2 + d^2} \tag{2-2}$$

Note that multiplying a complex number by its conjugate yields a real number. Our final result (2-2) shows that (2-1) may be expressed as a complex number having real and imaginary parts. The real part, since it has no j term in front of it, is

$$\frac{ac - bd}{c^2 + d^2}$$

and the imaginary part, identified by the j term, is

$$j\frac{bc + ad}{c^2 + d^2}$$

2-3. A Compensated Attenuator. As an exercise in manipulating complex numbers, try this very practical though tedious problem.

Figure 2-9 shows the circuitry of an RC voltage divider. Develop an expression for E_o, and show that when $R_1C_1 = R_2C_2$, the output is independent of the input frequency. This circuit is used in scope probes.

[1] *Ibid.*

[2] Chester H. Page, "Algebra of Electronics," D. Van Nostrand Company, Inc., Princeton, N.J., 1958.

R_1 and C_1 are in the probe, while C_2 is the combined capacitance of the coaxial cable, the wiring, and the input capacitance of the input stage. R_2 is the grid return resistor of the input stage, or the resistance of a stepped attenuator.

Let the R_1C_1 combination be called Z_1. Where

$$Z_1 = \frac{R_1(-jX_{C_1})}{R_1 - jX_{C_1}}$$

multiplying by the conjugate of the denominator gives

$$\frac{R_1(-jX_{C_1})}{R_1 - jX_{C_1}} \frac{R_1 + jX_{C_1}}{R_1 + jX_{C_1}} = \frac{R_1X_{C_1}{}^2}{R_1{}^2 + X_{C_1}{}^2} - j\,\frac{R_1{}^2X_{C_1}}{R_1{}^2 + X_{C_1}{}^2}$$

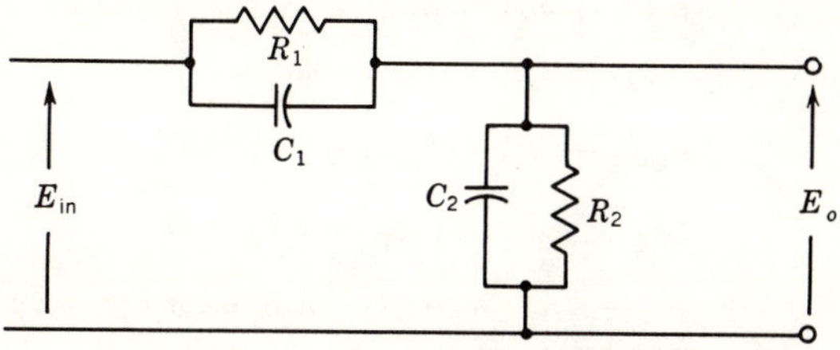

FIG. 2-9. A compensated attenuator.

Therefore $\qquad Z_1 = a_1 - jb$

where $\qquad a_1 = \dfrac{R_1X_{C_1}{}^2}{R_1{}^2 + X_{C_1}{}^2}$

and $\qquad b_1 = \dfrac{R_1{}^2X_{C_1}}{R_1{}^2 + X_{C_1}{}^2}$

Similarly, $\qquad Z_2 = a_2 - jb_2$

where $\qquad a_2 = \dfrac{R_2X_{C_2}{}^2}{R_2{}^2 + X_{C_2}{}^2}$

and $\qquad b_2 = \dfrac{R_2{}^2X_{C_2}}{R_2{}^2 + X_{C_2}{}^2}$

Therefore $\qquad \dfrac{E_o}{E_{in}} = \dfrac{Z_2}{Z_1 + Z_2} = \dfrac{a_2 - jb_2}{(a_1 - jb_1) + (a_2 - jb_2)}$

$$= \frac{a_2 - jb_2}{(a_1 + a_2) - j(b_1 + b_2)}$$

Multiplying by the conjugate of the denominator,

$$\frac{E_o}{E_{in}} = \frac{a_2 - jb_2}{(a_1 + a_2) - j(b_1 + b_2)} \frac{(a_1 + a_2) + j(b_1 + b_2)}{(a_1 + a_2) + j(b_1 + b_2)}$$

$$= \frac{a_2(a_1 + a_2) + b_2(b_1 + b_2)}{(a_1 + a_2)^2 + (b_1 + b_2)^2} + j\frac{a_2(b_1 + b_2) - b_2(a_1 + a_2)}{(a_1 + a_2)^2 + (b_1 + b_2)^2}$$

Now if E_o is to be independent of frequency, the reactive term must at least be equal to zero. Otherwise, there would be some ratio of X/R which results in a phase shift. We shall therefore set the j term equal to zero and mathematically investigate the relationship between the parameters, for this condition to be true:

$$\frac{a_2(b_1 + b_2) - b_2(a_1 + a_2)}{(a_1 + a_2)^2 + (b_1 + b_2)^2} = 0$$

Therefore,
$$a_2(b_1 + b_2) - b_2(a_1 + a_2) = 0$$

$$a_2b_1 + a_2b_2 - a_1b_2 - a_2b_2 = 0$$

$$a_2b_1 = a_1b_2$$

(this step is used further on); or

$$\frac{a_2}{b_2} = \frac{a_1}{b_1}$$

Now we can substitute for a and b:

$$\frac{\dfrac{R_2X_{C_2}^2}{R_2^2 + X_{C_2}^2}}{\dfrac{R_2^2X_{C_2}}{R_2^2 + X_{C_2}^2}} = \frac{\dfrac{R_1X_{C_1}^2}{R_1^2 + X_{C_1}^2}}{\dfrac{R_1^2X_{C_1}}{R_1^2 + X_{C_1}^2}}$$

Therefore,
$$\frac{R_2X_{C_2}^2}{R_2^2X_{C_2}} = \frac{R_1X_{C_1}^2}{R_1^2X_{C_1}}$$

$$\frac{X_{C_2}}{R_2} = \frac{X_{C_1}}{R_1}$$

But
$$X_C = \frac{1}{\omega C}$$

therefore,
$$\frac{1}{\omega C_2R_2} = \frac{1}{\omega C_1R_1}$$

and
$$C_2R_2 = C_1R_1$$

If this condition is met, we see that the network exhibits zero phase shift. This is also true of a pure resistive voltage divider, in which the output magnitude is also independent of frequency. To be sure that the amplitude of the output is independent of ω when $R_1C_1 = R_2C_2$, we can investigate the real part of the expression for E_o/E_{in}, which is

$$\frac{E_o}{E_{in}} = \frac{a_2(a_1 + a_2) + b_2(b_1 + b_2)}{(a_1 + a_2)^2 + (b_1 + b_2)^2} + j0 \qquad \text{for } R_1C_1 = R_2C_2$$

when $R_1C_1 = R_2C_2$, $\quad a_2b_1 = a_1b_2$, $\quad$ or $\quad a_1 = \dfrac{a_2b_1}{b_2}$

$$\frac{E_o}{E_{in}} = \frac{a_2[a_2(b_1/b_2) + a_2] + b_2(b_1 + b_2)}{[a_2(b_1/b_2) + a_2]^2 + (b_1 + b_2)^2}$$

$$= \frac{a_2^2[(b_1/b_2) + 1] + b_2(b_1 + b_2)}{a_2^2[(b_1/b_2) + 1]^2 + (b_1 + b_2)^2}$$

$$= \frac{a_2^2[(b_1 + b_2)/b_2] + b_2(b_1 + b_2)}{a_2^2[(b_1 + b_2)/b_2]^2 + (b_1 + b_2)^2}$$

$$= \frac{(b_1 + b_2)[(a_2^2/b_2) + b_2]}{(b_1 + b_2)^2[(a_2^2/b_2^2) + 1]}$$

$$= \frac{1}{b_1 + b_2} \frac{(a_2^2 + b_2^2)/b_2}{(a_2^2 + b_2^2)/b_2^2}$$

$$= \frac{b_2}{b_1 + b_2} = \frac{1}{b_1/b_2 + 1}$$

now

$$b_1 = \frac{R_1^2 X_{C_1}}{R_1^2 + X_{C_1}^2} = \frac{R_1^2/\omega C_1}{R_1^2 + 1/(\omega C_1)^2}$$

$$= \frac{\omega C_1 R_1^2}{(\omega C_1 R_1)^2 + 1} = \frac{R_1(\omega C_1 R_1)}{(\omega C_1 R_1)^2 + 1}$$

Similarly,

$$b_2 = \frac{R_2(\omega C_2 R_2)}{(\omega C_2 R_2)^2 + 1}$$

since

$$C_2R_2 = C_1R_1 = \tau$$

$$\frac{b_1}{b_2} = \frac{R_1\omega\tau/[(\omega\tau)^2 + 1]}{R_2\omega\tau/[(\omega\tau)^2 + 1]} = \frac{R_1}{R_2}$$

Therefore,

$$\frac{E_o}{E_{in}} = \frac{1}{R_1/R_2 + 1} = \frac{R_2}{R_1 + R_2}$$

This indicates that the output-to-input ratio depends solely upon the resistances when $R_1C_1 = R_2C_2$ and therefore is independent of frequency.

We now have conclusively shown that both the phase and amplitude characteristics of the network are independent of frequency when $R_1C_1 = R_2C_2$. We also see that the output is related to the input as in a simple resistance divider.

2-4. A Low-pass Filter. The first circuit we shall analyze is that of an *RC* low-pass filter, as shown in Fig. 2-10*a*. In servo systems this circuit is referred to as a phase-lag network. Each of these terms is

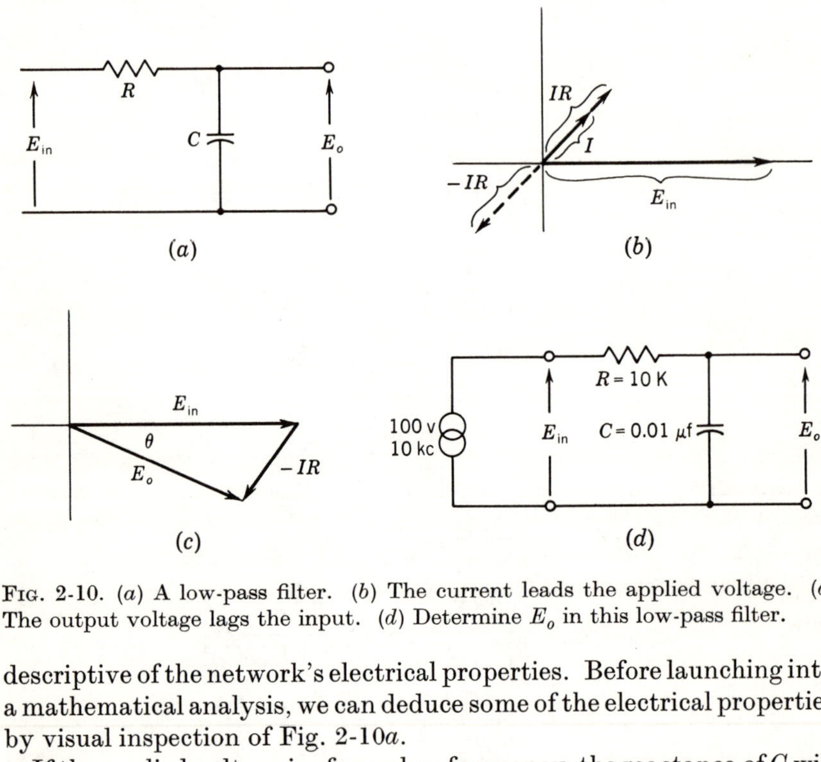

FIG. 2-10. (*a*) A low-pass filter. (*b*) The current leads the applied voltage. (*c*) The output voltage lags the input. (*d*) Determine E_o in this low-pass filter.

descriptive of the network's electrical properties. Before launching into a mathematical analysis, we can deduce some of the electrical properties by visual inspection of Fig. 2-10*a*.

If the applied voltage is of very low frequency, the reactance of *C* will be very high compared with *R*, and *C* may be considered an open circuit. Therefore, at low frequencies the input will appear virtually unattenuated at the output. Hence, we have the name *low-pass filter*. As the input frequency increases, X_C becomes smaller, causing the input to be increasingly attenuated. At an infinitely high frequency, $X_C = 0$ and therefore $E_o = 0$.

At low frequencies, X_C will be very large. The current will therefore lead the applied voltage by a large angle, but by less than 90°. Remember that the current leads the voltage in a capacitive circuit and lags the voltage in an inductive circuit. The voltage (IR) across the resistor will be in phase with the leading current and must therefore lead E_{in}, as shown in Fig. 2-10*b*. Now E_o is the difference between E_{in} and IR; that is, $E_o = E_{in} - IR$. Remember, these are directed quantities and must be added and subtracted by the rules of vector algebra. To subtract IR from E_{in}, rotate the IR vector by 180°, which is the same as multiplying it by -1. The vector's new position is shown in dotted lines, and is a $-IR$ in this position.

In Fig. 2-10*c*, we add vectorially the $-IR$ vector to our reference vector E_{in}; that is, $E_{in} + (-IR) = E_{in} - IR = E_o$. This is done by placing the tail of the $-IR$ vector on the head of the E vector, and completing the polygon. We see that E_o will lag E by an angle θ. Hence, we have the name *phase-lag network.*

We could also have determined the relative phase of the output by reasoning that if the current leads the voltage by some angle less than 90°, the I vector will lead the E_{in} vector. Assume that I leads E_{in} by 50°. The voltage across C, however, lags the current through it by 90°. Therefore, E_o, the capacitor voltage, must in this example lag the input E_{in} by 40°.

To analyze the circuit mathematically, we would use our voltage-divider relationship and write

$$E_o = E_{in} \frac{-jX_C}{R - jX_C}$$

We shall, however, solve the ratio of E_o to E_{in}, since we frequently wish to express the network gain or loss. This ratio is called the *network transfer function.* Later we shall analyze the properties of amplifiers and oscillators by investigating their transfer functions.

As a transfer function, we then have

$$\frac{E_o}{E_{in}} = \frac{-jX_C}{R - jX_C} = \frac{1/j\omega C}{R + 1/j\omega C} = \frac{1}{1 + j\omega RC}$$

Let $RC = \tau_h$, the circuit time constant. Then

$$\frac{E_o}{E_{in}} = \frac{1}{1 + j\omega\tau_h} \qquad (2\text{-}3)$$

This form of the solution is adaptable to slide-rule evaluation. The transfer function can be expressed in polar form by applying the Pythagorean theorem to the denominator. That is,

$$\frac{E_o}{E_{in}} = \frac{1}{\sqrt{1 + (\omega\tau_h)^2}\ \angle \tan^{-1}\omega\tau_h}$$

$$= \frac{1}{\sqrt{1 + (\omega\tau_h)^2}} \angle -\tan^{-1}\omega\tau_h \qquad (2\text{-}4)$$

In the circuit of Fig. 2-10*d*, find E_o when $E_{in} = 100$ volts at 10 kc, $R = 10$ kilohms and $C = 0.01$ μf.

We could solve this network directly by elementary a-c methods or use the previously derived expressions for the transfer function. Both methods will be applied in order to verify the algebra used in deriving Eqs. (2-3) and (2-4). Let us evaluate ω and X_C first since these quantities will be needed.

$$\omega = 2\pi f = 6.28 \times 10^4 \text{ radians/sec}$$

$$X_C = \frac{1}{\omega_C} = \frac{1}{6.28 \times 10^4 \times 10^{-8}} = 1{,}590 \text{ ohms}$$

$$E_o = \frac{E_{in}(-jX_C)}{R - jX_C} = \frac{(100\angle 0^\circ \text{ volts})(1.59\angle -90^\circ \text{ kilohms})}{10 - j1.59 \text{ kilohms}}$$

$$= \frac{159\angle -90^\circ \text{ kilohms}}{10.2\angle -9^\circ \text{ kilohms}} = 15.6\angle -81^\circ \text{ volts}$$

If we use Eq. (2-3),

$$E_o = \frac{1}{1 + j\omega\tau_h} E_{in}$$

where $\tau_h = RC = (10^4 \text{ ohms})(10^{-2}\ \mu f) = 10^{-4}$ sec

$$\omega\tau_h = 6.28 \times 10^4 \text{ radians/sec} \times 10^{-4} \text{ sec} = 6.28 \text{ radians}$$

The term radians actually has no dimensions, since it is the ratio of a length to a length and therefore need not be carried:

$$E_o = \frac{100\angle 0^\circ}{1 + j6.28} = \frac{100\angle 0^\circ}{6.4\angle 81^\circ} = 15.6\angle -81^\circ \text{ volts,}$$

which checks. If we use Eq. (2-4),

$$E_o = \frac{(100\angle 0^\circ)\angle -\tan^{-1} 6.28}{\sqrt{1 + (6.28)^2}} = \frac{100\angle -\tan^{-1} 6.28}{\sqrt{1 + 39.2}}$$

$$= \frac{100\angle -\tan^{-1} 6.28}{6.4} = 15.6\angle -81^\circ \text{ volts}$$

which also checks.

Problem 1. If $C = 0.01$ μf in Fig. 2-10*a*, what must R be if E_o is to lag E_{in} by 30° at $f = 10$ kc? What is E_o for this phase shift?

Solution. We inspect Eq. (2-4) and see that the phase angle is the angle whose tangent is $\omega\tau_h$ and that this angle is to be 30°. That is,

$$\tan^{-1} \omega\tau_h = 30^\circ$$

Therefore

$$\omega\tau_h = \tan 30^\circ$$

and $$\tan 30° = 0.578$$

$$\omega\tau_h = 0.578, \quad \tau_h = \frac{0.578}{6.28 \times 10^4} = 0.092 \times 10^{-4}$$

but $$RC = \tau_h = 0.092 \times 10^{-4}$$

Therefore $$R = \frac{0.092 \times 10^{-4}}{0.01\ \mu f} = 9.2 \times 10^2 \text{ ohms} = 0.92 \text{ kilohms}$$

The amplitude of E_o for this value of R is

$$E_o = \frac{1}{\sqrt{1 + (\omega\tau_h)^2}} E_{in} = \frac{100}{\sqrt{1 + (0.578)^2}} = \frac{100}{\sqrt{1.33}}$$
$$= \frac{100}{1.16} = 86.5 \text{ volts}$$

In Fig. 2-10*a*, is there a possibility of varying the network amplitude characteristic by juggling the values of R and C, while holding the phase shift constant at a fixed frequency?

This can be determined by inspection of Eq. (2-4). The phase shift is determined by $\omega\tau_h$, and since f is fixed, so is ω. Therefore, the phase shift will depend upon the RC product, which is τ_h. An infinite variety of RC combinations will result in the same phase shift. For example, suppose we were to vary RC so that R becomes $10R$, and C becomes $C/10$. Then τ_h will remain constant; and since the amplitude characteristic depends on τ_h only, it too will remain constant. This means that if we design for a specific phase shift at some frequency, the attenuation is fixed, and vice versa.

2-5. The Bode Plot. A most useful means of displaying the amplitude and phase characteristics of a network is to plot the magnitude of the transfer function versus frequency on one curve and the phase characteristic as a separate curve but with the same frequency axis. The curves may be drawn on semilog paper, so that several decades of frequency are included. The amplitude characteristic, which may vary over a wide range, can be conveniently plotted in terms of decibels. There are other advantages to plotting the transfer function in decibels. Curves with this type of display are known as Bode plots and find wide application in the analysis of amplifiers and servo systems.

Consider the low-pass filter of Fig. 2-10*a* again. The transfer function is

$$\frac{E_o}{E_{in}} = \frac{-jX_C}{R - jX_C} = \frac{1}{1 + j\omega RC}$$

Now RC is a time constant and has the dimensions of seconds.[1] If we

[1] $RC = \frac{\text{volt}}{\text{amp}} \times \frac{\text{charge}}{\text{volt}} = \frac{\cancel{\text{volt}}}{\cancel{\text{amp}}} \times \frac{\cancel{\text{amp}}\text{-sec}}{\cancel{\text{volt}}} = \text{sec.}$

substitute τ_h for RC, then τ_h may be considered as the period of some angular frequency ω_h. That is, $RC = \tau_h = 1/\omega_h$. The transfer function may then be written

$$\frac{E_o}{E_{in}} = \frac{1}{1 + j\omega/\omega_h} = \frac{1}{\sqrt{1 + (\omega/\omega_h)^2}\,\underline{/\tan^{-1}(\omega/\omega_h)}}$$

$$= \frac{1\underline{/-\tan^{-1}(\omega/\omega_h)}}{\sqrt{1 + (\omega/\omega_h)^2}}$$

In any given circuit such as Fig. 2-10a, we can immediately determine ω_h from $\omega_h = 1/RC$; or if frequency in cycles rather than radians per seconds is desired, we have

$$f_h = \frac{\omega_h}{2\pi} = \frac{1}{2\pi RC}$$

and $$\frac{E_o}{E_{in}} = \frac{1}{1 + jf/f_h} = \frac{1}{\sqrt{1 + (f/f_h)^2}\,\underline{/\tan^{-1}(f/f_h)}} = \frac{1\underline{/-\tan^{-1}(f/f_h)}}{\sqrt{1 + (f/f_h)^2}}$$

Once ω_h is identified, the gain at any frequency relative to ω_h can readily be determined. For example, if the network is driven by a source whose frequency ω happens to equal ω_h, the gain will be

$$\frac{E_o}{E_{in}} = \frac{1}{\sqrt{1 + (\omega/\omega_h)^2}}\,\underline{/-\tan^{-1}(\omega/\omega_h)} = \frac{1}{\sqrt{1 + (\omega_h/\omega_h)^2}}\,\underline{/-\tan^{-1}(\omega_h/\omega_h)}$$

$$= \frac{1}{\sqrt{1 + 1}}\,\underline{/-\tan^{-1} 1} = \frac{1}{\sqrt{2}}\,\underline{/-45^\circ}$$

To plot the phase and amplitude over a wide frequency range, four- or five-cycle semilog paper should be used. To accommodate the large gain variation which occurs at frequencies above ω_h, the gain ratio may be plotted in decibels. Let us assume ratios of ω/ω_h, calculate E_o/E_{in}, and then express the result in decibels. For example, if $\omega = \omega_h$, we just showed that

$$\left|\frac{E_o}{E_{in}}\right| = \frac{1}{\sqrt{2}} = \frac{1}{1.414}$$

Expressing this result in decibels, we have

$$\text{db} = 20 \log \left|\frac{E_o}{E_{in}}\right| = 20 \log \frac{1}{1.414} = 20(\log 1 - \log 1.414)$$

Since the log of 1 to any base is zero,

$$\text{db} = 20(-\log 1.414) = 20(-0.15) = -3$$

The results of Table 2-1 are plotted in Fig. 2-11. The exact amplitude curve is shown by the continuous line, while an approximation is shown by the dotted line. Notice that the amplitude characteristic for this

TABLE 2-1

$\dfrac{\omega}{\omega_h}$	$\dfrac{1}{1 + j\omega/\omega_h}$	db
0.01	$1.0\underline{/-0.6°}$	0
0.04	$1.0\underline{/-2.3°}$	0
0.10	$1.0\underline{/-5.7°}$	0
0.40	$0.93\underline{/-22°}$	−0.7
1.0	$0.707\underline{/-45°}$	−3.0
4.0	$0.24\underline{/-76°}$	−12.3
10.0	$0.1\underline{/-84°}$	−20.0
40.0	$0.03\underline{/-88.5°}$	−32.0
100.0	$0.01\underline{/-89.4°}$	−40.0

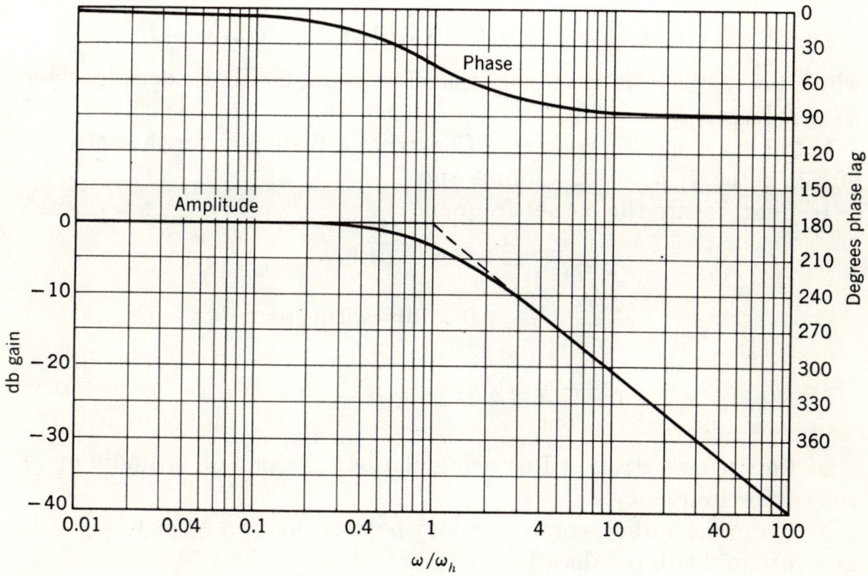

FIG. 2-11. The amplitude and phase response of the low-pass filter may conveniently be sketched on semilog paper.

single-time-constant low-pass RC filter could very well be constructed by drawing two straight lines which intersect at the frequency at which $\omega = \omega_h$. This frequency is called the *corner*, or *break*, frequency because of the abrupt change in the slope of the amplitude response. The slope of the straight line approximation for values of ω greater than ω_h is seen to be 6 db per octave, or 20 db per decade. The greatest error between the approximate and actual curve occurs at the corner frequency, where the gain is really 3 db away from the point of intersection. At an octave above or below the corner frequency, the actual response will be 1 db away from the straight-line approximation. Similarly, at plus or minus 2 octaves, the deviation is only ¼ db.

The validity of the straight-line approximations may be determined as follows:

$$\text{db} = 20 \log \left| \frac{E_o}{E_{in}} \right| = 20 \log \frac{1}{\sqrt{1 + (\omega/\omega_h)^2}}$$

$$= 20\{\log 1 - \log [1 + (\omega/\omega_h)^2]^{1/2}\}$$

$$= 20\{\log 1 - \tfrac{1}{2}\log [1 + (\omega/\omega_h)^2]\}$$

But $\log 1 = 0$; thus,

$$\text{db} = -10 \log [1 + (\omega/\omega_h)^2]$$

and when $\omega = 10\omega_h$

$$\text{db} = -10 \log (1 + 100) \approx -10 \log 100 = -10(2) = -20$$

which proves that the response falls off at about 20 db per decade when ω/ω_h is large.

Let us try a specific problem. Construct a Bode plot for the network of Fig. 2-10*a*, if $R = 9.2$ kilohms and $C = 0.01\ \mu$f.

1. First, locate the corner frequency:

$$\tau_h = RC = 0.092 \times 10^{-3} \text{ sec}$$

$$\omega_h = \frac{1}{\tau_h} = 10.9 \times 10^3 \text{ radians/sec}$$

$$f_h = \frac{\omega_h}{2\pi} = 1.74 \text{ kc}$$

2. Construct a straight-line segment along the zero db ordinate up to the corner frequency.

3. From the 0-db, corner-frequency point, construct a line which falls at a rate of 20 db per decade.

4. If greater accuracy is desired, draw a point 3 db below the corner frequency.

(*a*) Draw a point 1 db below the straight line segment, an octave above f_h, or 34.8 kc.

(*b*) Draw a point 1 db below the straight line segment, an octave below f_h, or 8.7 kc.

(*c*) Connect these points with a smooth curve, as shown in Fig. 2-12.

5. Plot the phase characteristic. This is readily accomplished by examining Fig. 2-11, or the tabulated calculations for plotting Fig. 2-11. At a frequency equal to $\frac{1}{100}$ of 17.4 kc, the phase lag is $-0.6°$. At $\frac{4}{100}$ of ω_h, it is $-2.3°$. At $\frac{1}{10}$, it is $-5.7°$. At ω_h, it is $-45°$; and so on. The resulting plot is shown in Fig. 2-12. Then, as shown by the dashed lines in Fig. 2-11, a smooth curve can be sketched between the three points. The amplitude characteristic should then appear as shown in Fig. 2-11.

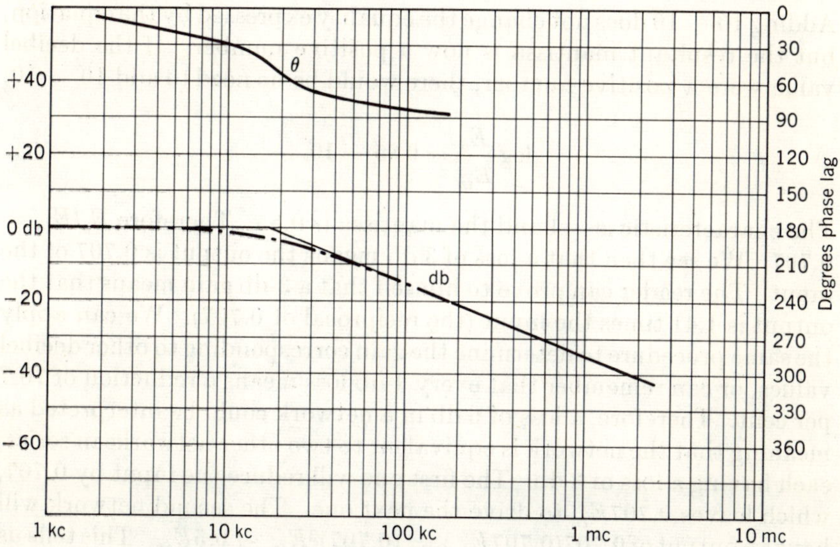

FIG. 2-12. A Bode plot for the circuit of Fig. 2-10*a*.

If it is desired to convert from gain in decibels to actual numerical gain, take the decibel value, which may be plus or minus, and substitute in the expression

$$20 \log \frac{E_o}{E_{in}} = \text{db}$$

Solve for E_o/E_{in}:

$$\log \frac{E_o}{E_{in}} = \frac{\text{db}}{20}$$

$$\frac{E_o}{E_{in}} = \log^{-1} \frac{\text{db}}{20}$$

This last expression is read, "E_o/E_{in} is the number whose logarithm is db/20." This is analogous to saying that 45° is the angle whose tangent is 1, which is written $45° = \tan^{-1} 1$. Let us see how this works out in an actual case. Suppose a Bode plot tells you that at a certain frequency the gain is −3 db. What is the actual numerical gain?

$$20 \log \frac{E_o}{E_{in}} = -3 \text{ db}$$

$$\log \frac{E_o}{E_{in}} = \frac{-3}{20} = -0.15 = (10.00 - 0.15) - 10$$

Adding 10 − 10 does not change the equality expressed by the equation, but the resultant mantissa is now a positive number. If the decibel value were a positive number, there would be no need to add 10 − 10:

$$\log \frac{E_o}{E_{in}} = 9.85 - 10$$

The characteristic is −1 and the mantissa is 0.85. Therefore $E_o/E_{in} = 0.707$. We see then that a loss of 3 db means the output is 0.707 of the input. The reader can prove to himself that a 3-db gain means that the output is 1.41 times the input (the reciprocal of 0.707). We can apply the same procedure to determine the gain corresponding to other decibel values, or can remember that every 3-db loss means a reduction of 70.7 per cent. Therefore, a loss of 6 db in a network could be interpreted as meaning that the network is equivalent to two other networks in series, each having a loss of 3 db. The first one will reduce the input by 0.707, which leaves $0.707E_{in}$ to drive the next one. The second network will have an output of $0.707(0.707E_{in}) = (0.707)^2E_{in} \approx 0.5E_{in}$. This tells us that a loss of 6 db approximately halves the input. Conversely, a gain of 6 db would double the input. Similarly, a loss of 12 db would halve a half of the input, which means that $E_o \approx \frac{1}{4}E_{in}$. A gain of 12 db would quadruple the input, and so forth. Doubling decibels means squaring the gain. That is, if we remember that 3 db = 20 log A, where $A = 0.707$, or $1/\sqrt{2}$, then 6 db = 2(3 db) = 2(20 log A) = 20 log A^2, where $A^2 = (0.707)^2$, or ½; and 12 db = 4(3 db) = 4(20 log A) = 20 log A^4, where $A^4 = 1/(\sqrt{2})^4 = \frac{1}{4}$, and so on.

2-6. A High-pass Filter. Analyze and construct a Bode plot for the network of Fig. 2-13. The circuit is apparently a type of high-pass filter, since the output must increase with increasing frequency because X_C decreases. The current will lead the voltage at some angle between zero and 90°. The leading current through R develops a leading output voltage. The network transfer function is

$$\frac{E_o}{E_{in}} = \frac{R}{R - jX_C} = \frac{R}{R - j/\omega C} = \frac{1}{1 - j/\omega RC}$$

Let $RC = \tau_l = 92\ \mu\text{sec}$:

$$\frac{E_o}{E_{in}} = \frac{1}{1 - j/\omega\tau_l}$$

If we define a frequency ω_l whose period is τ_l, then $\omega_l = 1/\tau_l$ and we may write

$$\frac{E_o}{E_{in}} = \frac{1}{1 - j\omega_l/\omega} = \frac{1}{1 - jf_l/f} = \frac{1\angle\tan^{-1}(f_l/f)}{\sqrt{1 + (f_l/f)^2}}$$

The characteristics of this transfer function are opposite to those of the low-pass filter. The lower the input frequency, the larger is the denominator of the transfer function, and therefore, the lower the output becomes. This is because the reactance of the capacitor increases as frequency decreases. The phase shift also increases as f decreases.

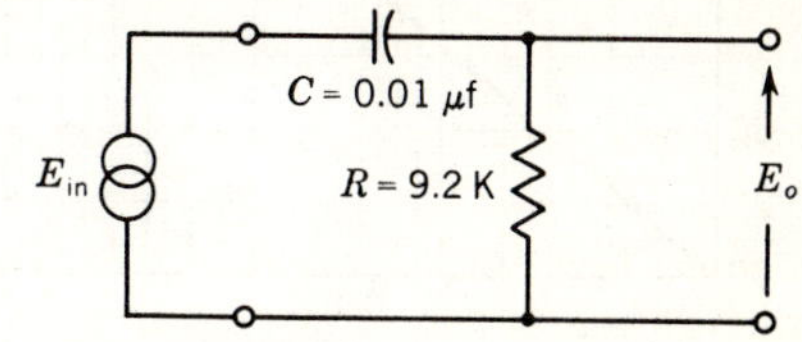

FIG. 2-13. A high-pass filter.

A general (normalized) single-time-constant, high-pass, *RC*-filter Bode diagram may be prepared by assuming ratios of ω/ω_l (Table 2-2). We are plotting ω/ω_l rather than ω_l/ω, in order to show clearly the effect of frequencies above and below the corner frequency. The resultant plot is shown in Fig. 2-14.

TABLE 2-2

$\dfrac{\omega}{\omega_l}$	$\dfrac{\omega_l}{\omega}$	$\dfrac{1}{1 - j\omega_l/\omega}$	db
100	0.01	$1.0\angle 0.6°$	0
25	0.04	$1.0\angle 2.3°$	0
10	0.10	$1.0\angle 5.7°$	0
2.5	0.40	$0.93\angle 22°$	−0.7
1	1.0	$0.707\angle 45°$	−3.0
0.25	4.0	$0.24\angle 76°$	−12.3
0.10	10.0	$0.1\angle 84°$	−20
0.025	40.0	$0.03\angle 88.5°$	−32
0.01	100.0	$0.01\angle 89.4°$	−40

Since $\omega_l = 2\pi f_l$, and $\omega = 2\pi f$, the transfer function may be expressed in terms of actual rather than radian frequency. At the frequency f_l, the output voltage will be 3 db below the input, and this fixes a break point on the Bode plot. The response will fall off at a rate of 20 db per decade below this break point. First, let us sketch the Bode plot on this

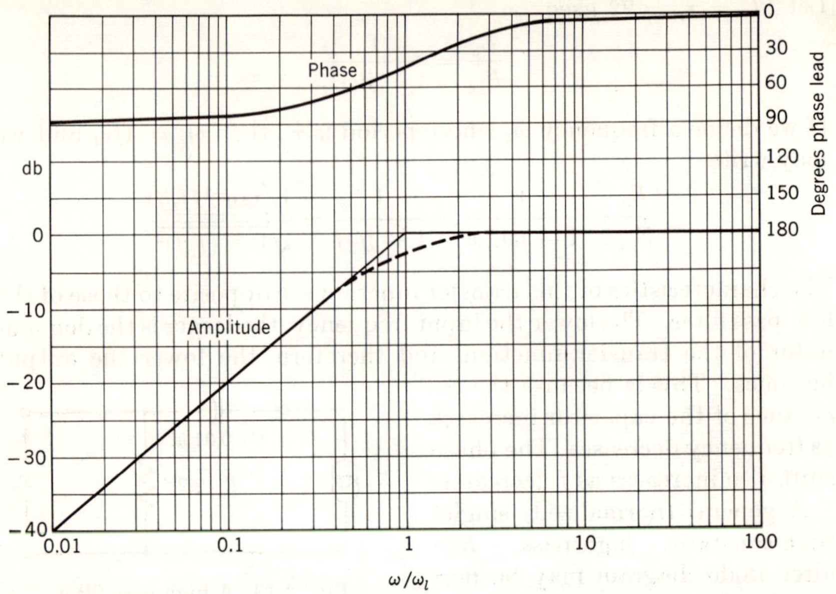

FIG. 2-14. The normalized Bode plot of the high-pass filter.

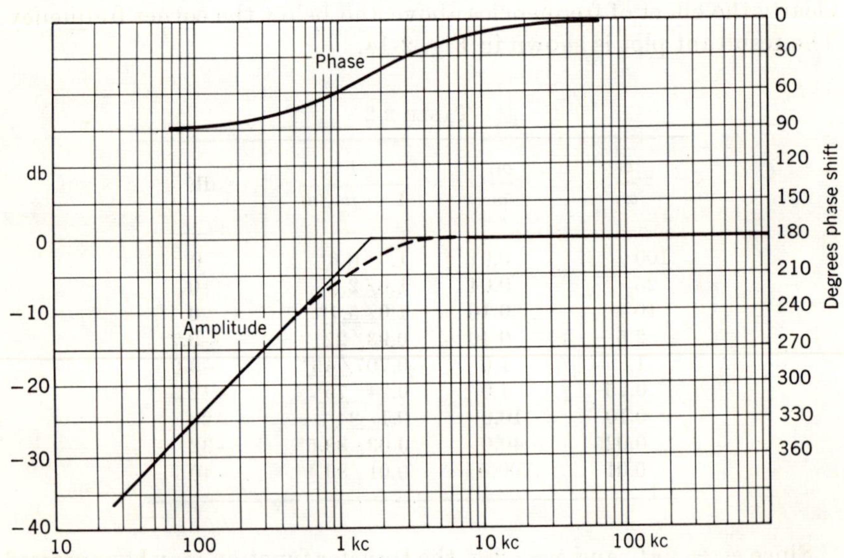

FIG. 2-15. The Bode plot for the circuit of Fig. 2-13.

basis, and then make some actual substitutions to verify the resultant curve:

1. Calculate f_l:

$$f_l = \frac{\omega_l}{2\pi} = \frac{1}{2\pi\tau_l} = 1.73 \text{ kc}$$

2. Locate f_l on the 0-db line.
3. At frequencies above f_l the output is essentially equal to the input.
4. At frequencies below f_l, the output falls at a rate of 20 db per decade. Therefore, at one-tenth of 1.73 kc, or 173 cps, mark a point at -20 db. Connect these points with a straight line.
5. If a little more accuracy is desired, the break point can be dropped 3 db and a smooth curve sketched in, as shown in Fig. 2-15. A point to remember is that at one octave above or below the break frequency, the actual response will be 1 db below the straight-line approximations. If we make actual substitutions, and locate the resultant points, we find close agreement with the sketched output-magnitude curve (Table 2-3).

TABLE 2-3

f, cps	f_l/f	$\sqrt{1+(f_l/f)^2}\angle\theta$	$1/\sqrt{1+(f_l/f)^2}\angle\theta$	$\log E_o/E_{in}$	$20 \log E_o/E_{in}$, db
173	10	$10\angle -84.6°$	$0.1\angle 84.6°$	-1	-20
433	4	$4.14\angle -76°$	$0.243\angle 76°$	-0.685	-13.7
1,730	1	$\sqrt{2}\angle -45°$	$0.707\angle 45°$	-0.15	-3
6,920	0.25	$1.03\angle -14°$	$0.98\angle 14°$	-0.012	-0.24

2-7. A Bandpass Filter. Figure 2-16*a* illustrates an *RC*-coupled type of bandpass filter. Suppose we wished to sketch the response curve in the form of a Bode plot. Here we must exercise caution because C_2 and R_2 act as a load upon C_1. Since C_2 is one-tenth of C_1, however, X_{C_2} will be 10 times as great as X_{C_1}. Therefore, we might assume a negligible loading effect of C_2 and R_2 upon C_1. Figure 2-16*a* can be redrawn as Fig. 2-16*b*. The dotted block containing R_1 and C_1 may be considered as an amplifier having a gain or transfer function of

$$\frac{E_o'}{E_{in}} = \frac{-jX_{C_1}}{R_1 - jX_{C_1}}$$

The output voltage of the first block E_o' is the input to the middle block which represents an ideal, unity gain amplifier having infinite input and zero output impedance. The isolation provided by this amplifier illustrates the fact that C_2 and R_2 are assumed as a negligible load upon C_1.

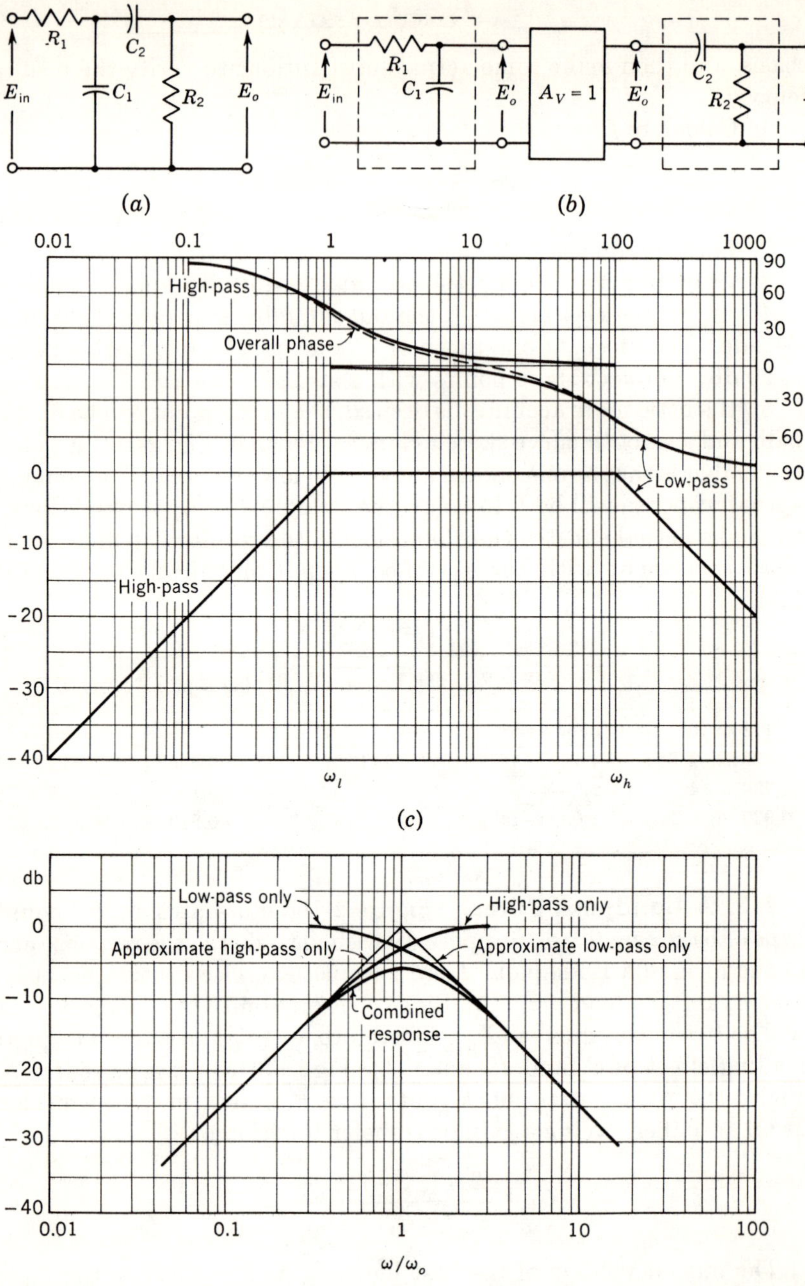

FIG. 2-16. (*a*) A bandpass filter. (*b*) The bandpass filter may be frequently analyzed in terms of high- and low-pass sections in series which are isolated by a unity gain amplifier. (*c*) A Bode plot of the output response of the bandpass filter. (*d*) The over-all frequency response of the bandpass filter is easily determined by sketching and combining the individual response curves.

The transfer function of the last block is

$$\frac{E_o}{E_o'} = \frac{R_2}{R_2 - jx_{C_2}}$$

The over-all transfer function (gain) is then the product of the individual gains:

$$\frac{E_o}{E_{in}} = \frac{E_o'}{E_{in}}\frac{E_o}{E_o'} = \frac{-jX_{C_1}}{R_1 - jX_{C_1}}\frac{R_2}{R_2 - jX_{C_2}}$$

$$= \frac{1/j\omega C_1}{R_1 + 1/j\omega C_1}\frac{R_2}{R_2 - j1/\omega C_2} = \frac{1}{1 + j\omega C_1 R_1}\frac{1}{1 - j1/\omega C_2 R_2}$$

Let $\tau_h = C_1R_1$, $\omega_h = 1/\tau_h$, $\tau_l = C_2R_2$, and $\omega_l = 1/\tau_l$;

then

$$\frac{E_o}{E_{in}} = \frac{1}{1 + j\omega/\omega_h}\frac{1}{1 - j\omega_l/\omega} = \frac{1}{(1 + j\omega/\omega_h)(1 - j\omega_l/\omega)}$$

In polar form,

$$\frac{E_o}{E_{in}} = \left[\frac{1}{\sqrt{1 + (\omega/\omega_h)^2}\,\underline{/\tan^{-1}(\omega/\omega_h)}}\right]\left[\frac{1}{\sqrt{1 + (\omega_l/\omega)^2}\,\underline{/-\tan^{-1}(\omega_l/\omega)}}\right]$$

$$= \left[\frac{1\underline{/-\tan^{-1}(\omega/\omega_h)}}{\sqrt{1 + (\omega/\omega_h)^2}}\right]\left[\frac{1\underline{/\tan^{-1}(\omega_l/\omega)}}{\sqrt{1 + (\omega_l/\omega)^2}}\right]$$

The above expression for E_o/E_{in} is of the form

$$\frac{E_o}{E_{in}} = (A\underline{/\theta})(B\underline{/\phi})$$

but the product of two vectors in polar form is

$$\frac{E_o}{E_{in}} = (AB)\underline{/\theta + \phi}$$

We see that the over-all phase angle is the algebraic sum of the individual phase angles. This tells us how to plot the phase response. The amplitude response in decibels is

$$\text{db} = 20 \log (AB) = 20(\log A + \log B) = 20 \log A + 20 \log B$$

Here we see that the over-all response in db is the algebraic sum of the individual decibel response curves. All we have to do is sketch the Bode plots for the individual blocks and then add them graphically.

Suppose we investigate a specific case in which $\omega_h = 100\omega_l$. On our graph paper we may make a mark at, say, $\omega_l = 1$, and then locate ω_h

two decades above, as shown in Fig. 2-16*c*. If we recognize that the first block in Fig. 2-16*b* is the previously analyzed low-pass filter, we can easily sketch its transfer function. The break frequency is ω_h. Below ω_h there is 0 db gain, and above it the amplitude falls off at 20 db per decade.

The second block may be recognized as a high-pass filter. The amplitude response of this block will be similar to that in Fig. 2-14. The over-all phase response may be determined by graphically adding the individual phase response curves with a pair of dividers. The resultant Bode plot is shown in Fig. 2-16*c*. Note how compactly the complete gain characteristics are displayed. We see that even though the amplitude response is fairly flat over two decades, the phase response varies from about 60° of lead to 60° of lag in this region.

Now let us consider a case in which $\omega_l = \omega_h$. In other words, $R_1C_1 = R_2C_2$. However, C_2 will still equal one-tenth of C_1 to afford negligible loading. Of course, this means that R_2 will have to equal $10 \times R_1$ if the break points, or corner frequencies, are to be identical. The general expression for the over-all transfer function was shown to be

$$\frac{E_o}{E_{in}} = \frac{1}{(1 + j\omega/\omega_h)(1 - j\omega_l/\omega)}$$

and since $\omega_h = \omega_l$ in this particular case, let us substitute ω_o for ω_l and ω_h:

$$\frac{E_o}{E_{in}} = \frac{1}{(1 + j\omega/\omega_o)(1 - j\omega_o/\omega)} = \frac{1 \angle \tan^{-1}(\omega_o/\omega - \omega/\omega_o)}{\sqrt{1 + (\omega/\omega_o)^2}\sqrt{1 + (\omega_o/\omega)^2}}$$

Let us calculate a few check points prior to sketching the response curve. At $\omega = \omega_o$, that is, $\omega/\omega_o = 1$,

$$\frac{E_o}{E_{in}} = \frac{1 \angle \tan^{-1}(1 - 1)}{\sqrt{1 + 1}\sqrt{1 + 1}} = \frac{1}{2} \angle 0° \text{ or } -6 \text{ db}$$

At $\omega/\omega_o = 10$,

$$\frac{E_o}{E_{in}} = \frac{1 \angle \tan^{-1}(0.1 - 10)}{\sqrt{1 + 100}\sqrt{1 + 0.01}} \approx \frac{\angle -\tan^{-1} 10}{\sqrt{100}} = 0.1 \angle -84.5° = -20 \text{ db}$$

At $\omega/\omega_o = 0.1$,

$$\frac{E_o}{E_{in}} = \frac{\angle \tan^{-1} 10}{\sqrt{1 + 0.01}\sqrt{1 + 100}} = 0.1 \angle 84.5° = -20 \text{ db}$$

At $\omega/\omega_o = 4$,

$$\frac{E_o}{E_{in}} = \frac{1 \angle \tan^{-1}(0.25 - 4)}{\sqrt{1 + 16}\sqrt{1 + 0.063}} \approx \frac{1 \angle -75°}{4.15}$$

or

$$\text{db} = 20(\log 1 - \log 4.15)$$
$$= -20(\log 4.15) = -20(0.618) = -12.5$$

In Fig. 2-16*d*, the response of the circuit is sketched. First, the high-pass response was approximated with two straight-line segments breaking at $\omega/\omega_o = 1$, and with the low-frequency segment sloping at 20 db per decade. Then the low-pass segments were sketched in a similar manner. This approximates the over-all response curve as a triangle with the apex at 0 db and the sides sloping at 20 db per decade. A better

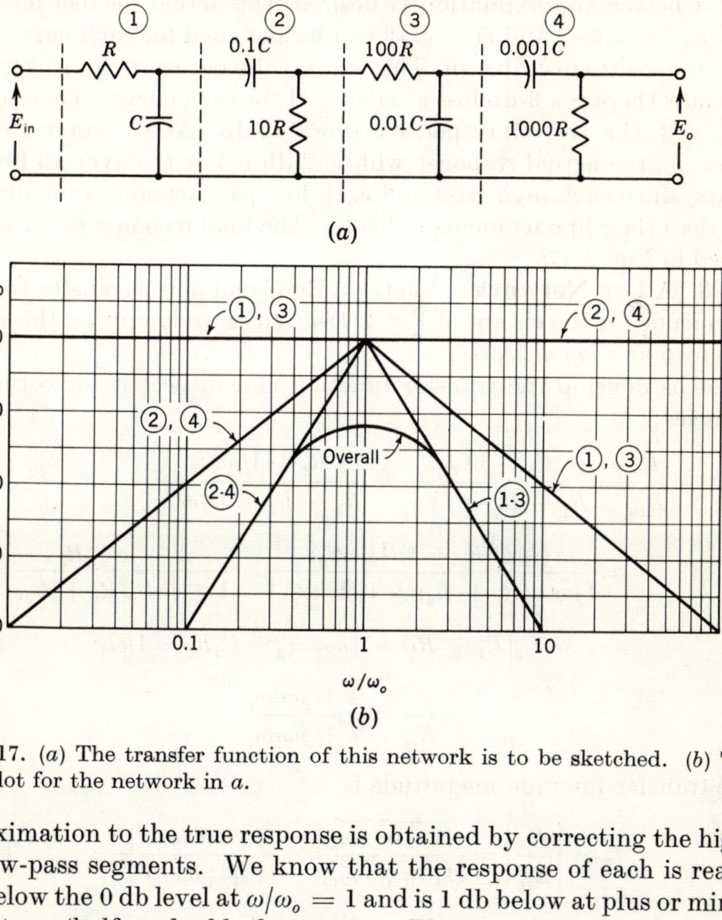

FIG. 2-17. (*a*) The transfer function of this network is to be sketched. (*b*) The Bode plot for the network in *a*.

approximation to the true response is obtained by correcting the high- and low-pass segments. We know that the response of each is really 3 db below the 0 db level at $\omega/\omega_o = 1$ and is 1 db below at plus or minus one octave (half or double frequency). Plotting and sketching these points, we obtain the high- and low-pass curves only. If these two curves are added with a pair of dividers, the combined response curve is obtained whose coordinates closely agree with the check points just calculated.

What do you think the response curve of the network in Fig. 2-17*a* would look like as a Bode plot? As each section barely loads the preceding section, we can sketch the approximate response of each individual section. Since a logarithmic scale is used, the over-all response is the sum of the individual response curves. Notice that the break points of each section are identical. Sections 1 and 3 have identical response curves, as do Secs. 2 and 4. We can add curves 1 and 3 with a pair of dividers to obtain curve 1-3, which slopes at 40 db per decade. Curve 1-2 is similarly obtained.

If a better approximation is desired, the actual decibel levels for $\omega = \omega_o$, $\omega = 2\omega_o$, and $\omega = \omega_o/2$ can be sketched for each curve and a new composite curve drawn. This is not really necessary, however, since we know there is a 3-db loss at $\omega/\omega_o = 1$ for each curve. Therefore, at $\omega/\omega_o = 1$, the over-all response is down 12 db. At an octave above or below ω_o, the actual response will be 2 db below the over-all line segments, since each high-pass and each low-pass section are really 1 db less than their line segments indicate. The final response curve is illustrated in Fig. 2-17*b*.

2-8. A Lag Network. Sketch a Bode plot of the transfer-function magnitude for the circuit of Fig. 2-18*a*. In servo language, this circuit is known as a *lag network*.

Let us develop the transfer function and inspect it for corner frequencies:

$$\frac{E_o}{E_{in}} = \frac{R_2 - jX_{C_2}}{R_1 + R_2 - jX_{C_2}} = \frac{R_2 + 1/j\omega C_2}{R_1 + R_2 + 1/j\omega C_2}$$

$$= \frac{(j\omega C_2 R_2 + 1)(1/j\omega C_2)}{[j\omega C_2(R_1 + R_2) + 1](1/j\omega C_2)} = \frac{1 + j\omega C_2 R_2}{1 + j\omega C_2(R_1 + R_2)} \tag{2-5}$$

Let $$\tau_1 = C_2(R_1 + R_2) = 1/\omega_1,\ \tau_2 = C_2 R_2 = 1/\omega_2: \tag{2-6}$$

$$\frac{E_o}{E_{in}} = \frac{1 + j\omega/\omega_2}{1 + j\omega/\omega_1} \tag{2-7}$$

The transfer-function magnitude is

$$\left|\frac{E_o}{E_{in}}\right| = \frac{\sqrt{1 + (\omega/\omega_2)^2}}{\sqrt{1 + (\omega/\omega_1)^2}} = \frac{[1 + (\omega/\omega_2)^2]^{1/2}}{[1 + (\omega/\omega_1)^2]^{1/2}} \tag{2-8}$$

$$\text{db} = 20\left\{\log\left[1 + \left(\frac{\omega}{\omega_2}\right)^2\right]^{1/2} - \log\left[1 + \left(\frac{\omega}{\omega_1}\right)^2\right]^{1/2}\right\}$$

$$= 20\left\{\tfrac{1}{2}\log\left[1 + \left(\frac{\omega}{\omega_2}\right)^2\right] - \tfrac{1}{2}\log\left[1 + \left(\frac{\omega}{\omega_1}\right)^2\right]\right\}$$

$$= 10\log\left[1 + \left(\frac{\omega}{\omega_2}\right)^2\right] - 10\log\left[1 + \left(\frac{\omega}{\omega_1}\right)^2\right] \tag{2-9}$$

Analyzing this final equation, we see that the total logarithmic response is the sum of two components. The first, $10 \log [1 + (\omega/\omega_2)^2]$, which is due to the numerator of Eq. (2-5) causes the response to increase as ω increases. A Bode plot of this component only, let us call it A, may be sketched in the following manner:

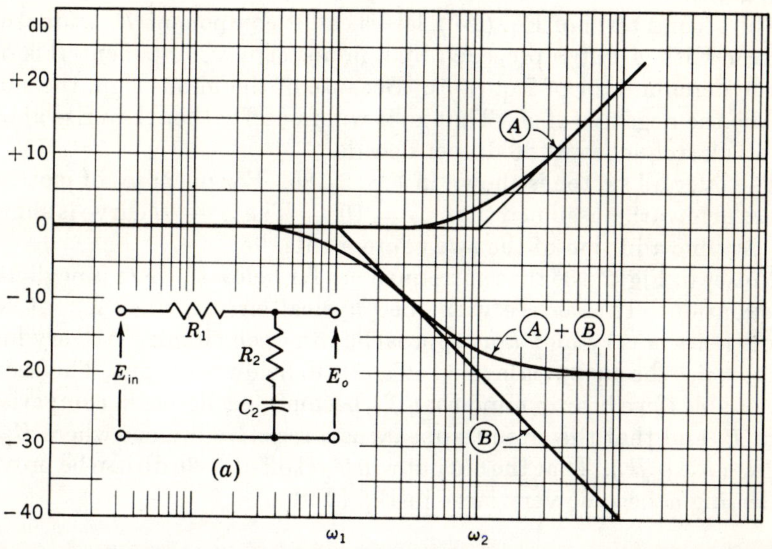

(b)

FIG. 2-18. (a) A lag network. (b) The A plus B curve is a Bode plot of the transfer function for the lag network of a.

When $\omega \ll \omega_2$, say one-tenth as much, we have

If $\omega/\omega_2 = 0.1$

$$A = 10 \log (1 + 1/100) \approx 10 \log 1 \approx 0$$

If $\omega/\omega_2 = 1$

$$A = 10 \log (1 + 1) = 3 \text{ db}$$

If $\omega/\omega_2 = 2$

$$A = 10 \log (1 + 4) = 10(0.7) = 7 \text{ db}$$

If $\omega/\omega_2 = 10$

$$A = 10 \log (1 + 100) \approx 10 \log 100 = 20 \text{ db}$$

For sketching purposes, we may conclude that a term of the form $1 + j\omega/\omega_2$ in the *numerator* of the transfer function has a break at ω_2, is flat at zero db below ω_2, and rises at 20 db per decade above ω_2. A slope of 20 db per decade is identical with 6 db per octave. In the vicinity of the break, say one octave above it, the slope varies and is not quite 6 db per octave. Actually, we find we are at 7 db rather than at 6. This is where we get the 1-db correction factor for the points which are $\pm$ one octave about the corner frequency. At the corner frequency the actual response is 3 db above the zero level. This is illustrated in Fig. 2-18*b*.

The second term of Eq. (2-9), let us call it component B, is identical in form with A but is preceded by a minus sign. Component B is due to the denominator of Eq. (2-5). Because of the minus sign, the slope will be the negative of A. That is, it will fall off rather than rise above its break frequency at 20 db per decade.

The over-all sketch is shown in Fig. 2-18*b*. For purposes of plotting, it was arbitrarily assumed that $\omega_2 = 10\omega_1$. The $A + B$ curve is simply the graphic addition of the two components.

Notice in Fig. 2-18*b* that at frequencies far below ω_1 there is negligible attenuation. This agrees with the physical circuit, since X_{C_2} at low frequencies is very high and approaches an open circuit. At very high frequencies the output is leveled off at 20 db below the input. This seems reasonable if we bear in mind that X_C becomes negligible in comparison with R_2 and that the circuit appears as a resistive divider where $E_o = E_{in}R_2/(R_1 + R_2)$. That the output will level off at -20 db can be proven by letting ω become very large in Eq. (2-9):

$$\text{db} = 10 \log \left[1 + \left(\frac{\omega}{\omega_2}\right)^2\right] - 10 \log \left[1 + \left(\frac{\omega}{\omega_1}\right)^2\right] \tag{2-9}$$

If $\omega \to \infty$

$$\text{db} \approx 10 \log \left(\frac{\omega}{\omega_2}\right)^2 - 10 \log \left(\frac{\omega}{\omega_1}\right)^2 = 20 \log \frac{\omega}{\omega_2} - 20 \log \frac{\omega}{\omega_1}$$

$$= 20\left(\log \frac{\omega}{\omega_2} - \log \frac{\omega}{\omega_1}\right) = 20 \log \frac{\omega/\omega_2}{\omega/\omega_1}$$

$$= 20 \log \frac{\omega_1}{\omega_2} = 20 \log \frac{1/C_2(R_1 + R_2)}{1/C_2R_2}$$

$$= 20 \log \frac{R_2}{R_1 + R_2}$$

2-9. A Lead Network. Construct a Bode plot for the lead network of Fig. 2-19*a*.

1. First develop the transfer function:

$$\frac{E_o}{E_{in}} = \frac{R_2}{R_2 + \dfrac{R_1(-jX_{C_1})}{R_1 - jX_{C_1}}} = \frac{R_2}{R_2 + \dfrac{R_1}{j\omega C_1}\,\dfrac{1}{R_1 + 1/j\omega C_1}}$$

$$= \frac{R_2}{R_2 + \dfrac{R_1}{j\omega C_1}\,\dfrac{1}{(j\omega R_1 C_1 + 1)/j\omega C_1}} = \frac{R_2}{R_2 + \dfrac{R_1}{1 + j\omega R_1 C_1}}$$

$$= \frac{R_2}{\dfrac{R_2 + j\omega R_1 R_2 C_1 + R_1}{1 + j\omega R_1 C_1}} = \frac{R_2(1 + j\omega R_1 C_1)}{R_1 + R_2 + j\omega R_1 R_2 C_1}$$

$$= \frac{1 + j\omega R_1 C_1}{R_1/R_2 + 1 + j\omega R_1 C_1} = \frac{1 + j\omega R_1 C_1}{(R_1 + R_2)/R_2 + j\omega R_1 C_1} \tag{2-10}$$

The next step is somewhat subtle. Since we wish to get the denominator into the form $1 + jX$, we must remove the $(R_1 + R_2)/R_2$ term in

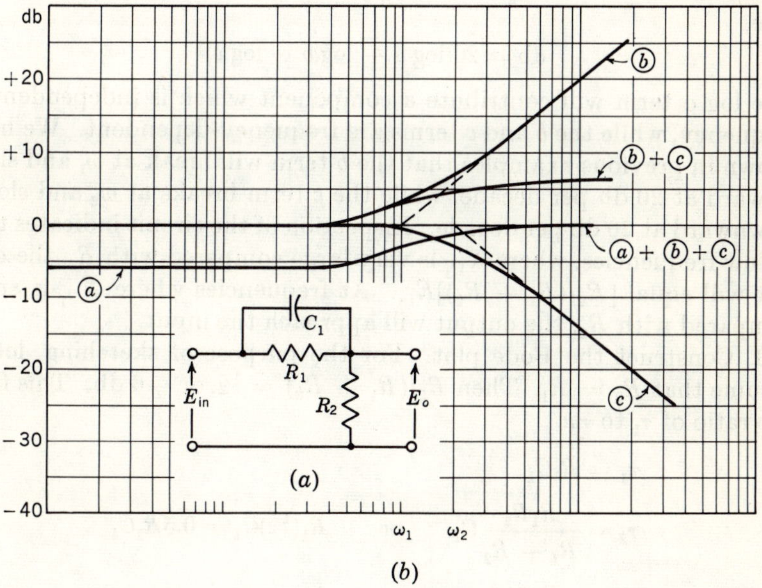

Fig. 2-19. (*a*) A lead network. (*b*) A Bode plot for the lead network of *a*.

some way without spoiling the numerator, which is in the desired form. Therefore, let us divide each term of the denominator by $(R_1 + R_2)/R_2$, and then multiply the whole denominator by this quantity, which will not change the value of the fraction:

$$\frac{E_o}{E_{in}} = \frac{1 + j\omega R_1 C_1}{\dfrac{R_1 + R_2}{R_2}\left(1 + j\omega C_1 R_1 \dfrac{R_2}{R_1 + R_2}\right)} = \frac{R_2}{R_1 + R_2} \frac{1 + j\omega R_1 C_1}{1 + j\omega R_{eq} C_1}$$

where
$$R_{eq} = \frac{R_1 R_2}{R_1 + R_2}$$

Now let $R_1C_1 = \tau_1 = 1/\omega_1$ and $R_{eq}C_1 = \tau_2 = 1/\omega_2$:

$$\frac{E_o}{E_{in}} = \frac{R_2}{R_1 + R_2} \frac{1 + j\omega/\omega_1}{1 + j\omega/\omega_2} \tag{2-11}$$

2. Analyze the transfer function to ascertain what contributions the various factors will make in the Bode plot. Eq.(2-11), the transfer function, is of the form

$$\frac{E_o}{E_{in}} = a \frac{b}{c}$$

where $a = R_2/(R_1 + R_2)$, $b = 1 + j\omega/\omega_1$, and $c = 1 + j\omega/\omega_2$. Therefore,

$$\text{db} = 20(\log a + \log b - \log c)$$

The log a term will contribute a component which is independent of frequency, while the b and c terms are frequency-dependent. We have shown in previous examples that the b term will break at ω_1 and slope upward at 20 db per decade, while the c term breaks at ω_2 and slopes downward at 20 db per decade. Inspection of the circuit indicates that at low frequencies, where X_{C_1} is very large compared with R_1, the output will equal $[R_2/(R_1 + R_2)]E_{in}$. At frequencies where X_{C_1} is small compared with R_2, the output will approach the input.

3. Construct the Bode plot. For the purpose of sketching, let us assume that $R_2 = R_1$. Then $R_2/(R_1 + R_2) = ½$, or -6 db. This fixes the ratio of τ_1 to τ_2:

$$\tau_1 = R_1 C_1$$

$$\tau_2 = \frac{R_1 R_2}{R_1 + R_2} C_1 \quad \text{or} \quad R_1(½)C_1 = 0.5 R_1 C_1$$

$$\frac{\tau_1}{\tau_2} = \frac{R_1 C_1}{0.5 R_1 C_1} = 2 = \frac{\omega_2}{\omega_1}$$

The Bode plot is sketched in Fig. 2-19*b*.

2-10. An Amplifier Equivalent Network. An equivalent circuit of an amplifier stage is shown in Fig. 2-20*a*. Let us analyze the circuit and try to sketch the frequency-response characteristic. The symbol A_v represents the mid-frequency gain of the stage. R_{eq} is the parallel combination of the tube plate resistance (r_p) and the plate load resistor (R_L). C_s represents the total wiring, output, and input capacitance hanging across the output. C_c is the coupling capacitor and R_g is the grid resistor of the following stage. Assume that $A_v = 40$, $r_p = 30$ kilohms, $R_L = 60$ kilohms, $C_s = 100\ \mu\mu$f, $C_c = 0.01\ \mu$f, and $R_g = 1$ meg.

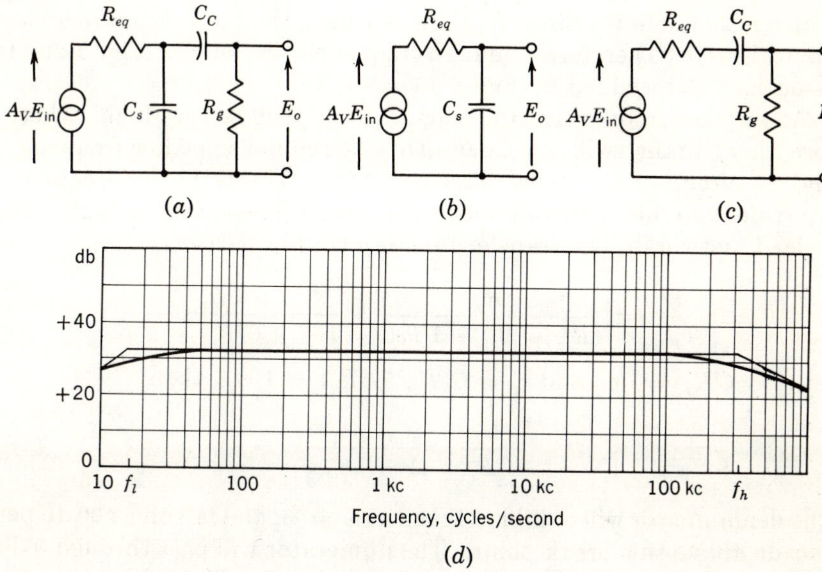

FIG. 2-20. (a) The equivalent circuit of a resistance-coupled amplifier. (b) The equivalent circuit at high frequencies. (c) The equivalent circuit at low frequencies. (d) The resistance-coupled amplifier is similar to a bandpass filter.

The most straightforward method of sketching the response curve is to develop the transfer function. The general form of the transfer function will contain factors in the numerator and denominator. The individual contributions of each factor are plotted and then graphically added to obtain the over-all response. It is not always necessary, however, to go through the labor of exactly evaluating the transfer function. We can make useful approximations in many cases. For example, in Fig. 2-20a, we might reason as follows. At some high frequency the reactance of C_s will be much less than the series combination of R_g and C_c. This reduces the circuit of Fig. 2-20a to that of 2-20b, which is known as the high-frequency equivalent circuit of a resistance-coupled amplifier.

The amplitude response for Fig. 2-20b will be identical with that of the lag network discussed in Sec. 2-4. The transfer function is $E_o/A_vE_{in} = 1/(1 + j\omega R_{eq}C_s)$. There will be a break at a high frequency $\omega_h = 1/R_{eq}C_s = 1/(20 \text{ kilohms})(100\ \mu\mu\text{f}) = 0.5(10^6)$ radians/sec, or 79.8 kc. The output will be 3 db down and will lag the input by 45°. At $\omega > \omega_h$ the output will fall at a rate of 20 db per decade.

Now we have said that at some frequency the circuit of Fig. 2-20*b* may be considered a valid approximation. We did not say it was the break frequency ω_h. To see if this circuit is valid at ω_h, let us investigate the loading effect of C_c and R_g upon C_s at ω_h. At ω_h, X_{C_s} will equal R_{eq}, which is 20 kilohms. Since R_g alone is 1 meg, the $C_c - R_g$ branch may be neglected. Therefore, the high-frequency response of Fig. 2-20*a* is essentially determined by Fig. 2-20*b*.

At very low frequencies, the reactance of C_s will be very high. Therefore, the C_s branch will appear as an open circuit at some low frequency, and the circuit of Fig. 2-20*a* reduces to 2-20*c*. This is the low-frequency equivalent circuit of an amplifier. The circuit is essentially a high-pass, or lead, network. The transfer function for Fig. 2-20*c* is

$$\frac{E_o}{A_vE_{in}} = \frac{R_g}{(R_{eq} + R_g) + 1/j\omega C} = \frac{j\omega CR_g}{1 + j\omega C(R_{eq} + R_g)}$$

let $\tau_1 = CR_g = 1/\omega_1$ and $\tau_2 = C(R_{eq} + R_g) = 1/\omega_2$; then

$$\frac{E_o}{A_vE_{in}} = \frac{j\omega/\omega_1}{1 + j\omega/\omega_2}$$

The denominator will exhibit a break at $\omega = \omega_2$ and fall off at 20 db per decade above this break point. The numerator will pass through 0 db at $\omega = \omega_1$ with a rising slope of 20 db per decade. That this is so may be determined by expressing the numerator magnitude in decibels. If db $= 20 \log \omega/\omega_1$ at $\omega = \omega_1$,

$$\text{db} = 20 \log 1 = 0$$

At $\omega = 0.1\omega_1$

$$\text{db} = 20 \log \tfrac{1}{10} = 20(\log 1 - \log 10) = -20 \text{ db}$$

At $\omega = 10\omega_1$,

$$\text{db} = 20 \log 10 = 20 \text{ db}$$

Since in this particular example $R_g \gg R_{eq}$, ω_1 will almost equal ω_2. If we assume that $\omega_1 = \omega_2 = \omega_l$, we may write the transfer function as

$$\frac{E_o}{A_vE_{in}} = \frac{j\omega/\omega_l}{1 + j\omega/\omega_l} = \frac{1}{1 + \dfrac{1}{j\omega/\omega_l}} = \frac{1}{1 - j\omega_l/\omega}$$

The denominator will exhibit a break point at $\omega = \omega_l = 1/1$ meg (0.01 μf) = 100 radians/sec, or 15.9 cps. The minus sign before the j term does not change the magnitude of the denominator. It signifies only that the output will lead the input. At $\omega = \omega_l$ the lead angle is 45°. Since the denominator contains ω_l/ω, increasing ω will cause the denominator to decrease. Therefore, the output will slope upward with increasing frequency, and by now we know this will be at 20 db per decade.

We have not yet shown that the low-frequency equivalent circuit is valid. To do this we must see if X_{C_s} is much larger than $R_g - jX_{C_c}$ at ω_l. At ω_l, $X_{C_s} = 1/\omega_l C_s = 1/10^2(10^{-10}) = 100$ meg, which justifies Fig. 2-20*c* at ω_l. The amplitude response is sketched in Fig. 2-20*d*. To plot the actual output-to-input ratio E_o/E_{in}, the transfer functions were multiplied by A_v. That is, at high frequencies

$$\frac{E_o}{E_{in}} = A_v \frac{1}{1 + j\omega/\omega_h}$$

where $\omega_h = 318$ kc; at low frequencies

$$\frac{E_o}{E_{in}} = A_v \frac{1}{1 - j\omega_l/\omega}$$

where $\omega_l = 15.9$ cps. The A_v term contributes

$$\text{db} = 20 \log A_v = 20 \log 40 = 32 \text{ db}$$

Notice how flat the amplitude response is between the break points. This region is referred to as the mid-frequency range. The mid-frequency equivalent circuit would not show any capacitors, since the effect, C_s, enters only above f_h, and C_c enters only below f_l.

Additional information pertaining to Bode plots may be found in servomechanism texts such as "Servomechanisms and Regulating System Design" by Chestnut and Mayer[1] or in "Servomechanism Analysis" by Thaler and Brown.[2]

2-11. The Concept of *Q*. When dealing with oscillatory waveforms, the term Q is frequently encountered. The Q of a component is a measure of its energy-storing ability. Since energy is stored in the electric field of a capacitor or in the magnetic field of an inductor, we associate Q with reactive elements. Having resistance in a circuit causes energy to be dissipated as heat.

[1] Harold Chestnut and Robert W. Mayer, "Servomechanisms and Regulating System Design," John Wiley & Sons, Inc., New York, vol. 1, 1951; vol. 2, 1955.

[2] George J. Thaler and Robert G. Brown, "Servomechanism Analysis," McGraw-Hill Book Company, Inc., New York, 1953.

Specifically, the Q of a circuit or component is defined as

$$Q = \frac{\text{reactive power}}{\text{real power}}$$

The reactive power is determined for the inductive and capacitive elements, while the real power is computed for resistive components. Capacitors as a rule have negligible resistive losses and consequently have very high Q values. Inductors may have appreciable resistive components because of winding resistance, skin effect, and core losses which result in lower Q values than for capacitors.

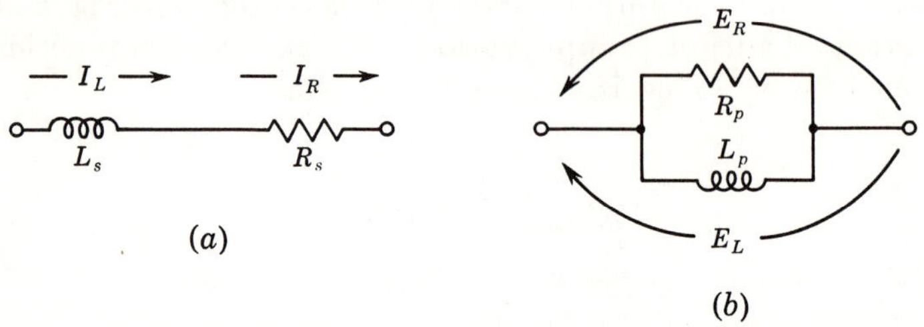

FIG. 2-21. (*a*) A resistance and inductance in series have the same current through them. $I_L = I_R$. (*b*) A resistance and inductance in parallel have the same voltage across them. $E_R = E_L$.

For a coil and resistor in series, such as in Fig. 2-21*a*,

$$Q_S = \frac{\text{reactive power}}{\text{real power}} = \frac{I_L{}^2 X_{Ls}}{I_R{}^2 R_S}$$

And, since $I_L = I_R$,

$$Q_S = \frac{X_{Ls}}{R_S} = \frac{\omega L_S}{R_S}$$

For a coil and resistor in parallel, as in Fig. 2-21*b*,

$$Q_P = \frac{\text{reactive power}}{\text{real power}} = \frac{E_L{}^2/X_{LP}}{E_R{}^2/R_P}$$

and, since $E_L = E_R$,

$$Q_P = \frac{R_P}{X_{LP}} = \frac{R_P}{\omega L_P}$$

Similarly, it may be shown that for a capacitor and resistor in series,

$$Q_S = \frac{X_{Cs}}{R_S} = \frac{1}{\omega C_S R_S}$$

and for a capacitor and resistor in parallel,

$$Q_P = \frac{R_P}{X_{CP}} = \omega C_P R_P$$

2-12. Parallel-to-series and Series-to-parallel Conversions. It is frequently expedient to convert the series circuit of Fig. 2-22 to its parallel equivalent, and vice versa. The impedance seen looking into the a-b terminals of either circuit at a specific frequency is to be the same. We can then write

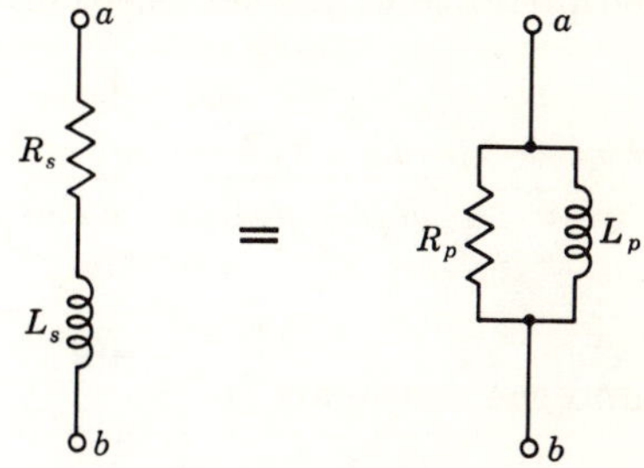

FIG. 2-22. At a given frequency the series RL circuit may be replaced by an equivalent parallel combination and vice versa.

$$Z_{abs} = Z_{abp}$$

$$R_S + j\omega L_S = \frac{R_P(j\omega L_P)}{R_P + j\omega L_P}$$

$$= \frac{R_P(j\omega L_P)}{R_P + j\omega L_P}\,\frac{R_P - j\omega L_P}{R_P - j\omega L_P}$$

$$= \frac{\omega^2 R_P L_P^2 + j\omega R_P^2 L_P}{R_P^2 + \omega^2 L_P^2}$$

$$= \frac{\omega^2 R_P L_P^2}{R_P^2 + \omega^2 L_P^2} + j\,\frac{\omega R_P^2 L_P}{R_P^2 + \omega^2 L_P^2}$$

If the parallel circuit is to be equivalent to the series circuit, the resistive part of the parallel impedance must equal the resistive part of the series circuit. Similarly, the reactive parts must equal each other. Therefore,

$$R_S = \frac{\omega^2 R_P L_P^2}{R_P^2 + \omega^2 L_P^2} = \frac{R_P}{R_P^2/(\omega L_P)^2 + 1}$$

If we define $R_P/\omega L_P = Q_P$,

$$R_S = \frac{R_P}{Q_P^2 + 1}$$

If Q_P is large (coil lightly shunted) we see that R_S will be quite small. We surely would expect R_S to be small if the coil has a high Q. Similarly,

$$j\omega L_S = j\,\frac{\omega R_P^2 L_P}{R_P^2 + \omega^2 L_P^2}$$

and
$$L_S = \frac{R_P^2 L_P}{R_P^2 + \omega^2 L_P^2} = \frac{L_P}{1 + (\omega L_P)^2/R_P^2} = \frac{L_P}{1 + 1/Q_P^2}$$

If Q_P is large, say, 10 or more, we see that $L_S \approx L_P$.

We now have developed the equations to convert the parallel circuit to its series equivalent. To convert the series circuit to its parallel equivalent, we must instead solve for R_P and L_P in terms of the series circuit parameters. The easiest way to do this is to equate the admittance seen looking into the a-b terminals of the parallel circuit to the admittance of the series circuit:

$$Y_{ab_P} = Y_{ab_S}$$

$$\frac{1}{R_P} + \frac{1}{j\omega L_P} = \frac{1}{R_S + j\omega L_S} = \frac{1}{R_S + j\omega L_S}\frac{R_S - j\omega L_S}{R_S - j\omega L_S}$$

$$= \frac{R_S - j\omega L_S}{R_S^2 + (\omega L_S)^2}$$

and the conjugate is

$$\frac{1}{R_P} - j\frac{1}{\omega L_P} = \frac{R_S}{R_S^2 + (\omega L_S)^2} - j\frac{\omega L_S}{R_S^2 + (\omega L_S)^2}$$

Equating real components,

$$\frac{1}{R_P} = \frac{R_S}{R_S^2 + (\omega L_S)^2}$$

or

$$R_P = \frac{R_S^2 + (\omega L_S)^2}{R_S} = \frac{1 + (\omega L_S/R_S)^2}{1/R_S}$$

but $Q_S = \omega L_S/R_S$, therefore $R_P = R_S(1 + Q_S^2)$. Here we see that R_S in the series circuit looks like a high value of R_P, if Q_S is high.

Equating reactive components,

$$\frac{1}{\omega L_P} = \frac{\omega L_S}{R_S^2 + (\omega L_S)^2}$$

or

$$\omega L_P = \frac{R_S^2 + (\omega L_S)^2}{\omega L_S}$$

$$L_P = \frac{R_S^2 + (\omega L_S)^2}{\omega^2 L_S} = \frac{R_S^2/(\omega L_S)^2 + 1}{\omega^2 L_S/\omega^2 L_S^2}$$

$$= \frac{(R_S/\omega L_S)^2 + 1}{1/L_S} = L_S\left[\left(\frac{R_S}{\omega L_S}\right)^2 + 1\right]$$

since $\omega L_S/R_S = Q_S$,

$$\frac{R_S}{\omega L_S} = \frac{1}{Q_S}$$

Therefore,

$$L_P = L_S\left(\frac{1}{Q_S^2} + 1\right)$$

Again, if Q_S is large, $L_P \approx L_S$. This is why the resonant frequency of a series circuit is essentially the same as that of the same elements in parallel.

In a similar manner, we can convert a capacitor and resistor in series to a parallel combination, and vice versa. For Fig. 2-23 we can write

$$R_S - j\frac{1}{\omega C_S} = \frac{R_P(-j1/\omega C_P)}{R_P - j(1/\omega C_P)}$$

$$= \frac{-j(R_P/\omega C_P)}{R_P - j(1/\omega C_P)}\frac{R_P + j(1/\omega C_P)}{(R_P + j1/\omega C_P)}$$

$$= \frac{R_P/(\omega C_P)^2}{R_P^2 + 1/(\omega C_P)^2} - \frac{j(R_P^2/\omega C_P)}{R_P^2 + 1/(\omega C_P)^2}$$

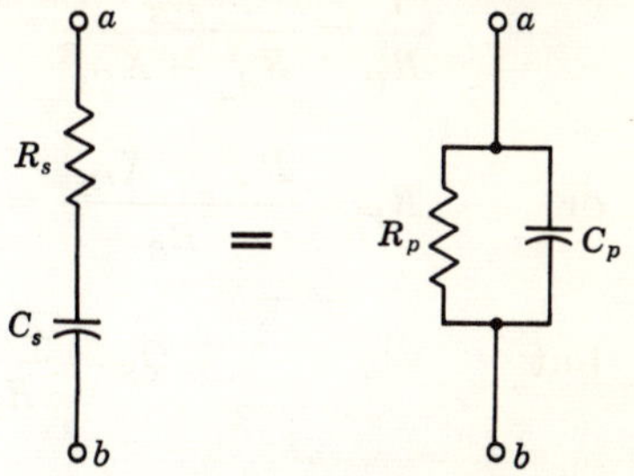

FIG. 2-23. At a given frequency we can equate the series RC combination to a parallel equivalent and vice versa.

Equating resistive components, we have

$$R_S = \frac{R_P/(\omega C_P)^2}{R_P^2 + [1/(\omega C_P)^2]}$$

and if we define

$$Q_P = \frac{R_P}{X_{C_P}} = \omega C_P R_P$$

multiplying numerator and denominator by $(\omega C_P)^2$,

$$R_S = \frac{R_P}{R_P^2(\omega C_P)^2 + 1} = \frac{R_P}{Q_P^2 + 1} = R_P\frac{1}{Q_P^2 + 1}$$

Equating reactive components, we have

$$\frac{1}{\omega C_S} = \frac{R_P^2/\omega C_P}{R_P^2 + [1/(\omega C_P)^2]}$$

or $$\omega C_S = \frac{R_P^2 + [1/(\omega C_P)^2]}{R_P^2/\omega C_P} = \frac{1 + [1/(\omega C_P R_P)^2]}{1/\omega C_P}$$

$$C_S = C_P\left[1 + \frac{1}{(\omega C_P R_P)^2}\right]$$

but $$Q_P = \frac{R_P}{X_{C_P}} = \omega C_P R_P$$

therefore, $$C_S = C_P\left(1 + \frac{1}{Q_P^2}\right)$$

To express R_P and C_P in terms of R_S and C_S, we equate admittances again:

$$Y_P = Y_S$$

$$\frac{1}{R_P} + \frac{1}{-jX_{C_P}} = \frac{1}{R_S - jX_S}$$

$$\frac{1}{R_P} - \frac{1}{jX_{C_P}} = \frac{1}{R_S - jX_{C_S}}\frac{R_S + jX_{C_S}}{R_S + jX_{C_S}} = \frac{R_S + jX_{C_S}}{R_S^2 + X_{C_S}^2}$$

$$\frac{1}{R_P} + \frac{j}{X_{C_P}} = \frac{R_S}{R_S^2 + X_{C_S}^2} + j\frac{X_{C_S}}{R_S^2 + X_{C_S}^2}$$

Equating resistive components,

$$\frac{1}{R_P} = \frac{R_S}{R_S^2 + X_{C_S}^2}$$

or $$R_P = \frac{R_S^2 + X_{C_S}^2}{R_S} = \frac{(X_{C_S}/R_S)^2 + 1}{1/R_S} = R_S\left[\left(\frac{X_{C_S}}{R_S}\right)^2 + 1\right]$$

but $$Q_S = \frac{X_{C_S}}{R_S} \qquad \text{or} \qquad \frac{1}{Q_S} = \frac{R_S}{X_{C_S}}$$

Therefore, $$R_P = R_S(Q_S^2 + 1)$$

To determine C_P, we equate reactive components:

$$\frac{1}{X_{C_P}} = \frac{X_{C_S}}{R_S^2 + X_{C_S}^2} = \frac{1/X_{C_S}}{(R_S/X_{C_S})^2 + 1}$$

$$\omega C_P = \frac{\omega C_S}{(1/Q_S)^2 + 1}$$

$$C_P = \frac{C_S}{(1/Q_S^2) + 1} = C_S \frac{1}{(1/Q_S^2) + 1}$$

It must be borne in mind that all these series-parallel equivalents are correct only at a particular frequency because of the fact that Q is a function of frequency. If Q is fairly constant over a range of frequencies, then the equivalent parameter values will hold over the same range. A few problems will illustrate the manner in which series-parallel conversions are performed.

Problem 1. The output frequency of a radio command tone generator is determined by an LC-series circuit. Nominally, $Q_L = 100$, $L = 0.5$ henry, and $C = 5{,}000\ \mu\mu\text{f}$ with negligible leakage or shunt resistance. Owing to environmental conditions, moisture penetrates the capacitor and produces an effective leakage resistance of 10 meg. Does a serious percentage change occur in the output frequency? Assume that C remains constant. What about the resultant circuit Q?

The original frequency was

$$f_r = \frac{1}{2\pi\sqrt{LC}} = \frac{1}{6.28\sqrt{0.5 \times 0.5 \times 10^{-8}}}$$

$$= \frac{0.159}{0.5} \times 10^4 = 3{,}180 \text{ cps}$$

Solution. With 10-meg leakage across C_P, the equivalent circuit changes from that of Fig. 2-24*a* to that of Fig. 2-24*b*. R_L is the coil resistance, and R_{CP} is the capacitor leakage resistance.

We shall now take C_P and convert it to C_S, as in Fig. 2-24*c*. This is a simple series-resonant circuit. We need not solve for R_{CS}, since it in no way affects the resonant frequency. It does, however, affect the Q of the circuit:

$$C_S = C_P\left(1 + \frac{1}{Q_P{}^2}\right)$$

where $Q_P = \omega C_P R_P$ or $\omega C_P R_{C_P}$ in Fig. 2-24*b*. Assuming that Q_P remains essentially constant, we have

$$Q_P = 2\pi f C_P R_{C_P} = 6.28(3.18 \times 10^3)(5 \times 10^{-9})(10^7) = 999 \approx 1{,}000$$

$$C_S = C_P\left(1 + \frac{1}{10^6}\right) \approx C_P$$

The frequency drift due to this factor is seen to be negligibly small.

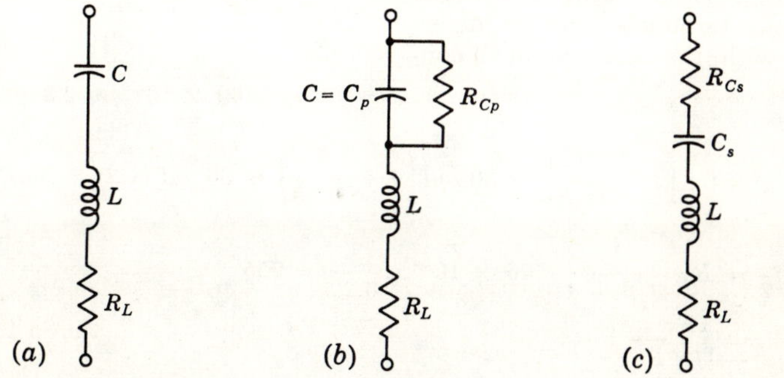

FIG. 2-24. The effect of leakage across capacitor C in *a* may be analyzed by the equivalent circuit of *b* and *c*.

To find the change in Q of the resonant circuit, we must first find R_{Cs}:

$$R_{Cs} = R_{CP}\frac{1}{Q_P{}^2 + 1} \approx 10^7\left(\frac{1}{10^6}\right) = 10 \text{ ohms}$$

The Q of the circuit in Fig. 2-24*a* is given by the coil Q, which is 100. Therefore,

$$R_L = \frac{\omega L}{Q_L} = \frac{6.28 \times 3.18 \times 10^3 \times 0.5}{100} = \frac{9.99 \times 10^3}{10^2} \approx 100 \text{ ohms}$$

The Q of the circuit in Fig. 2-24*c* is

$$Q_L = \frac{\omega L}{R_L + R_{Cs}} = \frac{9.99 \times 10^3}{110} = \frac{9.99 \times 10^3}{1.1 \times 10^2} = 90.6 \approx 90 \qquad \textit{Ans.}$$

The percentage change in circuit Q is about

$$\frac{100 - 90}{100} \times 100 = 10 \text{ per cent}$$

which is generally not serious.

Problem 2. At 2 Mc, the apparent input impedance of a certain triode amplifier consists of a 30-$\mu\mu$f capacitor in shunt with a 6-kilohm resistor, as shown in Fig. 2-25. Determine the inductance of a coil that must be placed in series with the ungrounded input terminal to make the circuit look purely resistive at this frequency.

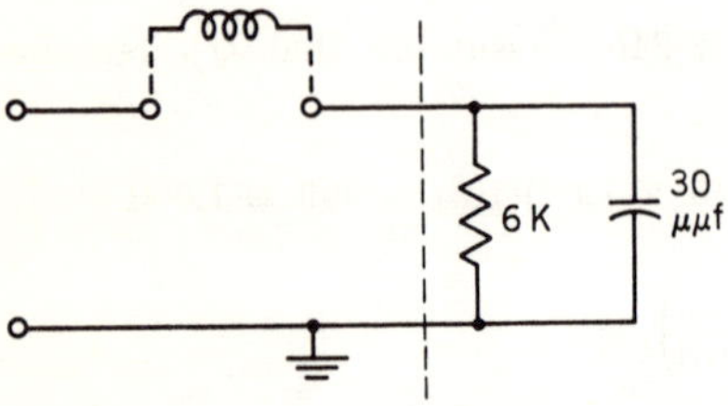

FIG. 2-25. The inductance required to make the circuit series-resonant at the input terminals may be determined with the series-parallel conversions.

Solution. We can convert the parallel RC combination into a series equivalent. The inductance required to make the circuit appear resistive must resonate with the series equivalent capacitor at 2 Mc:

$$\omega = 2\pi f = 12.55(10^6) \text{ radians/sec}$$

$$Q_P = \omega CR = 12.55 \times 10^6 \times 30 \times 10^{-12} \times 6 \times 10^3$$

$$= 2{,}260 \times 10^{-3} \approx 2.3$$

$$C_S = C_P\left(1 + \frac{1}{Q_P^2}\right) = 30\ \mu\mu\text{f}\left(1 + \frac{1}{5.29}\right) \approx 30\ \mu\mu\text{f}\ (1.2) = 35.7\ \mu\mu\text{f}$$

$$R_S = R_P \frac{1}{Q_P^2 + 1} = 6 \times 10^3 \times \frac{1}{6.29} = 955$$

Since $\omega_r = 1/\sqrt{LC}$,

$$L = \frac{1}{\omega_r^2 C} = \frac{1}{(12.55)^2(35.7)} = 175\ \mu\text{h} \qquad \textit{Ans.}$$

2-13. Series Resonance. In this section we shall investigate the well-known phenomena of series resonance. In the material which follows, the most fundamental concepts which are introduced in basic electronics will be expanded.

Consider the series circuit of Fig. 2-26a. The impedance of that part of the circuit to the right of the a-b terminals is

$$Z_{ab} = R_L + j\left(\omega L - \frac{1}{\omega C}\right) \tag{2-12}$$

At some angular frequency ω_r, the reactive term, will equal zero and the impedance will be purely resistive. This condition is known as *series resonance*, and ω_r or f_r is the series-resonant frequency. The series-resonant frequency may be expressed in terms of the circuit parameters by equating the reactive term of Eq. (2-12) to zero:

$$\omega L - \frac{1}{\omega C} = 0$$

$$\omega^2 LC - 1 = 0$$

$$\omega^2 = \frac{1}{LC} \tag{2-13}$$

$$\omega = \omega_r = \frac{1}{\sqrt{LC}} \tag{2-14}$$

$$f = f_r = \frac{1}{2\pi\sqrt{LC}} \tag{2-15}$$

From Eq. (2-12) we can see that Z_{ab} will exhibit a minimum impedance equal to R_L ohms. If the source impedance R_g is purely resistive as indicated, then the loop current is in phase with the applied voltage.

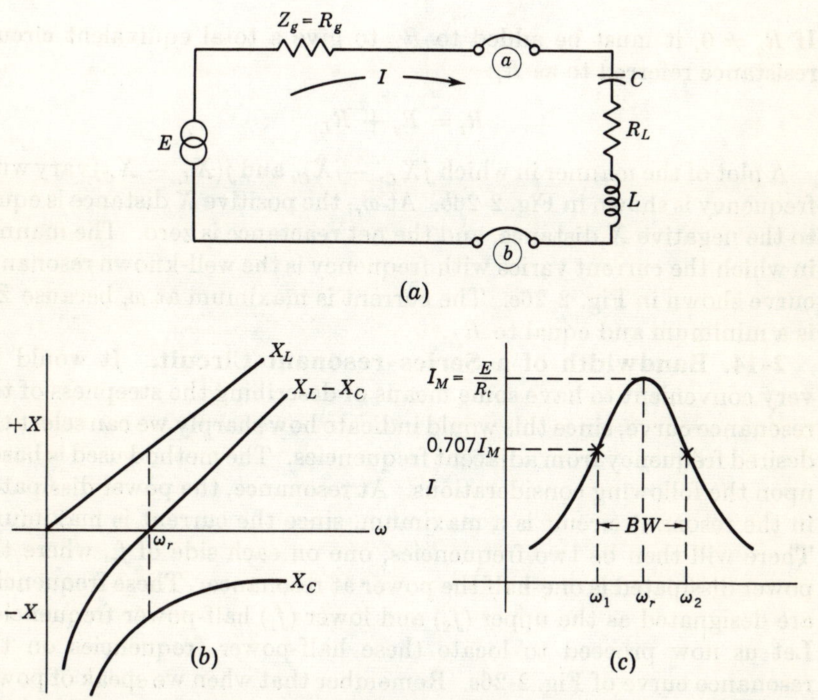

FIG. 2-26. (*a*) A series-resonant circuit. (*b*) A reactance sketch for the series-resonant circuit. (*c*) The current is maximum at the series-resonant frequency. We wish to describe the resonance curve in terms of ω_1 and ω_2, the two frequencies at which the current is 0.707 times the resonant current.

Notice that ω_r is independent of the circuit resistance and depends only upon L and C. Resistor R_L represents the total resistance between points b and a. This includes the d-c resistance of the coil winding, plus the a-c resistance which depends upon core losses and skin effect.

In polar form, the magnitude of the impedance seen looking into terminals a-b is

$$Z_{ab} = \sqrt{R_L^2 + \left(\omega L - \frac{1}{\omega C}\right)^2} \Big/ \tan^{-1} \frac{\omega L - (1/\omega C)}{R_L} \tag{2-12a}$$

and the current is

$$I = \frac{E}{Z_g + Z_{ab}} = \frac{E}{R_g + R_L + j[\omega L - (1/\omega C)]} \tag{2-16}$$

If the generator resistance ($R_g = 0$), then

$$I = \frac{E}{Z_{ab}}$$

If $R_g \neq 0$, it must be added to R_L to give a total equivalent circuit resistance referred to as R_t:

$$R_t = R_g + R_L$$

A plot of the manner in which jX_L, $-jX_C$, and $j(X_L - X_C)$ vary with frequency is shown in Fig. 2-26*b*. At ω_r, the positive X distance is equal to the negative X distance, and the net reactance is zero. The manner in which the current varies with frequency is the well-known resonance curve shown in Fig. 2-26*c*. The current is maximum at ω_r because Z_{ab} is a minimum and equal to R_L.

2-14. Bandwidth of a Series-resonant Circuit. It would be very convenient to have some means of describing the steepness of the resonance curve, since this would indicate how sharply we can select the desired frequency from adjacent frequencies. The method used is based upon the following considerations. At resonance, the power dissipated in the resonant circuit is a maximum, since the current is maximum. There will then be two frequencies, one on each side of f_r, where the power dissipated is one-half the power at resonance. These frequencies are designated as the upper (f_2) and lower (f_1) half-power frequencies. Let us now proceed to locate these half-power frequencies on the resonance curve of Fig. 2-26*c*. Remember that when we speak of power here, we mean real power that is dissipated in resistive elements. At f_r,

$$P_r = I_r^2 R_t \tag{2-17}$$

At f_2,

$$P_2 = \tfrac{1}{2} P_r$$

Therefore, $$I_2^2 R_t = \tfrac{1}{2} I_r^2 R_t$$

$$I_2 = \frac{I_r}{\sqrt{2}} = 0.707 I_r$$

In a similar manner we can show that

$$I_1 = \frac{I_r}{\sqrt{2}} = 0.707 I_r$$

Now it might be desirable to determine the bandwidth of the tuned circuit by inspection of the parameters, instead of by direct measurements upon an actual circuit. We can surely set up the following proportion, since we have developed the relationship between I at resonance and I at the half-power frequencies (I_{12}). Subscript 12 is used to designate a half-power point occurring at ω_1 and ω_2:

$$\frac{|I_{12}|}{|I_r|} = \frac{1}{\sqrt{2}} = \frac{E/\sqrt{R_t^2 + X_{12}}}{E/R_t} \tag{2-18}$$

$$\frac{1}{\sqrt{2}} = \frac{R_t}{\sqrt{R_t^2 + X_{12}^2}}$$

$$\frac{1}{2} = \frac{R_t^2}{R_t^2 + X_{12}^2}$$

Solving for the relationship between X_{12} and R_t, we obtain

$$X_{12} = \pm R_t$$

Notice that the net reactance is equal to the net resistance at the half-power points. This also tells us that the phase angle is plus or minus 45°. At ω_2 the circuit appears inductive and the phase angle is $-45°$, while at ω_1 the net reactance is capacitive and the current leads the voltage by 45°. The net reactance may be expressed in terms of L, C, and ω:

$$X_{12} = \omega_{12} L - \frac{1}{\omega_{12} C} = \pm R_t \tag{2-19}$$

$$\omega_{12}^2 LC - 1 = \pm \omega_{12} R_t C$$

$$LC\omega_{12}^2 \mp R_t C \omega_{12} - 1 = 0 \tag{2-19a}$$

which is analogous to the quadratic formula

$$ax^2 + bx + c = 0$$

whose solution is

$$x = \frac{-b}{2a} \pm \frac{\sqrt{b^2 - 4ac}}{2a}$$

Therefore $$\omega_{12} = \frac{\pm R_t C \pm \sqrt{R_t^2 C^2 + 4LC}}{2LC} \tag{2-19b}$$

Since $\sqrt{R_t^2C^2 + 4LC}$ is surely greater than R_tC, the case where the radical is preceded by a negative sign will result in negative frequencies. A negative frequency is meaningless to us, and this case is discarded. With just the positive sign before the radical, we have two possible frequencies:

$$\omega_{12} = \frac{\pm R_tC + \sqrt{R_t^2C^2 + 4LC}}{2LC} \tag{2-20}$$

The two roots are then

$$2\pi f_1 = \omega_1 = \frac{-R_tC + \sqrt{R_t^2C^2 + 4LC}}{2LC} \tag{2-20a}$$

and

$$2\pi f_2 = \omega_2 = \frac{+R_tC + \sqrt{R_t^2C^2 + 4LC}}{2LC} \tag{2-20b}$$

We now have developed three formulas that will enable us to determine the resonant and half-power frequencies in terms of circuit parameters. The frequency range between ω_1 and ω_2 is termed the bandwidth, BW. That is, BW $= \omega_2 - \omega_1$. A word of caution at this point: the quantity R_t includes both generator and coil resistance. The coil resistance varies with frequency, because of skin effect, etc., which means that R_t due to R_L also varies with frequency. The value of R_L will not necessarily be the same at f_1, f_r, or f_2. For these formulas to be most accurately used, we would require an R_2 and an R_1 corresponding to f_2 and f_1. If we had a table relating R_L to f, we would still be in poor shape, since we do not yet know the f_2 or f_1 with which to enter the table and to get the corresponding R_2 and R_1. A possible solution would be to use a method of successive approximations. Assume an f_2, read the corresponding value of R_2, that is, R_L at f_2 and from the table, substitute this in the formula for f_2, and see if this value of R_2 results in an f_2 which is the same as that assumed. If not, choose a different value of f_2, and repeat the process until the assumed f_2 equals the calculated f_2.

Fortunately, even though a-c coil resistance varies with frequency, the ratio of coil reactance to coil resistance stays reasonably constant over the bandwidth in most cases. As R_L increases with frequency, so does X_L, and the ratio of X_L to R_L stays fairly constant. The quantity X_L/R_L is referred to as the coil Q, or Q_L, and it affords us a convenient way of analyzing the tuned circuit. While coil manufacturers do not usually have charts of R_L versus frequency, curves of Q_L versus f are readily available (Fig. 2-27).

Later we shall develop a relation between Q_L and the half-power frequencies. But first let us see if we can relate the resonant and half-power frequencies directly to the circuit parameters. If we multiply ω_1 by ω_2, the result is

$$\omega_1\omega_2 = \frac{R_t^2C^2 + 4LC - R_t^2C^2}{4L^2C^2} = \frac{1}{LC}$$

but $\omega_r = 1/\sqrt{LC}$; therefore,

$$\omega_1\omega_2 = \omega_r^2 \qquad (2\text{-}21)$$

or $f_1f_2 = f_r^2$. This is the same as writing

$$\frac{f_1}{f_r} = \frac{f_r}{f_2} \qquad (2\text{-}22)$$

The term bandwidth, as we have been using it up to now, does not really tell us much unless the resonant frequency is also specified. For example, if you were told that the bandwidth of a series resonant circuit is 100 cps, could you tell if it were a sharply tuned circuit? Of course not. If the f_r is 500 cps, 100 cps would be a large percentage of f_r, resulting in a broad response curve. If f_r were 1 Mc, the tuning would be very sharp. Therefore, what we really need as a figure of merit for judging the selectivity of a tuned circuit is the ratio of the bandwidth to the resonant frequency. This ratio is sometimes referred to in the literature as the per unit bandwidth or just bandwidth (BW). We shall, however, define

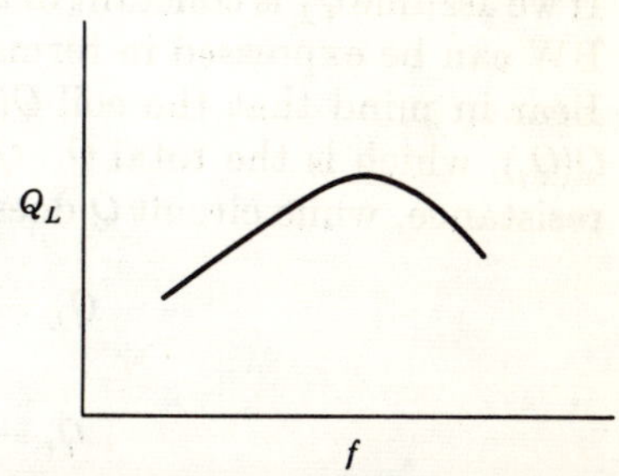

FIG. 2-27. The Q of a coil varies with the applied frequency. In the r-f range this variation is not too severe.

$$\frac{\omega_2 - \omega_1}{\omega_r} = \frac{f_2 - f_1}{f_r} = \text{per unit BW} \qquad (2\text{-}23)$$

and $\omega_2 - \omega_1 = \Delta\omega$ as the bandwidth.

Therefore, let us now develop a simple relationship between the per unit BW and the circuit parameters. We have already shown that

$$\omega_2 = \frac{R_tC + \sqrt{R_t^2C^2 + 4LC}}{2LC}$$

$$\omega_1 = \frac{-R_tC + \sqrt{R_t^2C^2 + 4LC}}{2LC}$$

$$\omega_r = \frac{1}{\sqrt{LC}}$$

By substituting in the expression for per unit BW,

$$\text{Per unit BW} = \frac{\dfrac{R_tC + \sqrt{R_t^2C^2 + 4LC}}{2LC} - \dfrac{-R_tC + \sqrt{R_t^2C^2 + 4LC}}{2LC}}{1/\sqrt{LC}}$$

$$= \frac{R_tC + \sqrt{R_t^2C^2 + 4LC} + R_tC - \sqrt{R_t^2C^2 + 4LC}}{2LC}\sqrt{LC}$$

$$= \frac{2R_tC\sqrt{LC}}{2LC} = \frac{R_t\sqrt{LC}}{L}$$

$$= \frac{R_t}{L}\sqrt{LC} \tag{2-24}$$

If we assume Q_L is constant in the interval between f_2 and f_1, the per unit BW can be expressed in terms of the coil and circuit Q's at resonance. Bear in mind that the coil $Q(Q_L)$ is normally higher than the circuit $Q(Q_t)$, which is the total Q. Q_L does not take into account generator resistance, while circuit Q does:

$$Q_L = \frac{\omega L}{R_L}$$

$$Q_t = \frac{\omega L}{R_t} = \frac{\omega L}{R_L + R_g}$$

The per unit BW depends on the over-all circuit Q rather than on just the coil Q.

Since per unit BW $= (R_t\sqrt{LC}/L)$, and $\omega_r = 1/\sqrt{LC}$ or $\sqrt{LC} = 1/\omega_r$, then

$$\text{Per unit BW} = \frac{R_t}{\omega_r L} = \frac{1}{Q_t} = \frac{\omega_2 - \omega_1}{\omega_r} \tag{2-25}$$

or

$$Q_t = \frac{\omega_r L}{R_t} = \frac{\omega_r}{\omega_2 - \omega_1} \tag{2-26}$$

The bandwidth is

$$\text{BW} = \omega_r/Q_t \tag{2-27}$$

We can bring coil Q (Q_L, that is) into the expression for total circuit Q by noting that

$$R = R_g + R_L \tag{2-82}$$

$$\frac{1}{Q_t} = \frac{R_g + R_L}{\omega_r L} = \frac{R_g}{\omega_r L} + \frac{R_L}{\omega_r L} = \frac{R_g}{\omega_r L} + \frac{1}{Q_L}$$

$$Q_t = \frac{1}{(R_g/\omega_r L) + 1/Q_L} \tag{2-29}$$

In general, $f_2 - f_r \neq f_r - f_1$. That is, the half-power frequencies are not equally spaced about the resonant frequency. If, however, the overall circuit Q (Q_t, that is) is ≥ 10, the error is negligible, and the half-power frequencies can be considered equally spaced about f_r. Therefore, if we know the circuit Q, we may write, when $Q_t \geq 10$,

$$\omega_2 = \omega_r + \frac{\mathrm{BW}}{2} = \omega_r + \frac{\omega_r}{2Q_t} \qquad (2\text{-}30)$$

$$\omega_1 = \omega_r - \frac{\mathrm{BW}}{2} = \omega_r - \frac{\omega_r}{2Q_t} \qquad (2\text{-}31)$$

If the circuit Q is about 10 or more, the voltage across either L or C will be a maximum at ω_r and will exhibit a frequency-response curve similar to that of the current. The same bandwidth, Q, and other relationships may be used. For example, the circuit Q could be evaluated by measuring 0.707 points of coil voltage. More about this later.

Let us summarize the results of our efforts up to this point.

1. The resonant frequency of a series RLC circuit is

$$\omega_r = \frac{1}{\sqrt{LC}}$$

2. The current at resonance is in phase with the applied voltage, and has a maximum value equal to E/R_t.

3. We have defined two half-power frequencies, ω_1 and ω_2, at which the current is 70.7 per cent of the resonant current and the phase angle between the current and the applied voltage is plus or minus 45°.

4. The resonant frequency is the geometric mean of ω_1 and ω_2. That is, $\omega_1\omega_2 = \omega_r^2$; therefore $\omega_r = \sqrt{\omega_1\omega_2}$. Sometimes it is difficult to determine a resonant point by direct measurement, since the current varies slowly in the immediate vicinity of resonance. By determining the half-power frequencies and by using the above relationship, ω_r can be readily determined.

5. The bandwidth, or spacing between the half-power frequencies, is

$$\mathrm{BW} = \omega_2 - \omega_1$$

6. The per unit bandwidth is a measure of the circuit selectivity, since it expresses what percentage of the resonant frequency is occupied by the bandwidth. The per unit bandwidth is related to the circuit Q by

$$\text{Per unit BW} = \frac{\omega_2 - \omega_1}{\omega_r} = \frac{1}{Q_t}$$

7. The above relationship (6) tells us that, if we know the circuit Q and ω_r, we can determine the bandwidth from

$$\text{BW} = \omega_2 - \omega_1 = \frac{\omega_r}{Q_t}$$

8. Conversely, the total Q of a series-resonant circuit can be determined by performing measurements to evaluate ω_1, ω_r, and ω_2:

$$Q_t = \frac{\omega_r}{\omega_2 - \omega_1}$$

9. If the source which drives the resonant circuit is purely resistive in nature, the circuit Q can be related to the coil Q by

$$Q_t = \frac{1}{(R_g/\omega_r L) + 1/Q_L}$$

10. If the circuit Q is known and it is desired to evaluate the coil Q, we can solve for Q_L and obtain

$$Q_L = \frac{1}{(1/Q_t) - (R_g/\omega_r L)}$$

11. If the circuit Q is known and is in the neighborhood of 10 or more, the half-power frequencies may be considered evenly spaced about ω_r. That is,

$$\omega_2 = \omega_r + \frac{\text{BW}}{2} = \omega_r + \frac{\omega_r}{2Q_t}$$

$$\omega_1 = \omega_r - \frac{\text{BW}}{2} = \omega_r - \frac{\omega_r}{2Q_t}$$

2-15. Analysis with Q Not Constant. Up to this point we have restricted ourselves to cases in which the coil Q is essentially constant over a range of frequencies. While this is the usual case, there are times, particularly when dealing with iron-core inductors at audio frequencies, when Q_L itself varies considerably with a slight change in frequency, as indicated in Fig. 2-28. Just how is this situation to be handled? Let us go back to fundamentals.

In the following derivation, we shall deal with the total circuit resistance only; therefore, the t subscript is to be understood. In a series RLC circuit, the current at any frequency is

$$I = \frac{E}{R + jX} \tag{2-32}$$

where R = total circuit resistance

$X = X_L - X_C$, the net reactance at any frequency

At resonance, $I = E/R_r$, where R_r = total circuit resistance at resonance. Therefore,

$$\frac{I}{I_r} = \frac{E/(R + jX)}{E/R_r} = \frac{R_r}{R + jX}$$

Dividing numerator and denominator by R_r,

$$\frac{I}{I_r} = \frac{1}{(R/R_r) + j(X/R_r)} \tag{2-33}$$

Since we wish to express I/I_r as a function of ω, we can write, for X,

$$X = X_L - X_C = \omega L - \frac{1}{\omega C} = \frac{\omega^2 LC - 1}{\omega C}$$

but $\omega_r = 1/\sqrt{LC}$; therefore $LC = 1/\omega_r^2$, and $C = 1/\omega_r^2 L$. Therefore,

$$X = \frac{\omega^2(1/\omega_r^2) - 1}{\omega/\omega_r^2 L} = \frac{(\omega/\omega_r)^2 - 1}{(\omega/\omega_r)1/\omega_r L}$$

Substituting in the expression for I/I_r, we have

$$\frac{I}{I_r} = \frac{1}{\dfrac{R}{R_r} + j\,\dfrac{(\omega/\omega_r)^2 - 1}{(\omega/\omega_r)R_r/\omega_r L}}$$

But $Q_r = \omega_r L/R_r$, where Q_r is the total circuit Q at resonance,

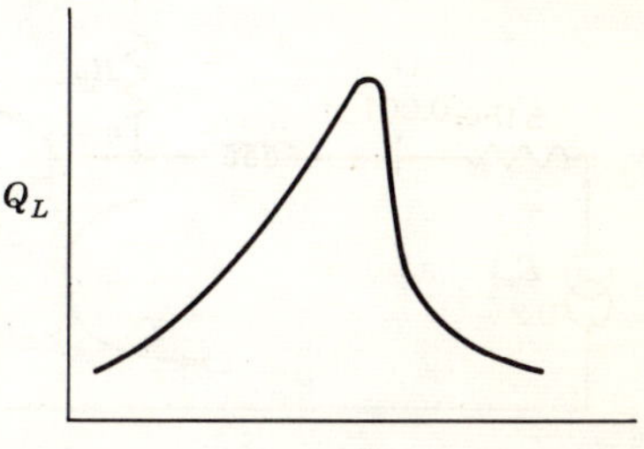

FIG. 2-28. The Q of this coil cannot be treated as constant.

$$\frac{I}{I_r} = \frac{1}{\dfrac{R}{R_r} + j\,\dfrac{(\omega/\omega_r)^2 - 1}{(\omega/\omega_r)1/Q_r}} = \frac{1}{\dfrac{R}{R_r} + jQ_r\,\dfrac{(\omega/\omega_r)^2 - 1}{\omega/\omega_r}}$$

$$= \frac{1}{\dfrac{R}{R_r} + jQ_r\left(\dfrac{\omega}{\omega_r} - \dfrac{1}{\omega/\omega_r}\right)}$$

$$= \frac{1}{\dfrac{R}{R_r} + jQ_r\left(\dfrac{\omega}{\omega_r} - \dfrac{\omega_r}{\omega}\right)} \tag{2-34}$$

Usually coil Q as a function of frequency is available from manufacturers' data. Therefore, it may be more convenient to express R/R_r in terms of Q:

$$R = \frac{\omega L}{Q}$$

$$R_r = \frac{\omega_r L}{Q_r}$$

and Q does not necessarily equal Q_r. Remember, R is the total circuit resistance; therefore,

$$\frac{R}{R_r} = \frac{\omega L/Q}{\omega_r L/Q_r} = \frac{\omega}{\omega_r}\frac{Q_r}{Q}$$

$$\frac{I}{I_r} = \frac{1}{\dfrac{\omega}{\omega_r}\dfrac{Q_r}{Q} + jQ_r\left(\dfrac{\omega}{\omega_r} - \dfrac{\omega_r}{\omega}\right)} = \frac{1}{\dfrac{f}{f_r}\dfrac{Q_r}{Q} + jQ_r\left(\dfrac{f}{f_r} - \dfrac{f_r}{f}\right)} \tag{2-35}$$

Equations (2-34) and (2-35) are exact and involve no assumptions other than L remaining constant. Let us see how Eq. (2-35) would be used in a typical problem.

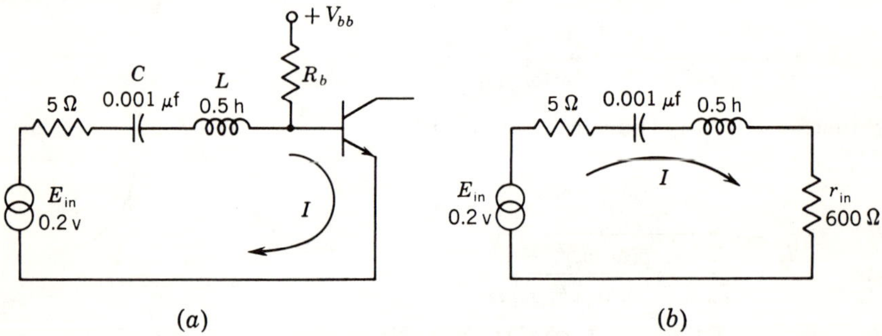

FIG. 2-29. The input current (I) which essentially flows into the transistor is to be determined. The inductor does not exhibit a constant Q over the frequency range of interest.

The radio command circuit of Fig. 2-29*a* operates in the following manner. As the input E_{in} approaches the vicinity of resonance, I increases. Since I also flows through the input resistance of the transistor, the resonant current rise will be amplified and in subsequent stages may be rectified and used to actuate a relay. The resistance seen looking into the transistor (r_{in}) terminals is about 600 ohms. Now R_b is in the order of 100 kilohms, which is so much larger than r_{in} that any shunting effect due to R_b may be neglected. The a-c equivalent circuit is that of Fig. 2-29*b*.

The system has been so designed that it takes a minimum of $I = 0.180$ ma for a relay to be actuated elsewhere. It has been found that the tone generator used to establish the frequency of E_{in} will vary by -10 per cent of the nominal frequency during operation. Manufacturers' data indicate that the coil Q (Q_L, that is) has a value of 40 at 6,400 cps, of 80 at 7,100 cps, and of 100 at 7,800 cps. Will the circuit operate with a minus 10 per cent variation of the nominal frequency?

First, let us determine the resonant frequency:

$$f_r = \frac{1}{2\pi\sqrt{0.5 \times 10^{-9}}} = 7{,}100 \text{ cps}$$

The tone generator varies about f_r by 10 per cent, or 710 cps. The lower limiting frequency (f_a) is then

$$f_a = 7{,}100 - 710 = 6{,}390 \text{ cps}$$

To see if the circuit will operate at f_a, we plug into Eq. (2-35). However, Q and Q_r in Eq. (2-35) are circuit Q's and not just coil Q's. In general,

$$Q = \frac{\omega L}{R_g + R_L}$$

or

$$\frac{1}{Q} = \frac{R_g + R_L}{\omega L} = \frac{R_g}{\omega L} + \frac{1}{Q_L}$$

At f_a,

$$\frac{1}{Q_a} = \frac{605}{2\pi \times 6{,}390 \times 0.5} + \frac{1}{40} = 0.0303 + 0.025 = 0.0553$$

Therefore $Q_a = 18.1$. Similarly,

$$\frac{1}{Q_r} = \frac{R_g}{\omega_r L} + \frac{1}{Q_{L_r}} = \frac{605}{2\pi \times 7.1 \times 10^3 \times 0.5} + \frac{1}{80}$$
$$= 0.0272 + 0.0125 = 0.0397$$

Therefore $Q_r = 25.2$. We can now plug into Eq. (2-35).

$$\frac{I}{I_r} = \frac{1}{\dfrac{6{,}390}{7{,}100} \times \dfrac{25.2}{18.1} + j25.2\left(\dfrac{6{,}390}{7{,}100} - \dfrac{7{,}100}{6{,}390}\right)}$$

$$= \frac{1}{0.897 \times 1.39 + j25.2(0.897 - 1.113)}$$

$$= \frac{1}{1.25 - j5.45} = \frac{1}{5.6\underline{/-77.1^\circ}} = 0.178\underline{/+77.1^\circ}$$

Since $$I_r = \frac{E_{in}}{R_g + R_{L_r}} = \frac{0.2}{R_g + (\omega_r L/Q_{L_r})} = \frac{0.2}{605 + (22 \times 10^3)/80}$$

$$= \frac{0.2}{605 + 276} = \frac{0.2 \text{ volt}}{881 \text{ ohms}} = 0.227 \text{ ma}$$

Therefore, $I = (0.178\underline{/77.1^\circ})(0.227 \text{ ma}\underline{/0^\circ}) = 0.041 \text{ ma}\underline{/77.1^\circ}$.

Since 180 μa is needed but only 41 μa is available at -10 per cent shift in command frequency, the relay will not actuate.

The 77.1° phase angle is positive, which means that the current is leading the voltage. This agrees with the fact that since f has shifted below f_r, the net circuit reactance will be capacitive.

2-16. The Universal Resonance Curve. At times we are faced with the problem of determining what frequency is required to give a specific current response. This is a difficult problem, since it means solving for ω in the equation

$$\frac{I}{I_r} = \frac{1}{\dfrac{\omega}{\omega_r}\dfrac{Q_r}{Q} + jQ_r\left(\dfrac{\omega}{\omega_r} - \dfrac{\omega_r}{\omega}\right)} \tag{2-36}$$

Furthermore, when ω is the unknown, we cannot determine Q since it depends on ω. Possibly Q can be expressed as a function of ω for a given coil, but this is a major problem in itself.[1] It would be much easier to develop a chart that plots the current ratio versus frequency. This, then, is our next undertaking.

To reduce the complexity of our problem, we shall first make the assumption that Q does not appreciably vary in the region about resonance. That is, $Q = Q_r$. Equation (2-36) therefore reduces to

$$\frac{I}{I_r} = \frac{1}{\dfrac{\omega}{\omega_r} + jQ_r\left(\dfrac{\omega}{\omega_r} - \dfrac{\omega_r}{\omega}\right)} \tag{2-37}$$

Let $\omega = \omega_r + k\omega_r$, where k may be positive or negative. To describe an ω greater than ω_r, say, twice as large, k would equal 1, since $\omega = 2\omega_r = \omega_r + k\omega_r = \omega_r + \omega_r$. This substitution may be simplified to

$$\omega = \omega_r(1 + k) \tag{2-38}$$

The parameter k carries its own sign. From (2-38) we see that

$$k = \frac{\omega - \omega_r}{\omega_r}$$

Substituting (2-38) in (2-37), we have

$$\frac{I}{I_r} = \frac{1}{\dfrac{\omega_r(1+k)}{\omega_r} + jQ_r\left[\dfrac{\omega_r(1+k)}{\omega_r} - \dfrac{\omega_r}{\omega_r(1+k)}\right]} \tag{2-38a}$$

[1] An excellent treatment of tuned circuits may be found in F. E. Terman, "Electronic and Radio Engineering," 4th ed., McGraw-Hill Book Company, Inc., New York, 1955.

Simplifying,

$$\frac{I}{I_r} = \frac{1}{1 + k + jQ_r\left(1 + k - \dfrac{1}{1 + k}\right)} = \frac{1}{1 + k + jQ_r\left[\dfrac{(1 + k)^2 - 1}{1 + k}\right]}$$

$$= \frac{1}{1 + k + jQ_rk\dfrac{2 + k}{1 + k}} \tag{2-39}$$

Equation (2-39) may be plotted as a universal resonance curve in the following manner. Assume values of Q_rk along the horizontal axis of a graph and plot the corresponding I/I_r values along the vertical axis. The range of values that Q_rk should encompass will be determined by the Q values expected and by deviation from the resonant frequency. Usually Q will lie between 10 and 500, so these two cases will be considered. Even for the case where $Q = 10$, we are seldom interested in the current response at frequencies 30 per cent higher or lower than ω_r. This fixes the range of k, which is determined as follows.

From (2-38) $\omega = \omega_r(1 + k)$. When ω is 30 per cent higher than ω_r,

$$\omega = \omega_r + 0.3\omega_r = 1.3\omega_r$$

Substituting in (2-38),

$$1.3\omega_r = \omega_r(1 + k)$$
$$0.3 = k$$

When ω is 30 per cent lower than ω_r

$$\omega = \omega_r - 0.3\omega_r = 0.7\omega_r$$

and for this case we find the lower limit of k, which is

$$0.7\omega_r = \omega_r(1 + k)$$
$$-0.3 = k$$

A satisfactory range along the horizontal axis can be determined by taking Q_rk equal to the product of the lowest Q considered, which is 10, and $k = \pm 0.3$, which is -3 to $+3$.

Let us go through a few illustrative calculations to show how the universal resonance curve of Fig. 2-30 is sketched. If $Q_rk = -3$, and $Q_r = 10$, $k = -0.3$:

$$\frac{I}{I_r} = \frac{1}{1 - 0.3 + j10(-0.3)\left(\dfrac{2 - 0.3}{1 - 0.3}\right)} = \frac{1}{0.7 - j3\,\dfrac{1.7}{0.7}}$$

$$= \frac{1}{0.7 - j7.3} = \frac{1}{7.3\underline{/-84.5^\circ}} = 0.14\underline{/84.5^\circ}$$

If $k = +0.3$, and $Q_r = 10$, $Q_rk = 3$:

$$\frac{I}{I_r} = \frac{1}{1 + 0.3 + j10(0.3)\left(\dfrac{2 + 0.3}{1 + 0.3}\right)} = \frac{1}{1.3 + j3\,\dfrac{2.3}{1.3}}$$

$$= \frac{1}{1.3 + j5.3} = \frac{1}{5.5\underline{/76.3^\circ}} = 0.18\underline{/-76.3^\circ}$$

Note that the curve is almost symmetrical about ω_r for these values of Q_rk. The higher the Q, the better the symmetry.

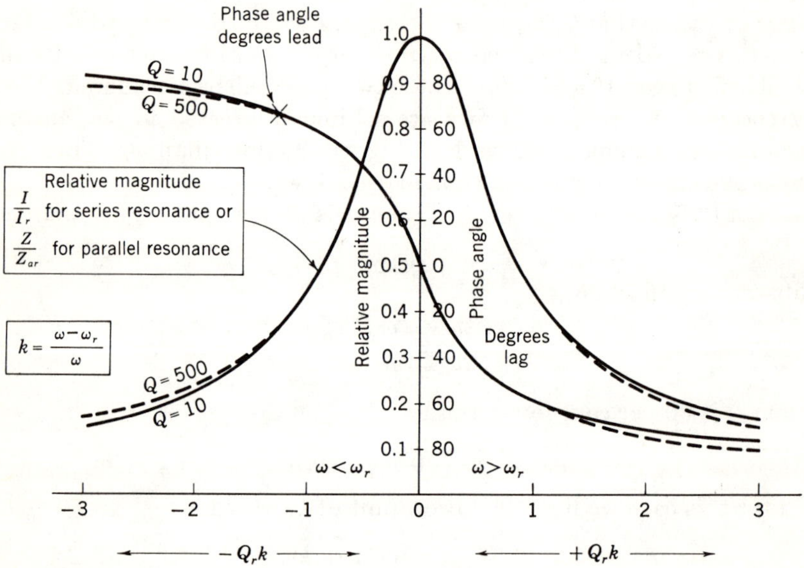

FIG. 2-30. Universal resonance curve. Leading phase angle for series resonance as shown will represent lag in parallel resonance.

Now suppose $Q_rk = -3$, but $Q_r = 500$. This means $k = -0.006$:

$$\frac{I}{I_r} = \frac{1}{1 - 0.006 + j500(-0.006)\left(\dfrac{2 - 0.006}{1 - 0.006}\right)}$$

$$\approx \frac{1}{1 - j3(2)} = \frac{1}{1 - j6} = \frac{1}{6.1\underline{/-80.6^\circ}} = 0.16\underline{/80.6^\circ}$$

If we were to continue the process and obtain sufficient points to sketch two resonance curves, one for $Q = 10$ and the other for $Q = 500$, we should find them virtually superimposed. The curve for $Q = 500$ may be used for values of Q ranging to infinity with very little error, as evidenced in Fig. 2-30. We shall see later that Fig. 2-30 may also be used for parallel-resonant circuits.

2-17. Resonant Voltage Rise. An interesting and useful phenomenon about series-resonant circuits is the large rise in voltage that occurs across L or C at ω_r when Q_t is large. We can prove this fact in the following manner.

The magnitude of the voltage across the capacitor is $E_C = IX_C$, but at resonance $I = I_r = E/R$. Therefore, $E_{C_r} = \frac{E}{R} X_{C_r}$. But at ω_r, $X_{C_r} = X_{L_r}$, or

$$E_{C_r} = \frac{EX_{L_r}}{R} = EQ_{t_r}$$

Q_{t_r} is the circuit Q at resonance.

Remarkably, the capacitor or inductor voltage at resonance may be Q_{t_r} times larger than the applied voltage. If 10 volts is applied to a series-resonant circuit having a $Q_{t_r} = 100$, the capacitor or inductor voltage will be 1,000 volts. When circuits of this type are designed, the capacitor working voltage must be determined on this basis. Actually ω_r is not the exact frequency at which E_L or E_C is a maximum, but the difference is slight if $Q_{t_r} = 10$. The exact frequency at which E_c is a maximum is

$$\omega = \omega_r \sqrt{1 - \frac{1}{2Q_{t_r}^{\ 2}}}$$

which lies slightly below ω_r. If $Q_{t_r} = 10$, this frequency is essentially the same as ω_r and the maximum capacitor voltage will approximately equal the capacitor voltage at resonance.

The voltage across the coil at resonance (E_{L_r}) is complicated by the fact that L has a series resistance (R_L) associated with it. Therefore, we shall start with Z_L rather than X_L:

$$Z_L = \sqrt{R_L^{\ 2} + (\omega L)^2}$$

Since R_L is usually unspecified, but Q_L is specified, we can write

$$Z_L = \sqrt{(\omega L)^2 \left[\frac{R_L^{\ 2}}{(\omega L)^2} + \frac{(\omega L)^2}{(\omega L)^2}\right]} = \omega L \sqrt{\frac{1}{Q_L^{\ 2}} + 1}$$

$$E_{L_r} = I_r Z_{L_r} = \frac{E}{R} \omega_r L \sqrt{\frac{1}{Q_{L_r}^{\ 2}} + 1}$$

$$= EQ_{t_r} \sqrt{\frac{1}{Q_{L_r}^{\ 2}} + 1} \approx EQ_{t_r} \qquad \text{when } Q_{L_r} \geq 10$$

The exact frequency at which the coil voltage is a maximum is slightly above ω_r, and is given by

$$\omega = \frac{\omega_r}{\sqrt{1 - 1/2Q_{t_r}^{\,2}}}$$

Again, if $Q_{t_r} \geq 10$, the coil voltage may be considered maximum at ω_r.

A note of caution—whenever you make any calculations involving the coil Q in the vicinity of ω_r, be sure to use the value of Q corresponding to ω_r. Q_L may vary widely over a broad frequency range, and therefore it is best either to measure the coil Q at the frequency of interest or to use manufacturer's data which should plot Q_L versus f.

Let us apply the material we have covered to some illustrative problems. In all cases, rms values of voltage and current are understood unless otherwise stated.

Problem 1. In Fig. 2-31, determine (*a*) the resonant frequency (f_r), (*b*) the circuit Q at f_r (Q_{t_r}), (*c*) the maximum voltage across the coil, (*d*) the bandwidth, (*e*) the upper and lower half-power frequencies, (*f*) the frequencies at which the current is down to one-half the resonant current, (*g*) the value of resistance that would have to be inserted in series with the rest of the circuit to double the existing bandwidth.

Solution

(*a*) $$f_r = \frac{\omega_r}{2\pi} = \frac{1}{2\pi LC} = \frac{0.159}{\sqrt{1(10^{-8})}} = \frac{0.159}{10^{-4}} = 1{,}590 \text{ cps}$$

(*b*) $$Q_{t_r} = \frac{\omega_r L}{R_t} = \frac{2\pi f_r L}{R_t} = \frac{6.28(1.59)10^3(1)}{10^2} = 100$$

$$\omega_r = 10^4 \text{ radians/sec}$$

(*c*) Since Q_{t_r} is greater than 10, the coil voltage will be considered maximum at ω_r, in which case

$$E_{L\,\max} = Q_{t_r} E_{in} = 100(10) = 1{,}000 \text{ volts}$$

(*d*) The bandwidth is given by

$$BW = \frac{\omega_r}{Q_{t_r}} = \frac{10^4}{10^2} = 100 \text{ radians/sec, or } 15.90 \text{ cps}$$

(*e*) Since Q_{t_r} is greater than 10, the half-power frequencies may be considered to be equally displaced about ω_r. Therefore,

$$f_2 = f_r + \frac{BW}{2} = 1{,}590 + \frac{15.90}{2} = 1{,}598 \text{ cps}$$

$$f_1 = f_r - \frac{BW}{2} = 1{,}582 \text{ cps}$$

(*f*) To find the frequencies at which the current is one-half the resonant current, we shall use the universal resonance curve of Fig. 2-31. In Fig. 2-31, we see that the current is 50 per cent below the resonant current at $Q_r k = \pm 0.85$. Since Q_r in Fig. 2-31 is total circuit Q or Q_{t_r}, we have

$$k = \pm \frac{0.85}{100}$$

but

$$k = \frac{\omega - \omega_r}{\omega_r}$$

Therefore $\omega = \omega_r(1 + k) = 10^4(1 + 0.0085)$ and $10^4(1 - 0.0085)$

$= 10{,}085$ radians/sec and 9,915 radians/sec

which respectively correspond to 1,605 cps and 1,578 cps.

(*g*) The present BW is 15.9 cps. The double of this amount is 31.8 cps. Since $\text{BW} = \omega_r/Q_{t_r}$, or $Q_{t_r} = \omega_r/\text{BW}$, and we are setting BW equal to 31.8 cps, the new Q_{t_r} must be $Q_{t_r} = 1{,}590/31.8 = 50$, which is to be expected, since Q and BW are inversely proportional. To reduce the Q by a factor of 2, we must double the resistance. Therefore, an additional 100 ohms must be inserted into the circuit.

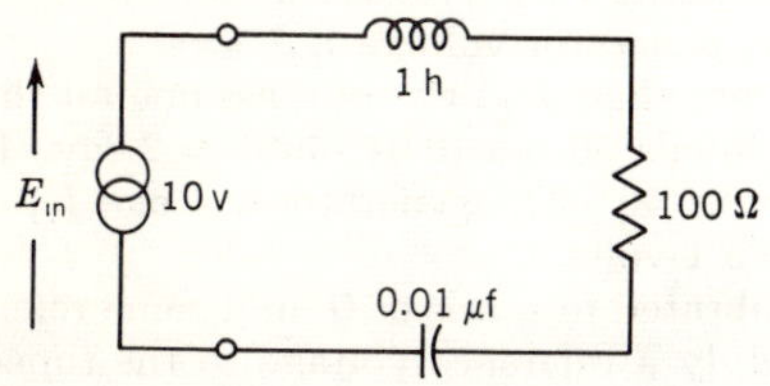

FIG. 2-31. Resonant circuit analyzed in detail.

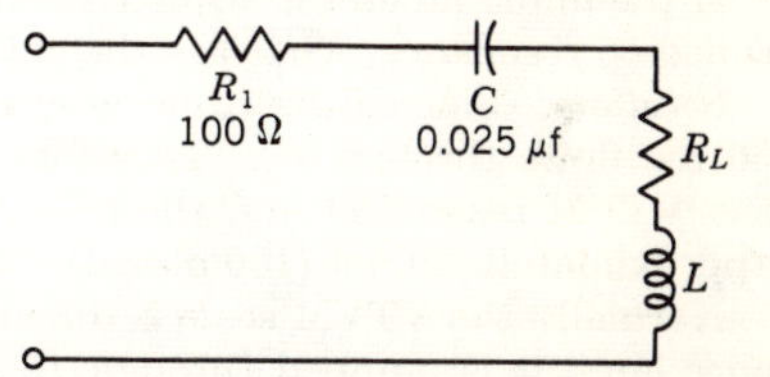

FIG. 2-32. The coil Q may be determined by measuring the selectivity of the circuit.

Problem 2. Measurements of current through the circuit of Fig. 2–32 show the response is down 3 db at $f_1 = 0.9$ kc and at $f_2 = 1.1$ kc. What is the coil Q—that is, Q_L?

Solution

$$Q_t = \frac{f_r}{f_2 - f_1} = \frac{1 \text{ kc}}{0.2 \text{ kc}} = 5$$

We shall need to know ωL. Since $\omega L = 1/\omega C$ at ω_r, and $1/\omega C = 6{,}280$ ohms we know that $\omega L = 6{,}280$. We previously showed that

$$Q_L = \frac{1}{1/Q_t - R_g/\omega_r L} = \frac{1}{1/5 - 100/6{,}280} = \frac{1}{0.2 - 0.016} = 5.42$$

The coil resistance R_L at ω_r is $R_L = \omega_r L/Q_L = 1{,}160$.

Problem 3. The basic circuit of a Q meter is shown in Fig. 2-33. Generator E_g, with source impedance R_g, pumps r-f current—measured by the r-f milliammeter—through R_1 and through the series $R_L LC$ circuit which shunts R_1. Capacitor C is adjusted until the $R_L LC$ branch is resonant at the generator frequency. A very high impedance VTVM measures the voltage across the capacitor, which at resonance is Q times the voltage applied to the tuned circuit. This voltage is the same as that across R_1, since the two branches are in parallel. Resistor R_1 is deliberately chosen at about 0.04 ohm, so that the source impedance that the series-resonant circuit looks into will be at least 0.04 ohm. Resistor R_g is about 100 ohms. With R_1 so low, the loading effect of the tuned circuit at resonance will be negligible, except for the very highest Q values, that is, lowest R_L values. Therefore, the generator amplitude may be adjusted to pump a known current through R_1, which produces a calculable driving voltage for the tuned circuit. This voltage will not vary significantly as C is adjusted.

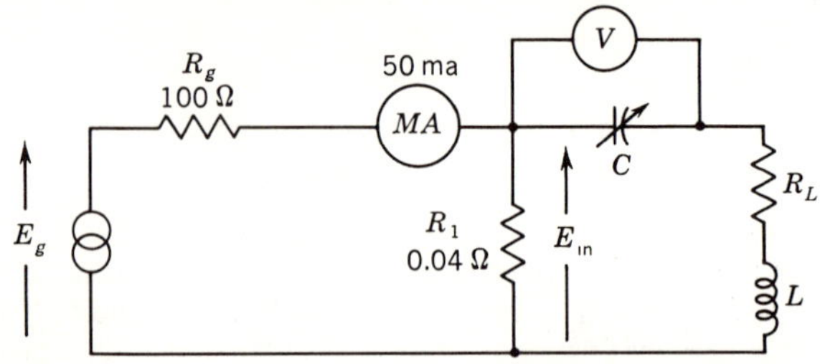

FIG. 2-33. Essentials of a Q meter.

If the milliammeter is adjusted to 50 ma and the VTVM reads 0.1 volt when C is tuned to resonance, what are the coil Q and generator voltage E_g?

Solution. Since R_1 is to be very much less than R_L, we may assume all the 50 ma flows through R_1. Therefore, E_{in} equals 50 ma (0.04 ohm) $= 2$ mv. If the VTVM reads 100 mv, the coil Q must be 50. The generator voltage E_g is approximately 50 ma (100 ohms) $+ E_{in} = 5.1$ volts.

Normally the VTVM scale is directly calibrated in terms of Q, and some reference current is pumped through R_1 to supply a reference voltage to the tuned circuit.

Problem 4. The circuit of Fig. 2-34*a* is part of a radio command system. The voltage across inductors rises as the resonant frequency of the branch they are located in is approached. This voltage is subsequently amplified and rectified and used to control current through a relay. Sketch, on conventional graph paper, a curve which describes the coil voltage as functions of the applied frequency. Assume that transformer T is ideal and serves solely as an impedance-transforming device. Assume also that the Q of each coil is 20.

Solution. The impedance seen looking through the secondary of T is obtained from

$$\frac{Z_s}{Z_p} = \left(\frac{N_s}{N_p}\right)^2$$

where N is the number of turns. If the reader does not recall this relationship from basic electronics, it will be found in Chap. 4. Therefore,

$$Z_s = \left(\frac{N_s}{N_p}\right)^2 Z_p = \left(\frac{1}{20}\right)^2 400 = 1 \text{ ohm}$$

Of course, the primary voltage will be stepped down from 20 volts to 1 volt. The simplified equivalent circuit is shown in Fig. 2-34*b*. The resonant frequency of the L_1C_1 branch is

$$f_{r_1} = \frac{1}{2\pi\sqrt{L_1C_1}} \approx 2 \text{ kc}$$

and that of the L_2C_2 branch is

$$f_{r_2} = \frac{1}{2\pi\sqrt{L_2C_2}} \approx 5 \text{ kc}$$

When E_{in} is at 2 kc, the impedance of the L_1C_1 branch is due solely to the coil resistance, which is

$$R_1 = \frac{\omega L_1}{Q_L} \approx \frac{1{,}200}{20} = 60 \text{ ohms}$$

When E_{in} is at 5 kc, the impedance of the L_2C_2 branch is $\omega L_2/Q_L \approx \dfrac{3{,}140}{20} = 157$ ohms. By investigating the minimum impedance—heaviest loads—that the gener-

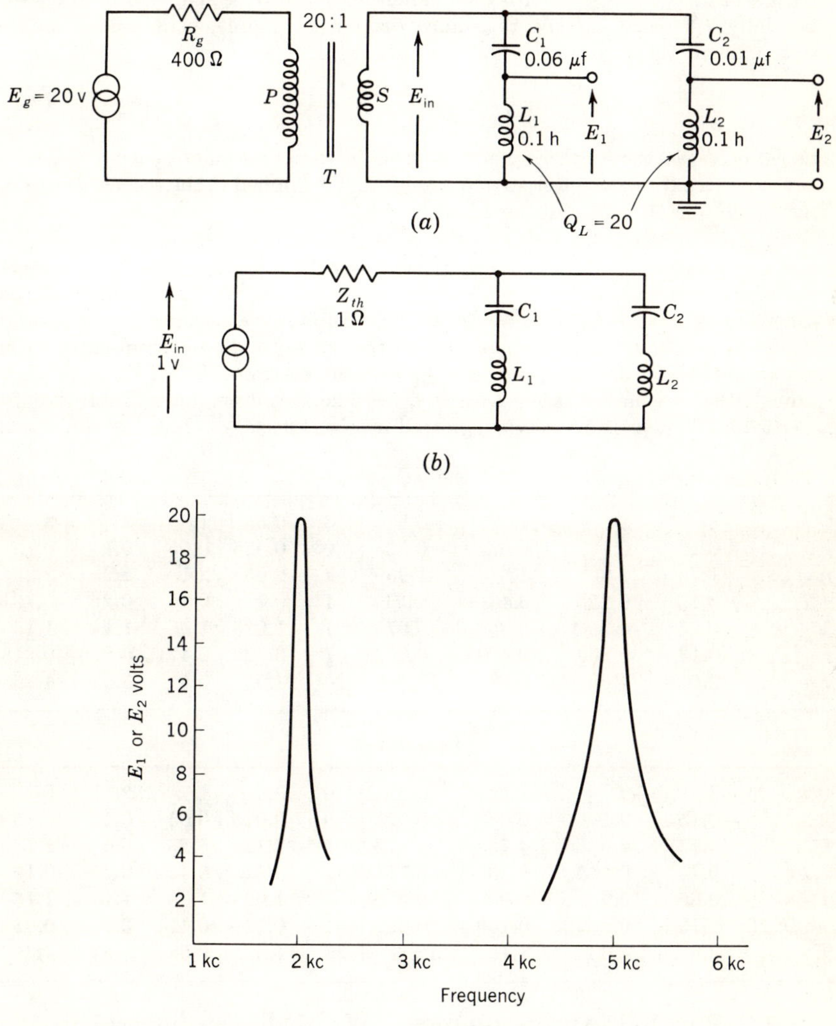

FIG. 2-34. (*a*) The channel-selecting portion of a radio command system. (*b*) The equivalent source that the tone selectors look into. (*c*) The frequency-response curves of the channel selectors.

ator must drive, we have determined that Z_{th} is negligible and that the tuned circuits may be considered as driven by a constant 1-volt source. Therefore, the responses of the two branches are entirely independent. This is usually desirable, and series-tuned circuits should be driven from a low-impedance, or constant-voltage, source.

At 2 kc, the voltage across L_1, designated by E_1 in Fig. 2-34*a*, will be $E_1 = E_{in}Q_t = 1$ volt(20) = 20 volts. Similarly, at 5 kc, E_2 will also equal 20 volts. To sketch E_1 and E_2 in detail, we can use the universal resonance curve as follows:

In general, the voltage across an inductor, if the Q is high—10 or more—is essentially $E_L = IX_L$. The resistance drop is negligible, compared with the reactive drop. Therefore,

$$\frac{E_L}{E_{L_r}} = \frac{IX_L}{I_rX_{L_r}} = \frac{I}{I_r}\frac{\omega L}{\omega_r L} = \frac{I}{I_r}\frac{\omega}{\omega_r} \tag{2-40}$$

The ratio I/I_r is easily determined from the universal resonance curve. Furthermore, we can express ω in terms of ω_r and k as it is applied in the resonance curve. That is, $\omega = \omega_r(1 + k)$ and therefore,

$$\frac{E_L}{E_{L_r}} = \frac{I}{I_r}(1 + k) \tag{2-41}$$

Remember, $k = (f - f_r)/f_r$, or $f = f_r(1 + k)$. Suppose we assume some values of Q_rk, determine k and ω, and then read off the corresponding current ratios from the curve. Substituting in (2-41) above, we can determine E_L. Taking the L_1C_1 channel first, we have Table 2-4 for $f_{r_1} = 2$ kc. And we have Table 2-5 for $f_{r_2} = 5$ kc. The resultant curve is plotted in Fig. 2-34*c*.

TABLE 2-4

Q_rk . . .	−3	−2	−1	−0.5	0	0.5	1	2	3
k	−0.15	−0.1	−0.05	−0.025	0	0.025	0.05	0.1	0.15
f_{kc}	1.7	1.8	1.9	1.95	2	2.05	2.1	2.2	2.3
I/I_r . .	0.15	0.225	0.43	0.71	1	0.71	0.45	0.25	0.19
$1 + k$	0.85	0.9	0.95	0.975	1	1.025	1.05	1.1	1.15
E_L/E_r	0.13	0.203	0.409	0.692	1	0.727	0.474	0.276	0.218
E_L . . .	2.6	4.06	8.19	13.8	20	14.5	9.5	5.53	4.36

TABLE 2-5

Q_rk . . .	−3	−2	−1	−0.5	0	0.5	1	2	3
k	−0.15	−0.1	−0.05	−0.025	0	0.025	0.05	0.1	0.15
f_{kc}	4.25	4.5	4.75	4.875	5	5.125	5.25	5.5	5.75
I/I_r . .	0.15	0.225	0.43	0.71	1	0.71	0.45	0.25	0.19
$1 + k$	0.85	0.9	0.95	0.975	1	1.025	1.05	1.1	1.15
E_L/E_r	0.13	0.203	0.409	0.692	1	0.727	0.474	0.276	0.218
E_L . . .	2.6	4.06	8.19	13.8	20	14.5	9.5	5.53	4.36

2-18. Parallel Antiresonance. We shall now investigate the phenomenon of parallel resonance, or antiresonance, as it is sometimes called. The circuit of Fig. 2-35*a* illustrates a completely general antiresonant circuit. The impedance seen looking into the *a-b* terminals may vary widely, depending upon the Q of the inductive and capacitive branches. At frequencies below the resonant frequency, the impedance of the inductive branch is low and a large current will flow through the coil. The current through the capacitive branch will be small because X_C is large at low frequencies. The line current flowing into the input terminals is therefore large. At high frequencies, the inductive branch offers a high impedance, but the capacitive branch has low impedance.

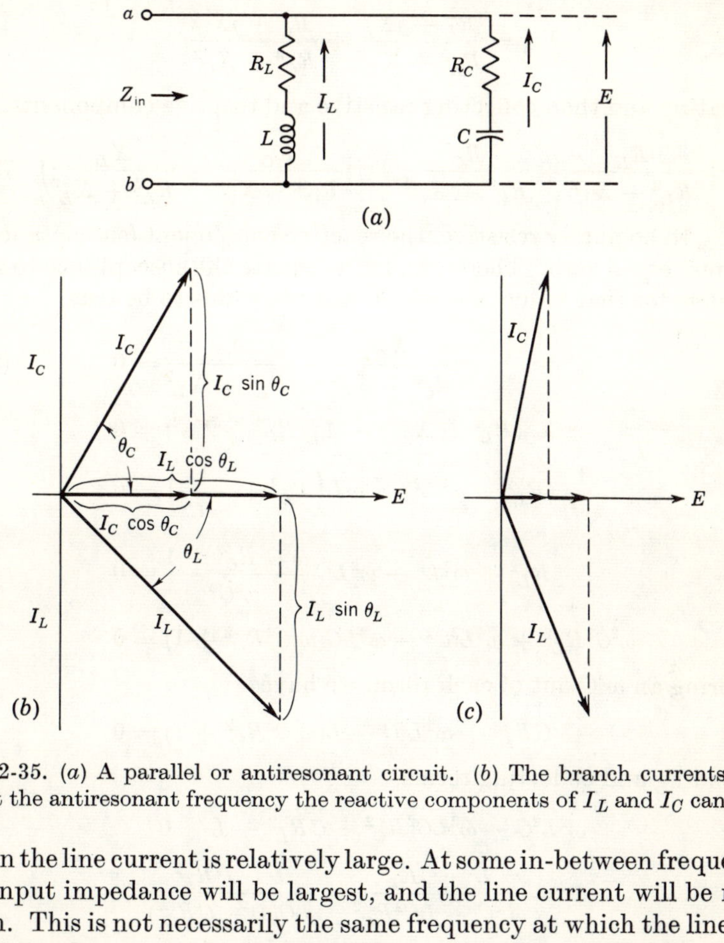

FIG. 2-35. (a) A parallel or antiresonant circuit. (b) The branch currents in a. (c) At the antiresonant frequency the reactive components of I_L and I_C cancel.

Again the line current is relatively large. At some in-between frequency, the input impedance will be largest, and the line current will be minimum. This is not necessarily the same frequency at which the line current is in phase with the applied voltage. If Q is even as low as 5, however, the error is in the neighborhood of 1 per cent, and therefore maximum impedance will be considered as occurring at the same frequency, which results in unity power factor. At some frequency which we shall define as the antiresonant frequency (f_{ar}), the impedance seen looking into terminals a-b is purely resistive. Our first objective is to determine how this frequency is related to the circuit parameters.

Since we are dealing with a parallel circuit, it is more convenient to work with admittances:

$$Y_{in} = \frac{1}{Z_{in}} = \frac{1}{Z_L} + \frac{1}{Z_C} = \frac{1}{R_L + jX_L} + \frac{1}{R_C - jX_C} \tag{2-42}$$

Rationalizing[1] each term, we obtain

$$Y_{in} = \frac{R_L - jX_L}{R_L^2 + X_L^2} + \frac{R_C + jX_C}{R_C^2 + X_C^2}$$

Separating and then collecting resistive and reactive components,

$$Y_{in} = \frac{R_L}{R_L^2 + X_L^2} + \frac{R_C}{R_C^2 + X_C^2} + j\left(\frac{X_C}{R_C^2 + X_C^2} - \frac{X_L}{R_L^2 + X_L^2}\right) \tag{2-43}$$

For Y_{in} to be purely resistive, the reactive component (susceptance) of Y_{in} must equal zero. Therefore, let us equate the susceptance to zero, and solve for that value of ω which will cause this to be true:

$$\frac{X_C}{R_C^2 + X_C^2} - \frac{X_L}{R_L^2 + X_L^2} = 0 \tag{2-44}$$

$$X_C(R_L^2 + X_L^2) - X_L(R_C^2 + X_C^2) = 0$$

$$\frac{1}{\omega C}(R_L^2 + \omega^2 L^2) - \omega L\left(R_C^2 + \frac{1}{\omega^2 C^2}\right) = 0$$

$$R_L^2 + \omega^2 L^2 - \omega^2 LC\,\frac{\omega^2 C^2 R_C^2 + 1}{\omega^2 C^2} = 0$$

$$\omega^2 C^2 R_L^2 + \omega^4 L^2 C^2 - \omega^2 LC(\omega^2 C^2 R_C^2 + 1) = 0$$

Factoring an $\omega^2 C$ out of each term, we have

$$CR_L^2 + \omega^2 L^2 C - L(\omega^2 C^2 R_C^2 + 1) = 0$$

Expanding and collecting terms,

$$\omega^2 L^2 C - \omega^2 LC^2 R_C^2 + CR_L^2 - L = 0$$

$$\omega^2 = \frac{L - CR_L^2}{L^2 C - LC^2 R_C^2} = \frac{L - CR_L^2}{LC(L - CR_C^2)}$$

$$\omega = \omega_{ar} = \frac{1}{\sqrt{LC}}\sqrt{\frac{L - CR_L^2}{L - CR_C^2}} \tag{2-45}$$

Notice that the parallel antiresonant frequency is actually dependent upon the branch resistances. In series circuits the resonant frequency

[1] Another way of expressing the fact that numerator and denominator are multiplied by the conjugate of the denominator.

was independent of the circuit resistances. Equation (2-45) is quite interesting. It indicates that resonance can be established not only by varying ω, L, or C but also by controlling R_L or R_C. This, however, is rarely done in practice, since R_L and R_C tend to deteriorate the selectivity of the circuit.

In most communications circuits, the resistance in the capacitive branch is negligible and that of the inductive branch is small if the coil Q is reasonably high. Then L will usually be much larger than CR_L^2 or CR_C^2, and Eq. (2-45) reduces to $\omega_{ar} = 1/\sqrt{LC}$, which is the same as the series-resonance case. If either CR_L^2 or CR_C^2 should be greater than L, the quantity under the radical will be negative, which results in an imaginary value of ω_{ar}. This is something we cannot physically generate and therefore has no meaning other than that no resonant condition will exist at any frequency. If $R_L = R_C$, the quantity under the radical is equal to 1, and therefore $\omega_{ar} = 1/\sqrt{LC}$ for this case. Should R_L equal R_C and also equal $\sqrt{L/C}$, ω_{ar} will just equal zero, and the circuit appears resistive at all frequencies.

In practical antiresonant circuits, the resistance in the capacitive branch is usually negligible, and Eq. (2-45) reduces to

$$\omega_{ar} = \frac{1}{\sqrt{LC}}\sqrt{\frac{L - CR_L^2}{L}}$$

which may be manipulated into

$$\omega_{ar} = \frac{1}{\sqrt{LC}}\sqrt{\frac{L^2 - LCR_L^2}{L^2}} \tag{2-45a}$$

Squaring both sides and substituting the series-resonant frequency ω_r for $1/\sqrt{LC}$, we obtain

$$\omega_{ar}^2 = \omega_r^2 \frac{L^2 - (1/\omega_r^2)R_L^2}{L^2} = \omega_r^2\left(1 - \frac{R_L^2}{\omega_r^2 L^2}\right)$$

and since $\omega_r L/R_L = Q_L$, we have

$$\omega_{ar} = \omega_r\sqrt{1 - \frac{1}{Q_L^2}} \tag{2-46}$$

which indicates that the series- and parallel-resonant frequencies are almost identical when Q_L is large in the vicinity of resonance.

Note that the value of Q_L in Eq. (2-46) is based upon the coil Q at ω_r and not at ω_{ar}. A slightly different expression for ω_{ar} is obtained if the coil Q at ω_{ar} is worked into Eq. (2-45a). Squaring both sides of (2-45a), we have

$$\omega_{ar}^2 = \frac{1}{LC}\left[\frac{L^2 - LCR^2}{L^2}\right]$$

Substituting ω_r^2 for $1/LC$, and separating the bracketed term into two terms,

$$\omega_{ar}^2 = \omega_r^2\left(\frac{L^2}{L^2} - \frac{1}{\omega_r^2}\frac{R^2}{L^2}\right)$$

Multiplying both numerator and denominator of the R^2/L^2 term by ω_{ar}^2,

$$\omega_{ar}^2 = \omega_r^2\left(1 - \frac{\omega_{ar}^2 R^2}{\omega_r^2\omega_{ar}^2 L^2}\right) = \omega_r^2\left(1 - \frac{\omega_{ar}^2}{\omega_r^2 Q_L^2}\right)$$

where Q_L this time is the coil Q determined at ω_{ar}. Solving for ω_{ar}, we obtain

$$\omega_{ar} = \frac{\omega_r}{\sqrt{1 + 1/Q_L^2}} \tag{2-46a}$$

Comparing Eqs. (2-46) and (2-46*a*), we see that even though the coil Q at ω_r may differ from that at ω_{ar}, the antiresonant frequency ω_{ar} is still essentially equal to ω_r, if Q_L is about 10 or more.

A physical interpretation of the circuit conditions at ω_{ar} may be obtained from Fig. 2-35*b*. The current in each branch is determined by the impedance of that branch. The current in the capacitive branch (I_C) will lead the applied voltage by an angle θ_C. The component of I_C which is in phase with the applied voltage (E) is $I_C \cos\theta_C$, while the quadrature (reactive) component is $I_C \sin\theta_C$. In the inductive branch, the current I_L will lag the applied voltage by an angle θ_L. We can also resolve I_L into an inphase and a quadrature component, $I_L \cos\theta_L$ and $I_L \sin\theta_L$, respectively. At ω_{ar}, the magnitudes and phase angles of I_L and I_C need not be the same, since R_L and R_C might be different. The quadrature components, $I_C \sin\theta_C$ and $I_L \sin\theta_L$, cancel, which leaves a total inphase current of $I_C \cos\theta_C$ plus $I_L \cos\theta_L$. The impedance seen looking into the *a-b* terminals of Fig. 2-35*a* at ω_r is then a finite quantity equal to the applied voltage divided by the net inphase current.

When the Q of each branch is high, so that the branch reactance is much larger than the branch resistance, the quadrature currents will be much larger than the inphase currents. This is illustrated in Fig. 2-35*c*. The net inphase current at ω_{ar} is therefore lower, which means that the input impedance at antiresonance is higher for the higher Q circuit. At ω_{ar}, the branch currents may be quite large, but their vector sum, if Q is high, results in a small line current.

2-19. A Practical Antiresonant Circuit. Figure 2-36 illustrates a practical antiresonant circuit commonly used in communications work. We have already developed Eq. (2-46) which expresses the resonant frequency of Fig. 2-35*a* with $R_C = 0$, which is the same as Fig. 2-36. Since one of our primary objectives is to gain skill in manipulating and interpreting complex-number equations, let us start from scratch in our analysis of Fig. 2-36. Let us first verify Eq. (2-46):

$$Y_{in} = \frac{1}{Z_{in}} = \frac{1}{R_L + jX_L} + \frac{1}{-jX_C} = \frac{R_L - jX_L}{R_L^2 + X_L^2} + j\frac{1}{X_C}$$

$$= \frac{R_L}{R_L^2 + X_L^2} + j\left(\frac{1}{X_C} - \frac{X_L}{R_L^2 + X_L^2}\right)$$

$$= G + jB = \frac{R_L}{R_L^2 + X_L^2} + j\frac{R_L^2 + X_L^2 - X_L X_C}{X_C(R_L^2 + X_L^2)} \tag{2-47}$$

For Z_{in} to be purely resistive the reactive component of Y_{in} must equal zero. That is,

$$\frac{R_L^2 + X_L^2 - X_L X_C}{X_C(R_L^2 + X_L^2)} = 0$$

The above will be true when the numerator is zero or

$$R_L^2 + X_L^2 - X_L X_C = 0 \tag{2-48}$$

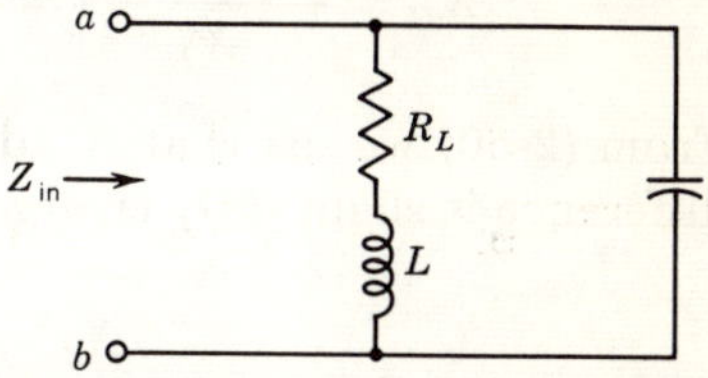

FIG. 2-36. A common antiresonant circuit.

Solving for the value of ω which makes the expression zero,

$$R_L^2 + \omega^2 L^2 - \frac{\omega L}{\omega C} = 0$$

$$\omega^2 = \frac{L}{L^2 C} - \frac{R_L^2}{L^2} = \frac{1}{LC} - \frac{R_L^2}{L^2}$$

$$\omega = \omega_{ar} = \sqrt{\frac{1}{LC} - \frac{R_L^2}{L^2}}$$

This can be manipulated into the form of Eq. (2-46) if we again let $\omega_r^2 = 1/LC$ and $Q_L = \omega L/R_L$:

$$\omega_{ar} = \sqrt{\omega_r^2 - \frac{\omega_r^2 R_L^2}{\omega_r^2 L^2}} = \sqrt{\omega_r^2 - \frac{\omega_r^2}{Q_L^2}} = \omega_r \sqrt{1 - \frac{1}{Q_L^2}}$$

which verifies Eq. (2-46).

2-20. Input Impedance at Antiresonance. The impedance at ω_{ar}, seen looking into terminals a and b of Fig. 2-36, is readily determined by examining Eq. (2-47). At ω_{ar}, the reactive component of Y_{in} is zero, which leaves the input admittance equal to G. Therefore, $Z_{in} = Z_{ar} = R_{ar} = 1/G$; where R_{ar} is the antiresonant impedance,

$$R_{ar} = \frac{R_L^2 + X_L^2}{R_L} = \frac{(R_L/X_L^2) + 1}{R_L/X_L^2} = \frac{(1/Q_L^2) + 1}{1/Q_L X_L}$$

All reactances are taken at ω_{ar}:

$$R_{ar} = X_L Q_L \left(1 + \frac{1}{Q_L^2}\right) \tag{2-49}$$

and if Q_L is large, say 10 or more,

$$R_{ar} \approx X_L Q_L$$

We can express R_{ar} in terms of X_C by solving (2-48) for X_C in terms of X_L and can thereby obtain Eq. (2-50):

$$R_L^2 + X_L^2 - X_L X_C = 0$$

$$X_C = \frac{X_L^2 + R_L^2}{X_L} = \frac{1 + R_L^2/X_L^2}{1/X_L} = X_L\left(1 + \frac{1}{Q_L^2}\right) \tag{2-50}$$

From (2-50) we see that X_C does not quite equal X_L at ω_{ar}, but the difference is slight if Q_L is large. Solving (2-50) for X_L, we obtain

$$X_L = X_C \frac{1}{1 + 1/Q_L^2} \tag{2-51}$$

and substituting in (2-49) yields

$$R_{ar} = X_C Q_L \tag{2-52}$$

Equation (2-52) can be used to express R_{ar} directly in terms of the circuit parameters, as follows:

$$R_{ar} = X_C Q_L = \frac{1}{\omega_{ar} C} \frac{\omega_{ar} L}{R_L} = \frac{L}{R_L C} \tag{2-53}$$

From (2-53), additional useful forms of expressing R_{ar} may be derived. For example,

$$R_{ar} = \frac{X_C X_L}{R_L} \approx \frac{X_C^2}{R_L} \approx \frac{X_L^2}{R_L} = Q_L^2 R_L \tag{2-54}$$

If resistance is present in the capacitive branch, it should be added to R_L when determining R_{ar}. For example, Eq. (2-53) and (2-54) would now be

$$R_{ar} = \frac{L}{(R_L + R_C)C} = \frac{X_C X_L}{R_L + R_C} = \frac{X_L^2}{R_L + R_C} = \frac{X_C^2}{R_L + R_C} \tag{2-55}$$

Equation (2-55) is not exact, but it is sufficiently accurate when the branch Q's are about 10 or more.

The total Q of the parallel circuit when resistance is present in both branches may be taken as

$$Q_T = \frac{X_L}{R_L + R_C} \tag{2-56}$$

when Q_T is 10 or more. An upper-case subscript (T) will be used to differentiate the total Q of the antiresonant circuit itself from the total *circuit* Q, designated Q_t. The total circuit Q, which includes the effect of source impedance, is developed in Sec. 2-21.

With the aid of the various expressions for R_{ar}, it may be shown that the branch currents are approximately Q_T times the line current at ω_{ar}. Let I_g equal the line current forced by some source E_g into the *a-b* terminals of Fig. 2-35*a*. The voltage across the *a-b* terminals is E_{ba}. We may write the following proportion:

$$\frac{E}{I_g} = R_{ar}$$

$$\frac{E}{I_L} = Z_L$$

If Q_T is high, therefore,

$$\frac{I_L}{I_g} = \frac{R_{ar}}{Z_L} \approx \frac{Q_T X_L}{X_L} = Q_T \tag{2-57}$$

Since $X_L \approx X_C$ at ω_{ar}, when Q_T is large, we see that

$$\frac{I_C}{I_g} = Q_T \tag{2-58}$$

2-21. Selectivity of an Antiresonant Circuit. The selectivity of a series-resonant circuit was described in terms of the bandwidth relative to the resonant frequency. This in turn was shown to be a function of the circuit Q, as expressed in the equation

$$\frac{\omega_2 - \omega_1}{\omega_r} = \frac{1}{Q_t}$$

It would be desirable to develop an analogous relationship for the selectivity of an antiresonant circuit.

Our first consideration will be how to represent the source which drives the antiresonant circuit. That is, shall a current or voltage source be used? The series-resonant circuit had to be driven from a low-impedance or constant-voltage source if the circuit was to be selective. If the series circuit were driven from a constant-current (high-impedance) source, the current would be independent of conditions in the tuned circuit, and no resonant current rise would manifest itself. An antiresonant circuit, on the other hand, should be driven from a constant current rather than from a constant voltage source. When the coil Q is high, the antiresonant circuit exhibits its maximum impedance at the frequency of antiresonance. If a constant-voltage source is connected across the antiresonant circuit, there will be no detectable change in voltage across it, since the voltage is held constant. The line current, it is true, will drop to a minimum value at ω_{ar}, but this is not discernible to a voltmeter across the input terminals of the antiresonant circuit. Now if some very high resistance is placed in series with the constant-voltage source, converting it to a constant-current source, the voltage across the antiresonant circuit will rise to a maximum at ω_{ar} because of the constant current flowing through the maximum impedance R_{ar} at ω_{ar}. Therefore, we may conclude that a current source is the ideal way to drive an antiresonant circuit.

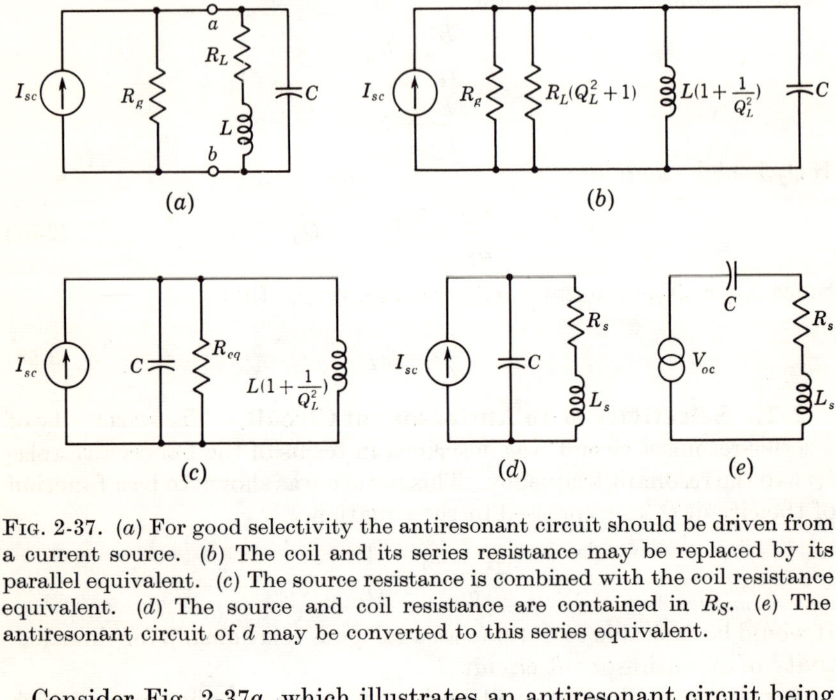

FIG. 2-37. (*a*) For good selectivity the antiresonant circuit should be driven from a current source. (*b*) The coil and its series resistance may be replaced by its parallel equivalent. (*c*) The source resistance is combined with the coil resistance equivalent. (*d*) The source and coil resistance are contained in R_S. (*e*) The antiresonant circuit of *d* may be converted to this series equivalent.

Consider Fig. 2-37*a*, which illustrates an antiresonant circuit being driven by a current source having an internal impedance R_g. Let us try to convert the circuit of Fig. 2-37*a* to some equivalent series circuit which we already know how to analyze.

The coil and its series resistor may be resolved into a parallel resistor and inductor, as shown in Fig. 2-37*b*. The two resistors form an equivalent resistor

$$R_{eq} = \frac{R_g R_L(Q_L^2 + 1)}{R_g + R_L(Q_L^2 + 1)}$$

where $Q_L = \omega_L/R_L$. It is also expedient to relocate the position of the capacitor to that shown in Fig. 2-37*c*. Now the inductor and resistor of Fig. 2-37*c* are reconverted to an equivalent series combination, as shown in Fig. 2-37*d*. Using the series-parallel conversions of Sec. 2-12 again, we have

$$R_S = \frac{R_{eq}}{Q_P^2 + 1}$$

where
$$Q_P = \frac{R_{eq}}{\omega L(1 + 1/Q_L^2)}$$

and
$$L_S = \frac{L_P}{1 + 1/Q_P^2}$$

With R_g absorbed into R_S, we can convert Fig. 2-37*d* into a voltage-source equivalent circuit which drives an equivalent series-resonant circuit, as shown in Fig. 2-37*e*. The selectivity of a series-resonant circuit is given by

$$\frac{\omega_2 - \omega_1}{\omega_r} = \frac{1}{Q_t}$$

where $Q_t = \omega L_S/R_S$, $\omega_2 - \omega_1$ is the bandwidth between the half-power points, and ω_r is the series-resonant frequency for Fig. 2-37*e*, which is equal to ω_{ar} of Fig. 2-37*a*. We must evaluate Q_t:

$$Q_t = \frac{\omega L_S}{R_S} = \frac{\omega L_P/(1 + 1/Q_P^2)}{R_{eq}/(Q_P^2 + 1)} = \frac{\omega L_P Q_P^2/(Q_P^2 + 1)}{R_{eq}/(Q_P^2 + 1)} = \frac{\omega L_P Q_P^2}{R_{eq}}$$

but
$$L_P = L\left(1 + \frac{1}{Q_L^2}\right) = L\,\frac{Q_L^2 + 1}{Q_L^2}$$

$$R_{eq} = \frac{R_g R_L(Q_L^2 + 1)}{R_g + R_L(Q_L^2 + 1)}$$

$$Q_P = \frac{R_{eq}}{\omega L(1 + 1/Q_L^2)} = \frac{R_{eq} Q_L^2}{\omega L(Q_L^2 + 1)}$$

Substituting for R_{eq} in Q_P,

$$Q_P = \frac{\dfrac{R_g R_L(Q_L^2 + 1)Q_L^2}{R_g + R_L(Q_L^2 + 1)}}{\omega L(Q_L^2 + 1)} = \frac{R_g R_L Q_L^2}{\omega L[R_g + R_L(Q_L^2 + 1)]}$$

Returning to the expression for Q_t, and substituting for L_P, R_{eq}, and Q_P,

$$Q_t = \frac{\omega L\,\dfrac{Q_L^2 + 1}{Q_L^2}\,\dfrac{R_g^2 R_L^2 Q_L^4}{\omega^2 L^2[R_g + R_L(Q_L^2 + 1)]^2}}{\dfrac{R_g R_L(Q_L^2 + 1)}{R_g + R_L(Q_L^2 + 1)}}$$

$$= \frac{R_g R_L Q_L^2}{\omega L[R_g + R_L(Q_L^2 + 1)]}$$

Since $R_L/\omega_L = 1/Q_L$,

$$Q_t = \frac{R_g Q_L}{R_g + R_L(Q_L^2 + 1)} \qquad \text{and} \qquad R_L = \frac{\omega L}{Q_L}$$

$$Q_t = \frac{R_g Q_L}{R_g + (\omega L/Q_L)(Q_L^2 + 1)} = \frac{R_g Q_L^2}{Q_L R_g + \omega L(Q_L^2 + 1)}$$

Therefore,

$$\frac{\omega_2 - \omega_1}{\omega_r} = \frac{1}{Q_t} = \frac{Q_L R_g + \omega L(Q_L^2 + 1)}{R_g Q_L^2} = \frac{1}{Q_L} + \frac{\omega L(Q_L^2 + 1)}{R_g Q_L^2}$$

and since we are considering ω at ω_{ar},

$$\frac{\omega_2 - \omega_1}{\omega_r} = \frac{1}{Q_t} = \frac{1}{Q_L} + \frac{\omega_{ar} L}{R_g} \frac{Q_L^2 + 1}{Q_L^2} \tag{2-59}$$

Equation (2-59) is exact. If $Q_L = 10$, $Q_L^2 + 1 \approx Q_L^2$, and $\omega_{ar} L Q_L \approx R_{ar}$,

$$\frac{\omega_2 - \omega_1}{\omega_r} = \frac{1}{Q_t} \approx \frac{1}{Q_L} + \frac{\omega_{ar} L}{R_g} \tag{2-59a}$$

$$\frac{\omega_2 - \omega_1}{\omega_r} = \frac{1}{Q_t} \approx \frac{1}{Q_L} + \frac{1}{Q_L} \frac{R_{ar}}{R_g} \tag{2-59b}$$

Equations (2-59), (2-59*a*), and (2-59*b*) indicate that the greater R_g is, the larger the circuit Q will be. Let us try some illustrative problems.

Problem 1. In the circuit of Fig. 2-38*a*, determine ω_{ar}, f_{ar}, R_{ar}, E_o, I_g, I_C, I_L at ω_{ar}, Q_t, the per unit bandwidth, and the upper and lower half-power frequencies.

Solution. As a first step, thevenize R_1, R_2, and E_g to obtain Fig. 2-38*b*. Since the Q of the antiresonant circuit to the right of the *a-b* terminals is high, $Q_L = 20$, the antiresonant frequency will almost equal the series resonant frequency of C and L:

$$\omega_{ar} \approx \frac{1}{\sqrt{LC}} = \frac{1}{\sqrt{10^{-8}(1)}} = 10^4 \text{ radians/sec}$$

therefore, $$f_{ar} = \frac{\omega_{ar}}{2\pi} = 1{,}590 \text{ cps}$$

The apparent resistance of the antiresonant circuit at ω_{ar} is

$$R_{ar} \approx \omega_{ar} L Q_L = 10^4(1)(20) = 200 \text{ kilohms}$$

Since the tuned circuit appears as a pure resistance, R_{ar} at ω_{ar}, E_o may be determined by the voltage-divider relationship

$$E_o = \frac{V_{oc}(R_{ar})}{Z_{th} + R_{ar}} = \frac{1(200)}{225} = 0.89 \text{ volt}$$

The generator current in Fig. 2-38*a* will be

$$I_g = \frac{E_g - E_o}{R_1} = \frac{2.00 - 0.89 \text{ volts}}{50 \text{ kilohms}} = 22.2\ \mu\text{a}$$

The capacitor current is

$$I_C = \frac{E_o}{X_C} = 0.89(10^4)(10^{-8}) = 0.089 \text{ ma} \approx 90\ \mu\text{a}$$

or, by Eq. (2-58) of Sec. 2-20 and Fig. 2-38*b*,

$$I_C = I_{th}Q_T = I_{th}Q_L$$

where I_{th} in Fig. 2-38*b* equals

$$I_{th} = \frac{V_{oc} - E_o}{Z_{th}} = \frac{1.00 - 0.89 \text{ volt}}{25 \text{ kilohms}} = 4.4\ \mu\text{a}$$

Therefore, $I_C = 4.4\ \mu\text{a}\ (20) = 88\ \mu\text{a}$, which checks closely. Notice that I_g and I_{th} are not the same. Thévenin's theorem must be used with caution if conditions

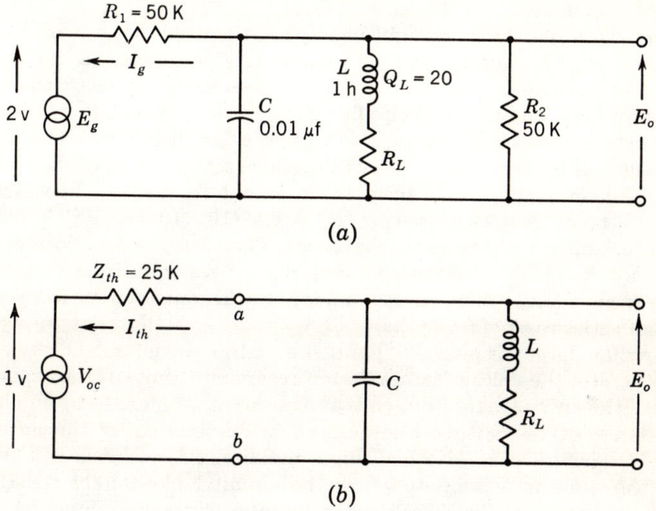

FIG. 2-38. (*a*) This circuit is to be analyzed in detail. (*b*) Simplified by Thévenin's theorem.

inside the equivalent circuit rather than the load are to be determined. The current in the inductive branch will essentially equal the capacitor current, since Q_T (Q_T same as Q_L, since there is no resistance in the capacitive branch) is large. The per unit bandwidth is, by Eq. (2-59*b*),

$$\frac{\omega_2 - \omega_1}{\omega_r} = \frac{1}{Q_t} = \frac{1}{Q_L} + \frac{1}{Q_L}\frac{R_{ar}}{R_g} = \frac{1}{20} + \frac{1}{20}\left(\frac{200}{25}\right)$$

where $R_g = Z_{th}$,

$$\frac{\omega_2 - \omega_1}{\omega_r} = \frac{1}{Q_t} = 0.05 + 0.05(8) = 0.45$$

The over-all circuit Q is $Q_t = 1/0.45 = 2.2$. The bandwidth is $\Delta\omega = \omega_2 - \omega_1 = 0.45(\omega_r) = 4{,}500$ radians/sec, or 718 cps. Since the circuit Q is only 2.2, the upper and lower half-power frequencies will not be symmetrically located about f_{ar}. As a first approximation, however, we may write

$$f_2 = f_{ar} + \frac{\Delta f}{2} = 1{,}590 + \frac{718}{2} = 1{,}950 \text{ cps}$$

and
$$f_1 = f_{ar} - \frac{\Delta f}{2} = 1{,}590 - \frac{718}{2} = 1{,}231 \text{ cps}$$

Problem 2. It is desired to measure the Q of a particular 100-mh inductor at a frequency of 10 kc. The only items of equipment available are a fixed frequency 10-kc oscillator which delivers 10 volts when open-circuited and whose internal resistance is 20 kilohms, an a-c VTVM, and a large variety of fixed and variable capacitors. Assume that the distributed capacitance of the coil windings is negligible and that the coil Q is at least 10. Describe a possible means of evaluating Q_L with this equipment.

Solution. Our first thought might be to place the coil in series with the capacitors and to vary the latter until a resonant rise in voltage across the coil is detected. At resonance, the coil voltage is essentially equal to the circuit Q multiplied by the generator open-circuit voltage. This method, however, is a poor one. Since the generator resistance is so high, the circuit Q for the series-resonant arrangement will be very low. This will produce a feeble voltage rise which will not even be a maximum at exactly the resonant frequency. Specifically, the circuit Q will be at least as low as $\omega L/R_g = 6.28(10^4)(10^{-1})/2(10^4) = 0.31$, if the inductor is assumed to have zero resistance. Therefore, let us consider an antiresonant circuit. It is because the majority of vacuum-tube and transistor amplifiers exhibit high source resistances that antiresonant circuits are more prevalent than series-resonant circuits.

Suppose that the coil is placed in parallel with the capacitors and the voltmeter is used to monitor the voltage across the antiresonant circuit. If the capacitance is varied until the antiresonant voltage is a maximum, we must be at a point where the impedance of the antiresonant circuit is maximum. If the antiresonant circuit Q—Q_L in this case—is at all high, maximum impedance will essentially occur at the same frequency at which the circuit looks purely resistive. The voltmeter reading at ω_{ar} is therefore the voltage across R_{ar}. Since R_{ar} and the generator resistance (R_g) are in series, they comprise a simple voltage divider, for which we may write

$$\frac{E_o}{E_g} = \frac{R_{ar}}{R_g + R_{ar}}$$

where E_o is the voltmeter reading. Solving for R_{ar},

$$E_oR_g + E_oR_{ar} = E_gR_{ar}$$

$$R_{ar} = \frac{E_oR_g}{E_g - E_o}$$

But if Q_L is 10 or more, $R_{ar} = Q_LX_L$, and X_L is a known quantity, since L and f are known. Therefore,

$$Q_L = \frac{R_{ar}}{X_L} = \frac{E_oR_g}{(E_g - E_o)X_L}$$

For this particular coil, $X_L = 6.28(10^4)(10^{-1}) = 6.28$ kilohms. If E_o, for example, were to read 8 volts,

$$Q_L = \frac{8(20)}{(10 - 8)(6.28)} \approx 12.8 \qquad \textit{Ans.}$$

Problem 3. Express the selectivity of an antiresonant circuit when R_{ar} is designed for maximum power transfer.

Solution. The selectivity is in general given by

$$\frac{f_2 - f_1}{f_r} = \frac{1}{Q_t} = \frac{1}{Q_L} + \frac{1}{Q_L}\frac{R_{ar}}{R_g}$$

But for the maximum power transfer $R_{ar} = R_g$. Therefore,

$$\frac{f_2 - f_1}{f_r} = \frac{1}{Q_t} = \frac{2}{Q_L}$$

and the circuit Q, that is, Q_t, is seen to be half of the coil Q.

Problem 4. What is the voltage across the antiresonant circuit at ω_{ar} when R_{ar} matches R_g?

Solution. Since $R_{ar} = R_g$, for maximum power transfer the output voltage will equal one-half of the generator voltage:

$$E_o = E_g \frac{R_{ar}}{R_g + R_{ar}}$$

and since $R_g = R_{ar}$,

$$E_o = E_g \frac{R_{ar}}{2R_{ar}} = \frac{E_g}{2}$$

2-22. Universal Resonance Curve for the Antiresonant Circuit. The universal resonance curve in Fig. 2-30 may also be applied to the antiresonant circuit of Fig. 2-39, if Q_L is high. The validity of this statement may be proven in the following manner.

The general expression for the input impedance in Fig. 2-39 is

$$Z_{in} = \frac{(R + jX_L)(-jX_C)}{R_L + j(X_L - X_C)} \tag{2-60}$$

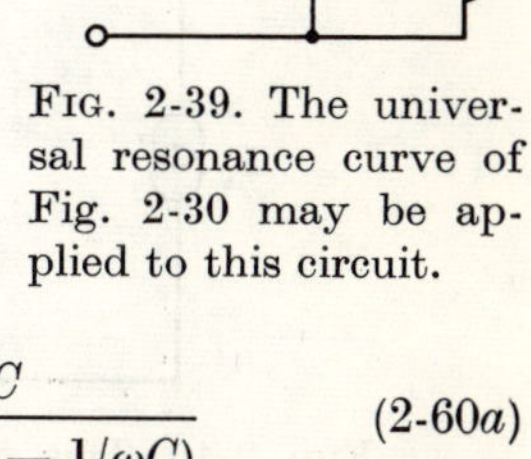

FIG. 2-39. The universal resonance curve of Fig. 2-30 may be applied to this circuit.

If Q_L is high, say 10 or more, $X_L \gg R$ and Eq. (2-60) simplifies to

$$Z_{in} = \frac{jX_L(-jX_C)}{R + j(X_L - X_C)} = \frac{L/C}{R + j(\omega L - 1/\omega C)} \tag{2-60a}$$

The R term in the denominator could not be discarded, since $X_L - X_C$ becomes very small in the vicinity of resonance. At $\omega = \omega_{ar} = \omega_r$, the reactive component in the denominator of (2-60a) equals zero, and we again derive $Z_{ar} = L/R_rC$, where R_r is the coil resistance at ω_r. Then

$$\frac{Z_{in}}{Z_{ar}} = \frac{L/C}{R + j(\omega L - 1/\omega C)} \cdot \frac{1}{L/R_rC} = \frac{R_r}{R + j(\omega L - 1/\omega C)} \tag{2-61}$$

For a series-resonant circuit we have

$$\frac{I}{I_r} = \frac{E/Z_{in}}{E/Z_r} = \frac{Z_r}{Z} = \frac{R_r}{R + j(\omega L - 1/\omega C)}$$

which is identical in form with Eq. (2-61). Therefore, the universal resonance curve may be applied to an antiresonant circuit, with the following considerations:

1. The Q is high, say 10 or more.
2. The vertical scale (ordinate values) indicates Z_{in}/R_{ar}.
3. The Q is constant over the frequency range of interest.
4. The phase angles for the antiresonant circuit are exactly the opposite of those shown for series resonance. For example, at frequencies below ω_{ar}, the reactance of the capacitive branch is high, and most of the current flows through the coil. The impedance is predominantly inductive, and therefore Z_{in}/Z_{ar} will have an angle between zero and 90°. On the other hand, a series-resonant circuit will exhibit capacitive reactance below ω_r, and therefore I/I_r is between zero and minus 90°. Since the phase angles in the universal resonance curve are shown for the series circuit, the reader must remember to invert the angles when dealing with antiresonant circuits.
5. The universal resonance curve in itself does not include the effect of the source resistance.

Problem 1. The circuit of Fig. 2-40 is to be used in a radio command system. The philosophy behind the proposal is as follows. Source E_g represents an

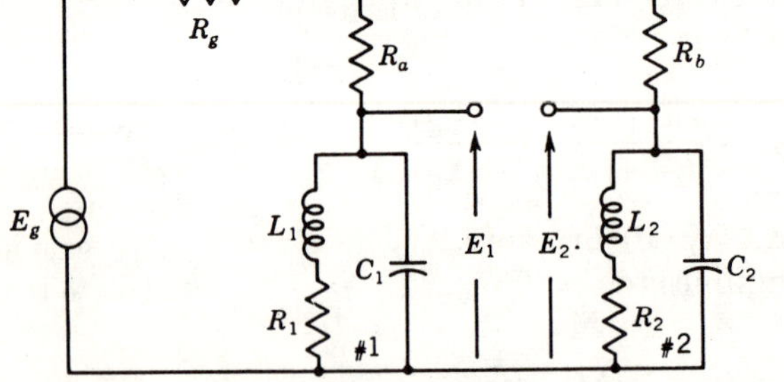

FIG. 2-40. This channel selector uses antiresonant circuits.

amplified tone signal obtained by demodulating a modulated r-f carrier. If the tone signal is used to drive a cathode follower which has a low output impedance, R_g will be low. Resistors R_a and R_b are used to isolate the number 1 and number

2 channels from each other and thereby prevent interaction through the common source impedance R_g. This is why a low R_g is desirable. Resistors R_a and R_b, which are relatively large, also serve to convert the voltage source E_g to more of a current source, which improves the selectivity of the antiresonant circuits.

When the modulating frequency for E_g is adjusted at the transmitter to that of the channel 1 antiresonant frequency, E_1 will be a maximum. Voltage E_1 may be subsequently amplified and rectified and used to drive a control relay. Channel 2 will behave in a similar manner.

With the aid of the universal resonance curve, roughly sketch on linear graph paper the values of E_1 and E_2 versus frequency. Assume the following parameters:

$$E_g = 20 \text{ volts} \qquad C_1 = 0.06\ \mu\text{f}$$
$$R_g = 100 \text{ ohms} \qquad L_1 = 0.1 \text{ henry}$$
$$R_a = 100 \text{ kilohms} \qquad C_2 = 0.01\ \mu\text{f}$$
$$R_b = 100 \text{ kilohms} \qquad L_2 = 0.1 \text{ henry}$$
$$Q_1 = Q_2 = 20$$

where Q_1 and Q_2 are the coil Q values.

Solution. Since the coil Q's are high, we may write

$$f_{r_1} = \frac{1}{2\pi\sqrt{L_1C_1}} = 2.18 \text{ kc} \approx 2 \text{ kc}$$

$$f_{r_2} = \frac{1}{2\pi\sqrt{L_2C_2}} = 5 \text{ kc}$$

$$R_{ar_1} = 2\pi f_{r_1} L_1 Q_1 = 6.28(2)10^3(0.1)(20) = 25 \text{ kilohms}$$

$$R_{ar_2} = 2\pi f_{r_2} L_2 Q_2 = 6.28(5)10^3(0.1)20 = 62.8 \text{ kilohms}$$

At f_{r_1}

$$E_1 = \frac{R_{ar_1}E_g}{R_{ar_1} + R_a} = \frac{25 \text{ kilohms}}{125 \text{ kilohms}} \times 20 \text{ volts} = 4.0 \text{ volts}$$

At f_{r_2}

$$E_2 = \frac{R_{ar_2}}{R_{ar_2} + R_b} E_g = \frac{62.8 \text{ kilohms}}{162.8 \text{ kilohms}} \times 20 \text{ volts} = 7.7 \text{ volts}$$

Table 2-6 may be compiled for channel 1.

An arbitrary value of Q_1k is chosen from the universal resonance curve, say $Q_1k = -3$. With $Q_1 = 20$, we find $k = -3/20 = -0.15$. Since $k = (f - f_r)/f_r$, we can solve for f_1 which yields $f = f_r(1 + k) = 2 \text{ kc } (1 - 0.15) = 1.7$ kc. From the curve, we see that $Z/Z_{ar} = 0.15\underline{/+84^\circ}$, or $Z = 25(0.15\underline{/84^\circ}) = 3.7\underline{/84^\circ} \approx j3.7$ kilohms:

$$E_1 = \frac{E_gZ}{R_a + Z} \approx \frac{20(3.7\underline{/84^\circ})}{100 + j3.7} \approx \frac{20(3.7\underline{/84^\circ})}{100}$$

$$|E_1| = 0.7 \text{ volt}$$

A similar table of values may be constructed for channel 2.

TABLE 2-6

Q_1k	−3	−2	−1	−0.5	0
k	−0.15	−0.1	−0.05	−0.025	0
f_{kc}	1.7	1.8	1.9	1.95	2
Z/Z_{ar} ..	$0.15\underline{/84°}$	$0.23\underline{/80°}$	$0.43\underline{/65°}$	$0.71\underline{/45°}$	1
Zk	$3.7\underline{/84°}$	$6.3\underline{/80°}$	$10.7\underline{/65°}$	$17.7\underline{/45°}$	25
$\lvert E_1\rvert$	0.7	1.3	2.1	3.16	4
Q_1k	0.5	1	2	3	0
k	0.025	0.05	0.1	0.15	0
f_{kc}	2.05	2.1	2.2	2.3	2
Z/Z_{ar} ..	$0.71\underline{/-45°}$	$0.43\underline{/-65°}$	$0.25\underline{/-73°}$	$0.19\underline{/-75°}$	1
Zk	$17.7\underline{/-45°}$	$10.7\underline{/-65°}$	$6.3\underline{/-73°}$	$4.75\underline{/-75°}$	25
$\lvert E_1\rvert$	3.16	2.1	1.3	0.91	4

The response of both channels is sketched in Fig. 2-41. What could be done to increase the output of channel 1 at ω_{r1}? Let us check the curves against calculated values for the half-power frequencies. For channel 1,

$$\frac{f_2 - f_1}{f_{r_1}} = \frac{1}{Q_t} = \frac{1}{Q_L} + \frac{1}{Q_L}\frac{R_{ar}}{R_a} = 0.05 + 0.05\left(\frac{25}{100}\right) = 0.0625$$

therefore,

$$Q_t = 1/0.0625 = 16$$

and

$$f_2 - f_1 = 0.0625(2 \text{ kc}) = 125 \text{ cps}$$

therefore, $f_2 \approx 2{,}060$ cps and $f_1 \approx 1{,}940$ cps, which checks closely with the 70.7 per cent level for channel 1.

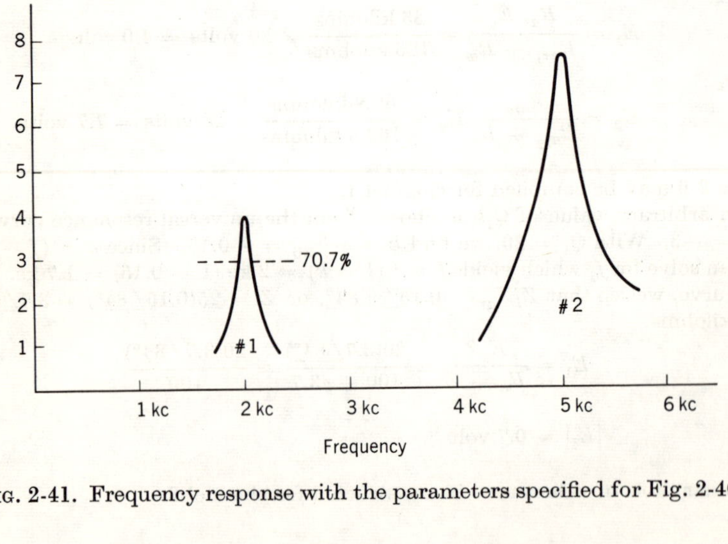

FIG. 2-41. Frequency response with the parameters specified for Fig. 2-40.

Problem 2. Figure 2-42*a* illustrates a frequency selective circuit in which a source E_g, having an internal impedance R_g, drives an antiresonant circuit (No. 1). The voltage developed across the antiresonant circuit is amplified by an amplifier A_v, which presents an infinite impedance at its input terminals and an output impedance of R_1 ohms at its output terminals. The open-circuit output voltage of

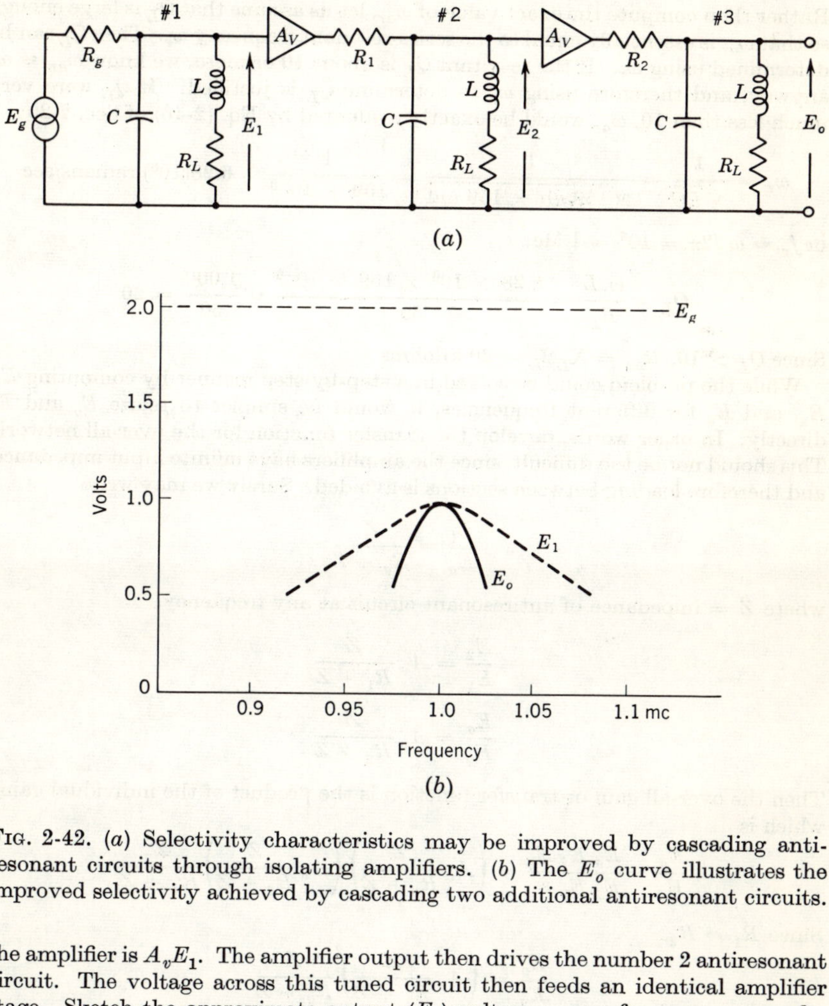

FIG. 2-42. (*a*) Selectivity characteristics may be improved by cascading antiresonant circuits through isolating amplifiers. (*b*) The E_o curve illustrates the improved selectivity achieved by cascading two additional antiresonant circuits.

the amplifier is A_vE_1. The amplifier output then drives the number 2 antiresonant circuit. The voltage across this tuned circuit then feeds an identical amplifier stage. Sketch the approximate output (E_o) voltage versus frequency curve for this tuned amplifier if

$$R_g = 20 \text{ kilohms} \qquad L = 159\ \mu\text{h}$$

$$R_1 = R_2 = 100 \text{ kilohms} \qquad R_L = 50 \text{ ohms}$$

$$A_v = 10 \qquad C = 159\ \mu\mu\text{f}$$

$$E_g = 2 \text{ volts}$$

Solution. First, determine the Q of the antiresonant circuit to see if the approximation formulas can be used. Since there is no resistance in the capacitive branch,

$$Q_t = Q_L = \frac{\omega_{ar} L}{R_L}$$

Rather than compute the exact value of ω_{ar}, let us assume that Q_L is large enough so that ω_{ar} is essentially equal to the series-resonant frequency ω_r. Then Q_L can be determined using ω_r. If the resultant Q_L is about 10 or more, we know $\omega_{ar} \approx \omega_r$ anyway, and therefore using ω_r to determine Q_L is justified. If Q_L were very much less than 10, ω_{ar} would be exactly evaluated by Eq. (2-46) of Sec. 2-20:

$$\omega_r = \frac{1}{\sqrt{LC}} = \frac{1}{\sqrt{159\ \mu\text{h} \times 159\ \mu\mu\text{f}}} = \frac{1}{159 \times 10^{-9}} = 6.28(10^6)\ \text{radians/sec}$$

or $f_r = \omega_r/2\pi = 10^6 = 1$ Mc:

$$Q_L = \frac{\omega_r L}{R_L} = \frac{6.28 \times 10^6 \times 159 \times 10^{-6}}{50} = \frac{1{,}000}{50} = 20$$

Since $Q_L > 10$, $R_{ar} = X_L Q_L = 20$ kilohms.

While the problem could be solved in a step-by-step manner by computing E_1, E_2, and E_o for different frequencies, it would be simpler to relate E_o and E_g directly. In other words, develop the transfer function for the over-all network. This should not be too difficult, since the amplifiers have infinite input impedance, and therefore loading between sections is avoided. Surely we may write

$$\frac{E_1}{E_g} = \frac{Z}{R_g + Z}$$

where Z = impedance of antiresonant circuit at any frequency.

$$\frac{E_2}{E_1} = A_v \frac{Z}{R_1 + Z}$$

$$\frac{E_o}{E_2} = A_v \frac{Z}{R_2 + Z}$$

Then the over-all gain or transfer function is the product of the individual gains, which is

$$\frac{E_o}{E_g} = \frac{E_o}{E_2}\frac{E_2}{E_1}\frac{E_1}{E_g} = \left(A_v \frac{Z}{R_1 + Z}\right)\left(A_v \frac{Z}{R_2 + Z}\right)\frac{Z}{R_g + Z}$$

Since $R_1 = R_2$,

$$\frac{E_o}{E_g} = \left(A_v \frac{Z}{R_1 + Z}\right)^2 \frac{Z}{R_g + Z}$$

and since an approximate rather than an exact result is desired, we may take advantage of the fact that $R_1 \gg Z$ because the maximum value of Z occurs at ω_r, which means that $Z = R_{ar} = 20$ kilohms and that R_1 is almost 10 times larger. Therefore, with this approximation, we have

$$\frac{E_o}{E_g} = \left(A_v \frac{Z}{R_1}\right)^2 \frac{Z}{R_g + Z}$$

At $\omega = \omega_r$, $Z = R_{ar}$, and

$$\frac{E_o}{E_g} = \left[\frac{10(20)}{100}\right]^2 \frac{20}{20 + 20} = 2$$

or $E_o = 0.5E_g = 1$ volt.

Let us now locate some other points with the aid of the universal resonance curve in Fig. 2-30. Choosing a frequency corresponding to, say, $Qk = -0.5$, we have $Z/Z_{ar} = 0.7\underline{/+45°}$. Therefore, $Z = 0.7\underline{/+45°}$ (20 kilohms) $= 14\underline{/+45°}$ $= 10 + j10$ kilohms, and

$$\frac{E_o}{E_g} = \left(\frac{A_v Z}{R_1}\right)^2 \frac{Z}{R_g + Z} = \left[\frac{10(14\underline{/+45°})}{100}\right]^2 \frac{14\underline{/+45°}}{20 + 10 + j10}$$

$$= \frac{8.5\underline{/+135°}}{30 + j10} = \frac{8.5\underline{/+135°}}{31.6\underline{/+18.4°}} = 0.27\underline{/+116.6°}$$

therefore $|E_o| = 0.27(2 \text{ volts}) = 0.54$ volt.

Since $k = (\omega - \omega_r)\omega_r$ and $Qk = -0.5$, we can determine the corresponding frequency at which the gain is $0.27\underline{/+116.6°}$.

$$k = \frac{-0.5}{Q} = -0.025$$

$$\frac{\omega - \omega_r}{\omega_r} = k$$

$$\omega - \omega_r = \omega_r k$$

$$\omega = \omega_r(k + 1)$$

or $f = f_r(k + 1) = 1$ Mc $(-0.025 + 1) = 0.975$ Mc, which is 0.025 Mc below the resonant frequency. Because of the symmetry of the universal resonance curve the output voltage magnitude at 1.025 Mc will also be 0.54 volt.

The output voltage at 975, 1,000, and 1,025 kc is plotted in Fig. 2-42*b*. Notice how selective the circuit seems to be. In order to gauge the improvement in selectivity due to the second and third antiresonant circuits, let us determine how E_1 varies with frequency. At ω_r,

$$E_1 = E_g \frac{R_{ar}}{R_g + R_{ar}} = 2 \text{ volts} \frac{(20 \text{ kilohms})}{20 + 20 \text{ kilohms}} = 1 \text{ volt}$$

Since at 0.975 and 1.025 Mc we found $Qk = -0.5$, and $Z/Z_{ar} = 0.7\underline{/\pm 45°}$, we can readily determine E_1 at these frequencies from

$$E_1 = E_g \frac{Z}{R_g + Z}$$

and, at 0.975 Mc,

$$E_1 = 2 \frac{14\underline{/+45°}}{20 + 10 + j10} = \frac{28\underline{/+45°}}{31.6\underline{/+18.4°}} = 0.885\underline{/+26.6°} \text{ volt}$$

or $|E_1| \approx 0.89$ volt. Again, because of the symmetry of the resonance curve, the output at 1.025 Mc will also equal 0.89 volt.

Problem 3. An antiresonant circuit is driven by an amplifier which develops an open-circuit voltage of 20 volts. The internal impedance is 50 kilohms. Design the antiresonant circuit so that 2 volts appears across it at a resonant frequency of 5 kc and only 1 volt at 4.8 and 5.2 kc.

Solution. If at f_r the tuned circuit voltage is to be 2 volts, we may write

$$\frac{E_o}{E_g} = \frac{2}{20} = \frac{R_{ar}}{R_g + R_{ar}}$$

Solving for R_{ar}, we have

$$R_{ar} = R_g \frac{E_o}{E_g - E_o} = \frac{50 \text{ kilohms}(2)}{20 - 2} = 5.55 \text{ kilohms}$$

Since the source impedance R_g, is about 10 times larger than R_{ar}, we may consider the generator as a constant current source of $I_g = 20$ volts/50 kilohms $= 0.4$ ma. The output voltage developed across the antiresonant circuit is, therefore, $E_o = I_g Z$. Then at 4.8 and 5.2 kc,

$$|E_o| = I_g|Z| = 1 \text{ volt} = 0.4 \text{ ma}|Z|$$

and therefore $|Z| = 2.5$ kilohms. This determines the $|Z|$-to-Z_{ar} ratio as $|Z|/Z_{ar} = 0.45$. From the universal resonance curve we see that the corresponding values of $Qk = \pm 1$, respectively. Since

$$k = \frac{f - f_r}{f_r} = \frac{4.8 - 5.0}{5} = -0.04$$

at 4.8 kc and $+0.04$ at 5.2 kc, we can determine Q, which is the same as Q_L if capacitor losses are negligible:

$$Q_L = \frac{Qk}{k} = \frac{1}{0.04} = 25$$

But $R_{ar} = X_L Q_L$, and therefore $X_L = 5{,}550/25 = 220$ ohms. Then,

$$L = \frac{X_L}{\omega_r} = \frac{220}{2\pi(5 \text{ kc})} = \frac{220}{31.4(10^3)} = 7 \text{ mh}$$

Since $X_C = X_L$,

$$C = \frac{1}{\omega_r X_C} = \frac{1}{31.4(10^3)(220)} = 0.144\ \mu\text{f}$$

The coil Q has already been specified at 25. The coil resistance is, therefore, $R_L = X_L/Q_L = 220/25 = 8.8$ ohms.

2-23. Impedance Transformation (the L Section). A problem which frequently arises is that of coupling a generator to a load where the load resistance is not what we want the generator to see. For example, a final r-f amplifier in a transmitter may require a 5-kilohm load for optimum operation. The transmission line which it is to feed may only be a 300-ohm line. The problem then is how the 300-ohm line can be matched to the final amplifier. It should be understood that impedance matching does not necessarily mean matching for maximum power transfer.

Consider the tuned circuit of Fig. 2-43. Source E_g of internal resistance R_g will look into a desired antiresonant resistance R_{ar} if L and C are correctly chosen for the frequency at which E_g operates. The load R_o, which is less than R_g, essentially looks into a low-impedance series-resonant circuit made up of L and C. By making the load a branch resistance of an antiresonant circuit, it will appear as a higher resistance at the input (1-2) terminals. That this is true may be shown in the following manner. Redrawing Fig. 2-43 as 2-44*a* clearly illustrates the antiresonant circuit seen looking into the 1-2 terminals. Usually the coil resistance is so much less than R_o that it may be neglected. If the coil resistance is not negligible, it can be lumped with R_o for purposes of analysis. The series combination of L and R_o can be converted

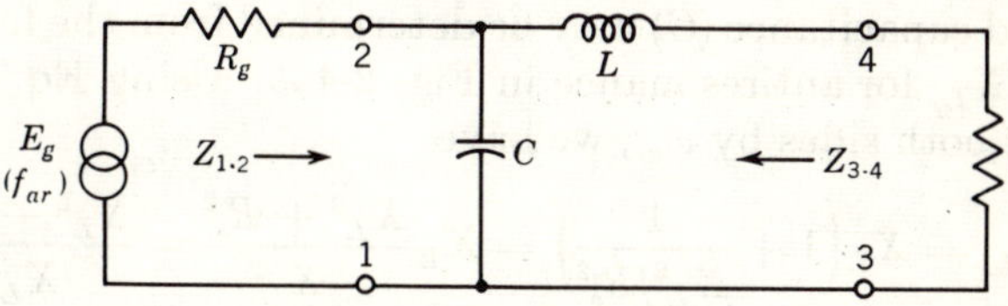

FIG. 2-43. At a single frequency L and C may be so chosen that R_g sees some desired impedance at the 1-2 terminals with a given value of R_o.

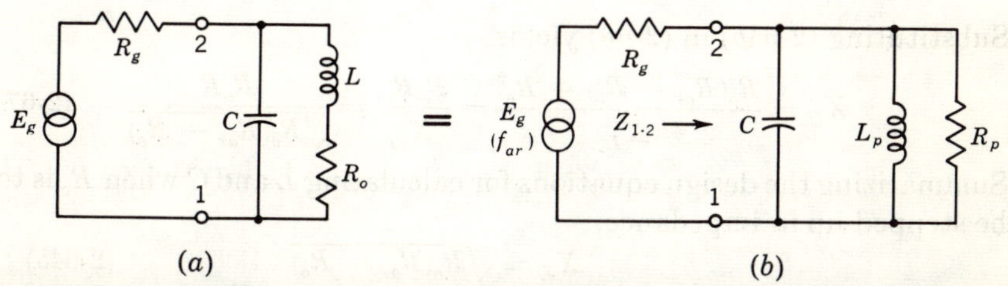

FIG. 2-44. (*a*) The idea is to make the load resistance (R_o) appear in a branch of an antiresonant circuit. (*b*) R_o is transformed into R_p with the aid of the series-parallel conversions.

to an equivalent parallel combination as shown in Fig. 2-44*b*. By means of the series-parallel conversions, we may write for Fig. 2-44*b*,

$$R_P = R_o(1 + Q_S^2)$$

where $$Q_S = \frac{X_L}{R_o} \tag{2-62}$$

and $$L_P = L\left(1 + \frac{1}{Q_S^2}\right) \tag{2-63}$$

At the frequency of E_g we want $Z_{1\text{-}2}$ in Fig. 2-44*b* to equal some desired resistance value, which is R_{ar}. If C and L are correctly chosen, X_C and X_{L_P} will cancel at ω_{ar}. The impedance at ω_{ar} seen looking into the 1-2 terminals is then R_P and by Eq. (2-62) we see that R_P must be greater than R_o. Therefore, at ω_{ar},

$$Z_{1\text{-}2} = R_{ar} = R_P = R_o\left(1 + \frac{X_L^2}{R_o^2}\right) = \frac{X_L^2 + R_o^2}{R_o} \tag{2-64a}$$

Solving (2-64*a*) for X_L

$$X_L^2 = R_o R_{ar} - R_o^2$$

$$X_L = \sqrt{R_o(R_{ar} - R_o)} \qquad (2\text{-}64b)$$

Since ω_{ar} is known, we may determine the required value of L from

$$L = \frac{X_L}{\omega_{ar}} \qquad (2\text{-}64c)$$

The required capacitance (C) may be determined from the fact that X_C must equal X_{L_P} for antiresonance in Fig. 2-44*b*. Using Eq. (2-63) and multiplying both sides by ω_{ar}, we have

$$X_{LP} = X_L\left(1 + \frac{1}{X_L^2/R_o^2}\right) = X_L \frac{X_L^2 + R_o^2}{X_L^2} = \frac{X_L^2 + R_o^2}{X_L} \qquad (2\text{-}65)$$

But at ω_{ar},

$$X_C = X_{LP} = \frac{X_L^2 + R_o^2}{X_L} \qquad (2\text{-}66)$$

Substituting (2-64*b*) in (2-66) yields,

$$X_C = \frac{R_o(R_{ar} - R_o) + R_o^2}{X_L} = \frac{R_o R_{ar}}{X_L} = \frac{R_o R_{ar}}{\sqrt{R_o(R_{ar} - R_o)}} \qquad (2\text{-}67)$$

Summarizing the design equations for calculating L and C when R_o is to be stepped up in impedance,

$$X_L = \sqrt{R_o(R_{ar} - R_o)} \qquad (2\text{-}64b)$$

$$X_C = \frac{R_o R_{ar}}{X_L} = \frac{R_o R_{ar}}{\sqrt{R_o(R_{ar} - R_o)}} \qquad (2\text{-}67)$$

If a situation arises where the load impedance is larger than that which we desire the generator to look into, some means of stepping the impedance downward must be effected. A convenient scheme to use this time is to place the load in parallel with an inductor. This parallel combination is equivalent to an inductor and resistor in series, where the series resistor is of lower value than the original load. The resultant series RL combination can then be resonated by a capacitor, to appear as a resistance to the generator.

Consider Fig. 2-45*a*, which may be converted to Fig. 2-45*b* by means of the series-parallel conversions:

$$L_S = L \frac{1}{1 + 1/Q_P^2}$$

where

$$Q_P = \frac{R_o}{X_L} \qquad (2\text{-}68)$$

$$R_S = R_o \frac{1}{Q_P^2 + 1} \qquad (2\text{-}69)$$

Inspection of Fig. 2-45*b* tells us that C must be chosen such that $X_C = X_{L_s}$ if $Z_{1\text{-}2}$ is to be resistive. From (2-68),

$$X_{L_s} = X_L \frac{1}{1 + X_L^2/R_o^2} = \frac{X_L R_o^2}{R_o^2 + X_L^2}$$

Therefore

$$X_C = \frac{X_L R_o^2}{R_o^2 + X_L^2} \tag{2-68a}$$

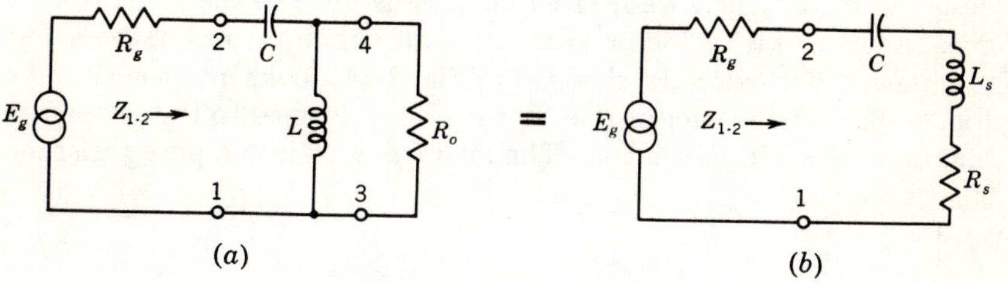

FIG. 2-45. Placing R_o in parallel with L makes it equivalent to a much smaller series resistor. (*b*) R_s is the transformed value of R_o.

Equation (2-69) may be used to relate R_o and X_L in terms of $Z_{1\text{-}2} = R_S$, where R_S is the lower resistance we want to see.

$$R_S = R_o \frac{1}{R_o^2/X_L^2 + 1} = \frac{R_o X_L^2}{R_o^2 + X_L^2} \tag{2-69a}$$

Solving for X_L,

$$R_o^2 R_S + R_S X_L^2 - R_o X_L^2 = 0$$

$$X_L^2(R_S - R_o) = -R_o^2 R_S$$

Multiplying both sides by -1 and solving for X_L,

$$X_L = \sqrt{\frac{R_o^2 R_S}{R_o - R_S}} = R_o \sqrt{\frac{R_S}{R_o - R_S}} \tag{2-69b}$$

Substituting (2-69*b*) into (2-68*a*) yields

$$X_C = \frac{\left(R_o \sqrt{\dfrac{R_S}{R_o - R_S}}\right) R_o^2}{R_o^2 + R_o^2 \dfrac{R_S}{R_o - R_S}} = \frac{R_o^3 \sqrt{\dfrac{R_S}{R_o - R_S}}\,(R_o - R_S)}{R_o^3 - R_o^2 R_S + R_o^2 R_S}$$

$$= (R_o - R_S)\sqrt{\frac{R_S}{R_o - R_S}} = (R_o - R_S)\frac{\sqrt{R_S}}{\sqrt{R_o - R_S}}$$

$$= \sqrt{R_o - R_S}\,\sqrt{R_S} = \sqrt{R_S(R_o - R_S)} \tag{2-68b}$$

Summarizing the design equations for calculating L and C when R_o is to be stepped down in impedance,

$$X_C = \sqrt{R_S(R_o - R_S)} \tag{2-68b}$$

$$X_L = R_o\sqrt{\frac{R_S}{R_o - R_S}} \tag{2-69b}$$

It should be understood that the L sections we have discussed were designed to present a desired value of resistance to the generator. It does not necessarily follow that the load simultaneously sees a pure resistance. For example, the load in Fig. 2-44*a* looks into the circuit of Fig. 2-46*a*. The capacitor and R_g may be converted to the series combination shown in Fig. 2-46*b*. The load will look into a pure resistance

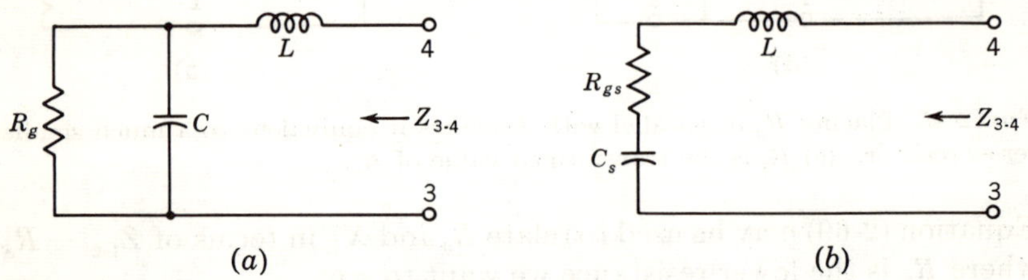

FIG. 2-46. (*a*) At times we are more concerned with what the load looks into. (*b*) The series equivalent that the load sees.

only if X_L equals X_{C_S}. There is no guarantee, however, that at the operating frequency of E_g this condition will exist. If R_g should be large compared with X_C, then C_S will essentially equal C; and ω_r, the series-resonant frequency looking into the 3-4 terminals, will be $1/\sqrt{LC}$. At the same time, if R_o is small compared with X_L looking into the 1-2 terminals of Fig. 2-44*a*, the value of L_P will approximately equal L looking into the 1-2 terminals of Fig. 2-44*b*. Then the antiresonant frequency ω_{ar} into the 1-2 terminals will essentially equal ω_r, the series-resonant frequency into the 3-4 terminals; and the network looks resistive from either side.

As an exercise, the reader might develop some design equations for assuring that the load, rather than the generator, always looks into a pure resistance. If it is necessary to match both R_g and R_o simultaneously, a transformer or more complicated network must be used. An excellent treatment is found in "Communication Engineering" by Everitt and Anner.[1]

Let us try a few illustrative problems.

[1] W. L. Everitt and G. E. Anner, "Communication Engineering," 3d ed., McGraw-Hill Book Company, Inc., New York, 1956.

Problem 1. The final amplifier of a 1-Mc transmitter is to work into a 10-kilohm resistive load. The antenna that it feeds looks like 100 ohms at the operating frequency. Design an L section to match the antenna to the final.

Solution. Since the 100-ohm load must be stepped up to 10 kilohms, the circuit of Fig. 2-44*a* will be used. Using Eqs. (2-64*b*) and (2-67), we have

$$X_L = \sqrt{R_o(R_{ar} - R_o)} = \sqrt{100(10^4 - 100)} = \sqrt{99(10^4)} \approx 1 \text{ kilohm} \qquad (2\text{-}64b)$$

$$L = \frac{X_L}{\omega_{ar}} = \frac{1 \text{ kilohm}}{6.28(10^6)} = 159\ \mu\text{h}$$

$$X_C = \frac{R_o R_{ar}}{X_L} = \frac{100(10 \text{ kilohms})}{1 \text{ kilohm}} = 1 \text{ kilohm} \qquad (2\text{-}67)$$

$$C = \frac{1}{\omega_{ar} X_C} = \frac{1}{6.28(10^6)(10^3)} = 159\ \mu\mu\text{f}$$

Problem 2. A 300-ohm transmission line is to feed a 72-ohm antenna at 100 Mc. It is desired that the transmission line look into a 300-ohm load. Design a suitable matching section.

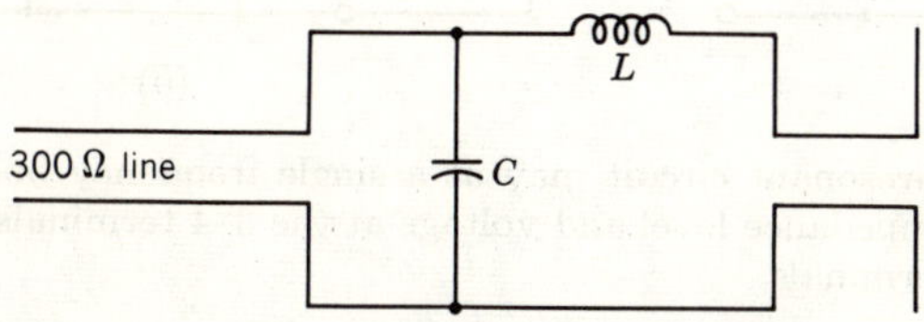

FIG. 2-47. A problem in impedance transformation.

Solution. The 72-ohm load must be stepped up to 300 ohms for the line to be terminated in the desired manner. An L section as shown in Fig. 2-47 will be used. By Eq. (2-64*b*),

$$X_L = \sqrt{72(300 - 72)} = 15.2 \text{ ohms}$$

$$L = \frac{X_L}{\omega_{ar}} = \frac{15.2}{6.28(10^8)} = 0.0241\ \mu\text{h}$$

By Eq. (2-67), $X_C = 300(72)/15.2 = 1{,}430$ ohms, and therefore $C = 1.1\ \mu\mu\text{f}$. Admittedly, such a matching section would be difficult to construct.

2-24. The Resonant Transformer.

At its antiresonant frequency, a tapped antiresonant circuit exhibits many characteristics associated with a transformer. For example, the driving voltage may be stepped up or down, and the impedance levels correspondingly varied.

Consider Fig. 2-48*a*, which illustrates a simple resonant transformer. To begin with, let's analyze the circuit for voltage and for impedance levels. Starting at the 1-2 terminals, the figure is redrawn as Fig. 2-48*b*. Looking into the 1-2 terminals, E_g sees an antiresonant circuit. The presence of capacitor C_1 in the inductive branch should not be disturbing to us. Surely there will be a frequency at which the net reactance in the inductive branch will be equal and opposite to that of capacitor C_2. If the circuit Q to the right of the 1-2 terminals is large, then, at the antiresonant frequency seen looking into the 1-2 terminals, I_g will be at a minimum, and I_1 and I_2 will be at a maximum. Furthermore, I_1 and I_2 will approximately equal each other, and be Q_t times larger than I_g. In many texts I_1 and I_2 are considered as a single current which circulates around the so-called tank circuit C_2, L, and C_1.

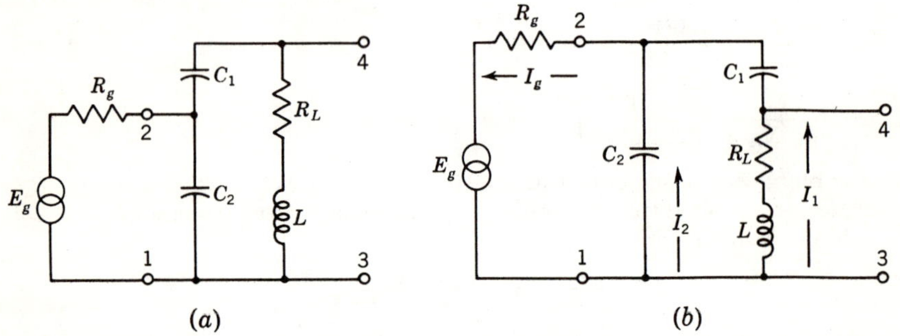

FIG. 2-48. (a) The antiresonant circuit may at a single frequency be used as a transformer. (b) The impedance level and voltage at the 3–4 terminals is greater than that at the 1–2 terminals.

Since I_g is much less than I_2 when Q_t is large, the voltage across C_2 is essentially

$$E_{1\text{-}2} = I_2 X_{C_2} \tag{2-70}$$

The voltage between the 3-4 terminals is essentially

$$E_{3\text{-}4} = I_1 X_L \tag{2-71}$$

But $I_1 = I_2$, for all practical purposes, when Q_t is large. Therefore,

$$E_{3\text{-}4} = I_2 X_L \tag{2-71a}$$

Now we have stated that if the circuit is to be resistive between the 1-2 terminals, that is, antiresonant, X_{C_2} must equal X_L minus X_{C_1}. In equation form,

$$X_{C_2} = X_L - X_C \tag{2-72a}$$

Therefore

$$X_L = X_{C_2} + X_{C_1} \tag{2-72b}$$

Substituting (2-72b) into (2-71a), we obtain

$$E_{3\text{-}4} = I_2(X_{C_1} + X_{C_2}) \tag{2-71b}$$

The output to input voltage ratio is then

$$\frac{E_{3\text{-}4}}{E_{1\text{-}2}} = \frac{I_2(X_{C_1} + X_{C_2})}{I_2 X_{C_2}} = \frac{X_{C_1} + X_{C_2}}{X_{C_2}} \tag{2-73a}$$

Since $X_{C_1} + X_{C_2}$ is greater than X_{C_2} only, $E_{3\text{-}4}$ must be greater than $E_{1\text{-}2}$. The voltage transformation ratio may be expressed in terms of the capacitors directly:

$$\frac{E_{3\text{-}4}}{E_{1\text{-}2}} = \frac{1/\omega_{ar}C_1 + 1/\omega_{ar}C_2}{1/\omega_{ar}C_2} = \frac{(C_1 + C_2)/C_1C_2}{1/C_2} = \frac{C_1 + C_2}{C_1} \tag{2-73b}$$

The impedance transformation ratio can be readily determined from the following considerations. The power put into the network at the 1-2 terminals is

$$P_{1\text{-}2} = \frac{(E_{1\text{-}2})^2}{Z_{1\text{-}2}} \tag{2-74}$$

If Q_t is large, the network losses are negligible, and an equal amount of power is made available at the 3-4 terminals. Remember, pure reactive elements cannot absorb power. Therefore,

$$P_{3\text{-}4} = \frac{(E_{3\text{-}4})^2}{Z_{3\text{-}4}} \tag{2-75}$$

but $P_{3\text{-}4} = P_{1\text{-}2}$, and therefore,

$$\frac{(E_{1\text{-}2})^2}{Z_{1\text{-}2}} = \frac{(E_{3\text{-}4})^2}{Z_{3\text{-}4}} \tag{2-76a}$$

$$\frac{(E_{3\text{-}4})^2}{(E_{1\text{-}2})^2} = \frac{Z_{3\text{-}4}}{Z_{1\text{-}2}} = \left(\frac{C_1 + C_2}{C_1}\right)^2 \tag{2-76b}$$

Equation (2-76*b*) may also be derived in the following manner. In Fig. 2-48*a*,

$$Z_{1\text{-}2} = \frac{-jX_{C_2}(R_L + jX_L - jX_{C_1})}{R_L + j(X_L - X_{C_1} - X_{C_2})} \tag{2-77a}$$

If Q_L is large, $X_L \gg R_L$:

$$Z_{1\text{-}2} = \frac{-jX_{C_2}j(X_L - X_{C_1})}{R_L + j(X_L - X_{C_1} - X_{C_2})} = \frac{X_{C_2}(X_L - X_{C_1})}{R_L + j(X_L - X_{C_1} - X_{C})} \tag{2-77b}$$

Rationalizing and simplifying (2-77*b*), we obtain

$$Z_{1\text{-}2} = \frac{X_{C_2}(X_L - X_{C_1})R_L}{R_L^2 + (X_L - X_{C_1} - X_{C_2})^2} + j\frac{X_{C_2}(X_L - X_{C_1})(X_L - X_{C_1} - X_{C_2})}{R_L^2 + (X_L - X_{C_1} - X_{C_2})^2} \tag{2-77c}$$

Since $Z_{1\text{-}2}$ is to be resistive at ω_{ar}, the reactive part of (2-77*c*) must equal zero. This will occur if the numerator of the j term is zero. That is,

$$X_{C_2}(X_L - X_{C_1})(X_L - X_{C_1} - X_{C_2}) = 0 \tag{2-77d}$$

Any value of ω which makes any of the factors of (2-77*d*) equal to zero will be a resonant frequency. A resonant frequency is one which makes $Z_{1\text{-}2}$ resistive. The frequency at which X_{C_2} is zero must be infinite. This solution has no physical significance other than that it is so high that C_2 is a short circuit.

For the factor $X_L - X_{C_1}$ to equal zero, X_L must equal X_{C_1}. This corresponds to some frequency at which the $C_1 - L$ branch in Fig. 2-48*b* is series resonant, and terminal 2 is virtually shorted to terminal 1 through this branch. This, however, is not the antiresonant point we are seeking.

Equating the last factor to zero yields the antiresonant condition, since $X_{C_2} = X_L - X_{C_1}$. A most interesting observation is that the third factor also represents the resonant condition when looking into the 3-4 terminals. That is, $X_L - X_{C_1} - X_{C_2} = 0$ can be written as $X_L = X_{C_1} + X_{C_2}$, which must be true for resonance at the 3-4 terminals. The following rule is therefore indicated. If a tapped parallel-resonant circuit appears antiresonant between any two taps, it will also appear resonant between any other pair of taps.

Dropping the j term in (2-77*c*) at ω_{ar} leaves

$$Z_{1\text{-}2} = \frac{X_{C_2}(X_L - X_{C_1})R_L}{R_L{}^2 + (X_L - X_{C_1} - X_{C_2})^2} \tag{2-77e}$$

and, since $X_L - X_{C_1} = X_{C_2}$, and $X_L - X_{C_1} - X_{C_2} = 0$ at ω_{ar}, (2-77*e*) simplifies to

$$Z_{1\text{-}2} = \frac{X_{C_2}{}^2}{R_L} \tag{2-77f}$$

Impedance $Z_{3\text{-}4}$ is readily determined, since it must equal R_{ar}, the antiresonant impedance across the total circuit. Therefore,

$$Z_{3\text{-}4} = R_{ar} = Q_t X_C = \frac{X_L X_C}{R_L} = \frac{X_C{}^2}{R_L} \tag{2-77g}$$

where $X_C = X_{C_1} + X_{C_2}$. Therefore,

$$\frac{Z_{3\text{-}4}}{Z_{1\text{-}2}} = \frac{X_C{}^2/R_L}{(X_{C_2})^2/R_L} = \frac{(1/\omega_{ar}C_1 + 1/\omega_{ar}C_2)^2}{(1/\omega_{ar}C_2)^2} = \frac{(C_1 + C_2)^2}{(C_1C_2)^2}C_2{}^2$$

$$= \left(\frac{C_1 + C_2}{C_1}\right)^2$$

which verifies (2-76*b*).

The resonant transformer of Fig. 2-48*a* utilized a capacitive divider. Essentially the same action may be obtained by dividing up the inductor as in Fig. 2-49. Again assuming a high-Q circuit in which the tank currents are Q times larger than the generator current and that L_1 and L_2 are not magnetically coupled, we may write

$$E_{3\text{-}4} = I_2 X_C = I_1 X_C \qquad (2\text{-}78)$$

and

$$E_{1\text{-}2} = I_1 X_{L_1} \qquad (2\text{-}79)$$

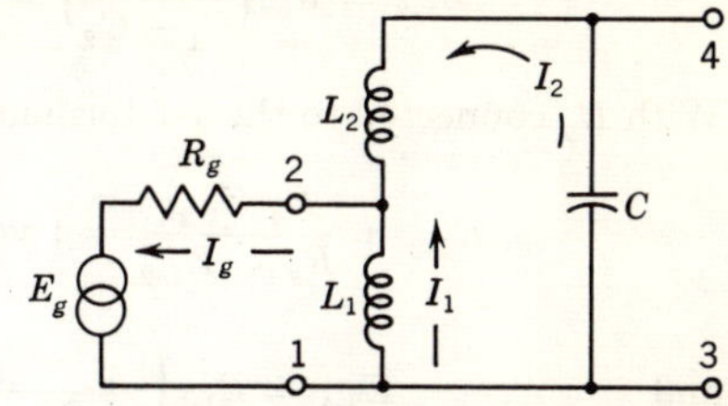

FIG. 2-49. Impedance transformation may also be achieved by tapping the inductor.

Since the circuit is resonant at the 1-2 terminals, it is also resonant at the 3-4 terminals, and

$$X_C = X_{L_1} + X_{L_2} \qquad (2\text{-}80)$$

Substituting (2-80) in (2-78) yields

$$E_{3\text{-}4} = I_1(X_{L_1} + X_{L_2}) \qquad (2\text{-}78a)$$

Therefore,

$$\frac{E_{3\text{-}4}}{E_{1\text{-}2}} = \frac{X_{L_1} + X_{L_2}}{X_{L_1}} = \frac{L_1 + L_2}{L_1} \qquad (2\text{-}81)$$

The impedance ratio will again be the square of the voltage ratio, if losses are negligible:

$$\frac{Z_{3\text{-}4}}{Z_{1\text{-}2}} = \left(\frac{L_1 + L_2}{L_2}\right)^2 \qquad (2\text{-}82)$$

The impedance of the generator can be stepped down by connecting it to the 3-4 terminals of either Fig. 2-48*a* or Fig. 2-49.

Problem 1. An r-f amplifier of 5-kilohm internal resistance drives the tapped resonant circuit shown in Fig. 2-50. If the open-circuit voltage of the amplifier is 1 volt, determine $E_{3\text{-}4}$ at ω_{ar}. What is the total circuit Q and bandwidth if the coil Q equals 20?

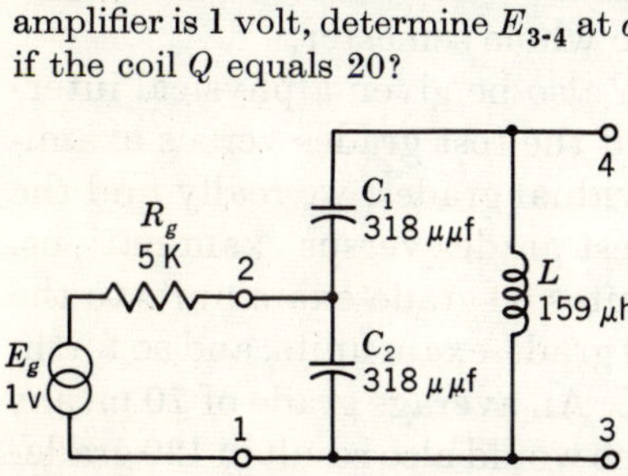

FIG. 2-50. The output voltage and selectivity at the 3-4 terminals are to be determined.

Solution. Looking into the 1-2 terminals,

$$f_{ar} = \frac{1}{2\pi\sqrt{LC_{eq}}}$$

$$\text{where } C_{eq} = \frac{C_1 C_2}{C_1 + C_2} = 159\ \mu\mu f$$

$$f_{ar} = \frac{0.159}{\sqrt{(159\ \mu h)(159\ \mu\mu f)}} = 1\ \text{Mc}$$

At ω_{ar},

$$X_L = \omega_{ar} L = 6.28 \times 10^6 \times 159 \times 10^{-6} = 1{,}000\ \text{ohms}$$

Therefore, $Z_{3\text{-}4} = R_{ar} = Q_L X_L = 20$ kilohms at the 3-4 terminals with the generator disconnected. Looking into the 1-2 terminals, we should see R_{ar} stepped down to

$$Z_{1\text{-}2} = R_{ar}\left(\frac{C_1}{C_1 + C_2}\right)^2 = 20 \text{ kilohms}\left(\frac{C_1}{2C_1}\right)^2 = 5 \text{ kilohms}$$

With E_g connected to the 1-2 terminals, the 1-2 terminal voltage will be, at ω_{ar},

$$E_{1\text{-}2} = \frac{E_g(Z_{1\text{-}2})}{R_g + Z_{1\text{-}2}} = 1 \text{ volt} \times \frac{5 \text{ kilohms}}{5 + 5 \text{ kilohms}} = 0.5 \text{ volt}$$

and

$$E_{3\text{-}4} = E_{1\text{-}2}\left(\frac{C_1 + C_2}{C_1}\right) = 0.5\left(2\frac{C_1}{C_1}\right) = 1.0 \text{ volt}$$

The 5-kilohm generator connected across the 1-2 terminals is equivalent to a 20-kilohm ($R_g{}^1$) generator across the 3-4 terminals, since the impedance ratio varies with the square of the voltage ratio (Eq. 2-76*b*). Therefore, looking into the 3-4 terminals, we have

$$\frac{1}{Q_T} = \frac{1}{Q_L} + \frac{1}{Q_L}\frac{R_{ar}}{R_g{}^1} = \frac{1}{20} + \frac{1}{20}\frac{20 \text{ kilohms}}{20 \text{ kilohms}} = 0.1$$

or $Q_T = 10$; also, $\Delta f = f_2 - f_1 = f_{ar}/Q_T = 100$ kc.

2-25. The Average-value Concept. When dealing with sine waves, we freely use the terms average and rms values. The true meaning of these terms is often obscure, however. In this section we shall therefore investigate the general meaning of average value and then see how it applies to sine waves.

Suppose you took six examinations in school and received grades of 80, 70, 90, 60, 40, and 80. Your average grade would then be determined by adding the individual scores together and dividing this sum by the number of exams. In this case, the sum total is 420, and 420 divided by six exams gives an average grade of 70. In the back of our minds lies the thought that if we had scored 70 on each of the six exams, we would have done just as well for the whole semester.

The concept of an average value can also be given a physical interpretation. In Fig. 2-51 we have plotted the test grades versus examinations. When we sum up all the individual grades, we really find the crosshatched area under the curve of test grades versus examinations. The first test with a score of 80 contributes 80 grade-exam units to the total area. The next test contributes 70 grade-exam units, and so forth. All six tests total 420 grade-exam units. An average grade of 70 means that a score of 70 on each of the six exams would also result in 420 grade-exam units. That is, if we draw a rectangle having a base equal to six exams and a height or grade of 70, the enclosed area would be equal to that bounded by the actual grade-versus-exam curve. We see, then, that if we had a curve relating y to x and wished to determine the average value of y in some interval, as between $X = X_1$ and $X = X_2$, as in Fig. 2-52, we must:

1. In some manner find the area bounded by the curve which lies between the limits $X = X_1$ and $X = X_2$. This is the shaded area of Fig. 2-52.

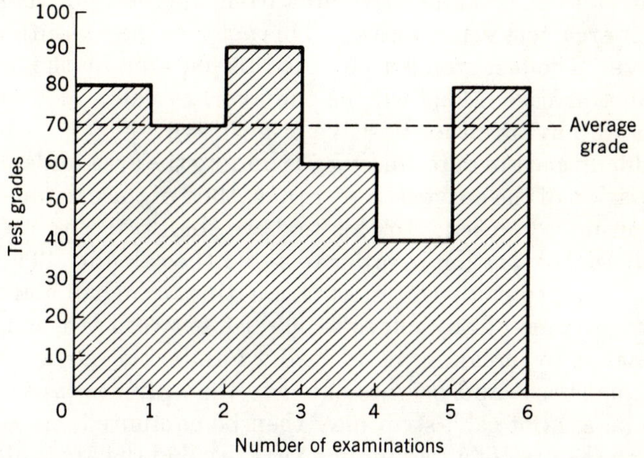

FIG. 2-51. The area bounded by the average grade line and number of examinations will equal the shaded area.

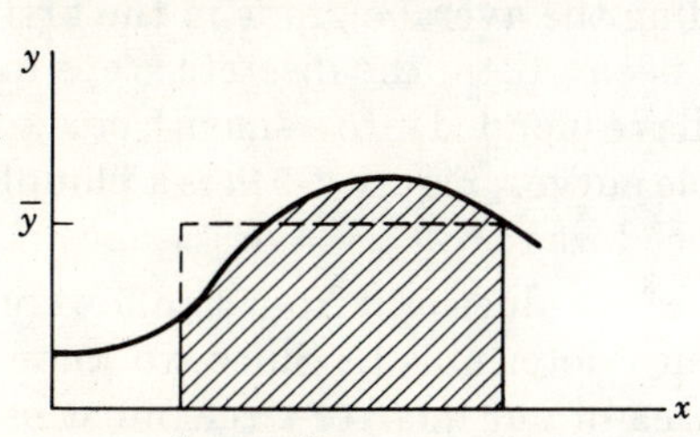

FIG. 2-52. The average value of the curve between X_1 and X_2 will bound an area equal to the shaded area.

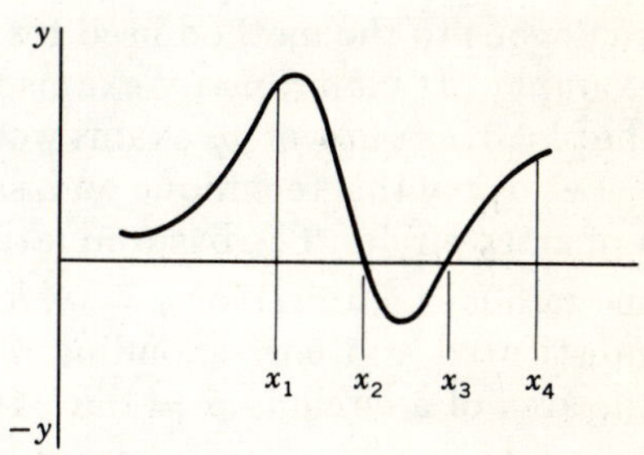

FIG. 2-53. A function with positive and negative areas.

2. Divide this area by the length of the base, which is $X_2 - X_1$ in Fig. 2-52. The result of this division is the average value of y, sometimes written $\bar{y}$. Remember, this is the average value in the interval $X_2 - X_1$.

3. Realize that the average value, $\bar{y}$, represents the height of a rectangle having the same base and an area equal to that bounded by the original curve.

In general, suppose we had a curve which plotted y as a function of x, as in Fig. 2-53. Let us say that we want to know the average y value in the interval between $x = x_1$ and $x = x_4$. To do this, we must, in some way, first find the area bounded by the curve and between the values of $X = x_1$ and x_2. This area is positive, since both x and y are positive during this interval. The next step is to find the area between x_2 and x_3. This area is negative, since the x values are positive and the y values are negative. Then, find the area between x_3 and x_4, which is again positive. Now, subtract the total negative area from the total positive area and get the net area between x_1 and x_4. This net area may be either positive or negative. The last step is to divide this net area by the base limits, $x_4 - x_1$ in this case. This will be the average value of y during this interval and may be either positive or negative.

The difficult part in determining $\bar{y}$ is finding the area under the curve. If the equation of the curve is known, a method of the calculus known as integration may be used. In this text we shall resort to a method of numerical approximation in order to avoid the calculus. The reader is urged, however, to study this most fascinating branch of mathematics.[1] The average value of any curve which lies between two limits can be approximated by breaking it up into a number of narrow rectangles having equal bases and heights equal to the ordinate (y) value of the curve. The area of each strip may then be computed, since we know how to find the area of a rectangle. Next, we add the areas of the strips to get the approximate area under the curve. Last, we divide this area by the number of strips considered to get the average height. This is analogous to the method used for finding the average grade in the first example. If many, many exams had been taken, the discrete steps in the plot of grades versus exams would have blended into a smooth curve.

Let us try this technique on a sample curve. Figure 2-54*a* is a plot of a quarter-circle. The basic equation used was $x^2 + y^2 = R^2$ (where R, the radius is 5 units), or $y = \sqrt{R^2 - x^2}$. Values of x from 1 to 5 were substituted and corresponding y values determined. Since we know the area of a circle is $A = \pi R^2$, the area of our quarter-circle must be

$$A = \frac{\pi R^2}{4} = \frac{3.14 \times 25}{4} = 19.6 \text{ units}$$

Now let us apply the approximation method and see if we arrive at the same area. In Fig. 2-54*b* we have divided the curve into rectangular strips, each with a base width interval (Δx) of 1 unit and a height equal to the ordinate value at the start of the interval. The area of the first rectangle is $y\,\Delta x = 5 \times 1 = 5$ units. The second rectangle area is $4.9 \times 1 = 4.9$ units, and so forth. Tabulating these calculations, we have Table 2-7.

[1] See A. E. Richmond, "Calculus for Electronics," McGraw-Hill Book Company, Inc., New York, 1958.

The total area computed in this manner is 21.5 units, as compared with the correct value of 19.6 units. This is fairly close and probably adequate for many purposes. Greater accuracy could be achieved by using more strips or by keeping the same number of strips but judiciously choosing the ordinate value within any strip interval. For

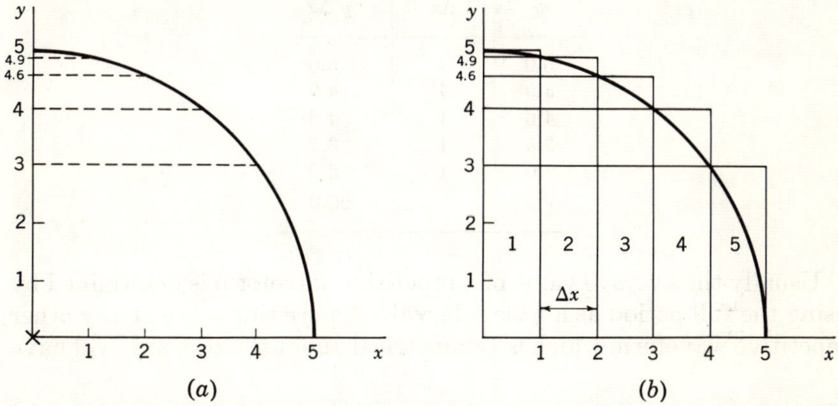

FIG. 2-54. (*a*) The area bounded by one-fourth of a circle is to be determined. (*b*) The area under a curve may be approximated by totaling the area of the rectangular strips.

example, in Fig. 2-54*b*, the ordinate values 5.0, 4.9, 4.6 used in the first three strips are fairly close to the average ordinate values during those intervals. In the fourth strip, however, a choice of $y = 3.5$ rather than 4

TABLE 2-7

y	Δx	$y\,\Delta x$
5.0	1	5.0
4.9	1	4.9
4.6	1	4.6
4.0	1	4.0
3.0	1	3.0
		21.5

would be closer to the average value of y during the fourth interval. Likewise, a choice of $y = 2$ would be a better choice than $y = 3$ during the fifth interval. With these choices of y, we have Table 2-8.

We now have 20.0 area units as compared with 19.6. This is an error of approximately

$$\frac{20.0 - 19.6}{19.6} \times 100 \approx 2 \text{ per cent}$$

The matter of average and effective values is discussed further in Chap. 3. Now let us return to the problem of finding the average value of a sine wave.

TABLE 2-8

y	Δx	$y\,\Delta x$
5.0	1	5.0
4.9	1	4.9
4.6	1	4.6
3.5	1	3.5
2.0	1	2.0
		20.0

Usually the average value of a repetitive waveform is determined by using the full period as a base interval. A pure sine wave or any other repetitive waveform which is symmetrical about the zero axis will have

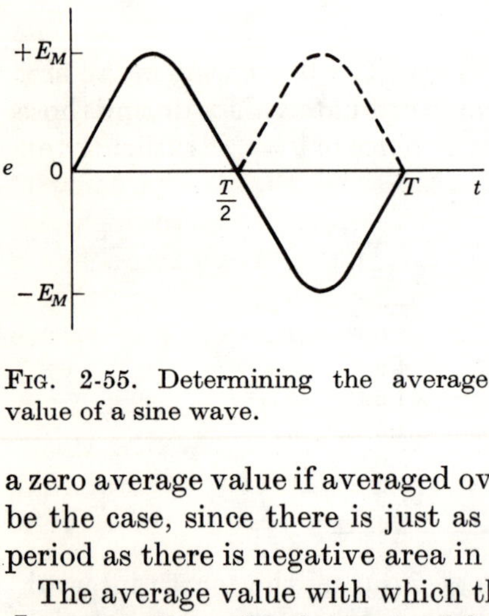

FIG. 2-55. Determining the average value of a sine wave.

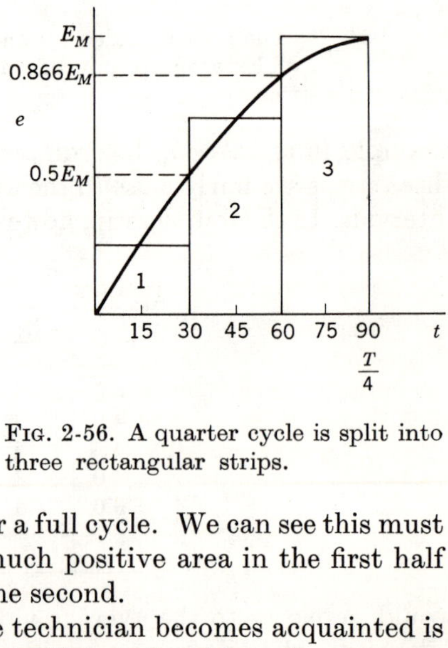

FIG. 2-56. A quarter cycle is split into three rectangular strips.

a zero average value if averaged over a full cycle. We can see this must be the case, since there is just as much positive area in the first half period as there is negative area in the second.

The average value with which the technician becomes acquainted is $\bar{E} = 0.637E_M$ and is determined for a half-period. This is also the same as the average value of a full-wave rectified sine wave, averaged over a full period (see Fig. 2-55). (The average value of a waveform is the d-c value.)

Let us then try to derive this relationship between the d-c and peak values of an unfiltered full-wave rectified sine wave, as in Fig. 2-55. We will first use our approximation method to find the area under a sine curve for one-fourth cycle (see Fig. 2-56) and then multiply this by four to get the total area under the full-wave rectified output. Let us divide the one-fourth cycle, 90°, into three 30° intervals, and judiciously choose our ordinate values.

TABLE 2-9

e	Δt	$e\,\Delta t$
$0.25E_M$	30°	$7.5E_M{}^\circ$
$0.72E_M$	30°	$21.6E_M{}^\circ$
$0.90E_M$	30°	$27.0E_M{}^\circ$
		$56.1E_M{}^\circ$

The area under a one-fourth cycle is then $56.1E_M{}^\circ$. The area under a full cycle is then $4 \times 56.1E_M{}^\circ = 224E_M{}^\circ$. The average voltage over a full cycle is then $224E_M{}^\circ$ divided by a base of 360°, or

$$\frac{224E_M{}^\circ}{360^\circ} = 0.62E_M$$

which is fairly close to the correct value of $0.637E_M$.

We could have expressed our angles in terms of radians rather than degrees. This is really the preferred way. The reader will recall that there are 2π radians in 360°. Therefore, to convert an angle in degrees to radians, let us use the conversion factor 2π radians/360°. For example, 30° in radians would be $30^\circ \times 2\pi$ radians/360° $= \pi$ radians/6.

Since there are 2π radians in 360°, every time the a-c generator sweeps through one cycle, 360°, it goes through 2π radians. If the frequency is f cps, the generator angular frequency in radians is $2\pi f$. The symbol ω is generally used to designate $2\pi f$. That is, $\omega = 2\pi f$. Radian measure is discussed in Cooke.[1]

2-26. The Effective (*RMS*) Value. Let us see if we can gain some additional insight into the significance of the effective, or root-mean-square, value. If we force a waveform of current through a resistor, that resistor becomes warmer. This will be true even though the current might be alternating in nature. If a direct current is passed through the resistor, which makes it just as hot as the alternating current, the power dissipated is the same in both cases. The value of direct current (or voltage) which produces a heating effect equal to the a-c value is equal to the effective value of the a-c waveform. It can be shown that the

[1] Nelson M. Cooke, "Basic Mathematics for Electronics," 2d ed., McGraw-Hill Book Company, Inc., New York, 1960.

effective value is related to the peak or maximum value, by

$$E_{\text{rms}} = \frac{1}{\sqrt{2}} E_M = 0.707 E_M$$

where E_M is the peak value. Usually the rms subscript is dropped and the meaning is understood. Let us try to derive this relationship.

The instantaneous power in a resistor is i^2R for any kind of waveform. For sine waves, $i = I_M \sin \theta$. The instantaneous power p is then

$$p = i^2R = (I_M \sin \theta)^2 R = (I_M{}^2 \sin^2 \theta)R$$

Let us investigate what the power waveform looks like as a function of the angle θ, or time. For simplicity, we could assume an R of 1 ohm and

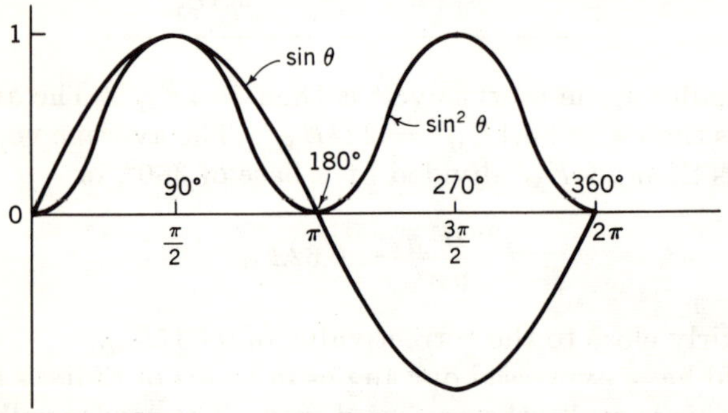

FIG. 2-57. A sine wave and the sine wave squared. The squared waveform has no negative area.

an I_M of 1 amp. This does not change the shape of the resulting waveform. Then $p = \sin^2 \theta$ is the equation which determines the shape of the power waveform. Values of R and I_M other than one unit will simply change the scale of our curve. One way to see what the $\sin^2 \theta$ function looks like is to assume values of θ from 0 to 360°. Look up the sine of each angle chosen, and then square it. A plot of the resulting values will look like Fig. 2-57. Note that the $\sin^2 \theta$ curve has no negative values, since the negative third- and fourth-quadrant sine functions become positive when squared. This is in agreement with the physical facts, since negative values of power would indicate that the resistor was generating power rather than absorbing it. Also, observe that the $\sin^2 \theta$ curve is likewise a sine wave but one having twice the frequency of the original $\sin \theta$ curve.

A neater way of arriving at the shape of the $\sin^2 \theta$ curve is to look up the $\sin^2$ function in a table of trigonometric identities. We would find that $\sin^2 \theta = \frac{1}{2}(1 - \cos 2\theta) = \frac{1}{2} - \frac{1}{2} \cos 2\theta$. From this identity we see that $\sin^2 \theta$ is made up of a constant quantity (the one-half term), from which is subtracted one-half of a double-frequency cosine curve. If we were to sketch this identity, we would draw the two terms separately and then combine them as shown in Fig. 2-58. The resultant sketch of $\sin^2 \theta$ is that of a sine curve, but it has an average value of one-half the peak values. The waveform varies about the value ½ rather than about zero, as in a sine curve.

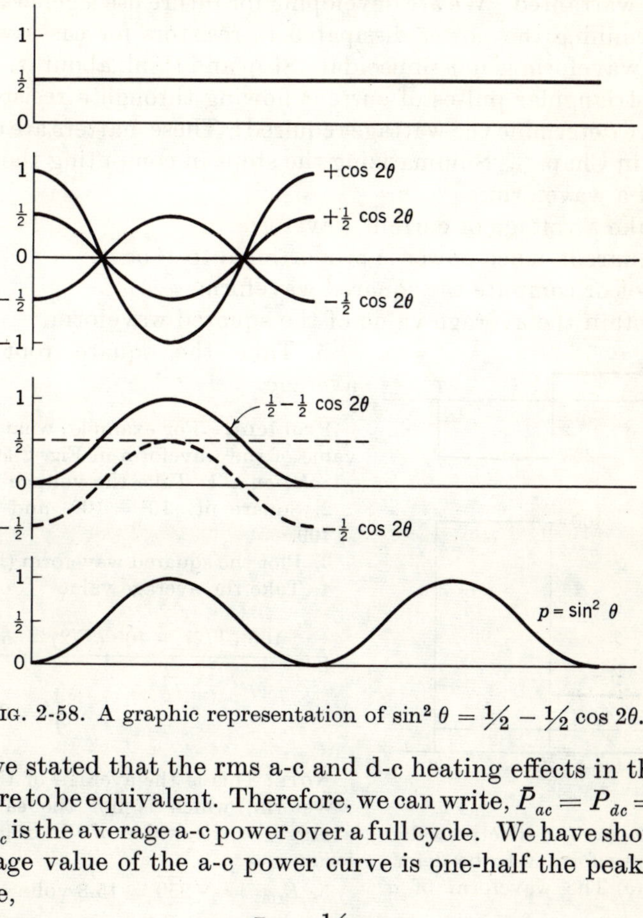

FIG. 2-58. A graphic representation of $\sin^2 \theta = \frac{1}{2} - \frac{1}{2}\cos 2\theta$.

Now we stated that the rms a-c and d-c heating effects in the same resistor are to be equivalent. Therefore, we can write, $\bar{P}_{ac} = P_{dc} = I_{dc}{}^2 R$, where $\bar{P}_{ac}$ is the average a-c power over a full cycle. We have shown that the average value of the a-c power curve is one-half the peak power. Therefore,

$$\bar{P}_{ac} = \tfrac{1}{2} P_M$$

for a sine wave only; and the maximum power is $P_M = I_M{}^2 R$; therefore, $\bar{P}_{ac} = \frac{1}{2} I_M{}^2 R$. Since the a-c and d-c powers are to be equal, we can write, $\frac{1}{2} I_M{}^2 R = I_{dc}{}^2 R$, or $\frac{1}{2} I_M{}^2 = I_{dc}{}^2$. Taking square roots,

$$\frac{I_M}{\sqrt{2}} = I_{dc}$$

$$0.707 I_M = I_{dc}$$

Since we have defined the rms value of a sinusoidal current as being equal in heating effect to a direct current, we can write

$$0.707I_M = I_{dc} = I_{\text{rms}}$$

or

$$I_{\text{rms}} = 0.707I_M$$

Perhaps this discussion of the rms value of a sine wave seems lengthy, but it is warranted. We are developing for future use a general method of determining the power dissipated in resistors for cases where the driving waveform is not sinusoidal. Stop and think about it. Suppose we had triangular pulses of current flowing through a resistor. How would we determine the wattage required? These matters are discussed further in Chap. 3. Summarizing the steps in computing the effective value of a waveform:

1. Take a voltage or current waveform.
2. Square it, since power is proportional to e^2 or i^2.
3. Plot or compute the squared waveform.
4. Obtain the average value of the squared waveform.
5. Take the square root of this average.

Problem. For example, what is the rms value of the waveform in Fig. 2-59*a*?

Solution. 1. Take the voltage waveform.

2. Square it. $10^2 = 100$ and $(-20)^2 = +400$.

3. Plot the squared waveform (Fig. 2-59*b*).

4. Take the average value

$$\bar{e} = \frac{100\,(T/2) + 400\,(T/2)}{T} = \frac{500T}{2} \times \frac{1}{T} = 250 \text{ volts}$$

NOTE: This is the average of the squared waveform, which is not the same as the average, or d-c, value of the original waveform.

5. $E_{\text{rms}} = \sqrt{250} = 15.8$ volts

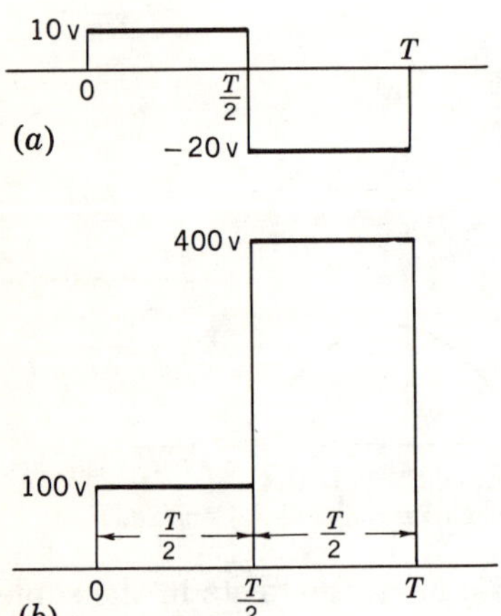

FIG. 2-59. (*a*) The effective value of this waveform is to be determined. (*b*) The waveform of *a* squared.

SUMMARY

In this chapter we have developed methods of a-c circuit analysis. The network theorems and other analytic tools of Chap. 1 provided the basic techniques. The *j* operator and other methods of complex algebra furnish the actual mathematical hardware needed. A variety of networks including low- and high-pass filters, attenuators, etc., was analyzed to illustrate the manner in which complex algebra is used.

The concepts of a transfer function and of the Bode diagram as a means of displaying it was introduced. The logarithmic amplitude-versus-frequency response of several networks was determined, since this powerful approach is most useful in the analysis of feedback amplifiers and servo systems. Section 2-12 developed a number of useful series-parallel conversions that frequently aid in the simplification of complex a-c networks.

A detailed analysis of resonance phenomena in both series and parallel circuits was undertaken. The universal resonance curve and certain relationships were developed in order to predict the response of tuned circuits. Sections 2-23 and 2-24 discussed the problem of impedance transformation and a number of solutions was investigated. The meaning of average and rms values was discussed in Secs. 2-25 and 2-26. Graphic techniques for approximating these values were developed.

PROBLEMS

2-1. What is the Thévenin's equivalent circuit seen looking into the 1-1 terminals of Fig. 2-60*a*?

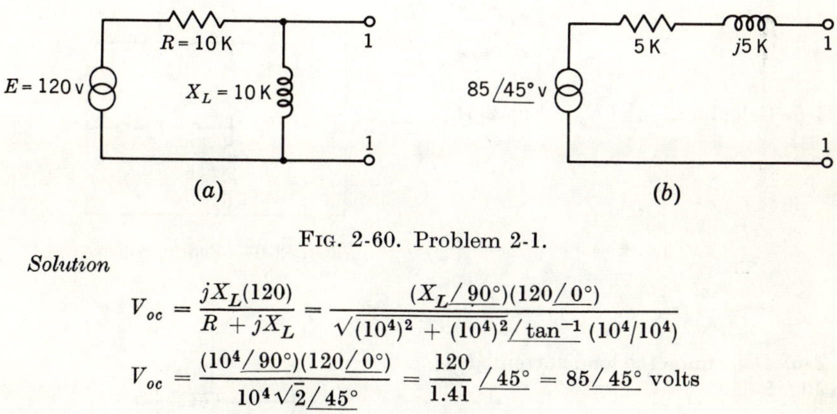

FIG. 2-60. Problem 2-1.

Solution

$$V_{oc} = \frac{jX_L(120)}{R + jX_L} = \frac{(X_L\underline{/90°})(120\underline{/0°})}{\sqrt{(10^4)^2 + (10^4)^2}\underline{/\tan^{-1}(10^4/10^4)}}$$

$$V_{oc} = \frac{(10^4\underline{/90°})(120\underline{/0°})}{10^4\sqrt{2}\underline{/45°}} = \frac{120}{1.41}\underline{/45°} = 85\underline{/45°} \text{ volts}$$

$$Z_{th} = \frac{(10^4)(10^4\underline{/90°})}{10^4 + j10^4} = 7{,}070\underline{/45°} = R + jX$$

$$= 7.071(\cos 45°) + j7.071(\sin 45°) = 5 + j5 \text{ kilohms}$$

The complete equivalent circuit is shown in Fig. 2-60*b*. The angle of V_{oc} is with respect to the original 120-volt generator taken at a reference of 0°.

2-2. Solve for the current in Fig. 2-61. *Ans.* 2.82$\underline{/-45°}$ amp.

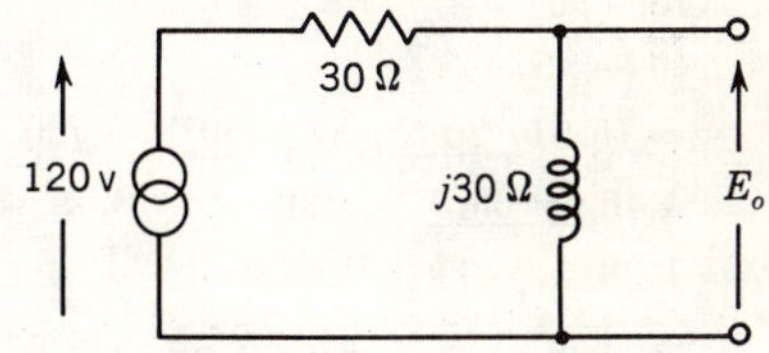

FIG. 2-61. Problem 2-2.

2-3. Determine E_o in Fig. 2-62.
Ans. $E_o = 0$ volts.

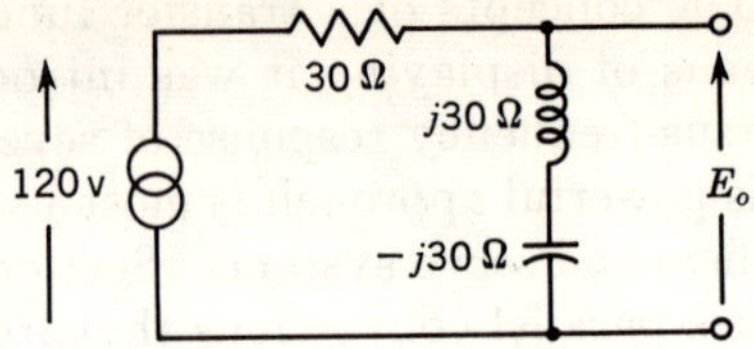

FIG. 2-62. Problem 2-3.

2-4. Determine E_o in Fig. 2-63.

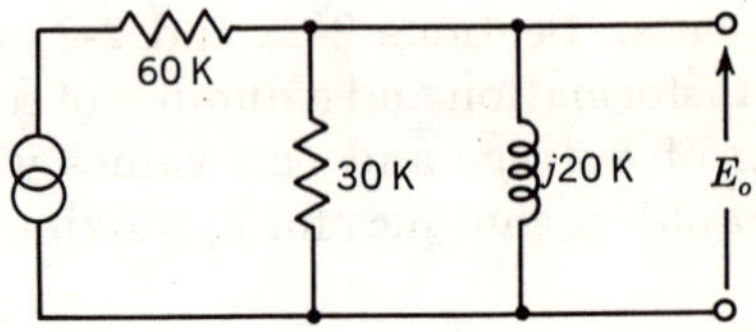

FIG. 2-63. Problem 2-4.

2-5. Calculate I_o and E_o in Fig. 2-64.

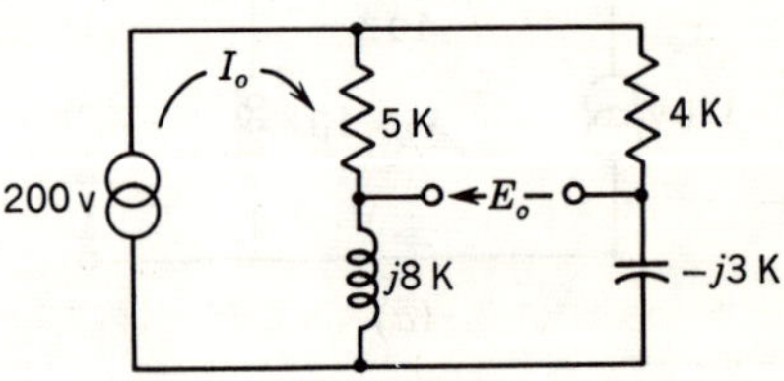

FIG. 2-64. Problem 2-5.

2-6. Determine the load current (I_L) in Fig. 2-65.

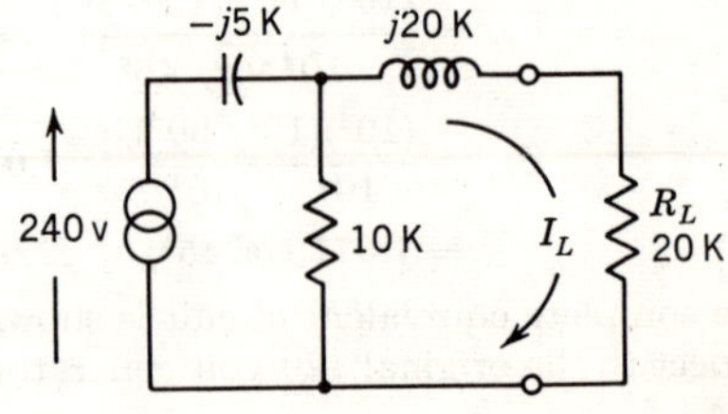

FIG. 2-65. Problem 2-6.

Solution

$$V_{oc} = \frac{240 \text{ volts (10 kilohms)}}{10 - j5 \text{ kilohms}}$$

$$= \frac{240(10)}{11.2\underline{/-26.5^\circ}} = 214\underline{/26.5^\circ} \text{ volts}$$

$$Z_{th} = \frac{10(-j5)}{10 - j5} + j20$$

$$= (8.91\underline{/26.5^\circ})(5\underline{/-90^\circ}) + j20$$

$$= 4.46\underline{/-63.5^\circ} + j20 = -j3.98 + 1.98 + j20$$

$$= 1.98 + j16 \text{ kilohms}$$

$$I_L = \frac{V_{oc}}{Z_{th} + R_L} = \frac{214\underline{/26.5^\circ} \text{ volts}}{21.98 + j16 \text{ kilohms}} = \frac{214\underline{/26.5^\circ}}{27.2\underline{/36.2^\circ}} = 7.8\underline{/-9.5} \text{ ma}$$

2-7. Compute E_o in Fig. 2-66.

Solution. The circuit is resonant, since $X_L = X_C$.

$$I = \frac{60\angle 0^\circ}{100 \text{ ohms} + X_L/Q_L} = \frac{60}{100 + 5{,}000/20} = \frac{60}{350} = 0.17\angle 0^\circ \text{ amp}$$

$$E_o = IZ_L = I(250 + j5{,}000)$$

$$\approx (5{,}000\angle 87.1^\circ)I$$

$$E_o = 855\angle 87.1^\circ \text{ volts}$$

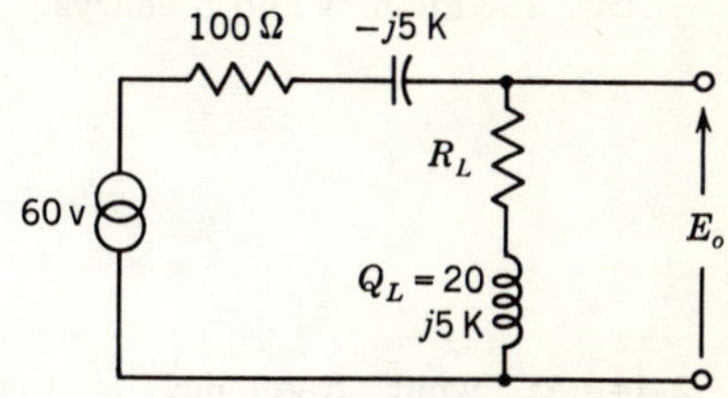

FIG. 2-66. Problem 2-7.

Alternative solution. Since the coil Q is high, the coil impedance is essentially its inductive reactance. The circuit Q is 5,000 ohms/350 ohms, or 14.3. The coil voltage will equal the circuit Q times the applied voltage. That is, $14.3(60) \approx$ 860 volts, which checks closely.

2-8. What is the series-resonant frequency of a 9-henry coil and 1-μf capacitor? *Ans.* 53 cps.

2-9. In trying to get selectivity out of the resonant circuit shown in Fig. 2-67, should the switch be in the 72- or in the 600-ohm position?

Ans. 72-ohm.

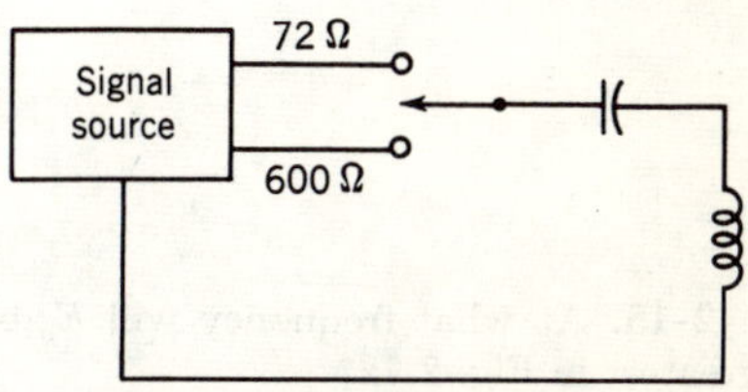

FIG. 2-67. Problem 2-9.

2-10. What effect, if any, is there upon the Q between two terminals of a coil if an identical coil is shunted across it?

2-11. Measurements upon the current in a series-resonant circuit resulted in the curve of Fig. 2-68. What is the circuit Q? *Ans.* 12.5.

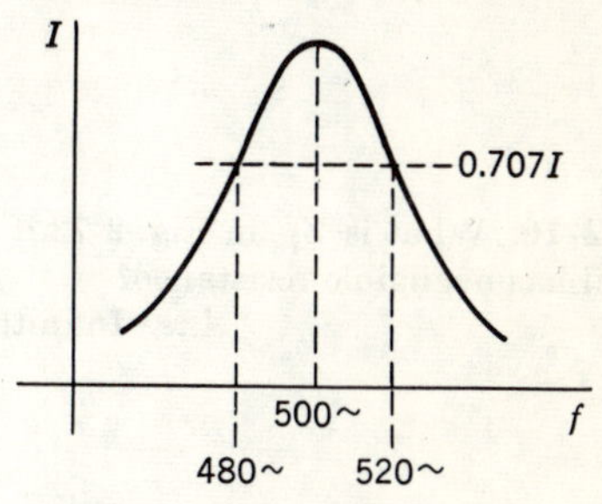

FIG. 2-68. Problem 2-11.

2-12. What is the maximum voltage between any two points in Fig. 2-69 at resonance? *Ans.* 100 volts.

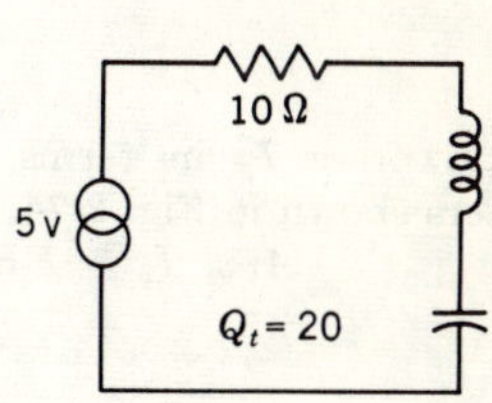

FIG. 2-69. Problem 2-12.

2-13. Approximately what values would the parameters have in the parallel equivalent of Fig. 2-70?

Ans. 185 kilohms and 5 henrys.

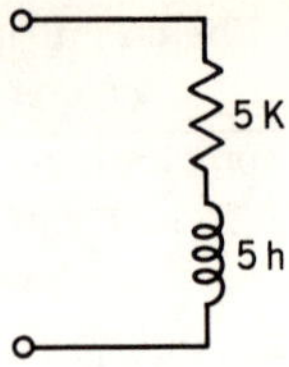

FIG. 2-70. Problem 2-13.

2-14. At what frequency is the magnitude of the output in Fig. 2-71 equal to $E_{in}/4$?

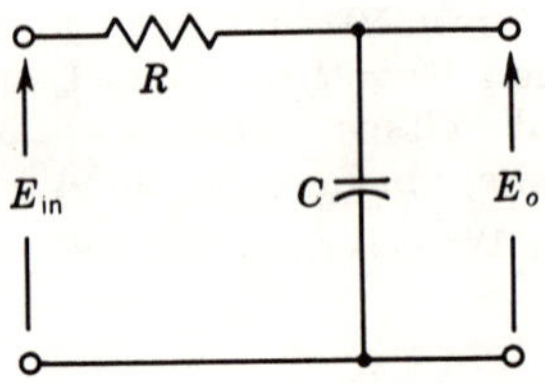

FIG. 2-71. Problem 2-14.

2-15. At what frequency will E_o be greatest in Fig. 2-72?

Ans. E_o will not change with frequency.

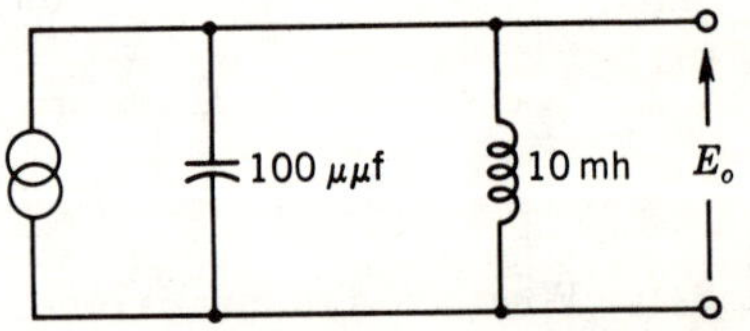

FIG. 2-72. Problem 2-15.

2-16. What is Z_{in} in Fig. 2-73 if the coil has negligible resistance?

Ans. Infinite.

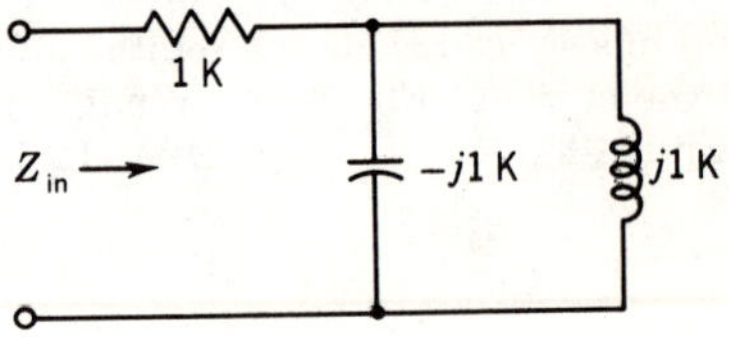

FIG. 2-73. Problem 2-16.

2-17. Express I_g in terms of the parameters shown in Fig. 2-74.

Ans. $I_g = ERC/L$.

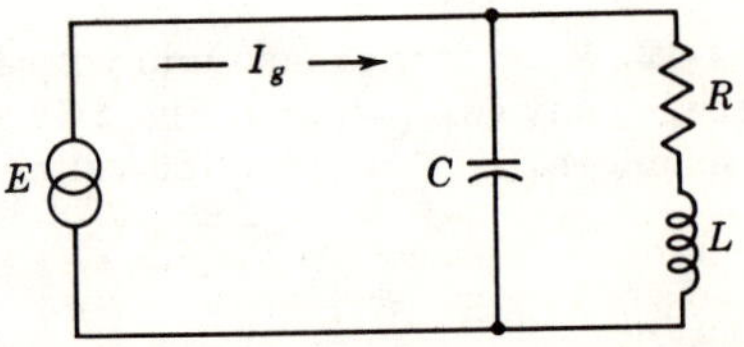

FIG. 2-74. Problem 2-17.

2-18. If the inductors and capacitors of Fig. 2-75*a* are lossless, sketch Z_{in} versus frequency.

Solution. First, let us reason how Z_{in} might vary with frequency and then verify our thinking with a mathematical analysis. At very low frequencies, X_{L_1} and X_{L_2} are very low, and so Z_{in} approaches zero. As the frequency increases, a point will be reached where X_{L_1} equals X_{C_1} and the antiresonant circuit presents an infinite input impedance. At frequencies above this antiresonant point, most of the current through the parallel branch will flow through C_1. The net impedance

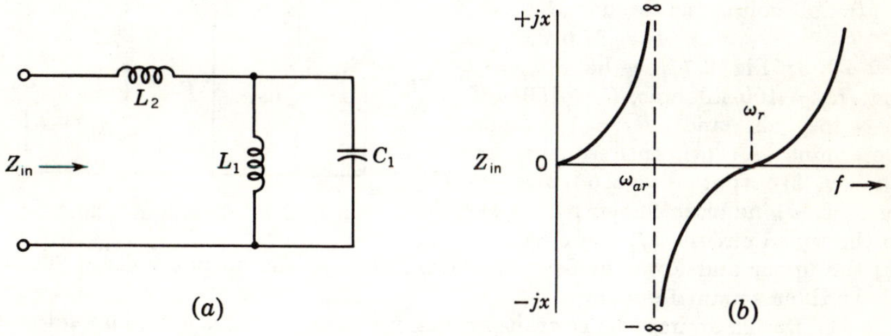

FIG. 2-75. Problem 2-18.

of the parallel branch is therefore capacitive. At some higher frequency, L_2 will be series resonant with this equivalent capacitance, and Z_{in} therefore equals zero. At frequencies above this series-resonant point, X_{C_1} approaches zero, but X_{L_2} approaches infinity, and therefore Z_{in} approaches infinity. The response should therefore appear somewhat as in Fig. 2-75*b*.

To verify Fig. 2-75*b* mathematically, we proceed as follows.

$$Z_{in} = jX_{L_2} + \frac{jX_{L_1}(-jX_{C_1})}{j(X_{L_1} - X_{C_1})} = \frac{-X_{L_2}(X_{L_1} - X_{C_1}) + X_{L_1}X_{C_1}}{j(X_{L_1} - X_{C_1})}$$

Z_{in} will go through points of zero impedance whenever the numerator is equal to zero. That is, $-X_{L_2}(X_{L_1} - X_{C_1}) + X_{L_1}X_{C_1} = 0$. Solving for X_C, to see how it must be related to the X_L values for this condition to be true, we obtain

$$X_{C_1} = \frac{X_{L_1}X_{L_2}}{X_{L_1} + X_{L_2}}$$

Inspection of the expression for X_C indicates it must equal the Thévenin's equivalent X_L that it looks back into. The only other possibility for the numerator of the expression for Z_{in} to equal zero, is when X_{L_1} and X_{L_2} equal zero, which occurs at d-c or zero frequency. To determine the number of antiresonant points where Z_{in} becomes infinite, we set the denominator equal to zero. By inspection, we see that this occurs when $X_{L_1} = X_{C_1}$.

For analyzing more complicated networks than that shown in Fig. 2-75*a*, the reader is referred to the literature on reactance sketches. See "Networks, Lines, and Fields" by Ryder[1] or "Communication Engineering" by Everitt and Anner.[2]

[1] John O. Ryder, "Networks, Lines, and Fields," 2d ed., Prentice-Hall, Inc., Englewood Cliffs, N.J., 1955.

[2] W. L. Everitt and G. E. Anner, "Communication Engineering," 2d ed., McGraw-Hill Book Company, Inc., New York, 1956.

2-19. In the circuit of Fig. 2-76, determine

(*a*) Resonant frequency.

Ans. $\omega_r = 20{,}000$ radians/sec.

(*b*) Current at resonance.

Ans. $I_r = 8.76$ ma.

(*c*) Approximate half-power frequencies.

Ans. 18,900 and 21,000 radians/sec.

(*d*) Coil voltage at resonance.

Ans. 87.6 volts.

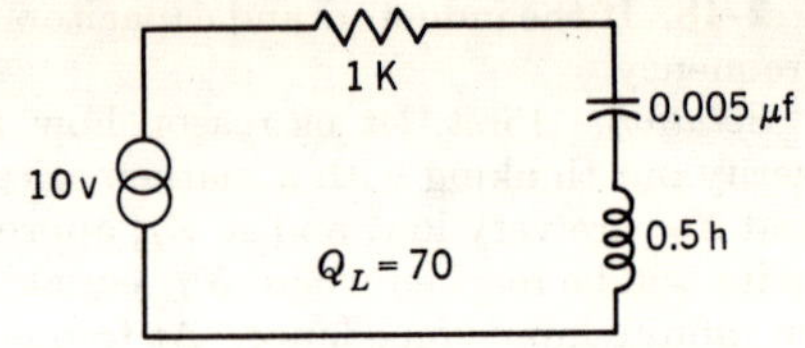

FIG. 2-76. Problem 2-19.

2-20. In Fig. 2-77, we have $I_{sc} = 1$ ma, $R_g = 100$ kilohms, $C = 159$ $\mu\mu$f, $L = 159$ μh, and $R_L = 10$ ohms. Determine the (*a*) antiresonant frequency, (*b*) output voltage, (*c*) capacitor current, (*d*) inductor current, (*e*) power in the tuned circuit, (*f*) the circuit Q, (*g*) the upper and lower half-power frequencies, (*h*) the output voltage 200 kc above the resonant frequency.

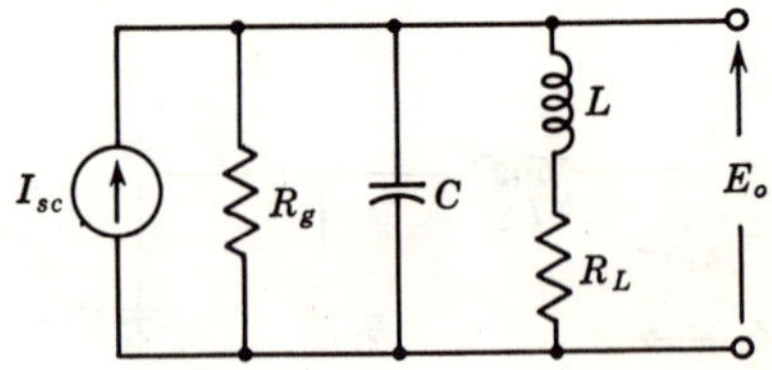

FIG. 2-77. Problem 2-20.

2-21. Design an impedance-transforming network to enable a 72-ohm antenna to see 72 ohms when looking into a 500-ohm line at 1 Mc.

2-22. Design the tuned circuit of Fig. 2-78 so that, at a resonant frequency of 500 kc, the output voltage magnitude will equal 60 per cent of the input. An inductor having a $Q_L = 50$ is available. Find the required values of L and C.

Solution

$$\frac{E_o}{E_{in}} = \frac{60}{100} = 0.6 = \frac{R_{ar}}{10^4 + R_{ar}}$$

Solving for the tuned circuit impedance at resonance (R_{ar}), we have 6 kilohms $+ 0.6R_{ar} = R_{ar}$, or $0.4R_{ar} = 6$ kilohms and $R_{ar} = 15$ kilohms. But $R_{ar} = Q_LX_L$, $X_L = R_{ar}/Q_L = 15{,}000/50 = 300$ ohms, since $\omega_L = X_L$ and $L = X_L/\omega$; and at $\omega = \omega_r = 2\pi$ (500 kc) $= 3.14 \times 10^6$, we have

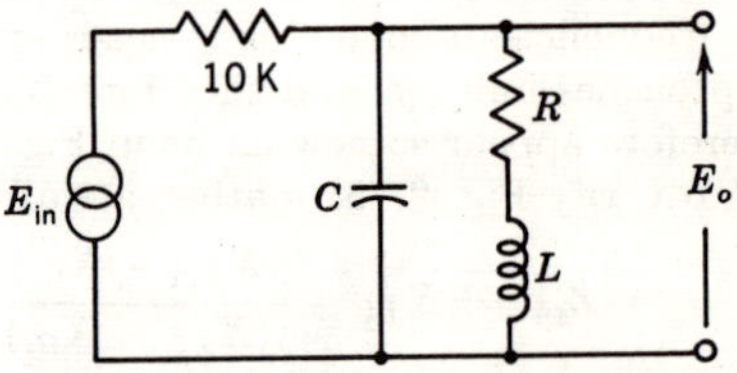

FIG. 2-78. Problem 2-22.

$$L = \frac{300}{3.14 \times 10^6} = 95.5\ \mu\text{h}$$

At ω_r, $X_L = X_C$ when $Q_L \approx 10$; therefore, $300 = 1/\omega_r C$,

$$C = \frac{1}{300\omega_r} = \frac{1}{3.14 \times 10^6 \times 3 \times 10^2} = \frac{1}{9.42 \times 10^8}$$

$$= 0.106 \times 10^{-8} = 0.00106\ \mu\text{f} = 1{,}060\ \mu\mu\text{f}.$$

Chapter 3

TRANSIENT ANALYSIS

MANY METHODS OF COMMUNICATION, remote guidance, automation, computation, and radar involve circuits which deal with electrical pulses. In this chapter we shall study the characteristics of pulses and the effect of various circuit elements upon them.

Pulsed waveforms can be analyzed by two methods. One is to consider the pulses as involving fast changes in waveforms which may or may not occur periodically. The other is to consider the pulses as being composed of a number of sine waves having different frequencies.

Both methods of analysis have their advantages and disadvantages. In this text we shall choose whichever viewpoint is better for the problem at hand. Generally, the first method, sometimes called transient analysis, will be used when we are dealing with timing or wave-shaping circuits. The second method of analysis, sometimes referred to as Fourier analysis, is, at our level, best adapted for determining required amplifier frequency-response characteristics for faithful pulse reproduction. If an amplifier is incapable of passing all the frequency components of a pulse, distortion results. In fact, we often test the frequency response of an amplifier by determining its ability to reproduce square waves.

3-1. The Square Wave. Take a look at Fig. 3-1, which illustrates a symmetrical square wave. By the first, or transient, method of analysis, the square wave can be considered as having a value of $+E$ volts during the interval from 0 to t_1. At t_1, an abrupt change in potential occurs, and the voltage drops to $-E$ volts. The voltage then stays at $-E$ volts until t_2, whereupon it instantaneously jumps back to $+E$ volts, and so forth. If the symmetrical square wave were to be analyzed for its sinusoidal components, it could be shown to consist of:

1. A fundamental sine wave having a frequency equal to the square wave repetition frequency.
2. Odd harmonics of the fundamental, each having an amplitude

equal to the fundamental amplitude divided by the number of the harmonic. For example, the third-harmonic amplitude would be one-third of the fundamental amplitude.

Figure 3-1 illustrates how a symmetrical square wave is built up of the fundamental and odd-harmonic frequencies. Notice that the greater the number of added harmonics, the closer is the square shape approached. If an amplifier had to pass square waves with perfect fidelity,

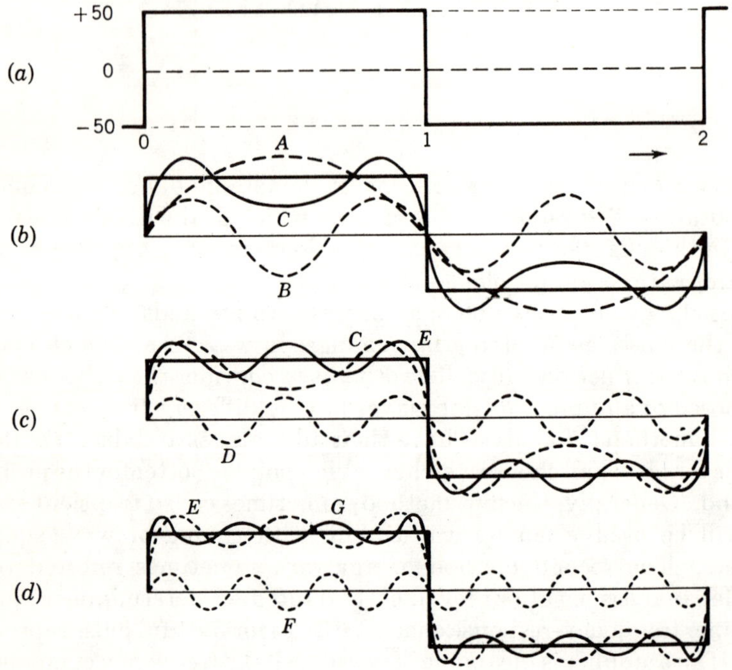

FIG. 3-1. Components of a symmetrical square wave. (a) Fundamental. (b) Third harmonic. (c) Fundamental plus third harmonic. (d) Fifth harmonic. (e) Fundamental plus third and fifth harmonic. (f) Seventh harmonic. (g) Fundamental plus third, fifth, and seventh harmonic.

it would have to be capable of amplifying uniformly an infinite number of harmonics, and also of preserving their proper phase relationships. Generally speaking, the more abrupt the changes in a waveform, the higher the harmonic content. Harmonics are always whole-number multiples of the fundamental. Also, the fundamental sine wave always has a frequency equal to the PRF (pulse repetition frequency).

Another point to observe about the symmetrical square wave of Fig. 3-1 is that it has no d-c component. This means that the average value of the waveform over a full cycle is zero. This can be readily seen, since there is just as much positive as negative area during a full cycle of the waveform. The average value of any waveform is its d-c component.

Figure 3-2*a* illustrates an unsymmetrical square wave which has a d-c component equal to one-third of the pulse amplitude. This is determined as follows:

The positive pulse area within the 1,000-μsec period is (100 volts)(333 μsec) = 33,300 volt-μsec. There is no negative pulse area, since the

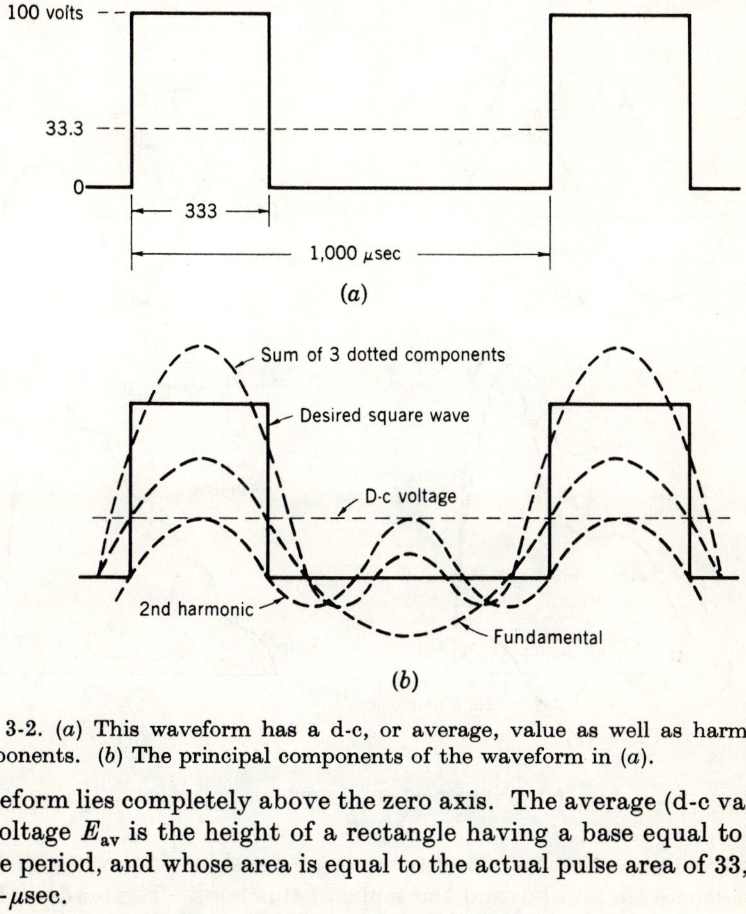

FIG. 3-2. (*a*) This waveform has a d-c, or average, value as well as harmonic components. (*b*) The principal components of the waveform in (*a*).

waveform lies completely above the zero axis. The average (d-c value) of voltage E_{av} is the height of a rectangle having a base equal to the pulse period, and whose area is equal to the actual pulse area of 33,300 volt-μsec.

Since the rectangle areas are to be equal, we can write

$$E_{av}\ (1{,}000\ \mu\text{sec}) = 33{,}300\ \text{volt-}\mu\text{sec}$$

$$E_{av} = \frac{33{,}300\ \text{volt-}\mu\text{sec}}{1{,}000\ \mu\text{sec}} = 33.3\ \text{volts}$$

Figure 3-2*b* shows how this unsymmetrical square wave could be made up of a d-c voltage which is one-third of the peak voltage, a fundamental

1,000-cps sine wave having an amplitude two-thirds of the peak value, a 2,000-cps sine wave having an amplitude one-third of the pulse amplitude, etc. Note that the fundamental frequency is equal to the reciprocal of the pulse period.

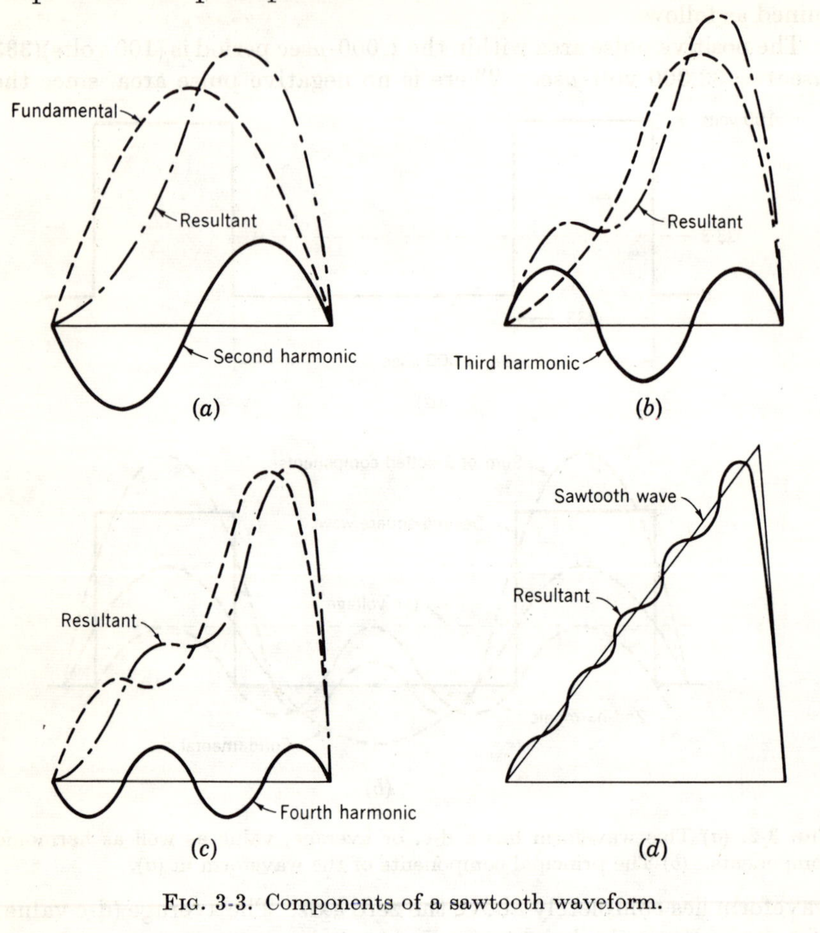

FIG. 3-3. Components of a sawtooth waveform.

Determining the harmonic content of nonsinusoidal waveforms is quite involved, and beyond the scope of this book. The reader who is interested in exploring the subject further is advised to study Fourier analysis in such books as "Calculus for Electronics" by Richmond[1] or "Mathematics for Radio Engineers" by Mautner.[2]

[1] A. E. Richmond, "Calculus for Electronics," McGraw-Hill Book Company, Inc., New York, 1958.

[2] Leonard Mautner, "Mathematics for Radio Engineers," Pitman Publishing Corporation, New York, 1947.

Some useful rules of thumb to follow when considering pulse waveforms are:

1. A waveform exhibiting sharp corners has many high-frequency components.

2. Waveforms which have a fast rise or fall have high-frequency components.

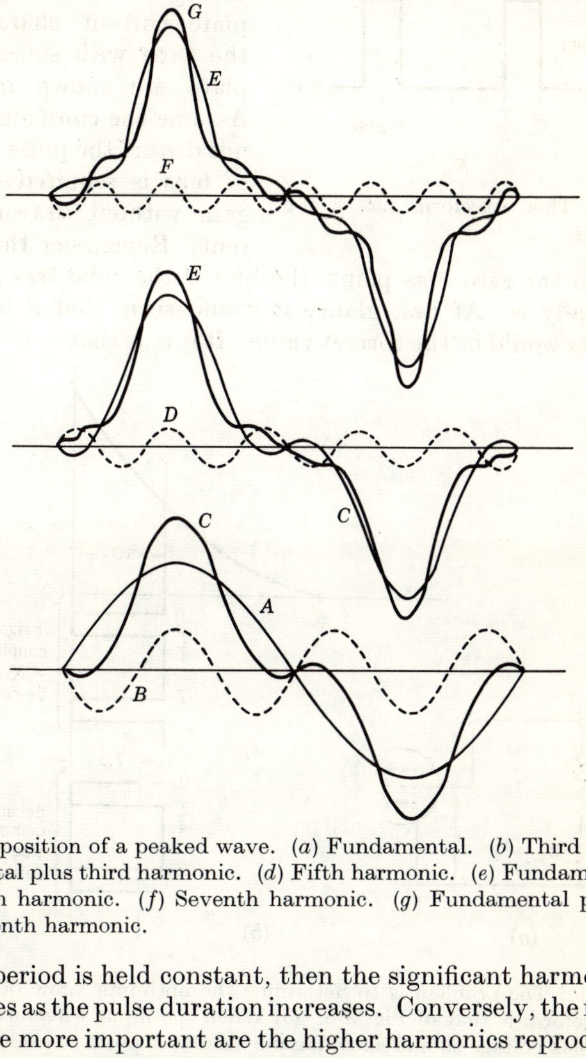

FIG. 3-4. Composition of a peaked wave. (*a*) Fundamental. (*b*) Third harmonic. (*c*) Fundamental plus third harmonic. (*d*) Fifth harmonic. (*e*) Fundamental plus third and fifth harmonic. (*f*) Seventh harmonic. (*g*) Fundamental plus third, fifth, and seventh harmonic.

3. If the period is held constant, then the significant harmonic content decreases as the pulse duration increases. Conversely, the narrower the pulse, the more important are the higher harmonics reproducing it.

4. The higher the PRF the higher the frequency components to be considered, since the fundamental is the PRF.

Figures 3-3 and 3-4 illustrate the composition of a sawtooth and a peaked wave.

3-2. Average Values. Frequently we shall encounter problems which necessitate a knowledge of the average (d-c) and effective (rms) values of a waveform. For an example in which a knowledge of the d-c, or average, value is required, consider the following problem.

Suppose the input signal of Fig. 3-5 were feeding the pulse amplifier of Fig. 3-6*a*. The grid voltage-plate current characteristics of the tube with some load in the plate are shown in Fig. 3-6*b*. Assume the coupling circuit does not distort the pulse. What value of bias is required for optimum gain without drawing grid current? Remember that the closer we are to the zero bias point, the higher the tube transconductance (g_m) usually is. At first glance it would seem that a bias of $E_{cc} = -10$ volts would be the correct value. It is true that -10 volts will not

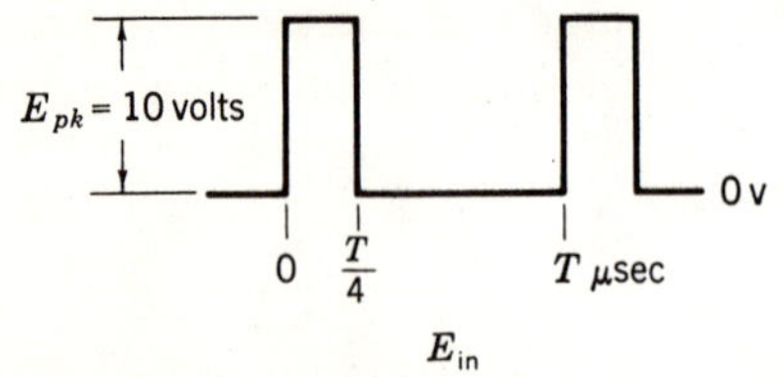

FIG. 3-5. This waveform has a d-c component.

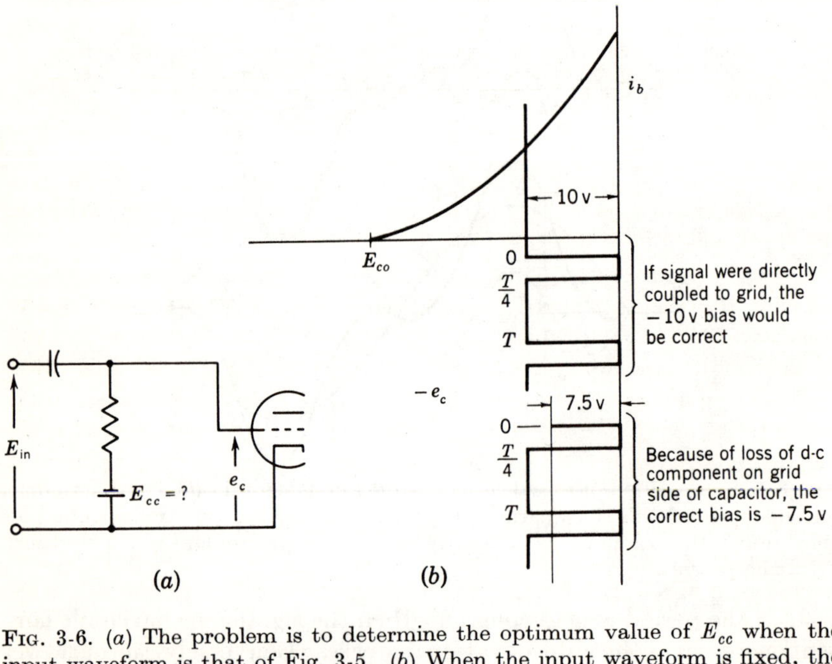

FIG. 3-6. (*a*) The problem is to determine the optimum value of E_{cc} when the input waveform is that of Fig. 3-5. (*b*) When the input waveform is fixed, the choice of bias voltage should allow for the decreased positive peak in the output.

permit the grid to go positive; but this is not the optimum bias. The input (E_{in}) surely has a d-c component, since the waveform exhibits a net positive area during a complete period (T). This positive area is

10 volts ($T/4$ μsec) = $10T/4$ volt-μsec. The average value is then the net positive area $10T/4$ divided by the period T, which is

$$E_{\text{av}} = \frac{10T/4 \text{ volt-}\mu\text{sec}}{T\ \mu\text{sec}} = 2.5 \text{ volts}$$

This 2.5-volt average value represents the height of a rectangle whose base is the total period T and whose area is equal to the actual pulse area in the waveform under consideration.

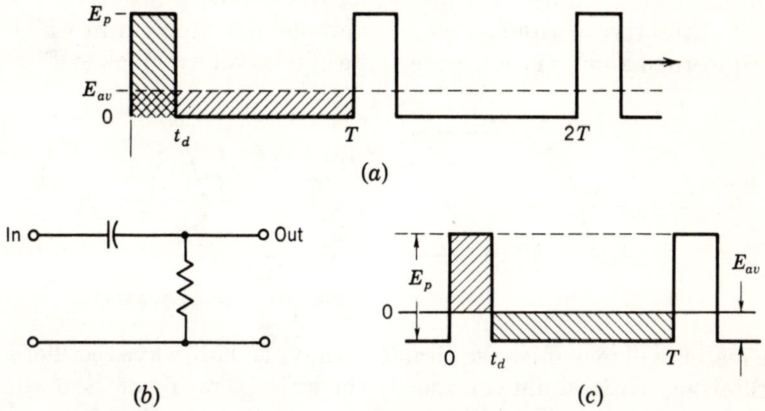

FIG. 3-7. The average value of waveform *a* will be lost at the output terminals in *b*, an *RC* coupling circuit. (*c*) The output voltage will swing positive and negative. The positive pulse area equals the negative pulse area.

Let us digress for a moment and look at Fig. 3-7*a*. The area of the pulse waveform is E_pt_d, where t_d is the pulse duration. When we find the average value of a waveform, we simply say that there must be a value E_{av} which, when multiplied by the period T, will give us an area equal to the pulse area. Therefore, we write $E_{\text{av}}T = E_pt_d$, or $E_{\text{av}} = E_pt_d/T$. Be sure you understand this concept of an average, or d-c, value of a pulse waveform.

Now we know from our basic electronics that a capacitor will not pass direct current. This means the waveform on the output side of the capacitor in Fig. 3-7*b* must have an average value of zero. The only way this can occur is for the zero axis of the waveform to shift, so that there is just as much negative area as positive area within one pulse period T. The amount of shift in the negative direction must be equal to E_{av}, and the resultant waveform is shown in Fig. 3-7*c*, which is identical with Fig. 3-7*a* except that the zero level of Fig. 3-7*c* replaces the E_{av} line of Fig. 3-7*a*.

Let us return to the problem of Fig. 3-6*a*. For maximum gain, we wish to operate as closely as possible to the zero bias point. From Fig. 3-7*c* we can see that the peak positive value of the output or grid side is the peak value E_p minus the average value E_{av}. Therefore, the magnitude of the bias is $E_p - E_{av}$, and since the bias is of necessity negative, let us place a minus sign in front of the entire expression. So we have

$$E_{cc} = -(E_p - E_{av}) = -(10 - 2.5) = -7.5 \text{ volts}$$

We see that because of the negative shift in the output waveform, we need bias the grid only 7.5 volts rather than 10 volts negative.

3-3. Effective Values. As an example of a problem in which we are concerned about the effective value of a waveform, look at Fig. 3-8.

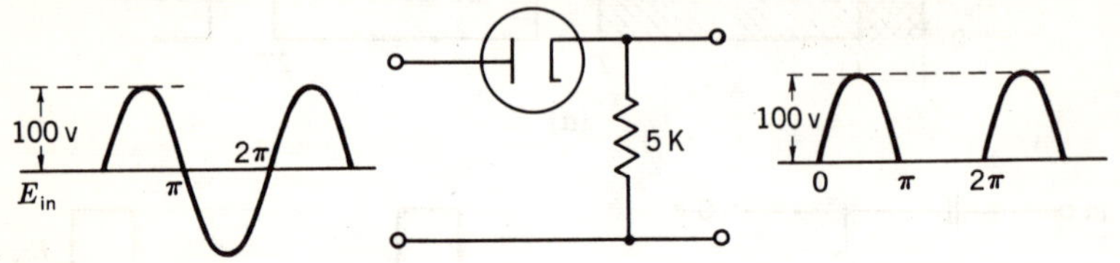

FIG. 3-8. The rectifier output voltage has a d-c component.

The reader will recognize the circuit as that of a half-wave rectifier with no filtering. How would you specify the wattage rating of the 5-kilohm load resistor if we assume that the tube, when conducting, has a negligible internal impedance? We will not solve this problem at the moment, since similar problems are discussed in other portions of the text. But we must bear in mind that rms values are necessary for determining power dissipation.

We can summarize the procedure for computing the effective value of an arbitrary waveform in four basic steps.

1. Express the current or voltage waveform as a function of time and peak value.
2. Square the current or voltage waveform.
3. Determine the average value of the current or voltage-squared waveform.
4. The square root of this average value is the effective, or rms, value.

Now let us familiarize ourselves with concepts of average and effective values by trying a few problems. The student must be sure he understands every detail here, since we shall use these principles again and again.

Problem 1. Find the average value of voltage for the waveform shown in Fig. 3-9*a*. Find the rms value.

Solution. To find the average value of a rectangular waveform:

1. First determine positive areas under the curve during the pulse period. $A^+ = E_m t_d$.

2. Determine negative areas during the pulse period. In this case, there are no negative areas, so $A^- = 0$.

3. Algebraically add the positive and negative areas. We have $A^+ + A^- = E_m t_d$.

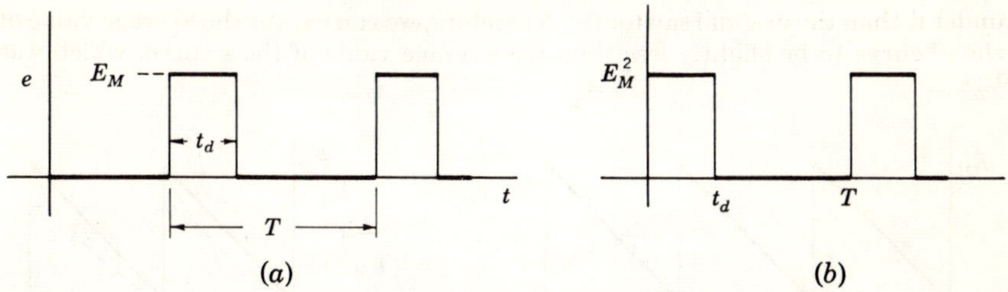

FIG. 3-9. (a) A rectangular waveform. (b) The square of the voltage waveform.

4. Divide the result by the pulse period to obtain the average value:

$$E_{\text{av}} = E_m t_d / T$$

The ratio of t_d/T is commonly referred to as the duty cycle.

To find the rms value:

1. Square the instantaneous values of voltage, since the power depends upon the square of the voltage. In this case, we would get a voltage-squared curve (see Fig. 3-9b) of similar shape, as follows:

2. Determine the average value of the voltage-squared curve. This is E^2_{rms}:

$$A^+ = E_m{}^2 t_d \qquad A^- = 0$$

$$A^+ - A^- = E_m{}^2 t_d$$

$$E^2_{\text{rms}} = E_m{}^2 \frac{t_d}{T}$$

3. The rms value is the square root of the average value of the voltage-squared curve; therefore,

$$E_{\text{rms}} = \sqrt{E_m{}^2 \frac{t_d}{T}} = E_m \sqrt{\frac{t_d}{T}}$$

Problem 2. Find the average and rms values of the sawtooth sweep waveform shown in Fig. 3-10a.

Solution. To find the average value:

1. Determine the positive area under the curve in the interval T. The area of a triangle is one-half the base times the altitude. Therefore, $A^+ = \frac{1}{2} T E_m$.

2. Determine the negative area, $A^- = 0$.

3. Determine the algebraic sum of the areas:

$$A^+ - A^- = \tfrac{1}{2} T E_m$$

4. Divide the resultant area by the period T:

$$E_{\text{av}} = \frac{\frac{1}{2} T E_m}{T} = \tfrac{1}{2} E_m$$

Now let us find the rms value. First of all we must generate an e^2 curve. This can be done in two ways. We can take the original sawtooth and square the instantaneous values along it, thus getting the resultant parabolic curve shown in Fig. 3-10*b*. For convenience we will choose E_m as 1 volt.

Plotting the results of the table we see that the e^2 parabolic curve has less area under it than the original sawtooth. Therefore, we can expect the average value of the e^2 curve to be slightly less than the average value of the e curve, which was $\frac{1}{2}E_m$.

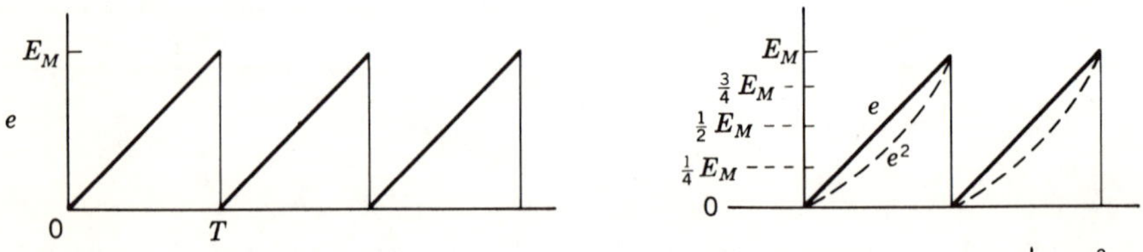

(*a*)

e	e^2
0	0
$\frac{1}{4}E_M$	$\frac{1}{16}E_M^2$
$\frac{1}{2}E_M$	$\frac{1}{4}E_M^2$
$\frac{3}{4}E_M$	$\frac{9}{16}E_M^2$
$1E_M$	$1E_M^2$

(*b*)

FIG. 3-10. (*a*) A sweep voltage with a d-c component. (*b*) A sweep waveform whose average and rms values are to be determined.

e . . .	0	$\frac{1}{4}E_M$	$\frac{1}{2}E_M$	$\frac{3}{4}E_M$	$1E_M$
e^2 . . .	0	$\frac{1}{16}E_M^2$	$\frac{1}{4}E_M^2$	$\frac{9}{16}E_M^2$	$1E_M^2$

The other way of generating the e^2 curve is to write an expression for the e curve, then square it. You will recall from algebra that the equation of a straight line is

$$y = \frac{\Delta y}{\Delta x}x + b$$

where $\Delta y/\Delta x$ is the slope of the line, and b is the y intercept. By y intercept we mean the value of y at $x = 0$.

In our sawtooth, the curve cuts the y axis at $y = 0$, so $b = 0$. The slope of the line is

$$\frac{\Delta y}{\Delta x} = \frac{E_m}{T}$$

The equation of the curve during the interval from 0 to T is then

$$e = \frac{E_m t}{T}$$

Therefore
$$e^2 = \left(\frac{E_m}{T}\right)^2 t^2$$

This is an equation of the form $y = ax^2$, which is standard for a parabola.

We surely know now that we have a parabolic curve. Our problem is to find the area under it. The only reason we bothered to sketch the curve was so that we could estimate by inspection the average value of the curve, which we decided was a little less than one-half of the peak value.

Table 3-1 is based upon the equation of the sawtooth, which is

$$e = \frac{E_m}{T} t$$

and shows corresponding values of t, e, and e^2.

TABLE 3-1

t	e	e^2
0	0	0
0.2	$0.2\ E_m/T$	$0.04(E_m/T)^2$
0.4	$0.4\ E_m/T$	$0.16(E_m/T)^2$
0.6	$0.6\ E_m/T$	$0.36(E_m/T)^2$
0.8	$0.8\ E_m/T$	$0.64(E_m/T)^2$
1.0	$1.0\ E_m/T$	$1.00(E_m/T)^2$

$$6\underline{|3.0}\,\frac{E_m}{T} \qquad 6\underline{|2.20}\left(\frac{E_m}{T}\right)^2$$

$$E_{av} = 0.5\,\frac{E_m}{T} \qquad E_{rms}^2 = 0.366\left(\frac{E_m}{T}\right)^2$$

Since we seek the average value of the e^2 curve, we need simply total the values in the e^2 column, and divide by the number of values taken. This is what you did when you figured your average grade in school, when given a number of tests. You added the scores and divided by the number of tests taken. Values for the e curve, when added and divided by the six readings, give

$$E_{av} = 0.5\,\frac{E_m}{T} = \frac{E_m}{2T}$$

which is the same result we derived analytically before. Remember, this is the d-c component.

The average value of the e^2 curve is $0.366(E_m/T)^2$ and this value is E_{rms}^2. Therefore,

$$E_{rms} = \sqrt{0.366\left(\frac{E_m}{T}\right)^2} = \sqrt{0.366}\,\frac{E_m}{T} = 0.61\,\frac{E_m}{T}$$

It can be shown by the methods of calculus that the exact average value of the e^2 curve is $(1/\sqrt{3})(E_m/T)$ for the sawtooth shown.

Let us see how close we came with our crude approximation method. $1/\sqrt{3} = 1/1.74 = 0.58$, as compared with our calculated value of 0.61. This is an error of

$$\frac{0.61 - 0.58}{0.58} = \frac{0.03}{0.58} \times 100 = 5.2 \text{ per cent}$$

Had we tabulated more values, the results would have been in closer agreement.

Problem 3. Let us try another problem. A vacuum tube's plate dissipation is rated at 2.5 watts maximum in the tube manual. This is an average, or steady-state, rating. For other reasons, the peak plate current cannot exceed 100 ma. If the plate current is to flow in pulses 10 μsec in width, at a repetition rate of 10,000 cps, what is the maximum permissible plate voltage during the plate-current pulse? What would a d-c milliameter read in the plate circuit? (See Fig. 3-11.)

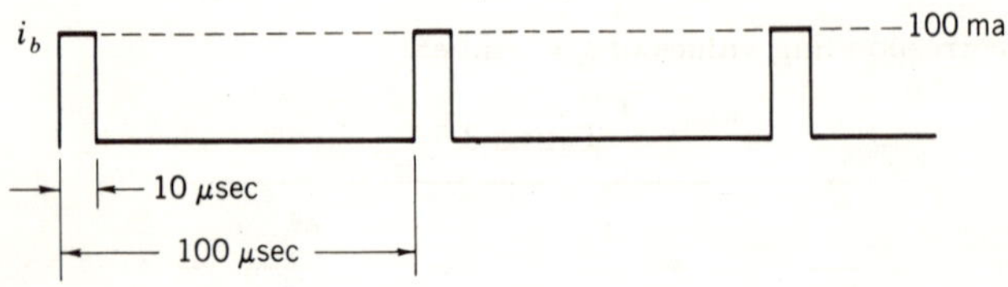

FIG. 3-11. The average value of the tube plate current is to be determined.

Solution. Even though 2.5 watts is the maximum plate dissipation, this is the maximum steady-state, or average, value. The peak plate dissipation can be greater, just so long as it averages 2.5 watts or less. We can drive the plate current with bursts having a peak value of 100 ma. The average value of plate current is then

$$I_{av} = 100 \text{ ma} \times \frac{10\ \mu\text{sec}}{100\ \mu\text{sec}} = 10 \text{ ma}$$

This is what a d-c milliameter in the plate circuit would read. Now since the direct, or average, current is 10 ma, the d-c plate-to-cathode voltage during the pulse cannot exceed

$$E_{dc} = \frac{2.5 \text{ watts}}{10 \text{ ma}} = 0.25 \times 10^3 = 250 \text{ volts}$$

Between pulses, the plate voltage may greatly exceed 250 volts. The plate current is cut off at this time, however, and therefore the power dissipated is zero.

The ratio of pulse presence to pulse period is sometimes referred to as the duty cycle. In this problem the duty cycle was 10 μsec/100 μsec = 0.1, or 10 per cent. If the duty cycle were reduced to 1 per cent, the plate voltage during the pulse could be raised to 2,500 volts if the tube did not arc internally.

3-4. The *RC* Circuit.

The waveform characteristics required by radar and pulse circuits, for example, duration, slope, repetition, amplitude, and frequency, are secured largely by controlling the variation of voltage with respect to time. Since the charging or discharging of a capacitor through a resistor requires time, the circuits commonly used for achieving these desired characteristics are of the resistance-capacitance (*RC*) type.

A fundamental point to bear in mind when analyzing *RC* circuits is that the charge on a capacitor cannot change instantaneously. The change in charge on a capacitor depends upon the number of electrons transferred into it. If the capacitor is initially uncharged, the actual charge depends upon the average rate at which charge is moving into it, multiplied by the time it takes to move it in. That is

$$q_{\text{electrons}} = \left(\frac{i_{\text{electrons}}}{\text{sec}}\right) t \text{ sec}$$

Now if we were to close a switch which causes some i to flow into a capacitor and consider the time at an infinitesimal instant after the switch closed, we would see that the charge must be very small, since the time t is small. If we were to consider smaller and smaller intervals of time, q would be less and less. In the limit, if t were zero, then q would be zero. This illustrates why the charge on a capacitor requires time to change. Sometimes it is helpful to imagine the capacitor as a bucket and the current as a flow of water.

The capacitance of a capacitor is defined as the charge per volt existing across it. That is, if some capacitor develops a voltage of 1 volt between its terminals when 1 coulomb of charge is forced into it, we say that this capacitor has a capacitance of 1 farad. In equation form, this can be written as

$$C = \frac{q}{e_C}$$

where e_C is the voltage across the capacitor. Therefore

$$e_C = \frac{q}{C}$$

Since capacitor charge q cannot change instantly, it follows that neither can the capacitor voltage e_C change immediately. Consequently, when analyzing *RC* circuits, we must be aware that the voltage across the capacitor cannot *change* instantly. The actual voltage across the capacitor can be any value or zero. All we are saying is that no matter what the voltage across the capacitor is at some instant of time, it will be the same value at any infinitesimally small instant of time later. Of course, if we consider the voltage at some time that is significantly different from the initial time, then the capacitor voltage will have changed by some amount. We are not making any restrictions concerning the *current* in an *RC* circuit. That may change instantaneously.

Most *RC* circuits, and the voltages applied to them, are reducible to the simple equivalent circuit of Fig. 3-12*a*. This shows a simple series circuit consisting of a battery, switch, capacitor, and resistor. When the d-c voltage is applied in the circuit of Fig. 3-12*a*, the waveforms of Fig. 3-12*b* will result.

Let us begin the analysis by assuming that the capacitor has no initial charge on it. This means—and the point is important—that the voltage across it is zero before the switch closes. Even at the instant when the switch closes, the capacitor voltage will still be at zero, for it takes time to build up a charge. This means that, at this time, there is no voltage drop across the capacitor, and so the full battery voltage is impressed across the resistor. To put it in other words, the peak value of the surge

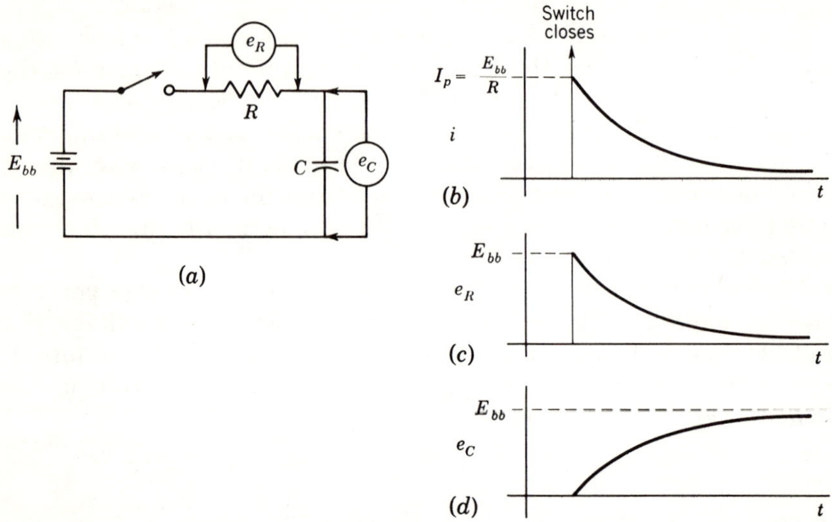

FIG. 3-12. (*a*) The capacitor will charge toward the supply voltage when the switch is closed. (*b*) The current waveform. (*c*) The resistor-voltage waveform. (*d*) The capacitor-voltage waveform.

of charging current which starts to flow is at first determined by the resistor. The formula for this peak value is as follows:

$$I_p = \frac{E_{bb}}{R}$$

But as the capacitor begins to charge, a voltage is built up across it which bucks the battery voltage, leaving less voltage for the resistor. As the charging continues, the current keeps decreasing, since the battery voltage is being bucked by an increasing voltage upon the capacitor.

In theory, the capacitor never quite charges to the battery voltage. Actually, however, after enough time has elapsed, the capacitor can be considered as having charged to E_{bb} volts. Now there will be no voltage left for the resistor, and the current should have decayed to zero.

Figure 3-12*b* shows how the resistor current i, the resistor voltage e_R, and the capacitor voltage e_C vary with time. The final value of capacitor voltage is determined solely by E_{bb}, and how long it takes to get there depends on the resistor and capacitor sizes. A large resistor, for example, holds the charging current down. This increases the time required to charge the capacitor. A small resistor permits a large initial charging current to flow into the capacitor, the voltage of which then builds up rapidly. Again, a large value of C increases the time required to charge the capacitor fully. This is apparent when we consider that the capacitor voltage e_C is at all times equal to $e_C = q/C$, where q is the charge on the capacitor.

The bigger C is, the smaller e_C will be for any given charge q. The exact relationship between the capacitor voltage and circuit parameters is

$$e_C = E_{bb}(1 - \epsilon^{-t/RC})$$

as a function of time, where $\epsilon = 2.718$. This formula is the solution to a differential equation which we can readily develop. We can write

$$E_{bb} = Ri + e_C$$

The latter in itself is simply a statement of Kirchhoff's voltage law, since the resistor drop, plus the capacitor voltage, must at all times equal the battery voltage.

This equation can be wrestled into a more meaningful form by showing that the instantaneous current flowing through a capacitor is at all times equal to

$$i_C = C\frac{de_C}{dt} = C\frac{\Delta e_C}{\Delta t}$$

We have discussed the fact that $q = Ce_C$. Let q change by some amount Δq. Then there must be a corresponding change in e_C, which we shall call Δe_C. Then

$$q + \Delta q = C(e_C + \Delta e_C)$$

$$\Delta q = Ce_C + C\Delta e_C - q$$

But $q = Ce_C$. Therefore, $\Delta q = C\Delta e_C$, and, if we divide both sides of the equation by Δt,

$$\frac{\Delta q}{\Delta t} = C\frac{\Delta e_C}{\Delta t}$$

but

$$\frac{\Delta q}{\Delta t} = i$$

Therefore,

$$i = C\frac{\Delta e_C}{\Delta t}$$

The quantity $\Delta e_C/\Delta t$ signifies a small change in e_C during the interval Δt. This does not mean, however, that i_C was exactly the same at the start of the interval under consideration as at the end. $\Delta e_C/\Delta t$ is really the average current throughout the interval Δt. As we make this interval Δt smaller and smaller, we approach the instantaneous value with greater accuracy. When we speak of instantaneous rates of change, the symbol d/dt is used. Now, in the series circuit we were analyzing, the same current must flow through R and through C. We may then write

$$E_{bb} = R\left(C\frac{de_C}{dt}\right) + e_C$$

Note that e_C is a function of its own rate of change, with respect to time. Equations involving rates of change are known as differential equations.

3-5. Normalizing a Curve. One way of solving RC problems is to set up and solve the differential equations involved. This is tedious and requires mathematical tools we have not yet mastered. The differential equation, however, is the most powerful approach and underlies all other methods.

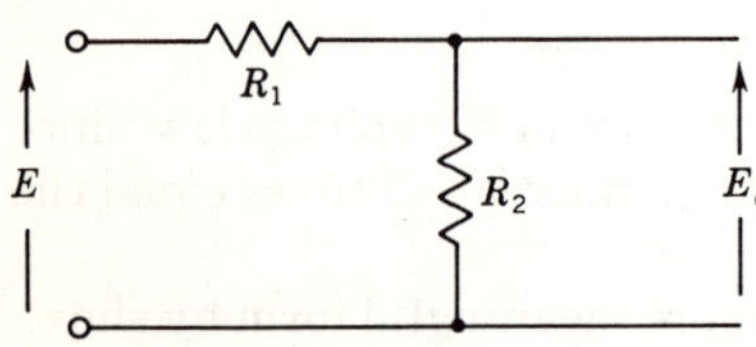

FIG. 3-13. A simple voltage divider. The output is to be graphically related to the input.

An alternate approach is to use the solution of the differential equation and plug in appropriate values wherever the formula calls for them. This is feasible, but requires extensive use of tables or slide rule. We may, however, modify this method as follows.

We will use the tables so as to generate a universal curve for solving RC problems. Let us first, however, get some practice in curve generation by posing a simpler problem.

Consider the voltage divider of Fig. 3-13. Suppose we were interested in plotting a graph which relates E_o to E for any specific values of R_1 and R_2. The relation between E_o and E is

$$E_o = \frac{R_2 E}{R_1 + R_2} = \frac{1}{R_1/R_2 + 1} E$$

Now, let
$$x = \frac{1}{R_1/R_2 + 1}$$

By inspection of the values for R_1 and R_2 in the circuit, we can determine x. And, therefore, $E_o = Ex$.

We could plot E_o against x for different values of E, and get a family of curves, as shown in Fig. 3-14a. This is very awkward, however, because the E applied to the circuit might be some value between E_2

and E_3 which is not plotted. We can solve this difficulty with a mathematical trick. Instead of plotting E_o against x, let us plot E_o/E against x. This is known as *normalizing a curve*. Since E_o can approach but never be greater than E, the maximum ordinate is 1. The curve now will look as shown in Fig. 3-14*b*. The ordinates could just as well have been expressed as percentages. Note that only one curve is necessary on the normalized graph.

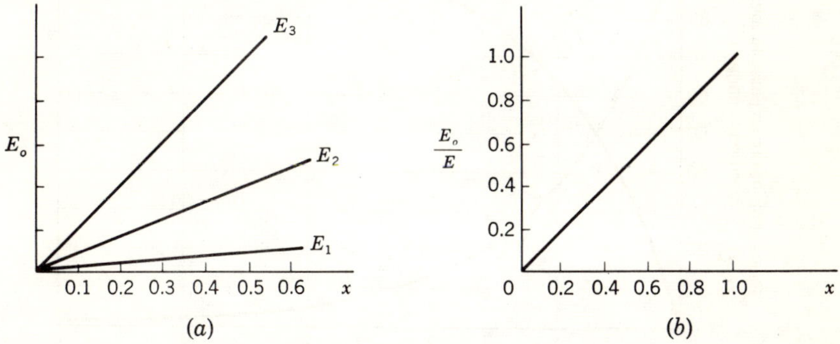

FIG. 3-14. (*a*) A family of curves relating the output to the attenuation ratio may be plotted. (*b*) It is more convenient to plot the ratio of the output to input versus the attenuation ratio.

Let us see how we can use this normalized curve to solve voltage-divider problems. Suppose, in Fig. 3-13*a*, that $R_1 = 10$ kilohms, $R_2 =$ 40 kilohms, and $E = 250$ volts. What is E_o? The first thing we do is evaluate x:

$$x = \frac{1}{10/40 + 1} = \frac{1}{1.25} = 0.8$$

Now we enter the chart in Fig. 3-14*b* at $x = 0.8$, and read the corresponding E_o/E, which is also 0.8. This means E_o is 80 per cent of E_{in}, or $E_o = 0.8(250) = 200$ volts.

Actually, it would have been more practical to plot E_o/E versus R_1/R_2 directly. But for a simpler voltage divider, a curve is not necessary.

3-6. The Time-constant Chart (Capacitor Charging). Suppose you were given that the equation for buildup of voltage across a capacitor with zero initial charge is

$$e_C = E(1 - \epsilon^{-t/RC})$$

For the moment, let e_C be the instantaneous voltage across the capacitor at any time (t). The capacitor is charging through some equivalent source resistance R. E is the open-circuit source voltage that the capacitor looks back into, and ϵ is 2.718, the base of the natural system of logarithms. If we could look at any specific circuit and evaluate R, C, and E, we could solve for e_C after any interval of time t. Working with such an exponential equation is somewhat awkward, and it would be worthwhile to investigate the possibility of developing a universal curve that can be applied to the exponential equation.

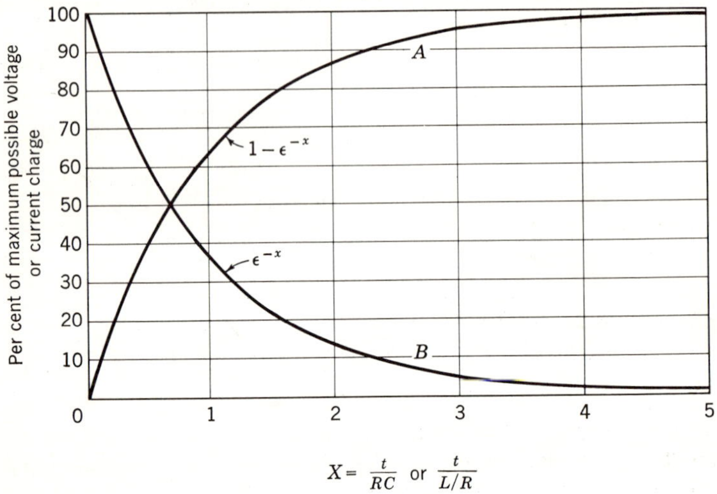

FIG. 3-15. Universal time-constant chart. Curve A: capacitor charge voltage; inductor charge current. Curve B: capacitor discharge voltage; capacitor charge current; inductor charge voltage.

First, let us normalize the equation so that we are not restricted to specific values of voltage. We can then write:

$$\frac{e_C}{E} = 1 - \epsilon^{-t/RC}$$

Then let $t/RC = x$, which gives us $e_C/E = 1 - \epsilon^{-x}$. We can then calculate or determine from tables the values of ϵ^{-x} for different values of x. If calculating, remember that $\epsilon^{-x} = \frac{1}{\epsilon^x}$. Next, let us take these tabulated values of ϵ^{-x} and subtract them from 1. This gives us the value of the function $1 - \epsilon^{-x}$ which is equal to e_C/E. We can now plot a curve of e_C/E vs. x, as shown in Fig. 3-15.

Let us see how we use the curve. Consider the circuit of Fig. 3-16. What will the capacitor voltage be in one second after the switch closes, if there is no initial charge on C? The first thing we do is evaluate x. We have set $t = 1$ sec, and from the circuit $RC = 1$ meg $\times$ 1 μf $= 1$ sec. The product of R times C is expressed in seconds (Table 3-2).

The exponent is then

$$x = \frac{t}{RC} = \frac{1 \text{ sec}}{(1 \text{ meg})(1\ \mu\text{f})} = \frac{1 \text{ sec}}{1 \text{ sec}} = 1$$

We then enter the chart at $x = 1$, and find the corresponding value of e_C/E at the curve intersection. We find that $e_C/E = 0.63$ for $x = 1$.

Since $E = 100$ volts, $e_C = 0.63(100) = 63$ volts in 1 sec. The value of the RC product determined by the actual circuit parameters is referred to as the time constant of the circuit. Whenever the time we are considering is equal in seconds to the circuit time constant, the capacitor will build up to 63 per cent of its final voltage. The value of the exponent x is equal to 1, since $t = RC$ under this condition.

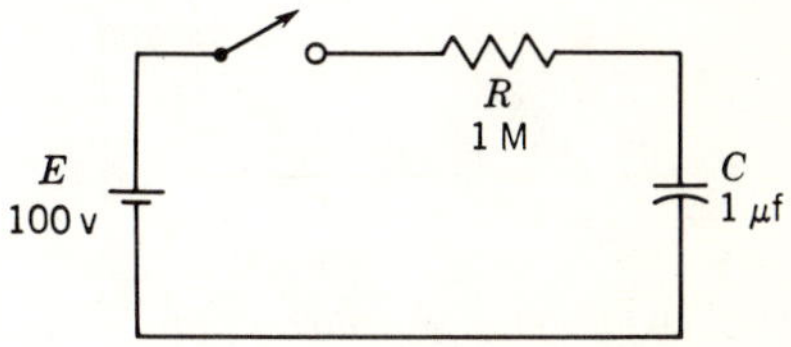

FIG. 3-16. The capacitor voltage at any instant of time is readily determined with the aid of the time-constant chart.

Suppose the switch were closed for 2 sec rather than for 1 sec. Now

$$x = \frac{t}{RC} = \frac{2 \text{ sec}}{1 \text{ sec}} = 2$$

We could say the switch is closed for 2 time constants in this case. The value e_C/E corresponding to $x = 2$ is about 0.87, which means that $e_C = 87$ volts, 2 sec after the switch closes.

TABLE 3-2

TIME CONSTANTS

R	C	t
Ohms	Farads	Seconds
Ohms	μf	μsec
Kilohms	μf	Milliseconds
Megohms	μf	Seconds
Megohms	μμf	μsec

Let us try a few sample problems involving the charging of capacitors.

Problem 1. In Fig. 3-17, estimate the capacitor voltage 100 μsec after the switch is closed.

Solution. First, determine the natural time constant of the circuit and then compare it with the number of time constants the switch has been closed. The natural time constant τ is

$$\tau = 0.1 \text{ meg } (100\ \mu\mu\text{f}) = 10\ \mu\text{sec}$$

The switch has been closed for 100 μsec, which is 10 time constants. We see from the time-constant chart that in four time constants, $x = 4$, the capacitor will be charged to 98 per cent of the final value. Therefore, in 10 time constants, the capacitor is surely charged to 120 volts.

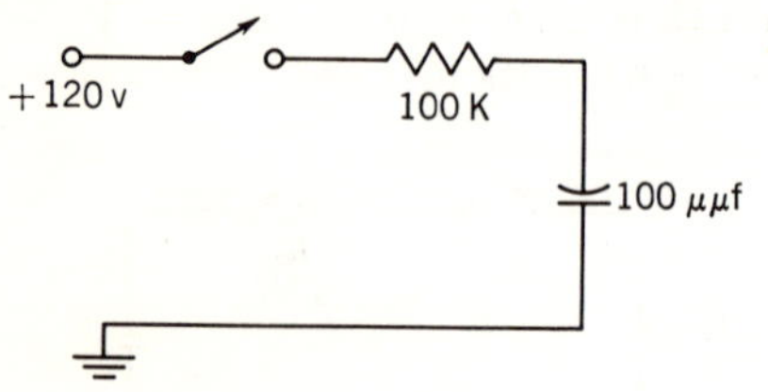

FIG. 3-17. The capacitor voltage in 7.5 μsec is to be determined.

Problem 2. In the circuit of Problem 1, what is the capacitor voltage 7.5 μsec after the switch closes?

Solution. The value of t/τ in this case is 7.5 μsec/10 μsec $= 0.75$. With the exponent value equal to 0.75, we see that the capacitor charges to 52 per cent of its full swing, which, in this case, is 120 volts. Therefore, the capacitor voltage is 0.52(120 volts) $= 62.4$ volts.

Up to this point we have considered only situations in which the capacitor is initially uncharged. What if the capacitor has some initial charge on it the moment a switch closes? To handle this case, let us expand the meaning of our equation for the capacitor-charging voltage, which is $e_C = E(1 - \epsilon^{-t/RC})$. The quantity E, in the broadest sense, is not necessarily the battery supply, but the difference in voltage between that across the capacitor and the battery at the instant of switching. This difference is the maximum possible change (swing) in capacitor voltage. Similarly, e_C is the actual change in capacitor voltage from the initial voltage. A few problems will best illustrate this.

Problem 3. In the circuit of Problem 1, suppose the switch were closed for 10 μsec, opened for 20 μsec, and reclosed for another 10 μsec. What is the capacitor voltage at the end of this interval?

Solution. In the first 10 μsec, the capacitor will charge to 63 per cent of its total swing, since $t/\tau = 1$. Therefore, $e_C = 0.63(120 \text{ volts}) = 75.5$ volts. With the switch then open for 20 μsec, the capacitor voltage will stay constant at this value of 75.5 volts, since there is no discharge path shunting the capacitor.

A total of 30 μsec has now elapsed. The switch then recloses. The capacitor has an initial voltage of 75.5 volts on it at the start of this 10-μsec interval. The maximum possible swing in capacitor voltage is now 120 volts $-$ 75.5 volts $= 44.5$ volts.

In 10 μsec the capacitor will have charged up to 63 per cent of 44.5 volts, which is 28 volts. Since we had an initial charge of 75.5 volts, the total capacitor voltage is now 75.5 volts $+$ 28 volts $= 104$ volts.

Problem 4. Suppose you had a 200-volt battery and wished to design an *RC* timing circuit with the following requirements. The output should reach 50 per cent of the applied input in 100 μsec, and the maximum current pulled from the supply is to equal 1 ma.

Solution. Since the output voltage is to lag the input, we know we are interested in taking the output off the capacitor. The circuit will look like Fig. 3-18. We also know that the maximum current flows the instant the switch is closed and that the capacitor looks like a short circuit. The only factors determining the peak current are E and R:

$$I_p = \frac{E}{R}$$

$$R = \frac{200 \text{ volts}}{1 \text{ ma}} = 0.2 \text{ meg}$$

Moreover, we know that e_o must be 100 volts (50 per cent of E) in 100 μsec. Therefore, entering the vertical scale on the time constant chart at 0.5, we see that

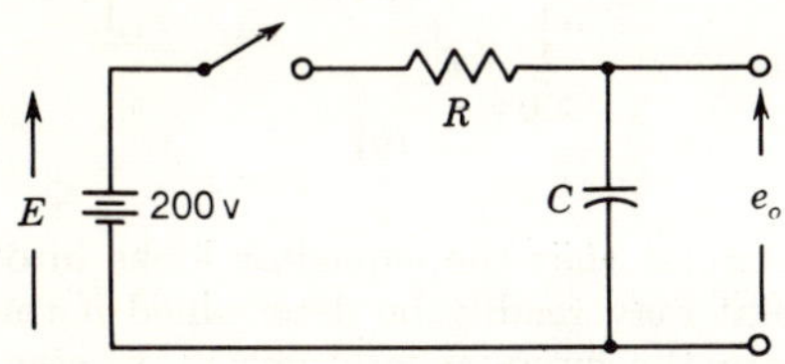

FIG. 3-18. The values of RC are to be chosen so that e_o will equal 50 per cent of E in 100 μsec.

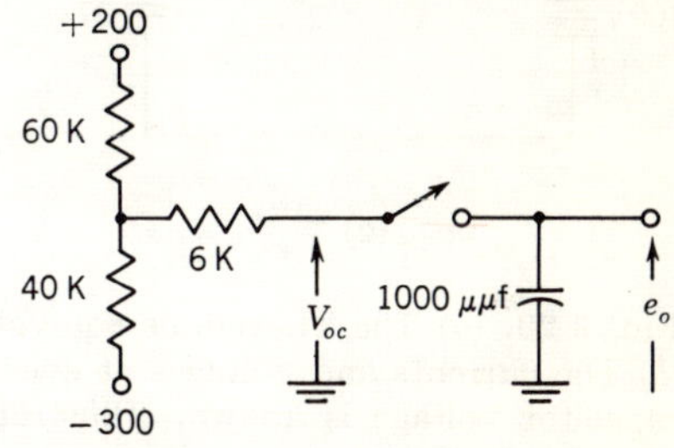

FIG. 3-19. A complex RC network in which the capacitor current and voltage are to be determined.

the corresponding value of the exponent is 0.7. That is, $t/\tau = 0.7$. We know that $t = 100$ μsec and, therefore, we can solve for τ:

$$\tau = RC = \frac{100\ \mu\text{sec}}{0.7} = 143\ \mu\text{sec}$$

$$R = 0.2 \text{ meg}$$

$$C = \frac{143\ \mu\text{sec}}{0.2 \text{ meg}} = 715\ \mu\mu\text{f}$$

Problem 5. Sometimes the charging or source resistance is not obvious. For example, consider Fig. 3-19. What will the capacitor voltage e_o and the capacitor current I_C be in 15 μsec? What will the currents in the 60- and 40-kilohm resistors be in 15 μsec?

Solution. The first thing we must determine is the time constant of the circuit. To do this, we must first determine the equivalent resistance that the capacitor looks back into. This is the Z_{th} of Thévenin's theorem:

$$Z_{th} = 6 + \frac{(60)(40)}{60 + 40} = 30 \text{ kilohms}$$

Therefore, $\tau = 30 \text{ kilohms} \times 1{,}000\ \mu\mu\text{f} = 30\ \mu\text{sec}$

The maximum voltage the capacitor will ever charge to is V_{oc}, the open-circuit voltage. We can easily determine this:

$$V_{oc} = 200 \text{ volts} - 500 \text{ volts} \times \frac{60 \text{ kilohms}}{100 \text{ kilohms}} = -100 \text{ volts}$$

In 15 μsec, $t/\tau = 15\ \mu\text{sec}/30\ \mu\text{sec} = 0.5$. Entering the horizontal axis of the curve for capacitor build-up voltage at $x = t/\tau = 0.5$, we find $e_C/E_{\max} = 0.37$. This means that in 15 μsec the capacitor will have charged to -37 volts with respect to ground. If the capacitor is charged to -37 volts and $V_{oc} = -100$ volts, then 63 volts must exist across Z_{th}. The current through Z_{th} must equal $I = 63$ volts/30 kilohms $= 2.1$ ma (see Fig. 3-20a).

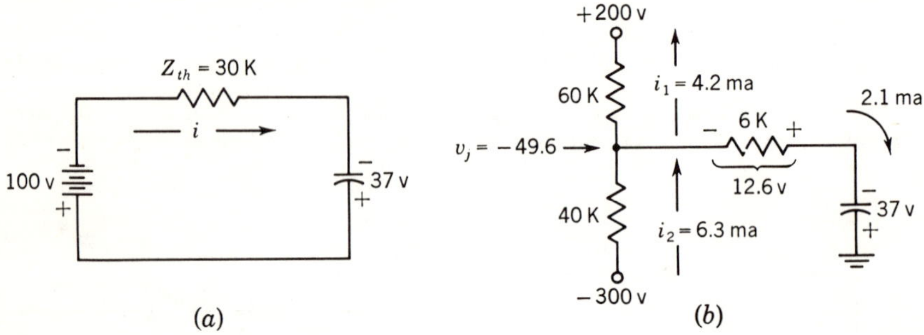

FIG. 3-20. (a) The Thévenin's equivalent circuit that the capacitor looks into. (b) The currents and voltages at every point may readily be determined if the capacitor voltage is known. Illustrated are the circuit conditions in 15 μsec.

The voltage across the 6-kilohm resistor is then 2.1 ma $\times$ 6 kilohms $= 12.6$ volts. The voltage at the junction of the 40- and 60-kilohm resistors is

$$v_j = -37 - 12.6 \text{ volts} = -49.6 \text{ volts}$$

(see Fig. 3-20b). Once the junction potential v_j is known, we can readily determine the current through the 60-kilohm resistor i_1. Since v_j is negative with respect to the 200-volt end of the 60-kilohm resistor, i_1 will flow as shown and be equal to

$$i_1 = \frac{200 + 49.6 \text{ volts}}{60 \text{ kilohms}} = \frac{250}{60} = 4.2 \text{ ma}$$

$$i_2 = 4.2 + 2.1 \text{ ma} = 63 \text{ ma}$$

3-7. Capacitor Discharging. When a capacitor discharges through a resistor, the same equation that was used for capacitor charging still holds true. That is,

$$e_C = E(1 - \epsilon^{-t/RC}) \tag{3-1}$$

where e_C is again the actual change in capacitor voltage and E is the maximum possible change. The actual capacitor voltage is then equal to the initial capacitor voltage minus the change in capacitor voltage.

An alternative expression which may be used in the analysis of RC circuits is

$$e_C = E\epsilon^{-t/RC} \tag{3-2}$$

In this case, e_C is *not* the actual change in capacitor voltage, but is the difference between the actual change which has occurred and the maximum possible change, E.

Now we can normalize and plot formula (3-2) in the same manner that we applied to formula (3-1). We can write

$$y = \frac{e_C}{E} = \epsilon^{-x}$$

where $x = t/RC$. A plot of $y = \epsilon^{-x}$ is illustrated in Fig. 3-21a. Notice that when $t = RC$ or $x = 1$, e_C/E equals 0.37. For formula (3-2) this is interpreted to mean that in a period of time equal to the natural time constant of the circuit, only 37 per cent of the maximum possible change remains to be traversed. This is another way of saying that 63 per cent of the maximum possible change has been traversed.

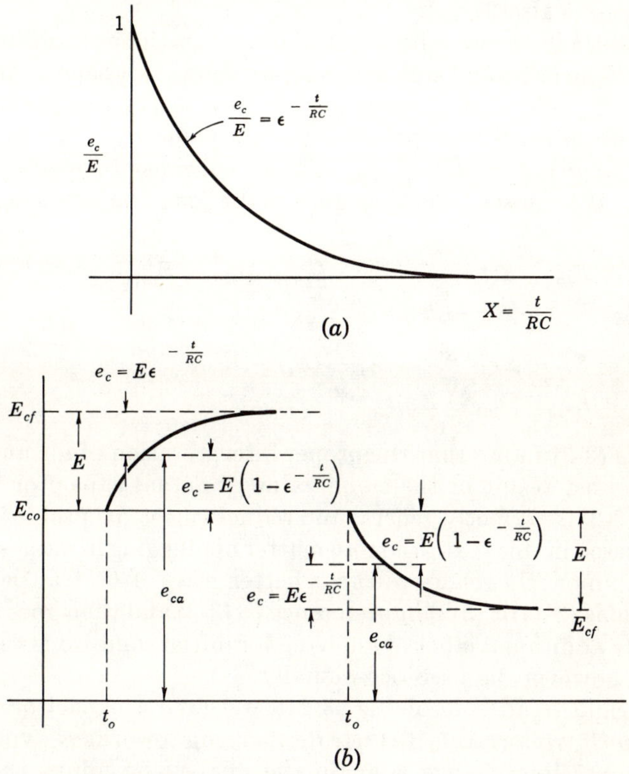

FIG. 3-21. (a) The capacitor voltage decays in an exponential manner. (b) Formulas (3-1) and (3-2),

where $E_{cf} - E_{c0} = E$ = maximum possible change in capacitor voltage during transient

e_{ca} = total instantaneous capacitor voltage

e_c = actual change in capacitor voltage from initial value

E_{c0} = initial (original) capacitor voltage at start of transient

E_{cf} = final capacitor voltage at end of transient.

Suppose we had a 1-μf capacitor charged to 100 volts and we suddenly switched a 1-meg resistor across it for 1 sec. The natural time constant of the circuit (RC) is 1 sec. By formula (3-1) or the curve of $y = 1 - \epsilon^{-x}$, which is the same thing, we would find that the change or decrease in capacitor voltage is 63 volts. This means that there are $100 - 63 = 37$ volts left across the capacitor. Formula (3-2) yields exactly the same answer for the capacitor voltage. In Fig. 3-21*a* we see that for $x = t/RC = 1$, $e_C/E = 0.37$. Since the maximum possible swing, E, is from 100 volts down to zero volts, $E = 100$ volts. Therefore, the remaining change in capacitor voltage (e_C) is 37 volts. In this case, since the capacitor is discharging towards zero volts, the actual voltage across the capacitor is also 37 volts.

Figure 3-21*b* illustrates the physical interpretations of formulas (3-1) and (3-2). The left-hand side of the figure shows a capacitor originally charged to E_{c0} volts. At t_0, switching occurs which causes the capacitor to charge up towards the final steady-state value E_{cf}. The maximum possible change is $E = E_{cf} - E_{c0}$. The actual capacitor voltage at any time is e_{ca}. We can see from the figure that e_{ca} may be expressed in two ways.

$$e_{ca} = E_{c0} + E(1 - \epsilon^{-t/RC}) = E_{c0} + (E_{cf} - E_{c0})(1 - \epsilon^{-t/RC})$$
$$= E_{cf} + (E_{c0} - E_{cf})\epsilon^{-t/RC}$$

or

$$e_{ca} = E_{cf} - E\epsilon^{-t/RC} = E_{cf} - (E_{cf} - E_{c0})\epsilon^{-t/RC}$$
$$= E_{cf} + (E_{c0} - E_{cf})\epsilon^{-t/RC} \tag{3-3}$$

Formula (3-3) shows that the proper interpretation of formulas (3-1) and (3-2) must result in the same value of actual capacitor voltage. Formula (3-3) is perfectly general and reduces most RC problems of the type discussed in this text to a mere matter of plugging in values. In the author's opinion, the reader will get a better grasp of RC transients if he uses formulas (3-1) or (3-2) in preference to (3-3) and then goes through the slightly additional labor of allowing for initial conditions. Formula (3-3) will, however, be used occasionally.

In the right-hand side of Fig. 3-21*b* we have a capacitor initially charged to E_{c0} volts. At t_0 it starts discharging towards E_{cf} volts. The maximum possible change is again the final value minus the initial value ($E = E_{cf} - E_{c0}$). As before, the actual $e_{ca} = E_{c0} + E(1 - \epsilon^{-t/RC})$ or $E_{cf} - E\epsilon^{-t/RC}$.

The quantity E in this case will be negative, since the magnitude of E_{cf} is less than that of E_{c0}. Both of these equations for e_{ca} in the discharge case are identical with those of the charging case on the left side of Fig. 3-21*b*. Hence, these equations will also simplify to Eq. (3-3) which is perfectly general.

A few sample problems will give us some background for handling capacitor discharge transients.

Problem 1. What will the capacitor voltage in Fig. 3–22 be 1 sec after the switch closes? The capacitor is initially charged to 120 volts.

Solution. The initial capacitor voltage is 120 volts. If the switch were indefinitely closed, the capacitor would discharge to zero. Therefore, the total capacitor voltage swing is 120 volts. The time constant of the circuit is

$$\tau = RC = 10 \text{ meg } (0.1\ \mu\text{f}) = 1 \text{ sec}$$

The equation we are going to solve graphically is

$$e_C = 120\epsilon^{-1\text{sec}/1\text{sec}}$$

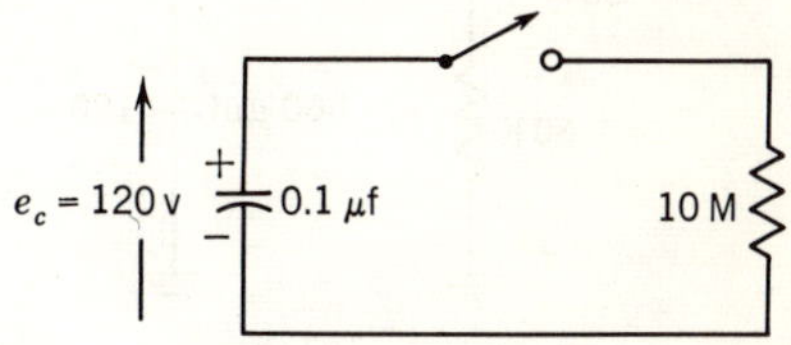

FIG. 3-22. Discharging a capacitor that is initially charged.

The designation e_C for capacitor voltage is used interchangeably with e_{ca} of Fig. 3-21*b*. There should be no confusion, however, as the problem discussion will make the meaning clear.

Entering the discharge curve at $t/RC = 1$, we see that the voltage across C will have fallen to 37 per cent of the maximum value. Therefore, in 1 sec, $e_C = 0.37 \times 120 = 44.4$ volts. We could just as well have reasoned that in one time constant the capacitor voltage will swing 63 per cent, or $0.63 \times 120 = 75.5$ volts. If the initial voltage is 120, the final voltage after one second is $120 - 75.5$ volts $= 44.5$ volts.

Problem 2. In the problem of Fig. 3–23*a*, the capacitor is initially charged to minus 120 volts with respect to ground. Plot the capacitor voltage and current as a function of time, starting at the instant the switch closes.

Solution. First thevenize the network to the left of the capacitor. We then have the equivalent circuit of Fig. 3-23*b*.

$$V_{oc} = 240\left(\frac{60 \text{ kilohms}}{90 \text{ kilohms}}\right) = 160 \text{ volts}$$

$$Z_{th} = \frac{60(30)}{90} = 20 \text{ kilohms}$$

The instant the switch closes, the capacitor will look like a 120-volt battery, since it cannot change its voltage instantaneously. The two sources are series-aiding, so that the peak current will be

$$I_p = \frac{120 + 160 \text{ volts}}{20 \text{ kilohms}} = 14 \text{ ma}$$

The capacitor must ultimately charge to +160 volts from −120 volts, so the total capacitor swing will be 120 + 160 volts = 280 volts. The natural time constant of the circuit is

$$\tau = 20 \text{ kilohms}(500\ \mu\mu\text{f}) = 10\ \mu\text{sec}$$

See Fig. 3-23*c* for an illustration of the capacitor voltage versus time and Fig. 3-23*d* for capacitor current. In 5 μsec, or one-half of a time constant, the capacitor will have swung 37 per cent of its swing, or 0.37(280 volts) = 104 volts. The actual capacitor voltage with respect to ground in 5 μsec is, then, $-120 + 104 = -16$ volts. And the capacitor current is

$$\frac{160 + 16 \text{ volts}}{20 \text{ kilohms}} = \frac{176}{20} = 8.8 \text{ ma}$$

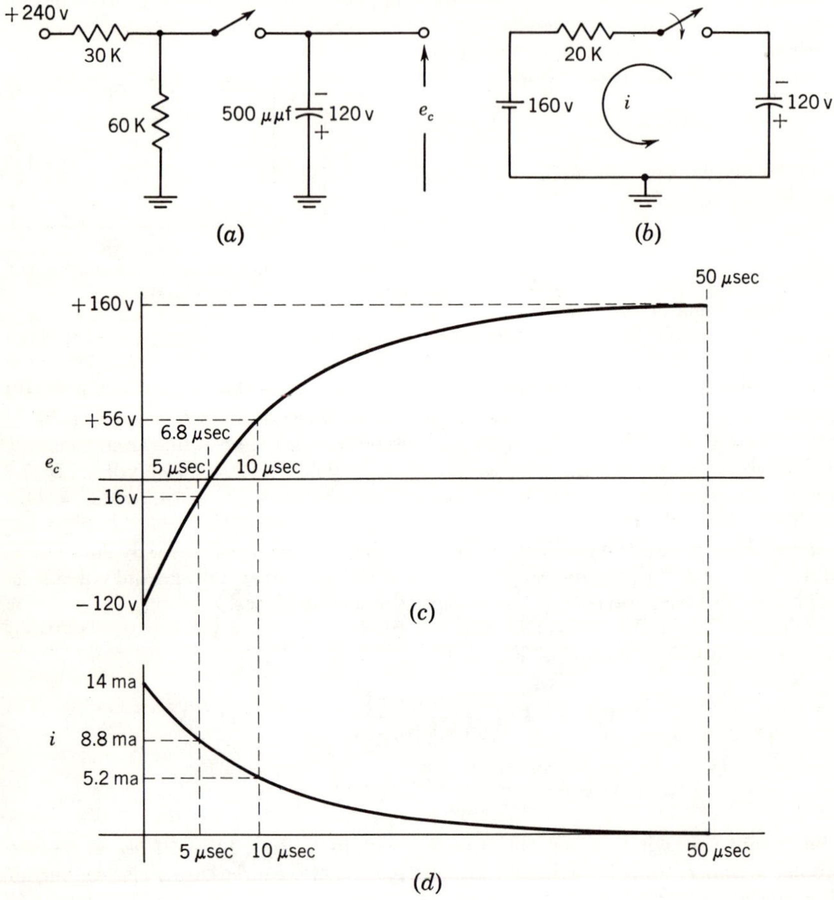

FIG. 3-23. (*a*) The charge or discharge path that the capacitor looks into is not obvious. (*b*) Thévenin's theorem simplifies the circuit that the capacitor looks into. (*c*) The capacitor voltage and (*d*) the capacitor current transient response for the circuit of *a*.

In one time constant, or 10 μsec, the capacitor voltage will have gone through 63 per cent of its total swing, which is $0.63(280) = 176$ volts. The capacitor voltage is then $-120 + 176 = +56$ volts. The current is now

$$\frac{160 - 56 \text{ volts}}{20 \text{ kilohms}} = \frac{104}{20} = 5.2 \text{ ma out of the capacitor}$$

It might be interesting to find the exact time it took for the capacitor to reach 0 volts. This means the capacitor must have swung through 120 volts to get to zero, since it had -120 volts to start with. We can therefore write

$$\frac{e_C}{E_{max}} = \frac{120}{280} = 0.43$$

and enter the curve for capacitor voltage buildup to find the corresponding $t/RC = 0.68$. Therefore, $t = 0.68 \times 10\ \mu\text{sec} = 6.8\ \mu\text{sec}$.

3-8. Periodic Switching. Frequently we encounter circuits in which the capacitor charges and discharges periodically. In some instances, the capacitor never gets a chance to charge or discharge fully before switching occurs. Let us investigate the simple RC switching circuit of Fig. 3-24*a*.

Problem 1. Assume that the switch is initially in position 1 and has been there for a long time. How does the output voltage vary with time, if the switch is in position 1 for 1 sec; in position 2 for 2 sec?

Solution. If the switch is in position 1 initially, $e_C = 100$ volts. Upon switching to 2, the capacitor starts to discharge through R_2 as in Fig. 3-24*b*. The time constant of the discharge path is $\tau_d = 0.5\ \text{meg} \times 1\ \mu\text{f} = 0.5$ sec. The switch is in position 2 for 2 sec, or $2/0.5 = 4$ time constants. C will then discharge by 98 per cent of 100 volts. Therefore, $e_C = +2$ volts in 2 sec. The switch now snaps back to position 1. Bear in mind that the capacitor now has an initial charge of 2 volts on it. The initial conditions are now as shown in Fig. 3-24*c*, and the capacitor starts to recharge through R_1. The switch stays in position 1 for 1 sec. The total swing that the capacitor could possibly go through is now 100 volts $-$ 2 volts $=$ 98 volts. Now $RC = \tau_C = 1$ sec. Since the switch is closed for 1 sec, therefore $t/\tau_C = 1\ \text{sec}/1\ \text{sec} = 1$.

We then enter the time constant chart at $t/\tau_C = x = 1$, and we see that the voltage swings through 63 per cent of its full swing. Therefore, at the end of 1 sec, C will have risen 98 volts (0.63) $=$ 62 volts. The total voltage on C will then be the increase of 62 volts, plus the initial voltage, 2 volts, which makes 64 volts. The switch now reverts to position 2 again, where it stays for 2 sec (Fig. 3-24*d*):

$$\tau_d = 0.5\ \text{sec} \qquad \text{and} \qquad x = \frac{t}{\tau_d} = \frac{2\ \text{sec}}{0.5\ \text{sec}} = 4$$

In 4 time constants, the capacitor will discharge 98 per cent of its initial charge. Another way of saying this is that the capacitor has 2 per cent of its initial charge left in 4 time constants. The maximum possible swing in this case is 64 volts. The final capacitor voltage in 2 sec is then 64 volts(0.02) $=$ 1.28 volts.

The switch again recloses in position 1 for 1 sec. $\tau_C = 1$ sec, and therefore $t/\tau_C = 1$. The capacitor will charge to 63 per cent of its possible swing, which is 100 volts $-$ 1.28 volts $=$ 98.7 volts, and 0.63(98.7) $=$ 62 volts. This brings us to $t = 6$ sec in Fig. 3-24*e*, where we have 62 volts across the capacitor when the switch flips back to position 2, where it stays for 2 sec. The exponent is

$$x = \frac{t}{\tau_d} = \frac{2\ \text{sec}}{0.5\ \text{sec}} = 4$$

Again the capacitor discharges to 2 per cent of its initial charge:

$$0.02(62) = 1.24\ \text{volts}$$

When the switch returns to position 1 at t_8, the capacitor starts to recharge. Since the initial voltage on the capacitor at $t = 8$ sec is virtually the same as at t_5, we can see that the final peak value will swing to about 62 volts.

Note that in this problem, a number of cycles had to be considered before we could determine what the capacitor voltage would look like after many cycles had elapsed.

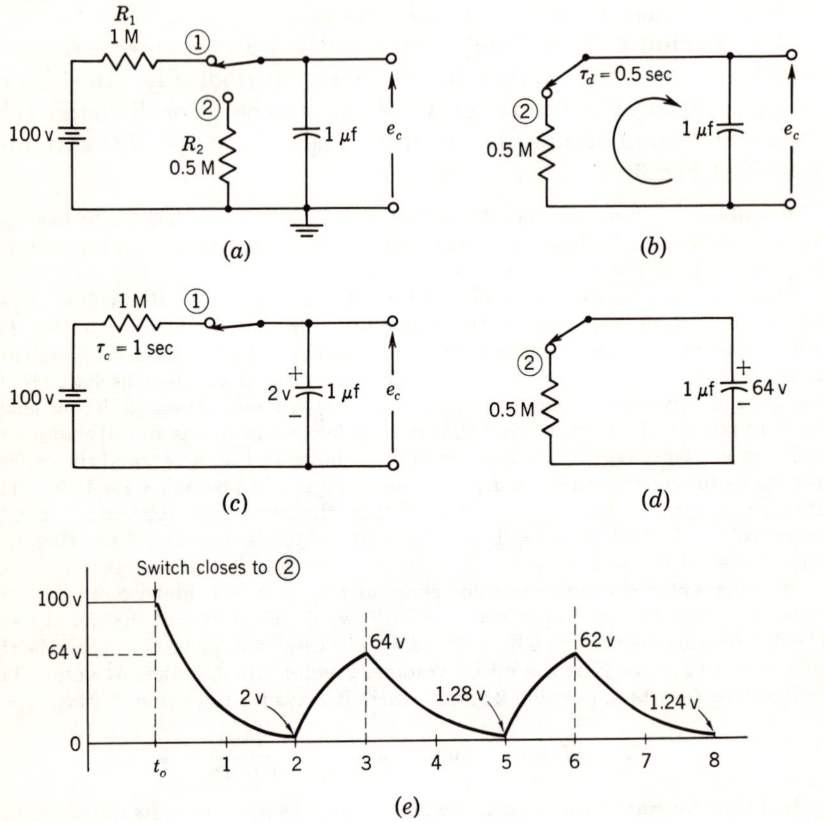

FIG. 3-24. (*a*) The switch repetitively swings between positions 1 and 2. (*b*) The discharge path for the capacitor. (*c*) The charge path for the capacitor. (*d*) The initial capacitor voltage is different at the start of the second cycle. (*e*) The output voltage is settling down to its final peak-to-peak swing at the third cycle.

Essentially the same result could have been derived by using the previously developed formula (3-3), $e_{ca} = E_{cf} + (E_{c0} - E_{cf})\epsilon^{-t/RC}$. We know that the capacitor voltage will have the somewhat sawtooth shape of Fig. 3-24*e* after many cycles. What we do not know is the upper limit e_{ca_2} and the lower limit e_{ca_1}. Using formula (3-3), we can write two simultaneous equations. With the switch having just transferred from position 2 to position 1, the capacitor starts charging from e_{ca_1} towards 100 volts for 1 sec. Therefore, e_{ca_1} is the initial voltage E_{c0} and 100 volts is E_{cf}. The final value reached at the end of this 1-sec interval is e_{ca_2}. Therefore, our first equation is

$$e_{ca_2} = 100 + (e_{ca_1} - 100)\epsilon^{-1\text{sec}/1\text{sec}} = 100 + (e_{ca_1} - 100)(0.37) \qquad (3\text{-}4)$$

By similar reasoning, the equation relating the discharge of the capacitor from e_{ca_2} to e_{ca_1} would be

$$e_{ca_1} = 0 + (e_{ca_2} - 0)\epsilon^{2\text{sec}/0.5\text{sec}} = e_{ca_2}(0.02) \qquad (3\text{-}5)$$

Substituting (3-5) in (3-4) and solving for e_{ca_2}, we find

$$e_{ca_2} = 63 \text{ volts}$$

and from (3-5),

$$e_{ca_1} = 1.3 \text{ volts}$$

This is essentially the same result as obtained previously. Formula (3-3) greatly simplifies calculations involving the ultimate response of a repetitively switched circuit.

Problem 2. Find e_o in 1.5 μsec after switch is closed in the circuit of Fig. 3-25.

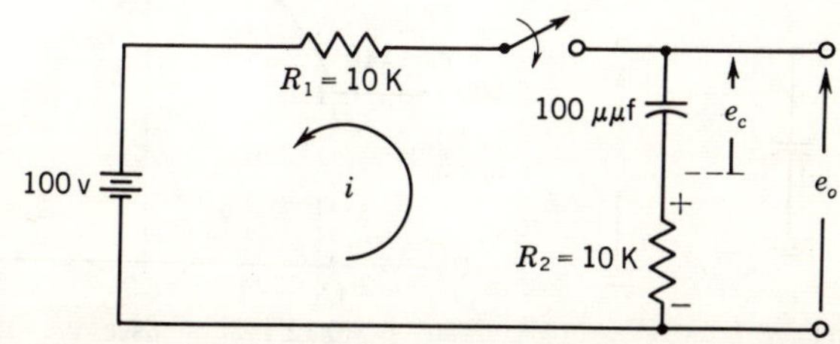

FIG. 3-25. The output voltage upon closing the switch will jump in value due to the voltage developed across R_2 by the capacitor-charging current.

Solution

$$RC = (20 \text{ kilohms})(100\,\mu\mu\text{f}) = (0.02 \text{ meg})(100\,\mu\mu\text{f}) = 2\,\mu\text{sec}$$

$$\frac{t}{RC} = \frac{1.5}{2} = 0.75 \qquad \text{and} \qquad \frac{e_c}{E} = 0.52$$

Therefore, $e_C = 52$ volts

The voltage left across the resistors is $100 - 52 = 48$ volts, and this is equally split between both. Therefore, $e_o = 24 + 52 = 76$ volts; or we could also have computed the current and then

$$i = \frac{100 - 52 \text{ volts}}{20 \text{ kilohms}} = \frac{48}{20} = 2.4 \text{ ma}$$

$$e_o = 2.4 \text{ ma } (10 \text{ kilohms}) + 52 \text{ volts} = 76 \text{ volts}$$

3-9. Rise and Fall Times. Now, consider the circuit of Fig. 3-26*a*, which is somewhat analogous to an overdriven amplifier with a capacitive load.

Problem. Assume that the switch is initially closed. At t_0 the switch opens for 5 μsec and then recloses. What will the output pulse look like, and what are the rise time t_r and the fall time t_f?

Solution. With the switch initially closed,

$$e_o = \frac{200(10 \text{ kilohms})}{20 \text{ kilohms}} = 100 \text{ volts}$$

When the switch opens, the output starts to rise toward 200 volts. The time constant during this 5-μsec interval is determined by R_L and by C:

$$\tau \text{ rise} = R_L C = 0.01 \text{ meg } (100\ \mu\mu\text{f}) = 1\ \mu\text{sec}$$

In 5 μsec, which is 5 time constants, we know the capacitor will have charged completely to 200 volts. The output (capacitor voltage) will have risen 63 per cent

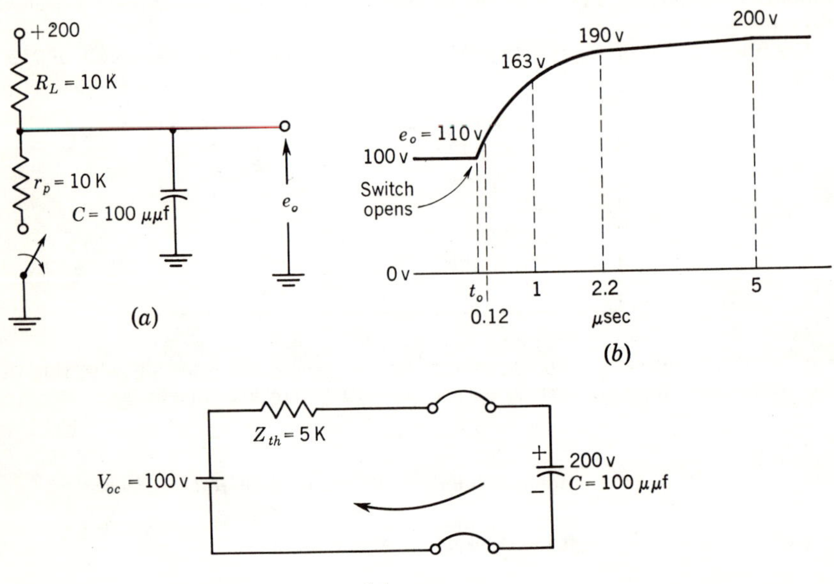

FIG. 3-26. (*a*) This circuit is analogous to an overdriven amplifier with a capacitive load. (*b*) A point-by-point analysis of the output voltage buildup. (*c*) The equivalent circuit when the switch recloses after being open for 5 μsec.

of its maximum possible swing, 100 volts, in one time constant, 1 μsec. This means that e_o in 1 μsec is

$$e_o = 100 + 0.63\ (100) = 163 \text{ volts}$$

This gives us a fair picture of the output rise. The rise time t_r is defined as the time between the 10 per cent and the 90 per cent levels of the output voltage swing. If we can find the time to swing 10 per cent and the time to swing 90 per cent, the interval between these levels is the rise time.

The time to swing 10 per cent of the 100-volt total swing is determined from

$$10 \text{ volts} = 100 \text{ volts } (1 - \epsilon^{-t/RC})$$
$$0.1 = 1 - \epsilon^{-t/RC}$$

Since $1 - \epsilon^{-t/RC}$ is plotted on the time-constant chart, we can enter the vertical axis at 0.1 and find the corresponding value of t/R_LC, which is 0.1. Now $R_LC = 1$ μsec. Therefore $t = 0.1\ (R_LC) = 0.1$ (1 μsec) = 0.1 μsec. The time to reach the 90 per cent level is determined from

$$90 = 100\ (1 - \epsilon^{-t/RC}) \qquad \text{and} \qquad 0.9 = 1 - \epsilon^{-t/RC}$$

Again we enter the vertical axis at 0.9 and intersect the capacitor buildup curve at $t/R_LC = 2.3$. Therefore, $t = 2.3$ (1 μsec) = 2.3 μsec. The rise time t_r is then $t_r = 2.3$ μsec − 0.1 μsec = 2.2 μsec. The pulse buildup then looks like Fig. 3-26*b*.

When the switch recloses, the capacitor is initially charged to 200 volts. The source voltage and impedance that the capacitor looks back into can be determined with the aid of Thévenin's theorem to yield Fig. 3-26*c*.

Considering the capacitor as a load and thevenizing, we have

$$V_{oc} = \frac{200 \text{ volts (10 kilohms)}}{10 + 10 \text{ kilohms}} = 100 \text{ volts}$$

$$Z_{th} = \frac{10(10)}{10 + 10} = 5 \text{ kilohms}$$

The capacitor will start to discharge through V_{oc} and Z_{th}. The effective time constant is

$$\tau_{\text{fall}} = Z_{th}C = 5 \text{ kilohms } (100\ \mu\mu\text{f}) = 0.5\ \mu\text{sec}$$

The time required to drop 10 per cent of the total swing (90 per cent remaining) of 100 volts is determined from

$$90/100 = \epsilon^{-t/\tau_{\text{fall}}}$$

Using the curve for capacitor discharge voltage, we have

$$\frac{t}{\tau_{\text{fall}}} = 0.1$$

Therefore, $t = 0.1$ (0.5 μsec) = 0.05 μsec. The time to drop 90 per cent of the 100 volts swing (10 per cent remaining) is obtained from

$$10/100 = \epsilon^{-t/\tau_{\text{fall}}}$$

and therefore

$$\frac{t}{\tau_{\text{fall}}} = 2.3$$

as read on the chart. Now, $t = 2.3 \times 0.5 = 1.15$ μsec. The fall time t_f is then $1.15 - 0.05 = 1.10$ μsec.

A general expression for rise (or fall) time can be developed. If a step of voltage is applied to an *RC* circuit, we know the response is exponential. From the time chart, we see that when

$$1 - \epsilon^{-x} = 0.1 \tag{3-6}$$

we shall have traversed 10 per cent of our total swing. The corresponding value of $X = t/RC = 0.1$, is therefore $t_{10\%} = 0.1RC$. After traversing 90 per cent of our swing, we have

$$1 - \epsilon^{-x} = 0.9 \tag{3-7}$$

and the corresponding value of X is $t/RC = 0.23$. Therefore, $t_{90\%} = 2.3RC$. The rise time is the difference in time between the 90 and 10 per cent levels. Therefore we can write

$$t_{\text{rise}} = t_{90\%} - t_{10\%} = 2.3RC - 0.1RC = 2.2RC$$

Perhaps the reader is wondering why the 10 and 90 per cent levels were chosen rather than the time to go from 0 to the maximum value. Consider Fig. 3-27*a* and

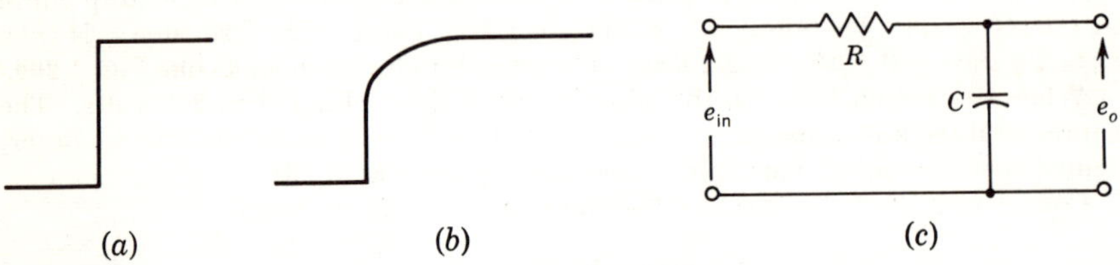

FIG. 3-27. (*a*) An input step voltage. (*b*) Rise time for this waveform might best be defined in terms of the intended application. (*c*) The rise time of the output waveform may be related to the corner (3 db) frequency of the network.

b. In Fig. 3-27*a*, we would not encounter much trouble if we defined the rise time as being between the 0 and 100 per cent levels. In Fig. 3-27*b*, however, the top of the waveform has a very gradual slope and if we were to use 0 and 100 per cent levels, the rise time would be tremendous, even though the upper part of the waveform might vary by only a few per cent. In some instances the 10 to 90 per cent definition of rise time might be wholly inapplicable.

3-10. Interrelation of Transient and Frequency Responses. Now think back to Chap. 2, where we drew Bode plots for some simple *RC* networks such as in Fig. 3-27*c*. The corner frequency (ω_c) for this low pass or integrating network was defined as the frequency at which the output was 3 db below the input. We also showed that the corner frequency was related to the circuit parameters by the relationship

$$\omega_c = \frac{1}{RC} \qquad \text{or} \qquad RC = \frac{1}{\omega_c}$$

Since $1/RC$ is the time constant of the *RC* network of Fig. 3-27*c*, we can begin to see a tie-in between the steady-state frequency response and the transient response. The rise time for the circuit of Fig. 3-27*c* is

$$t_r = 2.2RC = 2.2\left(\frac{1}{\omega_c}\right) = \frac{2.2}{2\pi f_c} = \frac{0.35}{f_c}$$

and $$f_c = \frac{0.35}{t_r}$$

Suppose in some way—and we soon shall do this—you boiled the equivalent circuit of a video amplifier down to Fig. 3-27c. You find that the equivalent of R is 5 kilohms and C, which might be stray, plus other capacitances is 20 $\mu\mu$f.

Problem. What is the rise time for a step of input voltage? At what frequency is $e_o = 0.707e_{in}$?
Solution

$$RC = 5 \text{ kilohms } (20\ \mu\mu\text{f}) = 0.1\ \mu\text{sec}$$
$$t_r = 2.2RC = 0.22\ \mu\text{sec}$$
$$f_c = \frac{0.35}{t_r} = \frac{0.35}{0.22\ \mu\text{sec}} = 1.59 \text{ Mc}$$

3-11. Differentiating a Waveform. Frequently we encounter pulse circuits referred to as differentiating or integrating circuits. In

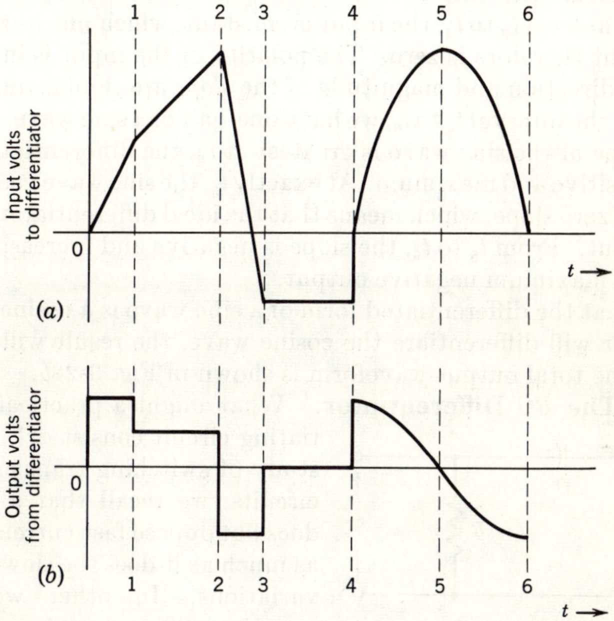

FIG. 3-28. The output waveform of an ideal differentiator when excited by the indicated input.

this section we shall endeavor to grasp the meaning of these terms and their significance in electronic circuitry.

A differentiating circuit is one whose output is proportional to the *rate of change* of the input signal. For example, consider Fig. 3-28a, which shows the input to an ideal differentiator. Let us try to predict what the output would look like.

During the interval from t_0 to t_1, the input is increasing in the positive direction and at a constant rate. Another way of expressing this is to say that the curve has a positive and constant slope during this interval. Now the output of our ideal differentiator depends upon the rate of change of the input only. Since the input is *changing* at a *constant* rate from t_0 to t_1, the output will therefore be some constant value from t_0 to t_1.

During the interval t_1 to t_2, the slope of the input is again constant but not so steep as from t_0 to t_1. The output therefore is again constant, but smaller than before, owing to the smaller slope. Note that even though the input is of higher value from t_1 to t_2 than from t_0 to t_1, the output is still less during t_1 to t_2. An ideal differentiator responds to the *rate of change, or slope*, of the input only.

During the interval t_2 to t_3, the slope is fairly steep and is constant and negative. The output, therefore, is constant and negative.

In the interval t_3 to t_4, the input is constant, which means zero slope. The output therefore is zero. The polarity of the input is immaterial. Only the direction and magnitude of the *slope* are significant.

During the interval t_4 to t_6, we have one-half of a sine wave. Since the rising slope of the sine wave is greatest at t_4, the differentiator output will be positive and maximum. At exactly t_5, the sine wave goes through a point of zero slope, which means that the ideal differentiator produces zero output. From t_5 to t_6, the slope is negative and increasing, which results in maximum negative output.

Note that the differentiated form of a sine wave is a cosine wave. If the reader will differentiate the cosine wave, the result will be a sine wave. The total output waveform is shown in Fig. 3-28*b*.

3-12. The *RC* Differentiator. What might a practical differentiating circuit consist of? From our study of switching transients in *RC* circuits, we recall that a capacitor does not oppose fast current changes as much as it does the slower current variations. In other words, the capacitor offers less impedance to the high-frequency content of fast current shifts than to the lower-frequency content inherent in slower variations. For example, consider Fig. 3-29.

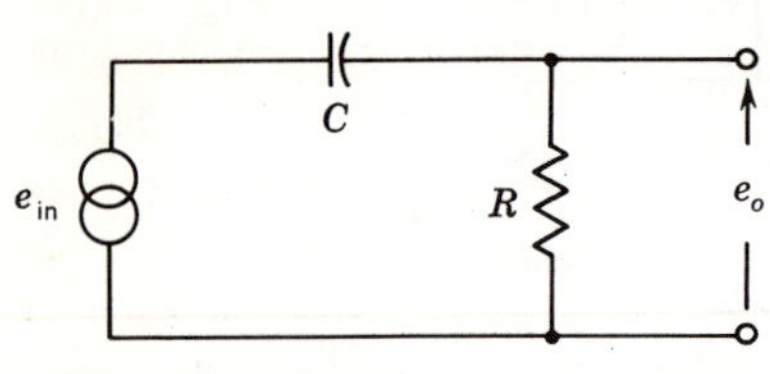

FIG. 3-29. For a constant input voltage the output increases as the input frequency is increased. Since the steep sides of a pulse contain high-frequency components, this circuit will develop more output voltage when an input pulse is changing most rapidly.

If we hold the source amplitude constant but increase the frequency the output will increase. This is because X_C becomes smaller as f increases, and therefore e_o increases by voltage-divider action.

Since pulsed waveforms are made up of many harmonically related sine waves, those having high-frequency components in steep slopes will produce higher outputs. This *RC* circuit would therefore seem to show the slope-discriminating property we would expect in a differentiator.

Another way of analyzing the simple circuit of Fig. 3-29 for differentiating action is to recall that the current through the capacitor is

$$i = C\frac{de_C}{dt} \approx C\frac{\Delta e_C}{\Delta t}$$

If we make R very, very small, the voltage across the capacitor will be essentially the same as the applied voltage. Then any changes in the applied voltage (Δe_{in}) will be impressed across the capacitor, and $\Delta e_c \approx \Delta e_{in}$. The current will then be

$$i \approx C\frac{\Delta e_{in}}{\Delta t}$$

The differentiated output voltage is taken off the resistor and by Ohm's law must be

$$e_o = Ri \approx RC\frac{\Delta e_C}{\Delta t} \approx RC\frac{\Delta e_{in}}{\Delta t}$$

Inspection of the expression for e_o shows that fast changes in voltage (large $\Delta e_{in}/\Delta t$) will result in large outputs compared to slow voltage shifts.

Now we know that R must be small, but how small? And what about the value of C? How do we choose them for a specific application as differentiating elements? To answer these questions, consider Fig. 3-30*a*, which is a differentiating network driven by a rectangular wave generator. Let us analyze this circuit by considering the PRF held constant, though R and C are variable. Assume the generator has zero internal impedance.

First consider the circuit response when the natural time constant of the circuit, $\tau = RC$, is large as compared with the pulse duration because of a large R, a large C, or both.

At t_0, the input instantaneously rises to $+E$ volts (Fig. 3-30*b* and *c*). Since $\tau = RC$ is large compared with t_1, the capacitor can charge only slightly, and therefore a large amount of charging current is still flowing at t_1. The voltage across the resistor is the input voltage minus the small amount that the capacitor has charged up to. At t_1, the input voltage drops to zero and our equivalent circuit is that of Fig. 3-30*d*. The capacitor discharges through R, driving the output negative as shown in Fig. 3-30*c*. With a large time constant compared with the pulse duration, the output very much resembles the input (see Fig. 3-30*c*). This is what we would want in a coupling circuit, not in a differentiator. The only resemblance between the output in this case and that of a differentiated output is output polarity. When the input rises in a positive direction, we get a positive output; when it falls, we

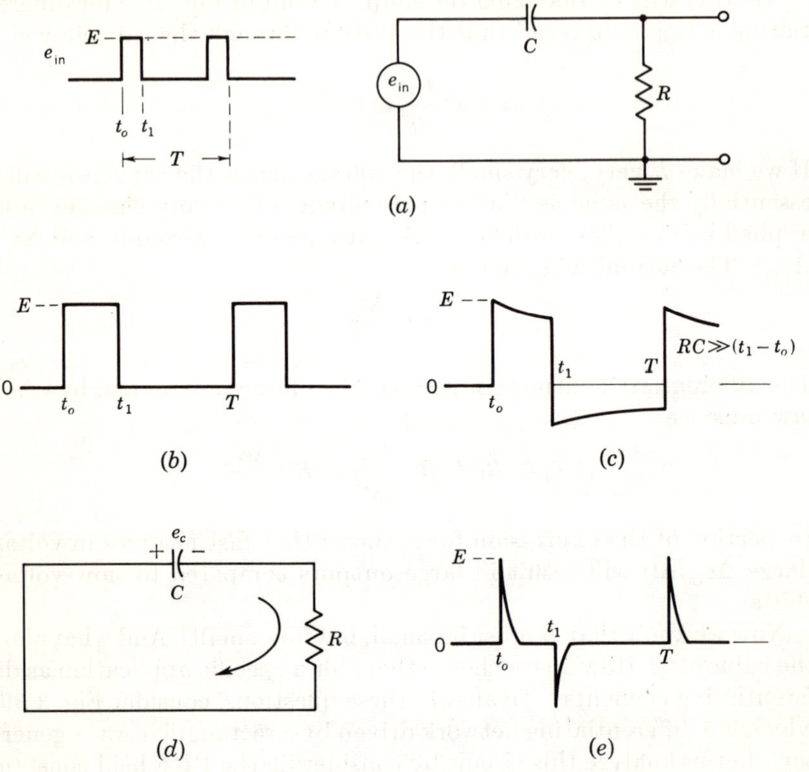

FIG. 3-30. (*a*) A rectangular waveform applied to an *RC* coupling circuit. (*b*) The input waveform. (*c*) The general shape of the output waveform resembles the input when the time constant is large relative to the pulse duration. (*d*) The equivalent circuit when the input waveform returns to zero volts. The driving generator is assumed to have zero internal impedance. (*e*) When the time constant is very small, an output is obtained only when the input waveform changes in amplitude.

get a negative output. The circuit fails us as a differentiator because it is developing an output during the interval t_0 to t_1, even though the input is not *changing*.

How do we surmount this difficulty? The reason we get an output during this interval is that there is still charging current flowing through *R*. Suppose we were to make $\tau = RC$ small as compared with the interval between t_1 and t_0. Then the capacitor would charge very rapidly up to E volts, which means that the charging current returns to zero, for all practical purposes, long before t_1 is reached. The output voltage must then be zero. As a rule of thumb, the time constant should be one-tenth the pulse duration for effective differentiation. The output will now appear as in Fig. 3-30*e*. Notice that now we get an output whenever the input changes, but none when it is constant.

Bear in mind that in many practical cases the generator resistance is not zero or negligible and therefore adds to R, which increases the time constant.

It would seem that by making R smaller and smaller we could approach extremely thin spikes having a magnitude equal to the shift in input voltage. This would be true were it not for inevitable stray shunting capacitance across R. This capacitance, in conjunction with the differentiating capacitor, forms a capacitive voltage divider.

3-13. The Capacitive Voltage Divider. In the differentiator, the capacitive voltage divider attenuates the magnitude of the output

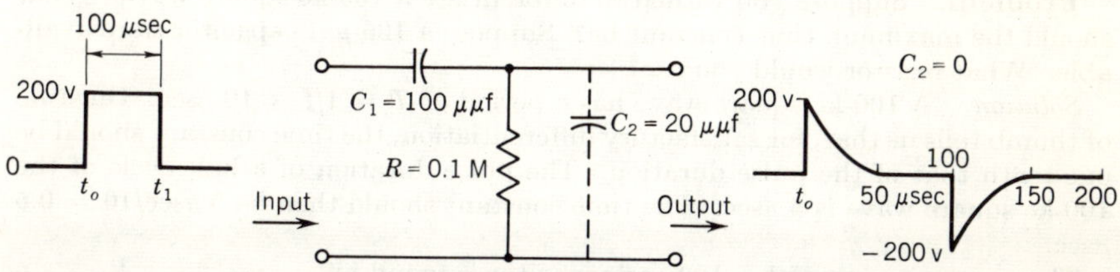

FIG. 3-31. Stray capacitance (C_2) across the output of the differentiator will reduce the peak output.

voltage shift due to an input voltage shift. For example, consider Fig. 3-31, and for the moment neglect C_2. The time constant is 10 μsec, which is one-tenth the pulse duration. We therefore have a good differentiator, since in 5 time constants, or 50 μsec, the capacitor is fully charged and the output voltage is zero. The output will shift by the full amount of input shift. Now suppose we have additional stray capacitance C_2 of 20 μμf shunting R. The peak output shift will now be

$$\Delta e_o = \Delta e_{in} \frac{C_1}{C_1 + C_2} = 200\left(\frac{100}{120}\right) = 167 \text{ volts}$$

Note that in a capacitive divider the capacitor in the numerator is *not* the one across which the output voltage appears.

For waveforms having infinitely steep sides such as we are considering, X_{C_2} will be very much smaller than R, and therefore we neglect R when considering the immediate shift in output. The input shift must then divide between C_1 and C_2. We can write, by the voltage-divider relationship,

$$\Delta e_o = \Delta e_{in} \frac{X_{C_2}}{X_{C_1} + X_{C_2}}$$

$$= \Delta e_{in} \frac{1/\omega C_2}{1/\omega C_1 + 1/\omega C_2} = \Delta e_{in} \frac{1/C_2}{1/C_1 + 1/C_2}$$

$$= \Delta e_{in} \frac{1/C_2}{(C_2 + C_1)/C_1 C_2} = \Delta e_{in} \frac{C_1}{C_1 + C_2}$$

If C_2 is an appreciable percentage of C_1, the effective time constant of the network will also be different.

If there is some leeway in the design of the differentiator elements, a larger value for C_1 could be chosen so that C_2 is a negligible percentage of C_1. R_1 of course would have to be reduced.

Problem. Suppose you wished to differentiate a 100-kc square wave. What should the maximum time constant be? Suppose a 100-$\mu\mu$f capacitor were available. What resistor would you use?

Solution. A 100-kc square wave has a period of $T = 1/f = 10$ μsec. Our rule of thumb tells us that, for satisfactory differentiation, the time constant should be one-tenth that of the pulse duration. The pulse duration of a half cycle of the 100-kc square wave is 5 μsec. The time constant should then be 5 μsec/10 = 0.5 μsec.

The resistor to use with a 100-$\mu\mu$f capacitor is found by

$$RC = 0.5\ \mu\text{sec}$$

$$R = \frac{0.5\ \mu\text{sec}}{100\ \mu\mu\text{f}} = 0.005\ \text{meg} = 5\ \text{kilohms}$$

An alternative method of regarding a differentiator is to consider it as a high-pass filter. The high-frequency components in fast changes readily go through the capacitor with little attenuation.

3-14. Ramp Input Applied to Differentiator. We have discussed the effect of shunt capacitance across R. It tended to reduce the peak output shift below the input shift, since a capacitive divider was present. Another factor which will tend to minimize the output shift is the finite rise time of the input. We have been discussing theoretical infinitely fast switching times, which in practice are only approximated. Let us therefore examine the circuit response to an input voltage of finite slope, as shown in Fig. 3-32.

Assuming that the capacitor is initially uncharged, as the input starts to rise the capacitor will start to charge. The voltage across the capacitor will surely lag the input during the interval t_r, since it takes time to charge. The rate at which C charges depends upon the time constant RC. The larger RC is, the greater the lag or difference between e_{in} and e_C. The output e_o, however, is the difference between e_{in} and e_C. Therefore, the larger RC is, the larger we would expect e_o to be. Let us go one step further. At t_0 there is no charge on C, so any increase in e_{in} will cause a proportionate increase in current. The output will then

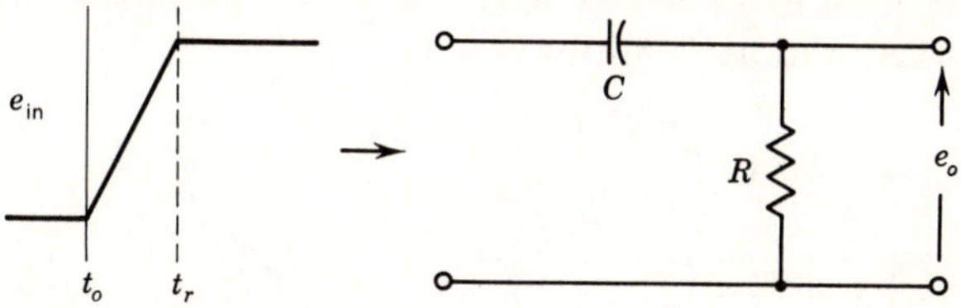

FIG. 3-32. An RC differentiator excited by a ramp input.

initially follow the input. As the capacitor starts to charge, the current starts to fall off. The input voltage increases, however, which forces more charging current into C. Since current is flowing into the capacitor, we will still get an output voltage across R. If the input were to continue rising at a constant rate, we would find that the current tapers off to some constant value, which causes the capacitor voltage to build up at the same *rate* as the input. The capacitor voltage will, however, be less than the input, because of the drop across R (see Fig. 3-33):

$$e_o = e_R = e_{in} - e_C$$

Another way of analyzing the circuit is to imagine the sloping or ramp input as being made up of many individual steps, as in a staircase (Fig. 3-34). Assume a time constant equal to the width of a step. That is, $RC = t_2 - t_1$. Let the height of each step be A volts. At t_1, e_o jumps to e_{in} and then tapers off as C charges. At t_2 minus,[1] e_o will be equal to 37 per cent of A. At t_2 plus, e_o jumps from $0.37A$ to $0.37A + A$ volts, or $1.37A$ volts. The equivalent circuit at t_2 plus is shown in Fig. 3-35.

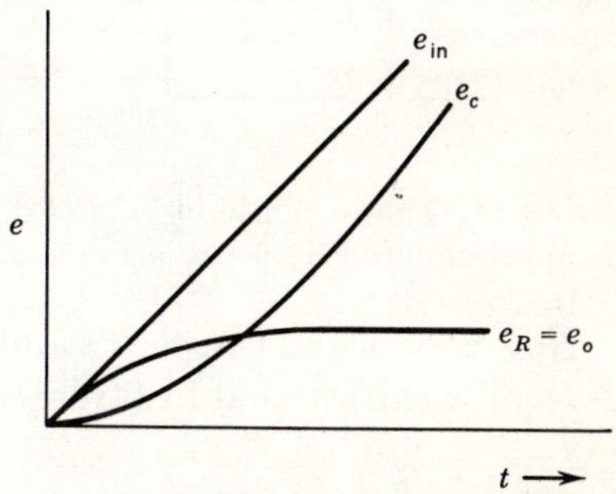

FIG. 3-33. The output voltage of the differentiating network approaches a constant value when the input voltage rises at a constant rate.

The capacitor is now trying to charge from $0.63A$ to $2A$, which is

[1] When a current or voltage rises or falls instantaneously at a time t, it is convenient to describe the voltage just before the jump by its value at t minus. Immediately after the jump we speak of the new level at t plus. Sometimes t minus and t plus are written t^- and t^+.

greater than from 0 to A volts. In one time constant, $t_3 - t_2$, the capacitor will swing through 63 per cent of the total swing. The total swing in e_C is $2A - 0.63A$, or $1.37A$:

$$0.63(1.37A) = 0.86A$$

At t_3 minus, e_C will then be $0.63A + 0.86A = 1.49A$. The output across R is then $2A - 1.49A = 0.51A$. At t_3 plus, the output jumps from $0.51A$ to $1.51A$, and at t_4 minus is

$$0.37(1.51A) = 0.56A$$

At t_4 plus, $e_o = 1.56A$.

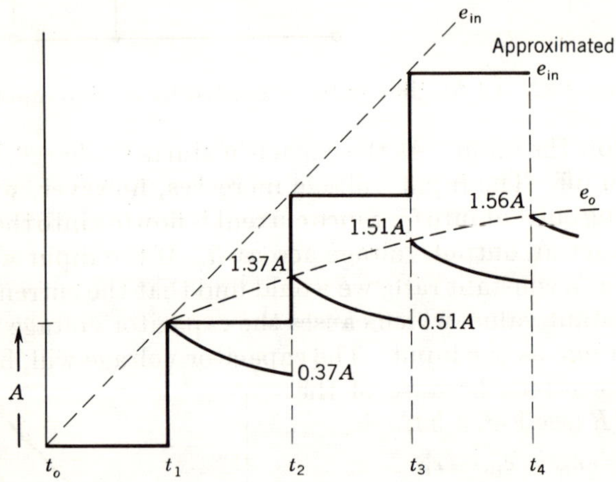

FIG. 3-34. The response of an RC differentiating network to a ramp input may be approximated by considering the ramp as a staircase.

Now connect all the peak values of e_o at t_1, t_2, t_3, and t_4 as shown in Fig. 3-34. We can see that in 4 time constants e_o has tapered off to a constant voltage, even though the input is changing. This is in agreement with our concept of a differentiating network. We get a constant output for a constant-input slope. In our illustration, we let the ramp duration equal 4 time constants before the output approaches a constant value. Now suppose we had to differentiate a ramp that lasted for 100 μsec. To get a constant output quickly, we would use a one-tenth, or 10-μsec, time constant. On the other hand, if we wished to couple the 100-μsec ramp faithfully to an amplifier, we would use a time constant equal to 1,000 μsec or more.

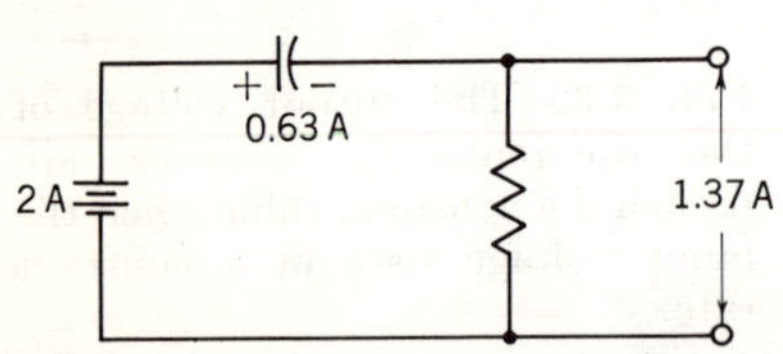

FIG. 3-35. The equivalent circuit at t_2 plus.

The exact expression for the current in the circuit is derived from the differential equation for the circuit, the solution for which is beyond the scope of this book. The results are all we shall concern ourselves with here.

The current at any time in a series-resistance single-capacitor circuit when excited by a ramp voltage is

$$i = C\left(\frac{\Delta e_{in}}{\Delta t}\right)(1 - \epsilon^{-t/RC})$$

$\Delta e_{in}/\Delta t$ is the slope of the ramp, which is a constant. Let $C\Delta e_{in}/\Delta t = I_M$, which is the maximum current that will flow through R when e_o has approached a constant value. Then

$$i = I_M(1 - \epsilon^{-t/RC})$$

This equation is in the same form as that for capacitor voltage buildup for a step of voltage. We can put this into the normalized form so that we can use our time constant chart for ramps as well as for steps of voltage.

$$\frac{i}{I_M} = 1 - \epsilon^{-t/RC}$$

From the time-constant chart we can see that when $t/RC = 1$, the current will have built up to 63 per cent of the maximum value. In 4 to 5 time constants the current can be considered as being at its maximum value. The output voltage, of course, is $e_o = iR$. The maximum output is

$$e_{o_M} = I_M R = RC\,\frac{\Delta e_{in}}{\Delta t}$$

Problem 1. A ramp voltage rising 128 volts in 2 μsec is impressed upon an RC circuit where $C = 100\ \mu\mu$f and $R = 100$ kilohms. What will the maximum output voltage across R be? What is e_R in 1 μsec?

Solution

$$RC = 10\ \mu\text{sec}$$

$$I_M = C\frac{\Delta e_{in}}{\Delta t} = 100\ \mu\mu\text{f}\left(\frac{128}{2\ \mu\text{sec}}\right) = 6{,}400\ \mu\text{a}$$

$$e_{RM} = 0.1\ \text{meg} \times 6{,}400\ \mu\text{a} = 640\ \text{volts}$$

In 1 μsec, $t/RC = 1/10$, and from the chart for $t/RC = 0.1$, the corresponding $i/I_M = 0.09$. Therefore,

$$i = 0.09 \times 6{,}400 = 576\ \mu\text{a}$$

$$e_o = 576\ \mu\text{a} \times 0.1\ \text{meg} = 57.6\ \text{volts in 1}\ \mu\text{sec}$$

As an illustration of how we can utilize our knowledge of RC circuit response to a ramp, consider the following.

Problem 2. The sawtooth sweep in an oscilloscope has the waveform shown in Fig. 3-36. During the long 95-μsec part of the sweep, the beam is traveling across the face of the tube from left to right. When the peak of the sweep is reached, the spot flies back to the left-hand side of the tube and again sweeps across. At high intensity levels, the 5-μsec flyback time can show up as an annoying return trace. To eliminate this effect, the sweep is sometimes fed into a cathode follower for isolation and low output impedance. The cathode follower drives a differentiating network whose time constant is about one-tenth of the flyback time, or 0.5 μsec. The narrow negative gate produced is amplified if

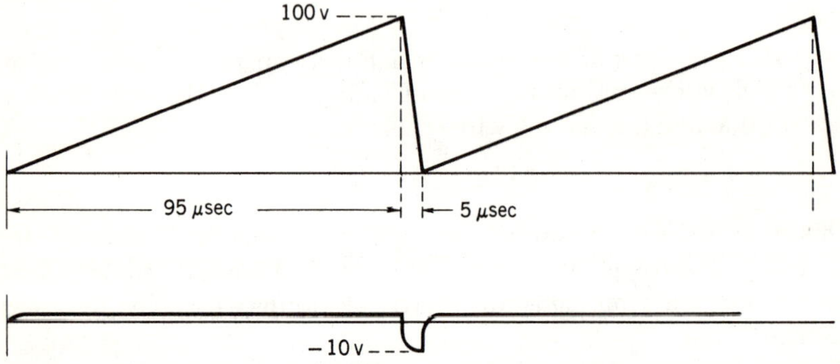

FIG. 3-36. The return trace of a sweep waveform may be differentiated to generate a blanking pulse.

necessary in a blanking amplifier and fed into the CRT grid, to cut off the beam during the return trace. Assuming unity gain and negligible output impedance for the cathode follower, determine the available blanking signal.

Solution. The peak negative output from the differentiator is $e_o = RI_M$ and $I_M = C\Delta e_{in}/\Delta t$.

$$RC = 0.5\ \mu\text{sec}$$

$$e_o = RI_M = RC\frac{\Delta e_{in}}{\Delta t} = 0.5\ \mu\text{sec}\left(\frac{100\text{ volts}}{5\ \mu\text{sec}}\right) = 10\text{ volts}$$

If R is chosen as 10 kilohms, or 0.01 meg, $C = 50\ \mu\mu$f.

The reader is urged to think up his own problems. Frequently the need is to conceive a circuit, rather than to analyze a given circuit. There is no reason why the technician who thoroughly masters this book cannot design modest circuits with engineering accuracy.

3-15. Integrating a Waveform. If we reverse the positions of R and C in our differentiating network, we have what is known as an integrator. The output voltage is taken off the capacitor, as shown in Fig. 3-37.

Just what do we mean by an integrator? To integrate means to sum up, and this is exactly what a capacitor does. Each electron that flows into the capacitor contributes to the capacitor voltage. At any time, the capacitor voltage is proportional to the sum or total of electrons that have flowed into or out of it. Recall that

$$e_C = \frac{q}{C}$$

If we had a constant current of 5 ma flowing into a 2-μf capacitor, the capacitor voltage in 10 msec would be (Fig. 3-38*a*):

$$e_C = \frac{5 \times 10^{-3} \times 10 \times 10^{-3}}{2 \times 10^{-6}}$$

$$= \frac{50 \times 10^{-6}}{2 \times 10^{-6}} = 25 \text{ volts}$$

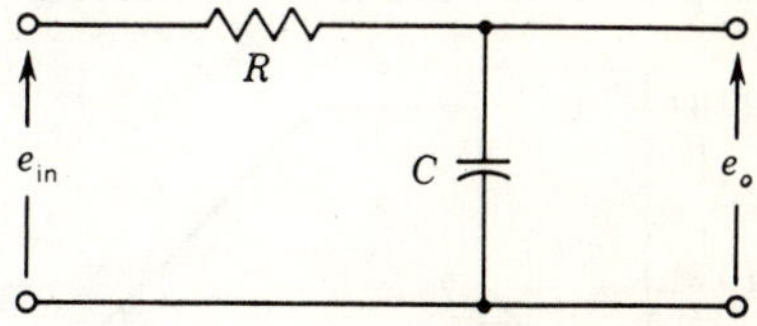

FIG. 3-37. An *RC* integrator.

Figure 3-38*b* graphically portrays the ramp, or sweep, voltage developed across the capacitor with a constant-current input. The capacitor voltage rises in a linear manner because the constant current con-

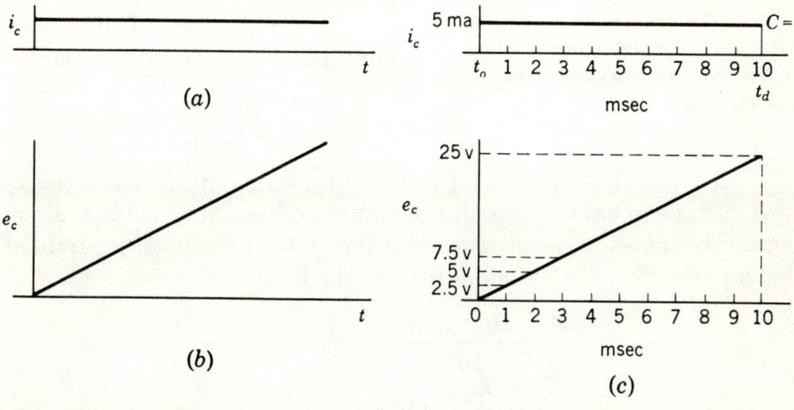

FIG. 3-38. (*a*) Constant charging current into a capacitor. (*b*) Ramp voltage across a capacitor with constant charging current. At the end of 10 msec a 2-μf capacitor driven by (*c*) a 5-ma constant-current source will have 25 volts across it.

tributed equal amounts of charge at every instant. We can gain a deeper insight into the meaning of integration by the graphical analysis of Fig. 3-38*a*.

Figure 3-38*a* tells us that a 5-ma step of constant current is applied to a 2-μf capacitor in the interval t_0 to t_d. Let us endeavor to plot e_C as a function of time. The technique we shall use is to break up the interval from t_0 to t_d into many smaller intervals, let us say 10. In the first millisecond, t_1, the charge q_1 moved into the capacitor is

$$q_1 = (5 \times 10^{-3} \text{ amp})(10^{-3} \text{ sec}) = 5 \times 10^{-6} \text{ amp-sec}$$

The corresponding voltage (e_{C_1}) is

$$\frac{q_1}{C} = \frac{5 \times 10^{-6}\text{ amp-sec}}{2 \times 10^{-6}\text{ farad}} = 2.5\text{ volts}$$

In case the reader is wondering how ampere-second/farad equals volts, he should review the definition of capacitance. We say that if one volt is developed across a capacitor when one coulomb of charge is forced into it, the capacitance is one farad. Therefore

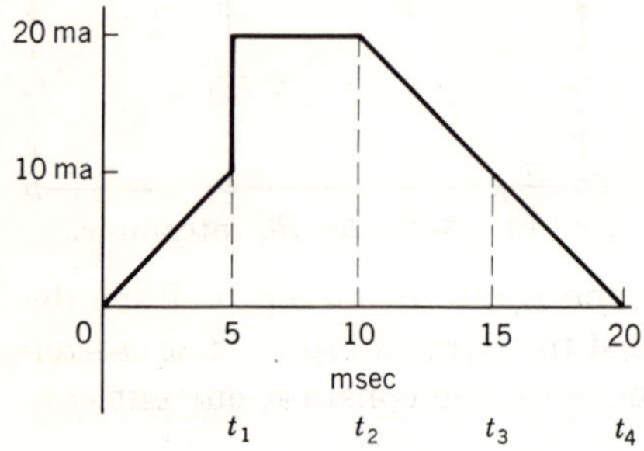

FIG. 3-39. The voltage across a 5-μf capacitor is to be determined at t_4 if this is the current waveform through the capacitor.

$$\text{Farads} = \frac{\text{coulombs}}{\text{volts}} = \frac{\text{amp-sec}}{\text{volts}}$$

or $$\text{Volts} = \frac{\text{amp-sec}}{\text{farads}}$$

In the interval from t_1 to t_2, the additional charge that flows into the capacitor is 5 ma(1 msec) = 5×10^{-6} amp-sec. The corresponding change in voltage is

$$\frac{5 \times 10^{-6}\text{ amp-sec}}{2 \times 10^{-6}\text{ farad}} = 2.5\text{ volts}$$

Since the current is constant, the capacitor will gain the same increment of voltage in each increment of time; and the final voltage is the sum of all the incremental charges that flow into the capacitor, divided by the capacitance. In our illustration, the final charge is

$$q_{\text{total}} = \frac{5 \times 10^{-6}\text{ amp-sec}}{1 \times 10^{-3}\text{ sec}}(10 \times 10^{-3}\text{ sec})$$
$$= 50 \times 10^{-6}\text{ amp-sec}$$

At t_d, $$e_C = \frac{50 \times 10^{-6}\text{ amp-sec}}{2 \times 10^{-6}\text{ farad}} = 25\text{ volts}$$

We know that charge has the dimensions of ampere-seconds. Looking at our curve of i_C versus time, we see a rectangle of height 5 ma and base 10 msec. The total charge of 50×10^{-6} amp-sec is the area of this rectangle. If the current were not constant, then the area bounded by the curve would be harder to evaluate, but would still give the total charge. From this we might infer that a device or circuit which integrates is giving an instant-by-instant evaluation of the area bounded by the curve of one variable plotted against another.

Let us try a simple problem in which the capacitor current i_C is not constant, as shown in Fig. 3-39.

Problem. What is the capacitor voltage at the end of 20 msec if $C = 5\ \mu$f?

Solution. If we can determine the final charge on the capacitor at t_4, we can find e_C, since $e_C = q/C$. Now the total charge on C is the area under the i_C versus the t curve. We can break up our irregular figure into geometric sections whose areas are readily determined. The area of a triangle is $\frac{1}{2}$(base) × (altitude), and therefore the charge accumulated from t_0 to t_1 is

$$\Delta q_1 = \tfrac{1}{2}(5 \times 10^{-3}\ \text{sec})(10 \times 10^{-3}\ \text{amp}) = 25 \times 10^{-6}\ \text{amp-sec}$$

From $(t_1 - t_2)$,

$$\Delta q_2 = (5 \times 10^{-3}\ \text{sec})(20 \times 10^{-3}\ \text{amp}) = 100 \times 10^{-6}\ \text{amp-sec}$$

From $(t_2 - t_4)$,

$$\Delta q_3 = \tfrac{1}{2}(10 \times 10^{-3}\ \text{sec})(20 \times 10^{-3}\ \text{amp}) = 100 \times 10^{-6}\ \text{amp-sec}$$

$$q_{\text{total}} = \Delta q_1 + \Delta q_2 + \Delta q_3 = 225 \times 10^{-6}\ \text{amp-sec}$$

Therefore, at t_4,

$$e_C = \frac{225 \times 10^{-6}\ \text{amp-sec}}{5 \times 10^{-6}\ \text{farad}} = 45\ \text{volts}$$

If at any time the current goes negative, this represents charge lost by the capacitor, which must therefore be subtracted.

3-16. The *RC* Integrator.

Now let us examine our simple *RC* network and see how effective it is as an integrator. We know that the capacitor voltage is the area under the i_C-versus-t curve. Sometimes we say the capacitor voltage is the integral of the capacitor current. In our circuit, however, we would like the capacitor voltage to be the integral of the input voltage—not the current. In other words, we want e_C to be proportional to the area under the input-voltage-versus-time curve. Suppose we apply the rectangular voltage pulse of Fig. 3-40*b* to the network of Fig. 3-40*a*.

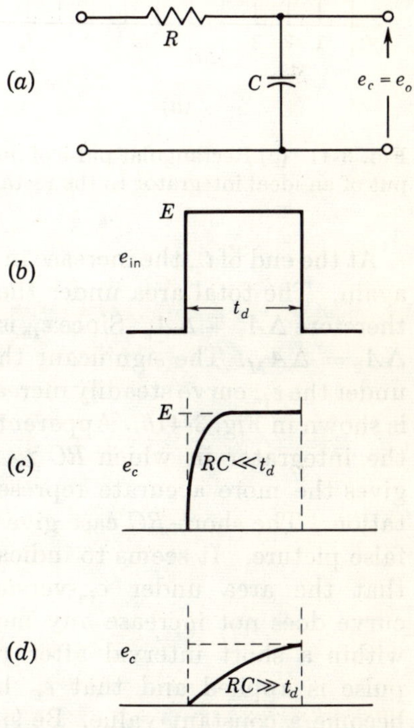

FIG. 3-40. (*a*) An integrating circuit. (*b*) Input waveform to the integrator. (*c*) Integrator output when the time constant is small relative to the pulse duration. (*d*) Integrator output when the time constant is large relative to the pulse duration.

Let us examine the output waveform for time constants much smaller and larger than the pulse duration. If *RC* is small compared with t_d, the capacitor charges rapidly to the applied input and will look as in Fig. 3-40*c*. When

$RC \gg t_d$, e_C will appear as in Fig. 3-40*d*. Now let us see what an ideal integrator would put out for the input pulse shown.

An ideal integrator would yield an output at all times which is proportional to the area under the e_{in}-versus-t curve. If we divide t_d into 10 equal increments Δt, the area of each increment will be $E(\Delta t)$ volt-sec. The increase in area in the first increment is $\Delta A_1 = E(\Delta t) = E(0.1t_d)$ volt-sec (see Figs. 3-41*a* and *b*).

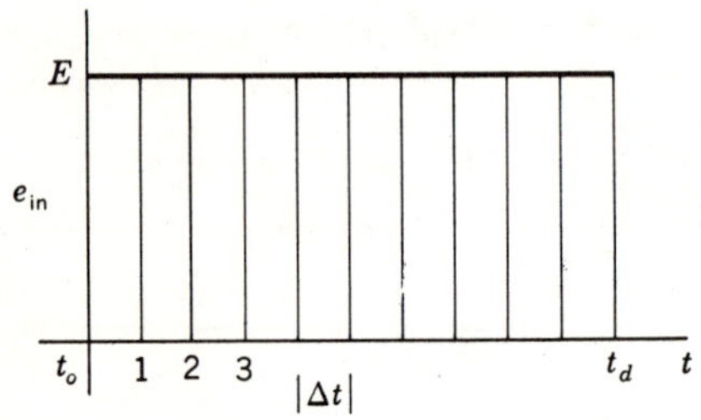

(*a*)

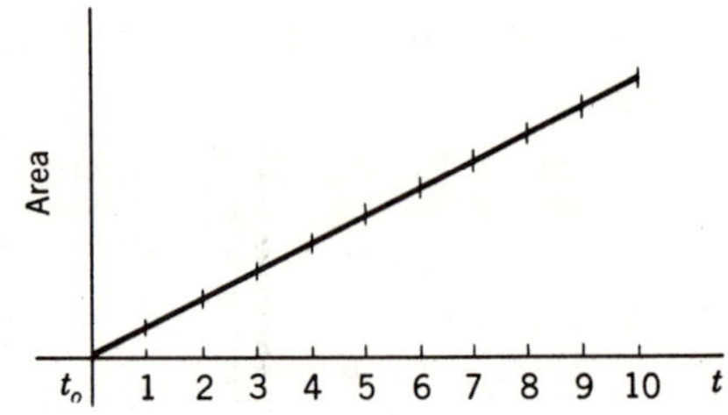

(*b*)

FIG. 3-41. (*a*) Rectangular pulse of duration t_d applied to an integrator. (*b*) Output of an ideal integrator to the rectangular pulse of *a*.

At the end of t_2, the increase in area $\Delta A_2 = E(\Delta t) = E(0.1t_d)$ volt-sec again. The total area under the e_{in}-versus-t curve at the end of t_2 is therefore $\Delta A_1 + \Delta A_2$. Since e_{in} is constant with respect to time, $\Delta A_1 = \Delta A_2 = \Delta A_M$. The significant thing is that as time goes by, the area under the e_{in} curve steadily increases with each increment of time. This is shown in Fig. 3-41*b*. Apparently the integrator in which $RC \gg T_d$ gives the more accurate representation. The short-*RC* case gives a false picture. It seems to indicate that the area under e_{in}-versus-t curve does not increase any more within a short interval after the pulse is applied and that e_C has become a constant value. Be sure you see that it is the actual value of e_o that has to be proportional, at every instant of time, to the area under the e_{in}-versus-t curve.

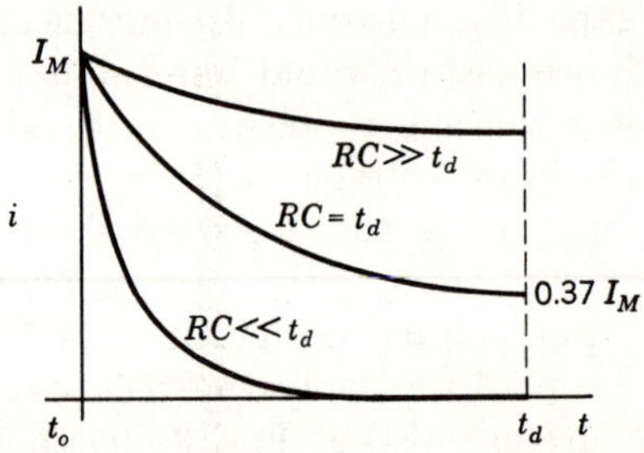

FIG. 3-42. The current response in an integrator as the time constant is varied relative to the pulse duration.

In our simple *RC* integrator, e_C does not rise linearly, but droops with time. This is because the capacitor voltage buildup bucks the applied pulse voltage, which in turn causes the charging current to decrease as time goes by. If the charging current were constant, then e_C would be a ramp or sweep voltage. This we have shown to be the required output of an ideal integrator for a rectangular input voltage. Increasing the time constant will tend to hold the charging current more constant during the pulse duration, since the capacitor is charging up slowly. Figure 3-42 illustrates this for different values of RC in which C is varied and R is held constant. If R were also varied, the initial current would vary.

3-17. *RC* Coupling. We shall briefly discuss the properties of a typical RC coupling circuit (Fig. 3-43), such as is used in vacuum-tube or transistor amplifiers. The primary purpose of the coupling circuit is to prevent the d-c component present on the input side from getting on the grid of the next stage. In some cases, such as in d-c amplifiers, we wish to couple the d-c component. However, d-c coupling means that any d-c drift in an amplifier stage will be amplified in subsequent stages. More will be said about this later.

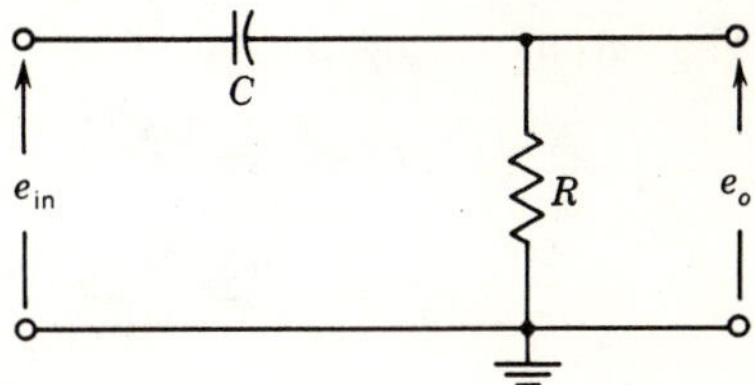

FIG. 3-43. *RC* coupling circuit.

Let us examine the behavior of the coupling circuit in Fig. 3-43 as regards sinusoidal inputs. Qualitatively, we can tell that if e_o is to bear a proportionate relationship in both phase and magnitude to e_{in}, the reactance of the capacitor must be negligible with respect to the resistance at the lowest frequency that we wish to pass undistorted. To minimize the phase displacement of e_o with respect to e_{in}, the circuit must look virtually resistive at the lowest frequency desired $R \gg X_C$. We can do this by making R large and X_C small, which means C large. So, for a satisfactory sinusoidal coupling circuit, we want a large time constant.

Problem. Suppose you had to design a coupling circuit for a 20-volt 10-cycle sine wave, with the provision that the output waveform follow the input with a maximum of 10° phase shift. The generator source impedance is 0.5 meg. From other considerations, the maximum permissible value for the output resistor is 1.5 meg. Find C and e_o.

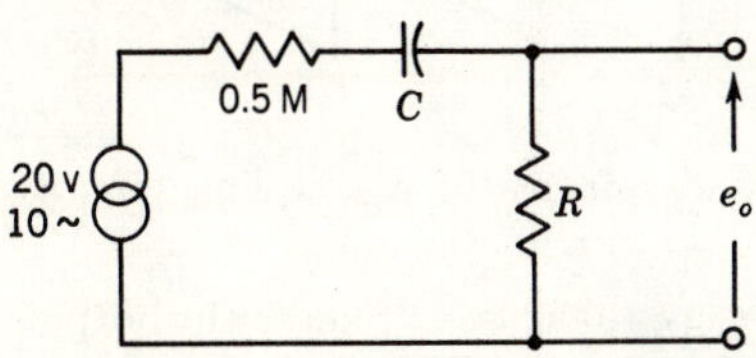

FIG. 3-44. Designing a coupling circuit for a specified minimum phase shift.

Solution. The circuit would be as shown in Fig. 3-44. To start our design, let us use the maximum permissible value for R. The total circuit resistance R_t is then $R_t = 1.5 + 0.5 = 2.0$ meg. From elementary a-c theory, we know that the capacitive reactance will produce a current which leads the voltage by an angle whose tangent is X_C/R_t (written as $\tan^{-1}[X_C/R_t]$). We can then write

$$\tan 10° = \frac{X_C}{2.0 \text{ meg}}$$

$$0.1763 = \frac{X_C}{2.0 \text{ meg}}$$

$$0.3526 \text{ meg} = X_C$$

$$353 \text{ kilohms} \approx X_C$$

$$353 \text{ kilohms} = \frac{1}{2\pi f C} = \frac{0.159}{10C}$$

$$C = \frac{0.159}{3.53 \text{ meg}} = 0.045 \ \mu\text{f}$$

The output voltage magnitude is

$$e_o = \frac{e_{in}R}{\sqrt{R_t^2 + X_C^2}} = \frac{20 \text{ volts}(1.5 \text{ meg})}{\sqrt{(2 \text{ meg})^2 + (0.35 \text{ meg})^2}}$$

$$= \frac{20 \text{ volts}(1.5 \text{ meg})}{2 \text{ meg}} = 15 \text{ volts}$$

3-18. Periodic Waveform Applied to *RC* Coupling Circuit.

We have already shown that to couple square, sweep, and sinusoidal waveforms with minimum distortion, a long time constant relative to the pulse duration is required. The principal points to remember are that the capacitor has a negligible impedance to abrupt changes in voltage and that the average value of the output waveform must be zero, since the capacitor will not pass the d-c component of the input waveform.

Problem. As an additional exercise in applying these principles, consider Fig. 3-45. The input waveform is a symmetrical square wave shifting between

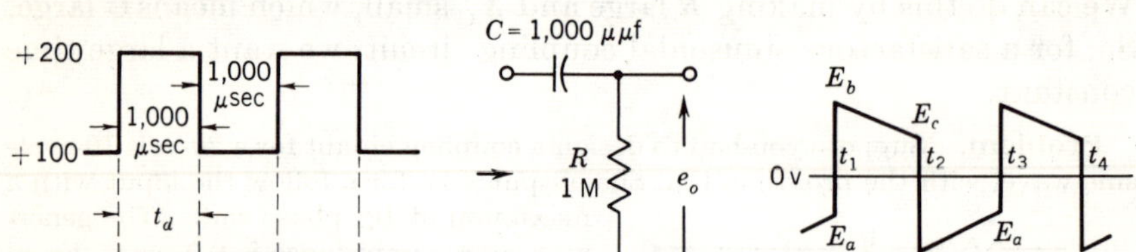

FIG. 3-45. Comparison of output and input waveforms of a coupling circuit.

+100 and +200 volts. The pulse duration is $t_d = 1{,}000$ μsec. Sketch the output waveform and determine the level of all break points (points of discontinuity). Assume that the generator has been on a long time.

Solution. We know that the average value of the output must be zero volts, since the capacitor cannot pass direct current. The positive pulse area must therefore equal the negative pulse area. Since the input is symmetrical, we know that the output waveform will also be symmetrical about the zero axis. The output waveform will have some tilt to it, as shown in Fig. 3-45, since the charging current is not constant. The longer the time constant, the less the tilt. The instantaneous 100-volt shifts which occur on the input side will pass through the capacitor and appear as 100-volt shifts on the output side. Let us try to set up some equations and see if we can evaluate points E_a, E_b, E_c, and E_d:

$$E_b - E_a = 100 \text{ volts} \tag{3-8}$$

$$E_c - E_d = 100 \text{ volts} \tag{3-9}$$

Equations (3-8) and (3-9) must be true, since the input shift appears unattenuated in the output.

$$E_c = E_b\epsilon^{-t/RC} = E_b\epsilon^{-1{,}000\,\mu\text{sec}/1{,}000\,\mu\text{sec}} = 0.37E_b \tag{3-10}$$

$$E_a = E_d\epsilon^{-t/RC} = E_d\epsilon^{-1{,}000\,\mu\text{sec}/1{,}000\,\mu\text{sec}} = 0.37E_d \tag{3-11}$$

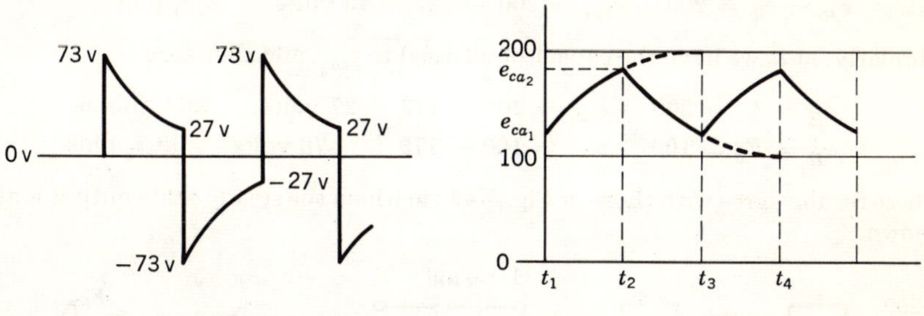

FIG. 3-46. (a) Exact values of the output waveform in the circuit of Fig. 3-45. (b) The capacitor voltage in Fig. 3-45.

Equations (3-8) to (3-11) may then be solved simultaneously by the substitution method, to yield

$$E_d = -73 \text{ volts}$$
$$E_c = +27 \text{ volts}$$
$$E_a = -27 \text{ volts}$$
$$E_b = +73 \text{ volts}$$

The solution is left to the reader (see Fig. 3-46a).

Another way to determine conveniently the output waveform after a long period of time is to use formula (3-3). We know that the input waveform of Fig. 3-45 will appear tilted in the output, and will have both positive and negative values, since the average d-c output must be zero. The problem is to determine points E_a, E_b, E_c, and E_d of the output waveform. It is easier to visualize the voltage across the capacitor (Fig. 3-46b), which will vary between e_{ca_1} and e_{ca_2}. We can write

$$e_{ca} = E_{cf} + (E_{c0} - E_{cf})\epsilon^{-t/\tau}$$
$$e_{ca_2} = 200 + (e_{ca_1} - 200)\epsilon^{-1{,}000\,\mu\text{sec}/1{,}000\,\mu\text{sec}}$$
$$= 200 + (e_{ca_1} - 200)(0.37) = 126 + 0.37e_{ca_1} \tag{3-12}$$
$$e_{ca_1} = 100 + (e_{ca_2} - 100)\epsilon^{-1{,}000\,\mu\text{sec}/1{,}000\,\mu\text{sec}}$$
$$= 100 + (e_{ca_2} - 100)(0.37) = 63 + 0.37e_{ca_2} \tag{3-13}$$

Solving (3-12) and (3-13) simultaneously,

$$\begin{aligned} e_{ca_2} &= 126 + 0.37(63 + 0.37e_{ca_2}) \\ &= 126 + 23.3 + 0.14e_{ca_2} \\ 0.86e_{ca_2} &= 149 \text{ volts} \\ e_{ca_2} &= 173 \text{ volts} \end{aligned}$$

and

$$e_{ca_1} = 63 + 0.37(1.73) = 127 \text{ volts}$$

Now that we have determined the capacitor voltage, we are in a position to compute the resistor voltage, since the sum of the two must always equal the input voltage. When the capacitor is at e_{ca_1} volts, and the input is 100 volts at t_1 minus, the resistor voltage must be

$$e_R = E_a = 100 - e_{ca_1} = 100 - 127 = -27 \text{ volts} \qquad \text{at } t_1 \text{ minus}$$
$$e_R = E_b = 200 - e_{ca_1} = 200 - 127 = 73 \text{ volts} \qquad \text{at } t_1 \text{ plus}$$

Similarly, at t_2 we have the capacitor charged to e_{ca_2}, and therefore

$$e_R = E_c = 200 - e_{ca_2} = 200 - 173 = 27 \text{ volts} \qquad \text{at } t_2 \text{ minus}$$
$$e_R = E_d = 100 - e_{ca_2} = 100 - 173 = -73 \text{ volts} \qquad \text{at } t_2 \text{ plus}$$

These results agree with those of Fig. 3-46*a* in which the steady-state output is also shown.

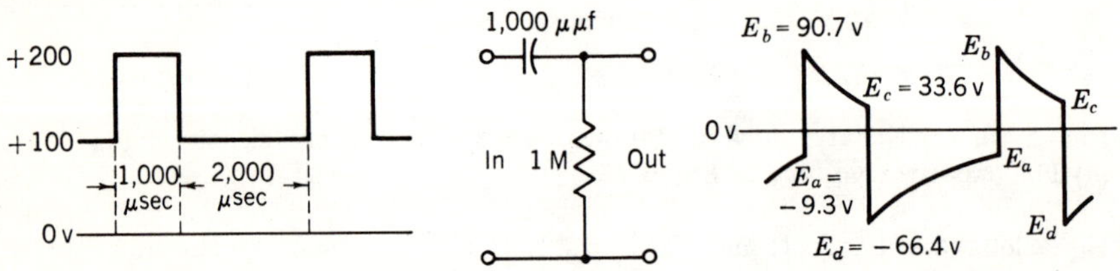

FIG. 3-47. Input and output waveforms for the coupling circuit shown.

Suppose our input waveform were not symmetrical and were as in Fig. 3-47. What would the output waveform now be if applied to the circuit of Fig. 3-47? The important clue in this case is that the output positive pulse area must equal the negative pulse area. We can therefore expect the positive peak to exceed the negative peak, since the duty cycle is less than 50 per cent. Let us again set up some equations:

$$E_b - E_a = 100 \text{ volts} \tag{3-14}$$

$$E_c - E_d = 100 \text{ volts} \tag{3-15}$$

Note that the same shift of the input waveform still occurs in the output. It is the levels that are different.

$$E_c = E_b \epsilon^{-1,000\,\mu\text{sec}/1,000\,\mu\text{sec}} = 0.37E_b \tag{3-16}$$

$$E_a = E_d \epsilon^{-2,000\,\mu\text{sec}/1,000\,\mu\text{sec}} = 0.14E_d \tag{3-17}$$

The simultaneous solution of these four equations yields $E_a = -9.3$ volts, $E_b =$ 90.7 volts, $E_c = 33.6$ volts, and $E_d = 66.4$ volts.

If formula (3-16) is used to determine the capacitor voltage, the equations will be

$$e_{ca_2} = 200 + (e_{ca_1} - 200)\epsilon^{-1{,}000\,\mu\text{sec}/1{,}000\,\mu\text{sec}} \tag{3-18}$$

$$e_{ca_1} = 100 + (e_{ca_2} - 100)\epsilon^{-2{,}000\,\mu\text{sec}/1{,}000\,\mu\text{sec}} \tag{3-19}$$

Solving these equations simultaneously yields

$$e_{ca_2} = 167 \text{ volts} \qquad \text{and} \qquad e_{ca_1} = 109.5 \text{ volts}$$

When the capacitor voltage is down at e_{ca_1} and the input is 100 volts, we have $E_a = 100 - 109.5 = -9.5$ volts. Then the input jumps up to 200 volts, but the capacitor voltage momentarily stays at e_{ca_1}. This condition gives $E_b = 200 - 109.5 = 90.5$ volts. The capacitor voltage starts building up from 109.5 toward 200 volts. At $e_{ca_2} = 167$ volts, with the input still at 200 volts, we have $E_c = 200 - 167 = 33$ volts. At this time, the input drops to 100 volts and the resistor voltage jumps down to $E_d = 100 - 167 = -67$ volts.

3-19. The Compensated Attenuator. We have another type of RC circuit to analyze which is commonly used in oscilloscope probes. Consider Fig. 3-48a, which is a simple resistive attenuator. C_2 is stray

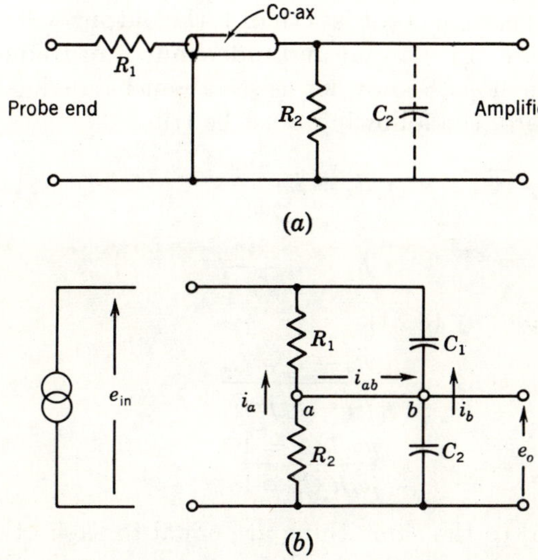

FIG. 3-48. (a) An uncompensated attenuator. (b) A compensated attenuator when $R_1C_1 = R_2C_2$.

or input capacitance which R_2 looks into. If R_1 and R_2 are separated physically, a co-ax cable is used to minimize stray pickup. The cable capacitance then appears in parallel with stray and input capacitances. R_1 tends to isolate C_2 from the probe and raises the impedance seen looking into the probe end. This means that there will be less loading effect upon whatever circuit we are measuring. This arrangement, however, is not very satisfactory. We wish an oscilloscope probe to reproduce faithfully the shape of any input waveform, even though it may be reduced in magnitude. The circuit of Fig. 3-48*a* will not do this. If we apply a step of voltage to the input, the output will not be a step. Instead, it will rise exponentially because C_2 is gradually charging.

We can circumvent this difficulty by putting a capacitor C_1 in parallel with R_1, to let fast shifts through and act as a capacitive divider with C_2. The question now is how large to make C_1, since C_2, R_2, and R_1 are usually known. The easiest way to solve this problem from a transient viewpoint is to imagine the elements connected as a bridge circuit (see Fig. 3-48*b*).

If we can find a way to balance the bridge at all frequencies, then the current in the middle arm, i_{ab}, must always equal zero. With $i_{ab} = 0$, the connection between terminals *a* and *b* could be removed with no change in circuit operation. The output would then be

$$e_o = e_{in} \frac{C_1}{C_1 + C_2}$$

Since $C_1/(C_1 + C_2)$ is a constant less than 1, the output will always be a constant percentage of the input and independent of frequency, since e_o is not a function of ω. So now let us see about balancing the bridge. For i_{ab} to equal zero, the following must be true:

$$i_a R_2 = i_b \frac{1}{j\omega C_2} \tag{3-20}$$

$$i_a R_1 = i_b \frac{1}{j\omega C_1} \tag{3-21}$$

If we divide (3-20) by (3-21),

$$\frac{i_a R_2 = i_b / j\omega C_2}{i_a R_1 = i_b / j\omega C_1}$$

or

$$\frac{j\omega R_2 C_2 = 1}{j\omega R_1 C_1 = 1}$$

Since things equal to the same thing are equal to each other, we can write $j\omega R_1 C_1 = j\omega R_2 C_2$ or $R_1 C_1 = R_2 C_2$. Apparently, our design criterion is that the two time constants must be equal.

We can prove the same result in another way. When the bridge is balanced, points *a* and *b* are at the same potential. When we considered the output as being tapped off the capacity divider, we wrote

$$e_o = e_{in} \frac{C_1}{C_1 + C_2}$$

We could just as well have written

$$e_o = \frac{R_2}{R_1 + R_2} e_{in}$$

since points a and b are connected. Again equating equalities, we have

$$\frac{C_1}{C_1 + C_2} = \frac{R_2}{R_1 + R_2}$$

thus
$$R_1C_1 + R_2C_1 = R_2C_1 + R_2C_2$$
or
$$R_1C_1 = R_2C_2$$

which is the result previously derived.

Problem 1. Let us try a problem. Assume you have an oscilloscope with an input impedance made up of 2 meg in parallel with 10 $\mu\mu$f. For some applications, this represents excessive loading, and we want the input impedance to be 10 meg in parallel with 12 $\mu\mu$f or less. A co-ax probe is to be designed having a length of 4 ft. The cable capacitance is 12 $\mu\mu$f per foot. What will the compensated circuit consist of? What is the attenuation factor?

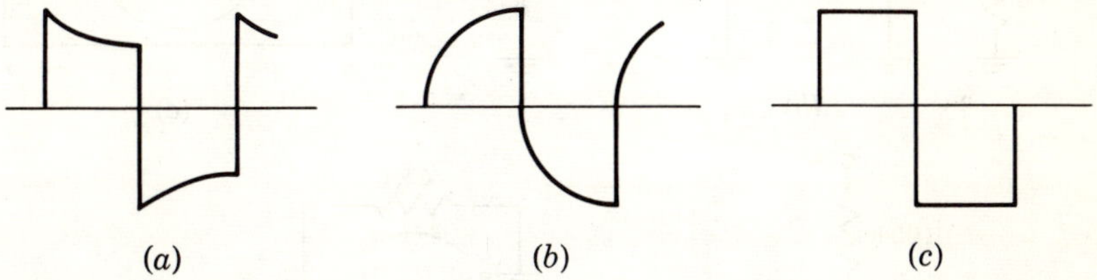

FIG. 3-49. (*a*) C_1 too large (overcompensated). (*b*) C_1 too small (undercompensated). (*c*) C_1 correct.

Solution. The total cable capacitance of 48 $\mu\mu$f will be in parallel with the 10-$\mu\mu$f scope input capacitance, making an effective C_2 of 58 $\mu\mu$f. R_2 is the 2-meg scope input resistance. To make the total input resistance 10 meg, R_1 must equal 8 meg. This immediately fixes the ratio of output to input, or attenuation factor:

$$\frac{e_o}{e_{in}} = \frac{R_2}{R_1 + R_2} = \frac{2 \text{ meg}}{2 + 8 \text{ meg}} = 0.2$$

To prevent frequency distortion, the probe must be compensated, such that

$$R_1C_1 = R_2C_2$$

or
$$C_1 = \frac{R_2C_2}{R_1} = \frac{2 \text{ meg} \times 58 \ \mu\mu\text{f}}{8 \text{ meg}} = 14.5 \ \mu\mu\text{f}$$

The total input capacitance is

$$\frac{C_1C_2}{C_1 + C_2} = \frac{14.5 \times 58}{72.5} = 11.6 \ \mu\mu\text{f}$$

In practice, C would be a small trimmer adjusted through a small hole in the probe. The adjustment is made by connecting the probe to a square-wave source and adjusting the trimmer until the output waveform is a square wave. If C_1 is too large, the output will be peaked as in Fig. 3-49*a*. If C_1 is too small, it will appear rounded as in Fig. 3-49*b*.

A convenient place to pick up a source of square waves (Fig. 3-49*c*) for adjusting C_1 is the voltage calibrator output on the scope itself.

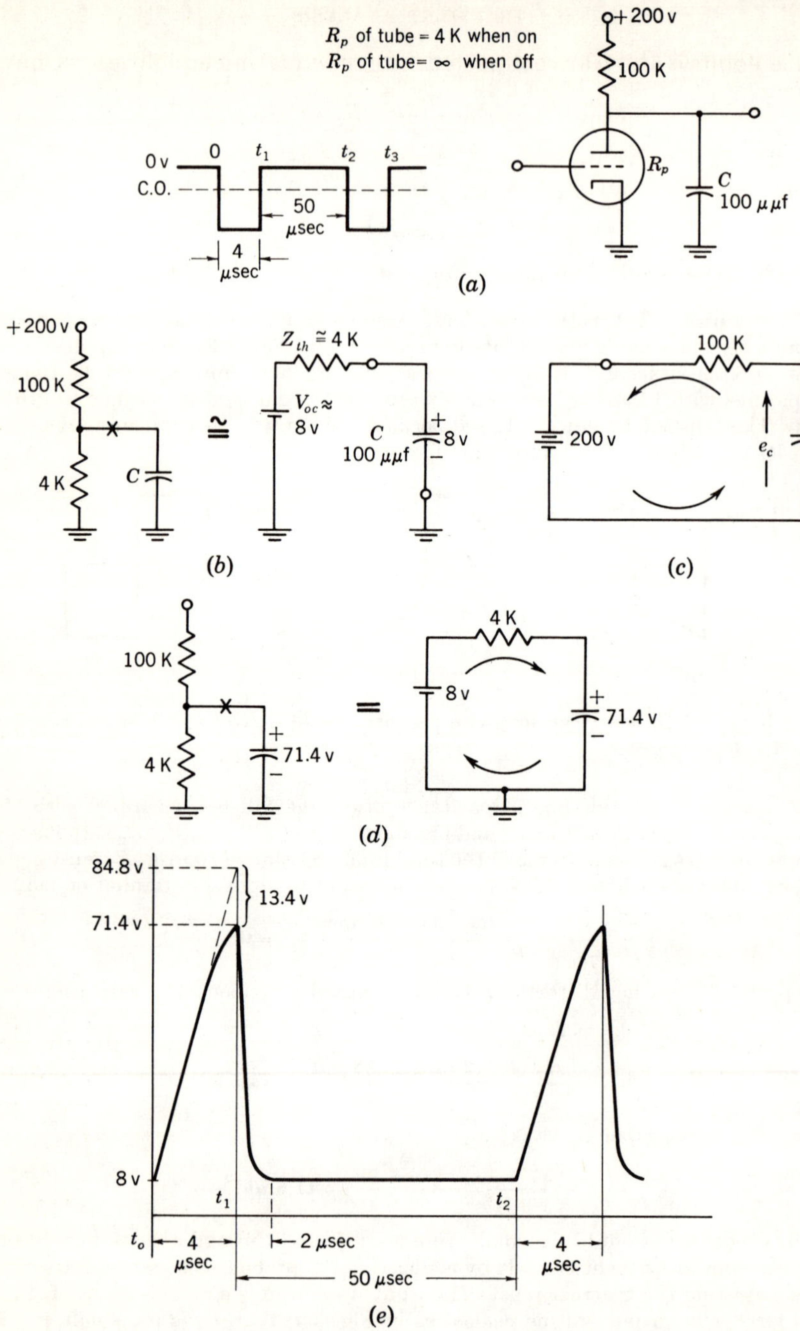

FIG. 3-50. (*a*) A sweep voltage generator. (*b*) Equivalent circuit at t_0 minus. (*c*) Equivalent circuit when the tube is cut off at t_0 plus. (*d*) Equivalent circuit when the tube is conducting at t_1 plus. (*e*) Comparison of an ideal linear sweep with the actual sweep.

Problem 2. Design a 10-to-1 probe attenuator for a 'scope with a 1-meg and 5-$\mu\mu$f input. Use a 4-ft co-ax with 10 $\mu\mu$f per foot of cable capacitance.

Solution

$$R_2 = 1 \text{ meg} \qquad \text{and} \qquad C_2 = \left(10\,\frac{\mu\mu\text{f}}{\text{ft}}\right)(4 \text{ ft}) + 5\ \mu\mu\text{f} = 45\ \mu\mu\text{f}$$

$$\frac{R_2}{R_1 + R_2} = 0.1$$

$$R_1 = 9 \text{ meg}$$

$$C_1 = \frac{R_2 C_2}{R_1} = \frac{1 \text{ meg}(45\ \mu\mu\text{f})}{9 \text{ meg}} = 5\ \mu\mu\text{f} \qquad \textit{Ans.}$$

3-20. Initial Slope of Capacitor Voltage. At times, we are interested in the rate with which the capacitor voltage builds up or decays at the instant of switching. We can determine this by writing a loop equation for a simple *RC* circuit:

$$E = RI + e_C = RC\,\frac{de_C}{dt} + e_C$$

where E is the maximum possible change and e_C is the actual change in capacitor voltage. We know that capacitor voltage cannot change instantaneously. Therefore, at the moment of switching, e_C must equal zero and the instantaneous slope of e_C is

$$\frac{de_C}{dt} = \frac{E}{RC}$$

The instantaneous slope of the capacitor-voltage-versus-time curve at the instant of switching is often useful in waveform sketching.

3-21. The Sawtooth Generator. The circuit of Fig. 3-50*a* is used to generate a sawtooth or sweep waveform such as oscilloscope time bases. When the input voltage on the grid cuts the tube off, the capacitor starts to charge from some low voltage towards the supply voltage. If the charging time constant is very large compared with the time the tube is held off, the capacitor voltage will rise almost linearly, because the capacitor does not have a chance to charge appreciably. This in turn means that the capacitor-bucking voltage will be negligible, and the initial charging current will have barely decreased. The capacitor current is then almost constant. In the section on integrators, we showed that constant current into a capacitor develops a ramp or sweep voltage across the capacitor. Of course, the price paid for holding down the capacitor voltage growth is a small output amplitude. This can, however, be remedied with amplifiers.

When the tube is switched on again, it forms a voltage divider with its load resistor. The capacitor discharges down to this steady-state value between the incoming pulses on the grid. We shall assume that the tube looks like a 4-kilohm resistor when switched on.

We wish to sketch the output waveform and determine its linearity. To do this, a detailed analysis is in order. We already have the required tools at our disposal.

Problem. Sketch the output waveform in Fig. 3-50*a*. By how many volts does the output depart from a linear rate of rise?

Solution. At t_0 minus, the equivalent circuit that the capacitor looks back into is obtained by applying Thévenin's theorem (see Fig. 3-50*b*). At t_0 plus, the tube is driven to cutoff and the equivalent circuit reverts to Fig. 3-50*c*. Note that the capacitor has an initial charge of 8 volts. At t_0 plus, e_c will then start rising from 8 to 200 volts, a swing of 192 volts. The charging path time constant is $RC = 100$ kilohms(100 $\mu\mu$f) $= 10$ μsec. Therefore, the exponent is $x = t/RC = 4$ μsec/10 μsec $= 0.4$. We use $t = 4$ μsec, since this is how long the tube is held off by the negative gate. Entering the chart at $x = 0.4$, we have a corresponding y value of 0.33. That is, the ratio of actual change in e_C to maximum possible change is 0.33. The maximum possible swing is 192 volts, however. The actual swing is then 33 per cent of the maximum possible, or

$$e_C = 0.33(192) = 63.4 \text{ volts}$$

Therefore, at the end of 4 μsec, or t_1 minus, the actual capacitor voltage is

$$e_C = 8 + 63.4 \text{ volts} = 71.4 \text{ volts}$$

At t_1^+, the tube is switched on again and the equivalent circuit is that of Fig. 3-50*d*. The capacitor now discharges. The maximum swing in e_C is from +71.4 volts down to +8 volts, or a 63.4-volt change. The discharge path time constant is $Z_{th}C = 0.4$ μsec. We know that in 5 time constants, or 2 μsec, the transient is over. That is, when $x = 5$, the ratio of actual to possible swing is 1. Since the tube is held on for 50 μsec, whereas 2 μsec is enough, the capacitor will surely have discharged down to 8 volts by t_2 minus. A plot of e_C, the output voltage, is shown in Fig. 3-50*e*.

To determine the departure from linearity,.determine the initial slope of the e_C curve at t_0 plus. In general, the slope at t_0^+ is given by E/RC, where E is the maximum possible swing in e_C. We have already established that at t_0 plus, $e_{C\,\max} = E$ is 192 volts, and $RC = 10$ μsec. The initial slope is then 192 volts/10 μsec $= 19.2$ volts/μsec. If the curve did not droop but kept rising linearly for 4 μsec, the linear swing in e_C would be (19.2 volts/μsec)(4 μsec) $= 76.8$ volts. Since we have an initial 8 volts on C, the actual e_C is $76.8 + 8$ volts $= 84.8$ volts. The departure from linearity is $84.8 - 71.4$ volts $= 13.4$ volts.

3-22. Energy of a Charged Capacitor. Our discussion of RC circuits would not be complete without mentioning the energy that a capacitor stores. It can be shown by methods of the calculus that the energy in a charged capacitor is

$$J = \tfrac{1}{2}Ce_C^{\,2}$$

where e_C is the capacitor voltage in volts, C is capacitance in farads, and J is the energy in joules. One joule is equal to one watt-second. Dimensionally, the equation can be written

$$J = \frac{q}{e_C} e_C{}^2$$

where $q/e_C = C$. Also, we know that q has the dimensions of ampere-seconds. Therefore,

$$J = \frac{\text{amp-sec}}{\text{volts}} (\text{volts})^2 = \text{volts-amp-sec}$$

Since volts $\times$ amp $=$ watts, we have $J =$ watt-seconds.

At times, the solution of engineering problems is greatly facilitated by considering the total energy within a system or circuit.

3-23. The *RL* Circuit. In this section we shall investigate the properties of resistance-inductance (RL) circuits. You will recall that in our discussion of RC circuits we stated that the voltage across a capacitor cannot change instantaneously. In RL circuits we have an analogous rule. The *current* through an inductor cannot change instantaneously! The voltage change across the coil, however, is unrestricted and may jump abruptly. It is a change in current that an inductance opposes.

In basic electronics you were told that the voltage across a coil is proportional to the rate at which current varies through the coil and the coil inductance. Mathematically, this is written as

$$e_L = -L \frac{di}{dt} \approx -L \frac{\Delta i}{\Delta t}$$

where di/dt is simply a shorthand way of writing $\Delta i/\Delta t$ when the time interval Δt is infinitesimally small. The reason why the $L\,di/dt$ term is preceded by a minus sign frequently confuses the student. If the reader will carefully work his way through the following discussion, the purpose of the minus sign will be clear.

Consider Fig. 3-51*a*. A current i has been assumed to flow in the direction of the arrow. Furthermore, we shall assume that the current is increasing in magnitude with the passage of time. When i increases with time, di/dt is positive. If i decreases with time, di/dt is negative. So di/dt may be interpreted as the instantaneous slope of the i-versus-t curve. When i is building up, the self-induced voltage of the coil will be of such polarity that it opposes an increase of current. To oppose the buildup of current, the coil must appear as a generator or source which is going to try to pump electrons in a direction opposite to that assumed for i.

The loop equation would be written as follows:

$$\text{Sources} = \text{sinks}$$

Starting at point A and following the current arrow around the loop, we first encounter source E, which is aiding the assumed direction of electron flow. Therefore, it is written on the source side as a positive source. The voltage drop across the resistor goes on the sink side. In going through the coil, we see a source which is trying to force electrons in a

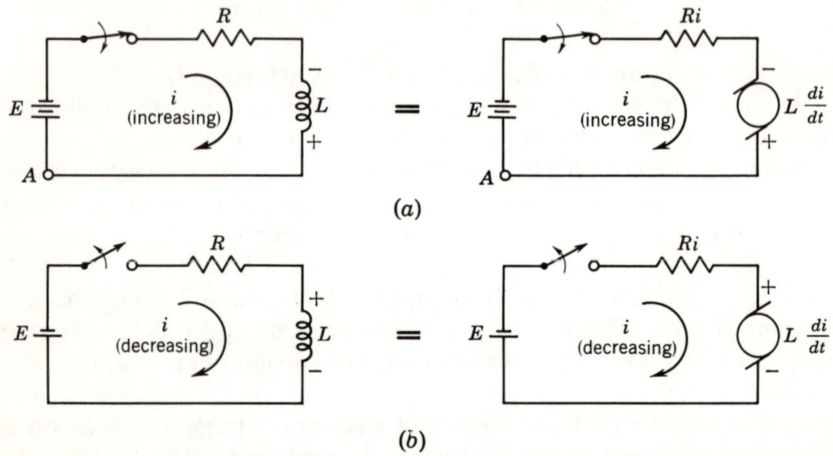

FIG. 3-51. The self-induced coil voltage opposes (*a*) the increase of current and (*b*) the decrease of current.

direction opposite to that assumed. Therefore, the coil voltage looks like a negative source when the current is increasing. The complete loop equation is

$$E - L\frac{di}{dt} = Ri$$

Now let us tie this in with the minus sign in our expression for e_L:

$$e_L = -L\frac{di}{dt}$$

The sign of e_L will determine whether it is a source that aids or opposes the assumed current direction. If e_L is positive, the source aids. If e_L is negative, as we shall see in this case, it opposes. When i is increasing with t, di/dt is positive and, therefore,

$$e_L = -L\left(+\frac{di}{dt}\right)$$

which makes e_L a negative source. This is the case in Fig. 3-51*a*. Were i decreasing with time, di/dt would be negative, and

$$e_L = -L\left(-\frac{di}{dt}\right)$$

makes e_L a positive source. This is the case in Fig. 3-51*b*.

In Fig. 3-51*b*, i is still assumed to be flowing in the same direction. However, i is now decreasing with respect to time, owing to a decrease in E. The self-induced coil voltage is always of such polarity that it opposes the change in current. If i is decreasing, as in Fig. 3-51*b*, the coil voltage will be in a direction to aid E and keep i going. The loop equation in this case is

$$E + L\frac{di}{dt} = Ri$$

Note that e_L is a positive source when di/dt is negative.

3-24. Universal Time-constant Chart and Equations. The general form of the loop equation for a simple RL circuit is

$$E - L\frac{di}{dt} = Ri \qquad \text{or} \qquad E = Ri + L\frac{di}{dt}$$

Remember that di/dt in itself may be plus or minus. As in the case of RC circuits, we are again confronted with a differential equation. We can deduce a few significant facts from the equation in this form. Consider Fig. 3-52. At t_0, let us close the switch. We know that current in an inductive circuit cannot change instantly. Therefore, if $i = 0$ at the instant before the switch closes (t_0 minus) it must be zero the instant after closure, or t_0 plus. The situation is analogous to a capacitor being unable to change its voltage instantly. The loop equation at t_0 plus is then

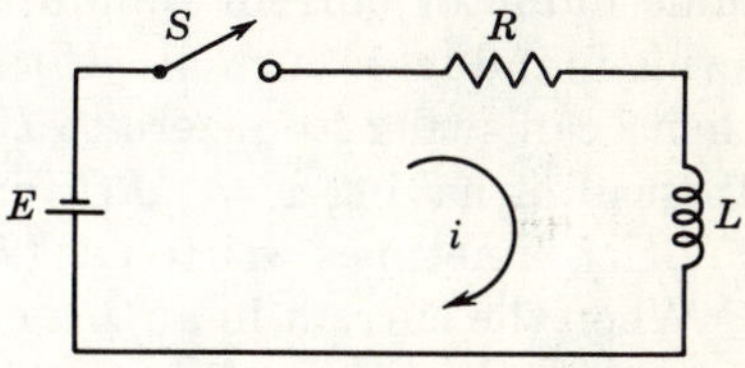

FIG. 3-52. An RL circuit.

$$E - L\frac{di}{dt} = R(0) \qquad \text{or} \qquad \frac{di}{dt} = \frac{E}{L}$$

In words, this tells us that, at t_0 plus, i is increasing at the rate of E/L amp/sec. At a later instant, di/dt will have some smaller value, since there is a drop across R due to i, which leaves less voltage for the coil.

If the reader is wondering how E/L has the dimensions of amperes per second, let him recall that a henry, the unit of inductance, is defined in the following manner. If a current, changing at a rate of one ampere per second, causes a self-induced coil voltage of one volt, that coil has an inductance of one henry.

We can also determine the final, or steady-state, current in the circuit after the switch has been closed a long time. In the steady state, the current is constant, not changing in value, which means $di/dt = 0$. We can then write

$$E - L(0) = Ri \qquad \text{or} \qquad I_{\text{max}} = \frac{E}{R}$$

This tells us that the steady-state current is limited only by the circuit resistance and source voltage.

The general solution of our loop equation involves advanced mathematics beyond the average technician's training. The exact expression relating the buildup of current as a function of time and circuit parameters in an LR circuit is

$$i = I[1 - \epsilon^{-t/(L/R)}] \tag{3-22}$$

where I is the maximum possible current change and the denominator of the exponent L/R is the time constant of an LR circuit. Note how analogous this expression is to that of capacitor voltage buildup:

$$e_C = E(1 - \epsilon^{-t/RC})$$

We can use the curve for $1 - \epsilon^{-x}$ on our time-constant chart to determine inductor current buildup. I represents the maximum possible swing in inductor current. If there is zero initial current in the circuit, then i can swing from zero to $I = I_{\text{max}} = E/R$. The y axis is now i/I. Instead of having $x = t/RC$ for the exponent of ϵ, we now have $x = t/(L/R)$, simetimes written as tR/L.

When the current in an LR circuit is decaying, we can use the curve for ϵ^{-x} on the time-constant chart:

$$i = I\epsilon^{-t/(L/R)} \tag{3-23}$$

Here again, I is the maximum possible change, but i is the amount of change remaining, and *not* the actual amount of change completed.

A general equation which allows for initial conditions is

$$i_{La} = I_{Lf} + (I_{L0} - I_{Lf})\epsilon^{-t/(L/R)} \tag{3-24}$$

While formula (3-24) is perhaps the simplest to use, we shall seldom apply it. The reader will get a better insight into RL circuits by keeping track of initial conditions as he goes through a problem.

The following RL problem should be carefully studied. A good approach is first to try the problem without examining the solution. Then verify your answer and compare methods of solution. There is often more than one way to solve the problem. The important thing is to be sure that you have a justifiable reason for each step.

Problem 1. The 1-henry coil of Fig. 3-53*a* has a winding resistance of 1 kilohm. If the coil is connected across a 100-volt d-c source having an internal resistance of 4 kilohms, find: (*a*) the current at t_0^+, or instant after switch closes; (*b*) the rate at which current is changing with respect to time at t_0^+; (*c*) the coil current in 200 μsec; (*d*) the current rise time; (*e*) the coil voltage in 200 μsec; (*f*) the coil voltage in 1,000 μsec; and (*g*) the coil voltage at $t = \infty$.

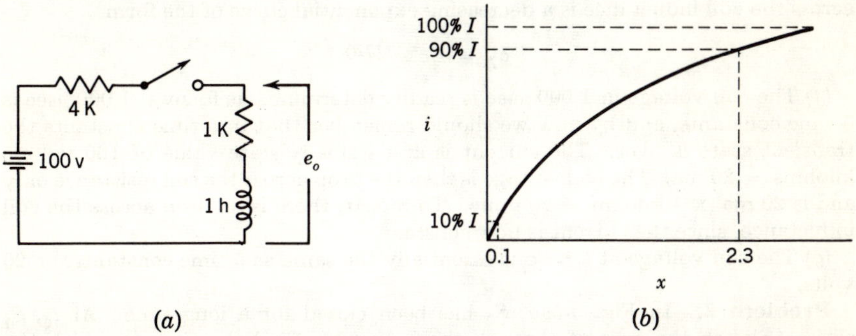

FIG. 3-53. (*a*) An *RL* circuit. (*b*) Rise time of current in an *RL* circuit.

Solution. (*a*) At t_0^+, the current must be zero, since it was zero the moment before.

(*b*) At t_0^+

$$\frac{di}{dt} = \frac{E}{L} = \frac{100 \text{ volts}}{1 \text{ henry}} = 100 \text{ amp/sec}$$

(*c*) To find i in 200 μsec, we must first determine the circuit time constant $L/R = 1$ henry/5 kilohms $= 200$ μsec.

The equation for coil buildup current is

$$i = I[1 - \epsilon^{-t/(L/R)}]$$

and the exponent

$$x = \frac{t}{L/R} = 1 \qquad \text{at} \qquad t = 200\ \mu\text{sec}$$

Since 200 μsec is the circuit time constant, we are simply determining i in 1 time constant. Entering the horizontal axis of the chart at $t/(L/R) = 1$, we see that $i/I = 0.63$. To find i at this time, we must determine I, the maximum possible current change. In this case, $I = 100$ volts/5 kilohms $= 20$ ma, since we are swinging from zero to the steady-state value. In 200 μsec, $i = 0.63I = 12.6$ ma.

(*d*) The coil-current-rise time is the time between $i = 10$ per cent I, and $i = 90$ per cent I (see Fig. 3-53*b*). The time to swing 10 per cent of maximum swing is $t/(L/R) = 0.1$, or $t = 0.1L/R$. The time to swing 90 per cent is $t/(L/R) = 2.3$, or $t = 2.3L/R$. The rise time t_r is then $t_r = 2.3L/R - 0.1L/R = 2.2L/R$. This derivation for rise time in an RL circuit is perfectly general. In this specific problem,

$$t_r = 2.2\frac{L}{R} = 2.2\left(\frac{1 \text{ henry}}{5 \text{ kilohms}}\right) = 2.2(200\ \mu\text{sec}) = 440\ \mu\text{sec}$$

(*e*) The coil voltage in 200 μsec is the sum of the drops across the coil inductance and the coil resistance. We have already established that the coil current is 12.6 ma at this instant of time. The coil resistance drop is then 12.6 volts. The source impedance drop is 4 kilohms $\times$ 12.6 ma $=$ 50.4 volts. The coil inductance drop is then 100 $-$ (50.4 $+$ 12.6) $=$ 37 volts. The total coil voltage is 12.6 $+$ 37 $=$ 49.6 volts. The same result could have been obtained by recalling that the voltage across the coil inductance is a decreasing exponential curve of the form

$$e_L = E\epsilon^{-t/(L/R)}$$

(*f*) The coil voltage in 1,000 μsec is readily determined as follows: 1,000 μsec is 5 time constants, and by now we should remember that in 5 time constants the transient state is over. The current is at its steady-state value of 100 volts/5 kilohms $=$ 20 ma. The coil voltage is then the drop across the coil resistance only and is 20 ma $\times$ 1 kilohm $=$ 20 volts. To repeat, there is no drop across the coil inductance, since the current is now constant.

(*g*) The coil voltage at $t = \infty$ is essentially the same as 5 time constants, or 20 volts.

Problem 2. In Fig. 3-54*a*, S_2 has been closed for a long time. At t_0, S_1 closes. Sketch the current through the inductor. Find the time at which the inductor current is zero.

Solution. At t_0, the inductor current is

$$I_{L0} = \frac{120 \text{ volts}}{20 + 60 \text{ kilohms}} = 1.5 \text{ ma}$$

At t_0, S_1 closes. At $t_0{}^+$, the circuit is that of Fig. 3-54*b*. Now I_{L0} cannot change instantaneously, since it is the current through the coil. We can break the circuit at x and determine the Thévenin's equivalent that the inductor looks back into. The V_{oc} at point x is -120 volts, and Z_{th} is 40 kilohms. We can now draw Fig. 3-54*c*. The coil current at $t_0{}^+$ is still I_{L0}. Right now, we do not know nor are we interested in what I_{L0} does in the rest of the network. What we are saying is that I_{L0} is the current in the coil at $t_0{}^+$. V_{oc} will start building up a current i, as shown in Fig. 3-54*c*. The total swing in inductor current is from $-I_{L0}$ to $+I_{Lf}$, where I_{Lf} is the final value of inductor current at $t = \infty$. We say $-I_{L0}$, because I_{L0} is opposite to the direction of i after S_1 closes. I_{Lf} as determined from Fig. 3-54*c* is

$$I_{Lf} = \frac{V_{oc}}{Z_{th}} = \frac{120 \text{ volt}}{40 \text{ kilohms}} = 3 \text{ ma}$$

The time constant is $L/R = 1$ henry/0.04 meg $= 25\ \mu$sec.

The total swing is 4.5 ma. In 25 μsec, the coil current will have swung through 63 per cent of 4.5 ma $=$ 2.8 ma. The actual coil current, however, is $-1.5 + 2.8$ ma $= +1.3$ ma. In 5 time constants, the current will have swung through 4.5 ma or have reached its steady-state value of 3.0 ma. In equation form, i could be written as

$$\begin{aligned} i &= -I_{L0} + [I_{Lf} - (-I_{L0})][1 - \epsilon^{-t/(L/R)}] \\ &= -1.5 + 4.5 \text{ ma}(1 - \epsilon^{-t/25\ \mu\text{sec}}) \end{aligned}$$

To find t when $i = 0$, we could set $i = 0$ in the equation and solve for the corresponding value of t. This need not be done if a time-constant chart is available. We see from Fig. 3-54d that for $i = 0$, i must have increased by 1.5 ma. This is a swing of 1.5 ma out of 4.5 total, or

$$\frac{1.5 \text{ ma}}{4.5 \text{ ma}} \times 100 = 33.3 \text{ per cent}$$

Entering the curve of inductor buildup at 33.3 per cent tells us that 0.4 time

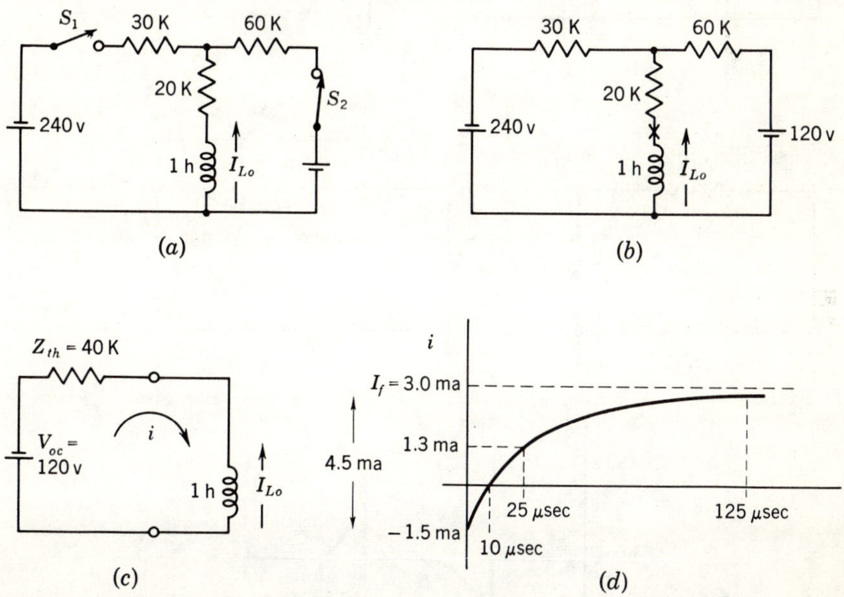

FIG. 3-54. (a) A complex RL circuit at t_0 minus. (b) S_1 closed in Fig. a at t_0. The inductor current momentarily remains constant. (c) The equivalent circuit that the inductor looks back into. (d) The inductor current time response.

constant are required. That is, the exponent of our equation must equal 0.4. Therefore,

$$\frac{t}{25\ \mu\text{sec}} = 0.4$$

$$t = 0.4 \times 25\ \mu\text{sec} = 10\ \mu\text{sec} \qquad \textit{Ans.}$$

Problem 3. In Fig. 3-55a, the switch has been open for a long time. Sketch the output voltage if S is closed at t_0 for 50 μsec, and then opened.

Solution. With S open, e_o must equal 300 volts, since there is no drop across the 30-kilohm resistor. The moment (t_0^+) that S closes, the inductor current will still stay at zero, and therefore e_o is simply

$$e_o = 300 \text{ volts} \times \frac{60 \text{ kilohms}}{60 + 30 \text{ kilohms}} = 200 \text{ volts}$$

As the coil current builds up, e_o will start to rise from 200 volts. This should be clear, since the increasing coil current must flow through the 60-kilohm resistor. If the switch were never reopened, e_o would climb back to 300 volts, since the coil looks like a short circuit in the steady state.

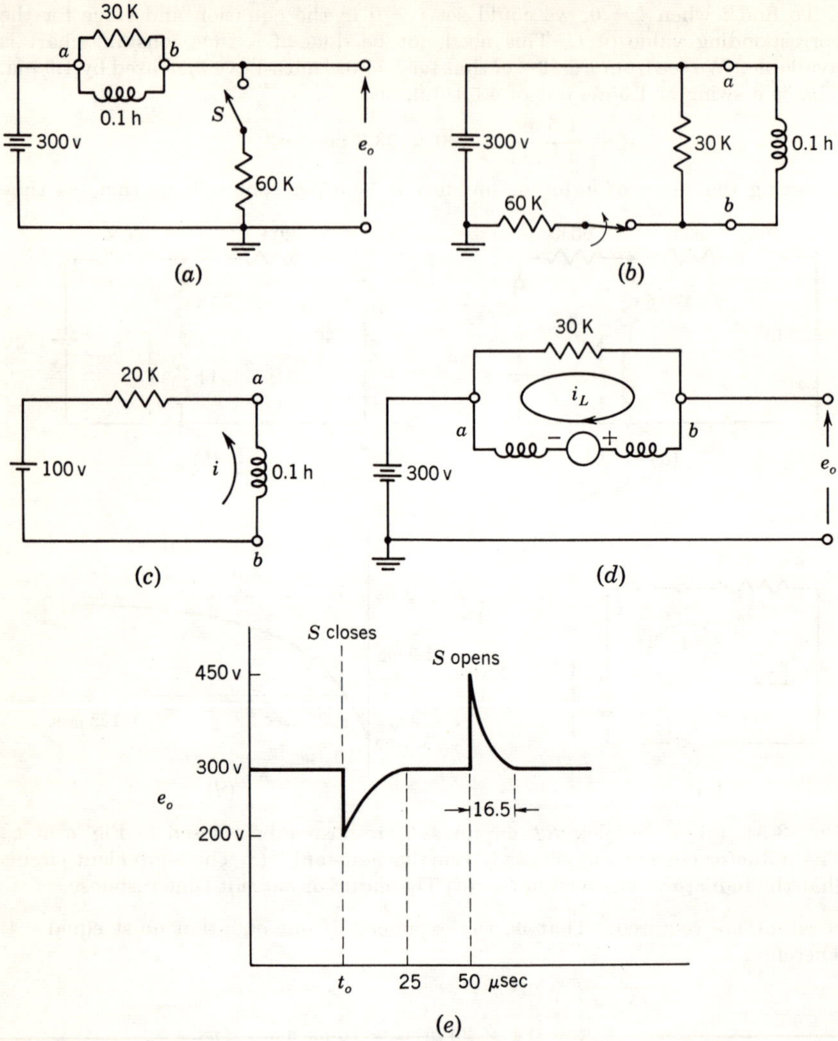

FIG. 3-55. (a) An RL switching circuit. (b) Circuit a redrawn. (c) The equivalent circuit that the inductor looks into. (d) The equivalent circuit when S reopens. (e) The output voltage time response for the circuit.

In 50 μsec, the switch reopens. To find e_o at this time we must first determine the coil current. Let us apply Thévenin's theorem to determine the equivalent circuit that the coil looks back into. Redraw the circuit, as shown in Fig. 3-55b and as shown in Fig. 3-55c.

$$V_{oc} = 300 \text{ volts} \times \frac{30 \text{ kilohms}}{90 \text{ kilohms}} = 100 \text{ volts}$$

$$Z_{th} = \frac{60 \times 30}{60 + 30} = 20 \text{ kilohms}$$

The charging time constant is

$$\tau_c = \frac{L}{R} = \frac{0.1 \text{ henry}}{20 \text{ kilohms}} = 5\ \mu\text{sec}$$

Since the switch is closed for 50 μsec, or 10 time constants, we can be sure that the inductor current has reached its steady-state value of 100 volts/20 kilohms = 5 ma. This is in perfect agreement with Fig. 3-55*a*, since the steady-state current, when L appears as a d-c short circuit, is 300 volts/60 kilohms = 5 ma. It will take only about 5 time constants, or 25 μsec, before the inductor current is 5 ma. The output voltage is 5 ma × 60 kilohms = 300 volts at this time.

In 50 μsec, S suddenly opens. The coil will develop a voltage because of the collapsing magnetic field. The polarity of this voltage will be such that it tries to keep the coil current unchanged. When the switch was closed, coil current flowed from the b to the a terminal. With the switch just opened as in Fig. 3-55*d*, the coil current must momentarily stay at 5 ma in the same direction. The polarity of the self-induced generator must then make the a side negative. The coil current then returns through the 30-kilohm resistor to the b side. The peak voltage developed across the coil is the same as that across the 30-kilohm resistor, since the two elements are in parallel. The magnitude of this peak voltage must be

$$e_{pk} = 5 \text{ ma} \times 30 \text{ kilohms} = 150 \text{ volts}$$

The output voltage will therefore have jumped to 300 + 150 volts = 450 volts. This is not an error. Switching transients in inductive circuits frequently exceed the supply voltage. If the 30-kilohm resistor had been 60 kilohms, the output would have jumped to 600 volts.

The discharge time constant is 0.1 henry/30 kilohms = 3.3 μsec. The jump in the output will then settle down to a steady-state value of 300 volts in 5 time constants, or 16.5 μsec (see Fig. 3-55*e*).

3-25. The *RL* Differentiator. Since the voltage across a pure inductance is proportional to the rate of change of current, the voltage has the properties of a differentiating element. Remember always that $e_L = -L\,di/dt \approx -L\Delta i/\Delta t$.

Problem. If the waveform of current through the inductor of Fig. 3-56*a* is as shown in Fig. 3-56*b*, sketch the voltage waveform read by e_L.

Solution. Assuming a positive or reference current direction, as shown, we see that during the interval from t_0 to t_1 the current is flowing in the assumed direction and increasing. Therefore, e_L is positive between t_0 and t_1. Also, the i_L curve is steepest at t_0, so that e_L will be maximum at this time (Fig. 3-56*c*). From t_1 to t_2, the current is still flowing in the assumed direction, since the sign of i_L is still positive here. However, i_L is decreasing, which means di/dt is negative. The coil, in an effort to maintain i_L at the value it had at t_1, will now reverse its polarity and act as a source trying to keep i_L flowing (Fig. 3-56*d*).

At t_2 the direction of i_L actually reverses, as evidenced by i_L being negative between t_2 and t_3. The slope di/dt is also negative during this interval, which indicates that the current is building up in this reversed direction. The coil will then look like a source which tends to keep the current from growing by developing the polarity shown in Fig. 3-56*e*. If you have any trouble seeing this, just remove the coil and replace it with a generator. Then ask yourself what polarity this generator must have to pump current in the direction opposing i_L when i_L is increasing in magnitude. When i_L decreases in magnitude, the generator polarity will try to aid the dwindling i_L.

Between t_3 and t_4, i_L is still flowing in the negative direction. The magnitude, however, is decreasing, and therefore the self-induced generator will have the polarity shown in Fig. 3-56*f*. Notice that the coil voltage aids the current.

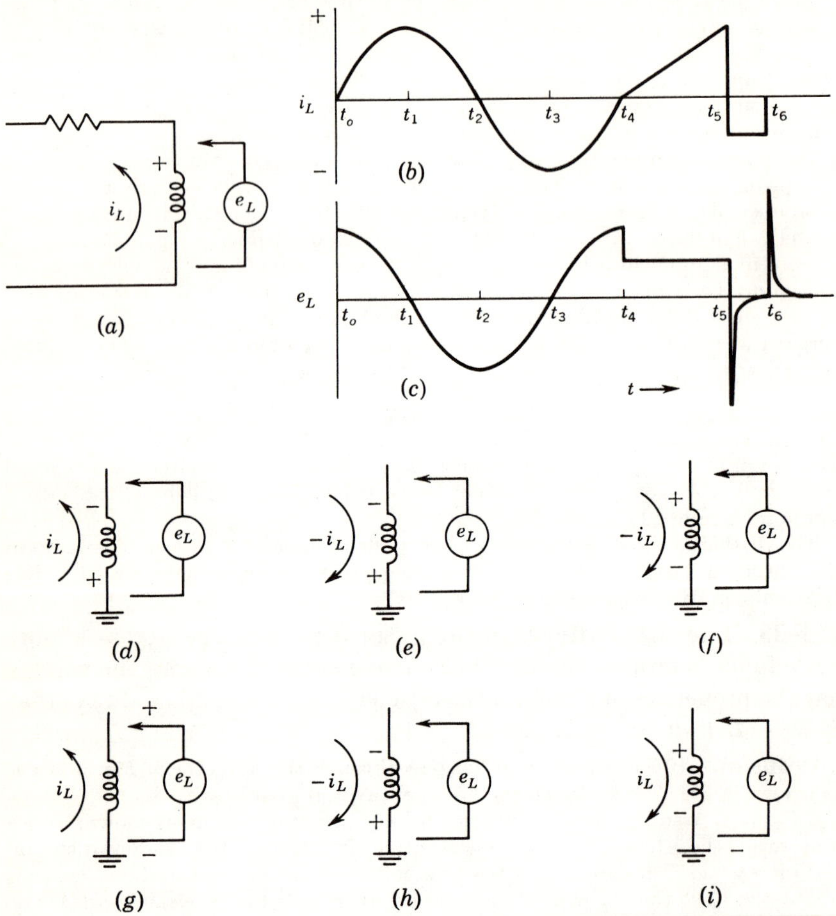

FIG. 3-56. (*a*) An *RL* differentiating circuit. (*b*) The input current waveform. (*c*) The differentiated output waveform. (*d*) The coil voltage between t_1 and t_2. (*e*) The coil voltage between t_2 and t_3. (*f*) The coil voltage between t_3 and t_4. (*g*) The coil voltage between t_4 and t_5. (*h*) The coil voltage between t_5 and t_6. (*i*) The coil voltage at t_6 plus.

The reader may already have noticed that the sinusoidal part of i_L results in a leading e_L. This is in agreement with our a-c theory, which tells us the current lags the voltage 90° in an inductor.

At t_4, the current is still flowing in the reverse direction. It has decreased to almost zero. In an effort to keep the current going, e_L will be as shown in Fig. 3-56*f*. And e_L looks like a positive source.

At t_4^+, i is infinitesimally small, flowing in the assumed positive direction and increasing in magnitude with a slope somewhat less than at t_4^-. In an effort to keep the current from building up, the coil voltage will be as shown in Fig. 3-56*g*. e_L is still positive, but jumps to a lower value because the slope is less at t_4^+ than at t_4^-.

At t_5^+, the current has suddenly reversed. A large negative spike will be produced because the slope is very steep between t_5^- and t_5^+. The output is negative because the self-induced coil voltage will try to keep the current flowing as in Fig. 3-56*g*, although i_L at t_5^+ is as shown in Fig. 3-56*h*.

Between t_5 and t_6, the current is constant (zero slope). Therefore, $e_L = 0$, since the coil is a pure inductance with no resistance.

At t_6, i_L is still flowing in the reverse direction and dropping in magnitude toward zero. To keep the reverse current going, e_L will jump in a positive direction (see Fig. 3-56*i*).

RL circuits are seldom used as diffentiating elements, because an *RC* differentiator is so much cheaper. As with *RC* circuits, the time constant L/R should be small compared with the pulse duration for good differentiating action. Choosing L/R equal to one-tenth the pulse duration is the usual rule of thumb.

3-26. The *RL* Integrator. In an *RL* integrator, the output voltage would be taken off the resistor, since the inductor smooths out current variations. As with *RC* circuits, the time constant should be 10 times the pulse duration.

The *RC* circuit is used in place of the *RL* circuit whenever possible because of the higher cost, weight, and volume of inductors compared with capacitors. Furthermore, when we consider a capacitor in some application, the capacitance is usually all we are concerned with. But with an inductor, such factors as the winding capacitance and Q must be carefully considered. These hidden properties of a practical inductor may result in oscillatory phenomena known as ringing. When we deal with inductors containing iron cores, the inductance becomes a function of the current, owing to saturation effects. When L is not a constant parameter, the inductor is said to be nonlinear. Often these nonlinear properties are put to good use in peaking circuits, magnetic amplifiers, and power supplies.

3-27. Energy of a Current-carrying Inductor. An inductor carrying a current of i amperes has an amount of energy stored in its magnetic field equal to $J = \frac{1}{2}Li^2$. The energy (J) is expressed in joules, or watt-seconds. One joule equals one watt-second.

When switching takes place in a current-carrying inductor, the sudden collapse of the magnetic field may cause high-voltage transients to appear in the circuit. This high induced voltage can cause arcing at switch contacts, and the stored energy in the coil's field will largely dissipate itself in the arc.

The circuit of Fig. 3-57*b* illustrates a moderately effective but inexpensive method of arc suppression for inductive circuits. R_L is the coil resistance, and R_S is the arc-suppressing resistor. As a rule of thumb, R_S is usually chosen to be 10 times larger than R_L.

First, consider Fig. 3-57*a*. The switch has been indefinitely closed. At t_0, S suddenly opens. This means that i must decrease from a steady-state value of 28 volts/28 ohms = 1 amp to zero in a negligible time. The inductor, in an effort to keep the same i flowing, will develop an enormous voltage of reversed polarity, as shown. This induced voltage, acting in series with the supply voltage, will usually cause an arc to be developed across the switch contacts. The magnitude of the induced voltage depends upon the rate at which the switch and the arc break

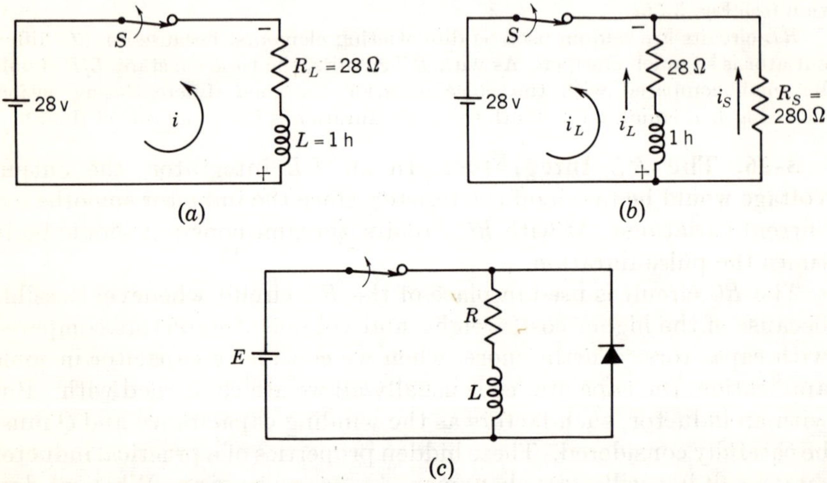

FIG. 3-57. (*a*) The inductive kick when S opens will form an arc at the switch contacts. (*b*) R_s will partially absorb the inductive kick and suppress the arc when S opens. (*c*) A diode used as an efficient arc suppressor.

the circuit. Also, the inductance and initial coil current are factors to be considered. As a hypothetical case, assume that the steady-state current of 1 amp drops to zero in 1,000 μsec when S opens. The jump in coil voltage will be, approximately,

$$e_L \approx -L\frac{\Delta i}{\Delta t} = 1 \text{ henry} \times \frac{1 \text{ amp}}{10^{-3} \text{ sec}} = 1{,}000 \text{ volts}$$

Now let us examine the circuit of Fig. 3-57*b*. The steady-state value of i_L is also 1 amp. A certain amount of current i_s is wasted in the arc-suppressing resistor R_s. This current is plainly 0.1 amp. Let us assume that S opens instantly, which is the worst case. What is the coil voltage now? Well, i_s will instantly drop to zero, since it flows through the pure resistance R_s. However, i_L cannot cease instantly. But we said that S opens instantaneously, which means that i_L cannot return through the switch. Therefore, i_L flows instead through the parallel branch comprising R_s. The peak voltage across the coil terminals must then be the same as the drop across R_s, since the two are in parallel. We can then write

$$e_{Lpk} = i_L R_S = \frac{E_b R_s}{R_L} = \frac{28 \text{ volts} \times 280 \text{ ohms}}{28 \text{ ohms}} = 280 \text{ volts}$$

This is much less than the previously determined value of e_L without suppression.

Frequently, the additional supply drain due to R_S is prohibitive, and a diode is used, as shown in Fig. 3-57*c*. With S closed, the diode current is negligible, since it is reverse-biased. If the source voltage is high, several diodes may be stacked in series. When S opens, the upper end of the coil jumps negative and forward-biases the diode. A forward-biased diode looks like a low value of resistance ranging from 10 to 100 ohms, depending upon the type used. The diode-forward resistance is then equivalent to R_S. The diode chosen must be capable of passing a peak current equal to the steady-state value of coil current.

Perhaps at this point some readers are questioning the relationship

$$e_{Lpk} = E_b \frac{R_s}{R_L}$$

since it does not contain L. Experience tells us that if L were 1 μh and everything else stayed the same, there would be no inductive kick worth mentioning. Yet mere substitution of values for E_b, R_S, and R_L yielded 280 volts for e_L, whether L was 1 μh or 10 henrys. The answer lies in the fact that Fig. 3-57*b* is incomplete.

There is always some stray and shunt capacitance across the coil. If L is small, it will not swamp out the effect of shunt capacitance. Since this capacitance is across the coil, it will keep the coil voltage from changing instantaneously.

An exact analysis is beyond the scope of this book. The thing to remember is that physical components such as R, L, and C never exist in the pure state.

3-28. Series *RLC* Circuit and Logarithmic Decrement. The response of *RLC* networks to nonsinusoidal waveforms is difficult to determine without the aid of calculus. Even with the aid of calculus, the problem can be quite formidable. For this reason we shall discuss just a few basic ideas and circuits.

Consider the circuit of Fig. 3-58*a*. We know that, because of L, the initial current must be zero. Because of C, the final current must be zero.

The current in Fig. 3-58*a* will assume the form of either a pulse as in Fig. 3-58*b* or a damped sine wave as in Fig. 3-58*c*. The response is that of Fig. 3-58*b* if $L/C < R^2/4$, and that of Fig. 3-58*c* if $L/C > R^2/4$. If the response is that of Fig. 3-58*c*, the angular frequency of the sine wave is

$$\omega = \sqrt{\frac{1}{LC} - \frac{R^2}{4L^2}} = \sqrt{\omega_r^2 - \frac{R^2\omega_r^2}{4\omega_r^2 L^2}} = \omega_r\sqrt{1 - \frac{1}{4Q_r^2}} \approx \omega_r$$

when $Q \geq 10$, ω_r is the series-resonant frequency, and Q_r is the circuit Q

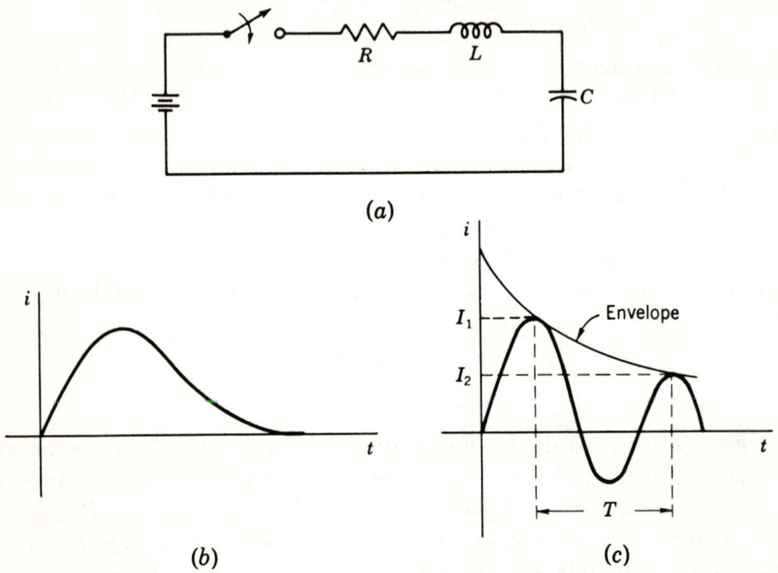

FIG. 3-58. (*a*) A series *RLC* circuit. (*b*) Overdamped response of the *RLC* circuit when $L/C < R^2/4$. (*c*) Underdamped response of the *RLC* circuit when $L/C > R^2/4$.

at this frequency. The ratio of the currents between successive cycles is

$$\frac{I_2}{I_1} = \epsilon^{-RT/2L}$$

The quantity $RT/2L$ is sometimes referred to as the logarithmic decrement. Since $T = 1/f$ and $Q = 2\pi fL/R$, or $1/Q = R/2\pi fL$, we can write

$$\frac{I_2}{I_1} = \epsilon^{-R/2fL} = \epsilon^{-\pi R/2\pi fL} = \epsilon^{-\pi/Q}$$

The value of Q is the total circuit Q, which involves all resistances.

We can also show that the time required for the peak current at some future cycle to equal 63 per cent of the peak current in the first cycle is $2Q/\omega$ sec.

Since $I_2 = I_1\epsilon^{-\pi/Q}$, $I_3 = I_2\epsilon^{-\pi/Q}$, $I_4 = I_3\epsilon^{-\pi/Q}$, $I_5 = I_4\epsilon^{-\pi/Q}$, we can readily write an expression for the current in any arbitrary cycle after the first. For example,

$$I_5 = I_4\epsilon^{-\pi/Q} = I_3\epsilon^{-\pi/Q}\epsilon^{-\pi/Q} = I_2\epsilon^{-\pi/Q}\epsilon^{-\pi/Q}\epsilon^{-\pi/Q} = I_1\epsilon^{-4\pi/Q}$$

Notice that the number of times π/Q appears in the exponent is equal to the number of cycles n between I at the $(n + 1)$st cycle and at the first cycle. That is,

$$I_{n+1} = I_1\epsilon^{-n\pi/Q} \qquad \text{or} \qquad \frac{I_{n+1}}{I_1} = \epsilon^{-n\pi/Q}$$

Now, I_{n+1} means I at any cycle. It might be interesting to see what percentage I_{n+1} is of I_1 when the exponent of ϵ is -1. That is,

$$\frac{n\pi}{Q} = 1 \qquad \text{or} \qquad n = \frac{Q}{\pi} \text{ cycles}$$

When $n = Q/\pi$,

$$\frac{I_{n+1}}{I} = \epsilon^{-1} = 0.63$$

In other words, in Q/π cycles the current peak will be down to 63 per cent of the peak in the first cycle. Since the time per cycle T is the period, or $1/f$, the time required for n cycles is $nT = n/f$. For Q/π cycles, the time is $Q/\pi f$. However, $\omega = 2\pi f$, or $f = \omega/2\pi$. Substituting in $Q/\pi f$, we have $2Q/\omega$ sec as the time for the oscillatory current peak to decay to 63 per cent of the initial peak.

The higher the circuit Q, the less decay there will be in current amplitude between successive cycles. If R were zero, so that no energy loss occurred in the form of heat, the oscillations would continue forever.

The circuit of Fig. 3-58*a* could be used to measure the circuit Q by counting the number of cycles on an oscilloscope that is required for the nth current peak to equal 63 per cent of the first. The Q is then $Q = n\pi$.

3-29. The Parallel *RLC* Circuit. Frequently we encounter the parallel-resonant circuit of Fig. 3-59*a*, whose equivalent is Fig. 3-59*b*. The circuit is generally operated so that at t_0, S opens. Again the response may or may not be a damped sinusoid. Figure 3-59*b* is usually easier to analyze and more informative than Fig. 3-59*a*. Therefore, let us first see how we can convert Fig. 3-59*a* to Fig. 3-59*b*.

You will recall from Chap. 2 that $R_P = R_S(Q_S^2 + 1) \approx R_SQ_S^2$. The value of Q_S is to be taken at the series-resonant frequency ω_r determined by L_S and C in Fig. 3-59a. L_P will essentially equal L_S if Q_S is moderately high, since

$$L_P = L_S\left(1 + \frac{1}{Q_S^{\,2}}\right) \approx L_S$$

The circuit of Fig. 3-59a is generally operated so that the switch is

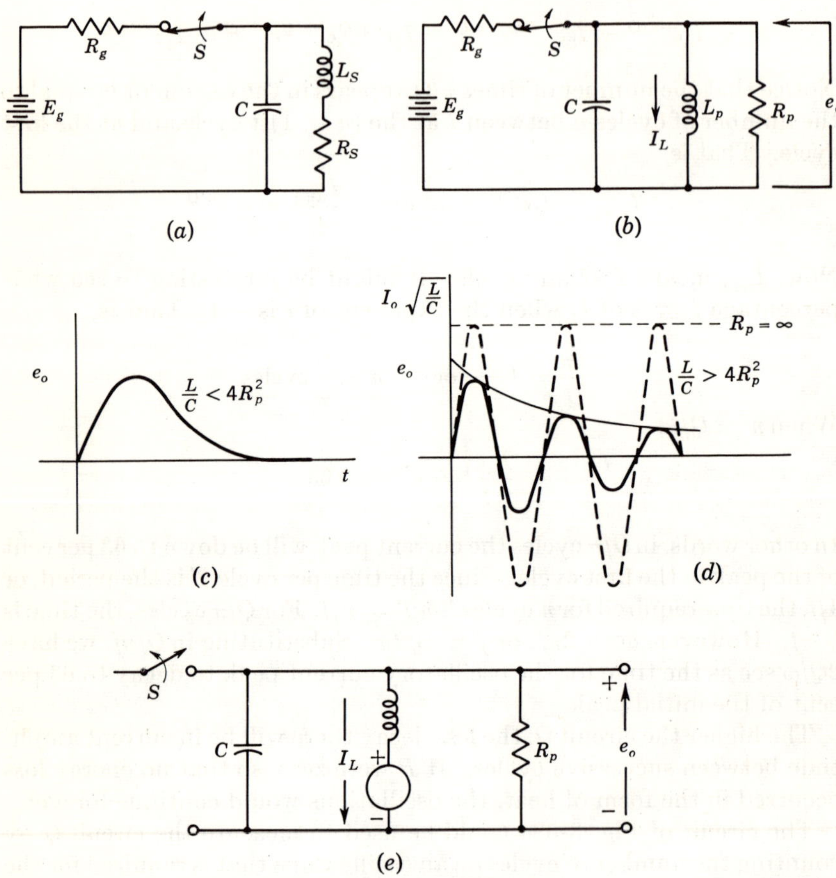

FIG. 3-59. (a) A parallel RLC circuit. (b) Equivalent circuit. (c) Overdamped response of the parallel RLC circuit. (d) Underdamped response of the parallel RLC circuit. (e) Equivalent circuit when S is opened.

opened at some instant of time t_0. The only losses in the series-resonant circuit of Fig. 3-59a are then due to R_S. Therefore, with S open, use $Q_S = \omega_r L_S / R_S$, where

$$\omega_r = \frac{1}{\sqrt{L_S C}}$$

Opening S in Fig. 3-59b will produce a voltage across the tank circuit, e_o, that will look like Figs. 3-59c or d. Figure 3-59c prevails if $L/C < 4R_P^2$ and Fig. 3-59d if $L/C > 4R_P^2$. The frequency of the damped sine wave in Fig. 3-59d is

$$\omega = \sqrt{\frac{1}{LC} - \frac{1}{4R_P^2C^2}} = \sqrt{\omega_r^2 - \frac{1\omega_r^2}{4\omega_r^2R_P^2C^2}} = \omega_r\sqrt{1 - \frac{1}{4Q_P^2}}$$

$$\approx \omega_r \qquad \text{if } Q_P \text{ is at all high}$$

Remember that $Q_P = 1/\omega C_P R_P$, as developed in Chap. 2.

If the switch in Fig. 3-59b has been closed a long time, the current through L_P will have reached a steady-state value I_0 equal to $I_0 = E_g/R_g$. Since L_P is a short circuit to direct current, there will be no initial voltage on C when S opens. The instant S opens, the voltage across the inductor will try to jump abruptly, owing to the collapse of the surrounding magnetic field. The self-induced voltage in L will try to keep I_L flowing in the same downward direction. The polarity of the coil voltage will appear as in Fig. 3-59e, and the upper end of the tank circuit will initially swing positive. The capacitor slows down the rate at which e_o rises, since it cannot change its potential instantaneously.

The energy stored in the magnetic field of the inductor is then transferred to the capacitor, minus some amount dissipated in the resistor. If the Q_P is high, most of the coil energy is transferred to the capacitor, and we can write the approximation

$$\tfrac{1}{2}LI_0^2 = \tfrac{1}{2}Ce_o^2$$

where $I_0 = \dfrac{E_g}{R_g}$

This relationship enables us to solve for the peak value of the continuous sine wave which would occur if $R_P = \infty$, which is approximately the same in a high-Q circuit:

$$e_{o\max} = I_0\sqrt{\frac{L}{C}}$$

The material in this discussion is often used in the design and analysis of time-mark generators and peaking circuits.

SUMMARY

In this chapter we learned that nonsinusoidal waveforms may be broken down into a number of harmonically related sinusoidal waves. The output waveform of a circuit depends upon the circuit response to these component frequencies. Techniques for evaluating average and rms values of nonsinusoidal waveforms were presented. The *RC* circuit was discussed, and two fundamental principles were stressed: namely, that the voltage across a capacitor cannot change instantaneously and that a capacitor will not pass direct current. An exact interpretation of the time-constant chart was presented, so that quantitative solutions of switching transients could be readily made. With the aid of tools developed in Chap. 1, multiple-resistor single-capacitor circuits were simplified to a simple series *RC* circuit to which the time-constant chart could be applied. A formula was presented that is especially useful for determining an output waveform due to a repetitive exciting waveform.

The waveshaping properties of *RC* differentiating and integrating networks were investigated. We learned that the differentiator could be used to generate spikes or trigger pulses, while the integrator could generate a sawtooth. The differentiator tends to produce an output proportional to the rate of change of input, while the integrator output is proportional to the area under the input-vs.-time curve. In discussing the *RC* coupling circuit, we learned that a long time constant relative to the input pulse period is required for undistorted transmission.

The analysis of *RL* circuits was also based upon two fundamental principles: namely, that the current through an inductor cannot change instantaneously and that the average voltage across an inductor is zero. We learned that the inductor may be regarded as a source whose polarity is such that it will always try to maintain a constant current through itself. Multiple-resistor single-inductor circuits were analyzed and simplified to equivalent circuits that could be quantitatively analyzed with the time-constant chart.

Series and parallel *RLC* circuits were also investigated to some extent. The discussion of *RLC* circuits was of necessity limited, owing to the complicated nature of the equations involved. The probable nature of the response to be expected with a given circuit was shown to be dependent upon certain relationships between the parameters.

PROBLEMS

3-1. Determine e_o, 40 μsec after the switch closes in Fig. 3-60.

Ans. 34.4 volts.

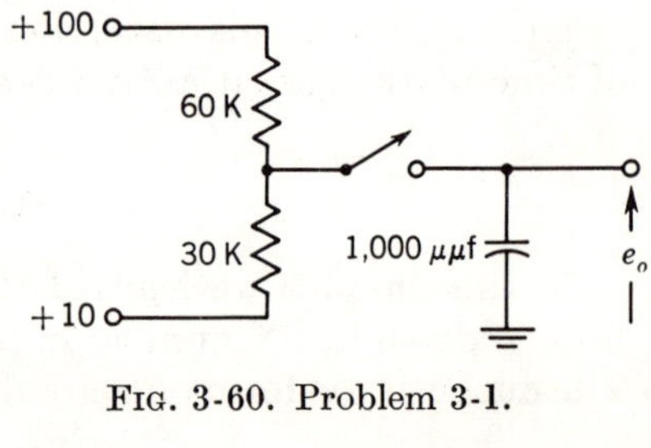

FIG. 3-60. Problem 3-1.

3-2. What is e_o, the instant the switch closes in Fig. 3-61? *Ans.* 120 volts.

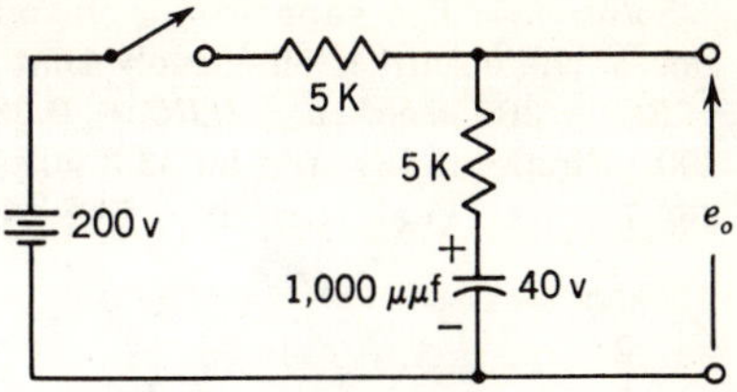

FIG. 3-61. Problem 3-2.

3-3. What is e_o, 15 μsec after t_0 in Fig. 3-62? *Ans.* −30.8 volts.

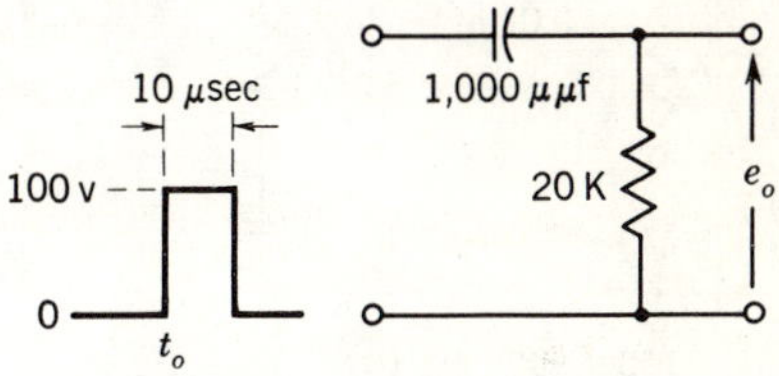

FIG. 3-62. Problem 3-3.

3-4. In Fig. 3-63, the switch is closed for 1 sec at t_0, and then opened at t_1. What will e_o be 1 sec after t_1?

Ans. −19.5 volts.

Solution. The capacitor-charging time constant is 1 sec. At t_1^-, the capacitor will be charged to 63 volts. Opening S at t_1, gives a discharging time constant equal to 2 sec. In 1 sec after t_1, the capacitor will have swung through 39 per cent of its maximum swing, which is 63 volts. This means the change in capacitor voltage at this time is 24.5 volts. The actual voltage left across the capacitor is 63 − 24.5 = 38.5 volts, and this is split equally between the resistors, which leaves −19.5 volts at the output terminals.

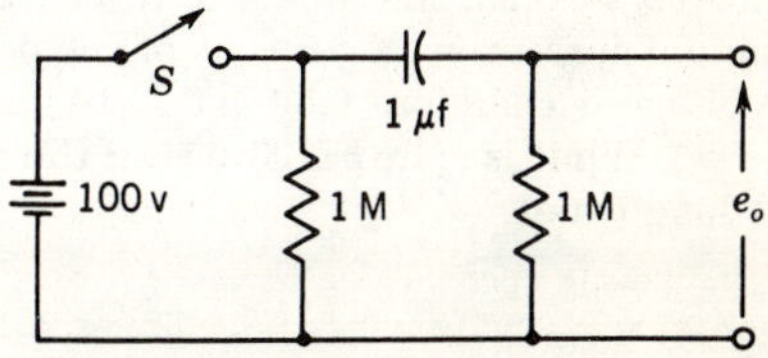

FIG. 3-63. Problem 3-4.

3-5. Prove that the circuit of Fig. 3-64*b* will reach its final amplitude in twice the time required by that of Fig. 3-64*a*.

FIG. 3-64. Problem 3-5.

3-6. (*a*) The switch is initially open. What is e_o, 100 μsec after the switch is closed in Fig. 3-65*a*? *Ans.* −65.6 volts.

Solution. The capacitor is initially charged to 300 volts. When the switch closes, the circuit is effectively that of Fig. 3-65*b*.

$RC = 300\ \mu$sec. $x = t/RC = 0.33$, when $t = 100\ \mu$sec. Initial value of $e_C =$ 300.· Final value would be 33.3 volts. Maximum possible swing is $300 - 33.3 =$ 266.7 volts. Peak current $= {}^{266.7}\!/_{30}$ kilohms $= 8.9$ ma.

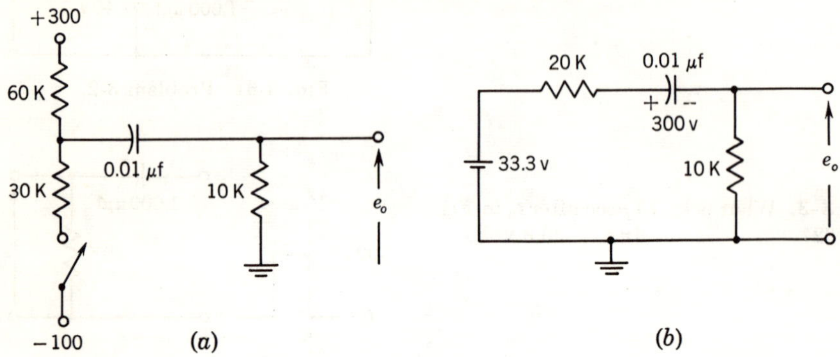

FIG. 3-65. Problem 3-6.

At $x = 0.33$, the capacitor discharge or charge current curve indicates that the actual current is 74 per cent of the peak value of 8.9 ma, or 6.56 ma. The output voltage is therefore 6.56 ma $\times$ 10 kilohms, or -65.6 volts.

(*b*) What is e_o in Fig. 3-65*a*, if the switch is opened 100 μsec after being closed a long time? *Ans.* 34 volts.

3-7. What will the capacitor voltage in Fig. 3-66 be 100 μsec after the switch closes? *Ans.* 0 volts.

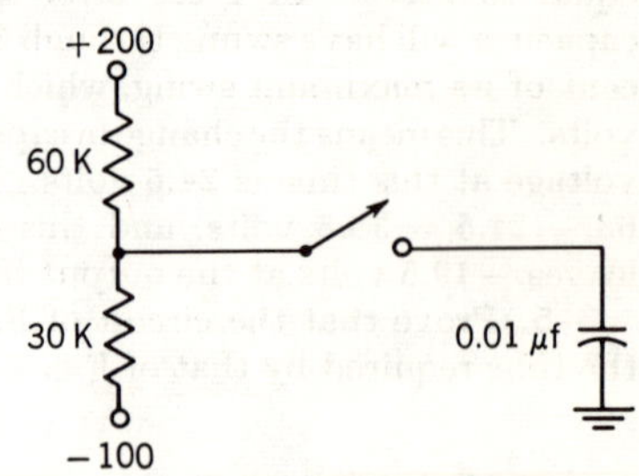

FIG. 3-66. Problem 3-7.

3-8. (*a*) What will e_o be at t_1^- in Fig. 3-67? *Ans.* 37 volts.

(*b*) Would increasing C make the output more or less like the input? *Ans.* More.

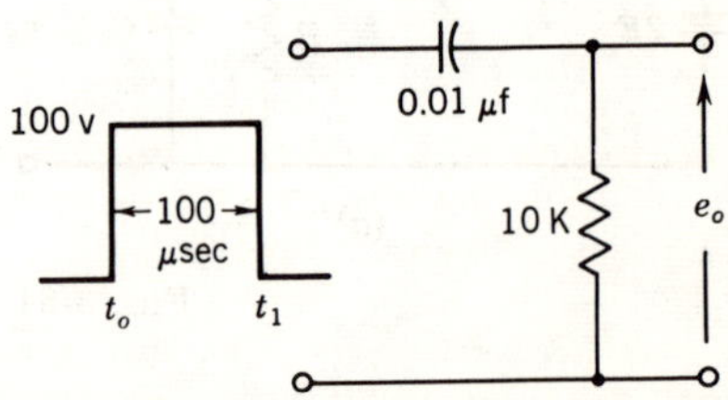

FIG. 3-67. Problem 3-8.

3-9. How long will it take for e_C to reach 157.5 volts after the switch is opened in Fig. 3-68?

Ans. 100 μsec.

Solution

$$RC = 100\ \mu\text{sec} \qquad \frac{157.5\ \text{volts}}{250\ \text{volts}} = 0.63$$

therefore $\quad x = 1 = \dfrac{t}{RC} \quad$ or $\quad t = 100\ \mu\text{sec}$

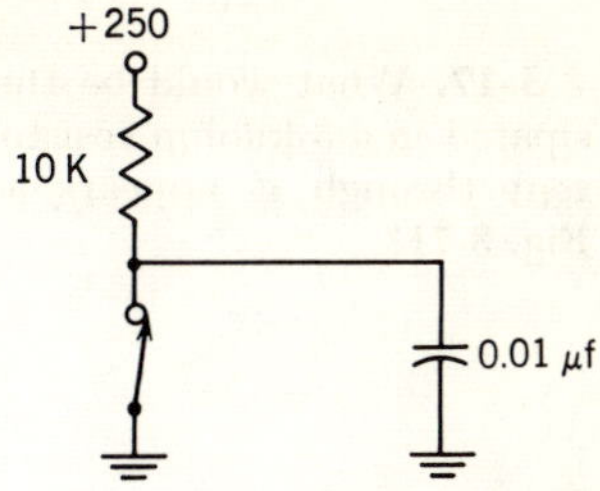

FIG. 3-68. Problem 3-9.

3-10. What is E_{ab}, 500 μsec after the switch in Fig. 3-69 closes?

Ans. 11.7 volts.

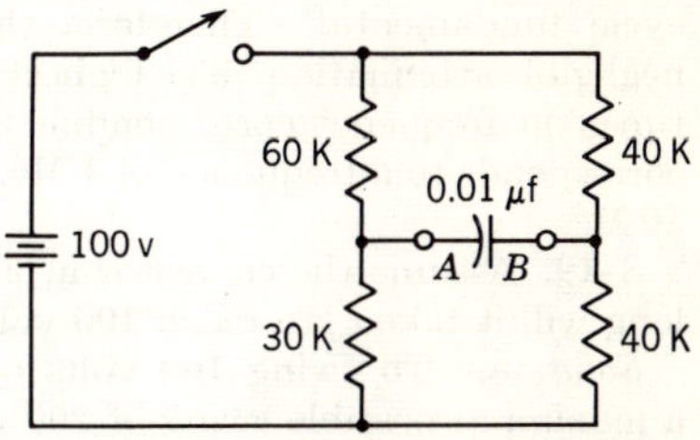

FIG. 3-69. Problem 3-10.

3-11. (*a*) What is the initial current when the switch closes in Fig. 3-70? Why?

Ans. 0 amp.

(*b*) What is the final current if the switch has been closed for a long time?

Ans. 0 amp.

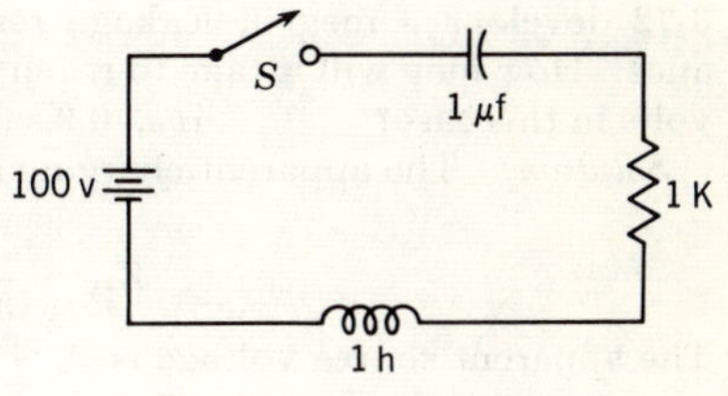

FIG. 3-70. Problem 3-11.

3-12. Design a circuit in which a capacitor charges to 25 per cent of the charging voltage in 100 μsec. The resistor should be 50 kilohms for other reasons.

3-13. Design a capacitive divider having an output equal to 20 per cent of the input signal. The circuit is to draw 10 ma from a 10-kc sine-wave generator developing 160 open-circuit volts. The generator internal impedance is negligible.

3-14. What nominal value of trimmer capacitor would be used to compensate a 'scope probe working into 3 ft of co-axial cable having a shunt capacitance of 5 $\mu\mu$f per foot and terminated in an amplifier having 20 $\mu\mu$f of input capacitance shunted by 2 meg? The attenuation ratio due to the probe is to be 0.2.

Ans. 8.8 $\mu\mu$f.

3-15. What effect would a leaky capacitor have in an *RC* integrator?

3-16. What is the power dissipated in a 5-kilohm resistor if a 500-volt rectangular waveform is impressed upon it for 0.1 sec, once every second?

3-17. What would be the power dissipated in a 5-kilohm resistor if the current through it appears as shown in Fig. 3-71?

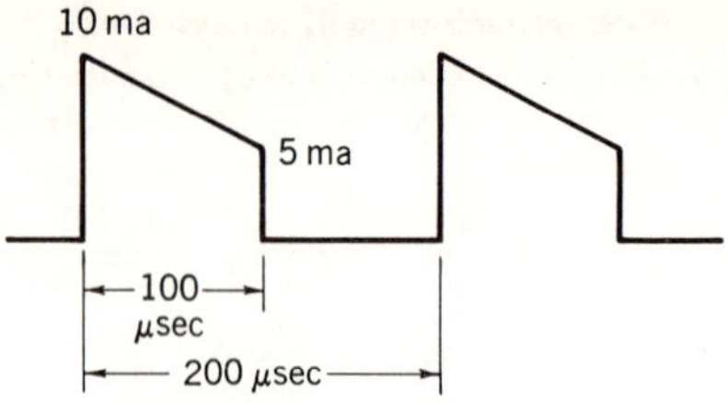

FIG. 3-71. Problem 3-17.

3-18. Approximately what minimum bandwidth should an amplifier have in order to pass 1-μsec pulses that occur at a repetition rate of 1,000 cps?

Solution. The lowest frequency component present in the pulse is the 1,000-cycle fundamental. Therefore, the amplifier should pass 1,000 cycles with negligible attenuation. The highest frequency component to consider is at least 10 times the frequency corresponding to the pulse duration. A 1-μsec pulse duration corresponds to a frequency of 1 Mc. Therefore, the amplifier should be flat up to 10 Mc.

3-19. Assume the capacitor in Fig. 3-72 has infinite leakage resistance. How long will it take e_o to reach 100 volts?

Solution. To swing 100 volts out of a maximum possible swing of 200 volts, or 50 per cent, locates $t/RC = 0.7$ on the time-constant chart. Therefore, $t = 0.7RC = 0.7(1 \text{ sec}) = 0.7$ sec.

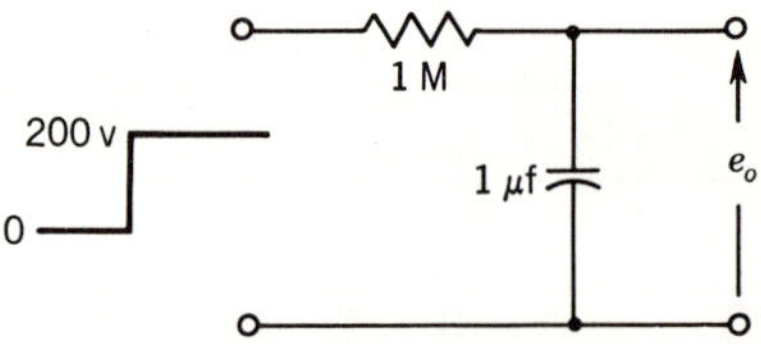

FIG. 3-72. Problems 3-19 and 3-20.

3-20. Suppose the capacitor in Fig. 3-72 develops 3 meg of leakage resistance. How long will it take to reach 100 volts in this case? *Ans.* 0.9 sec.

Solution. The apparent charging resistance seen by the capacitor is now

$$Z_{th} = \frac{3(1)}{3+1} = 0.75 \text{ meg}$$

The apparent source voltage is

$$V_{oc} = 200 \text{ volts}\left(\frac{3 \text{ meg}}{4 \text{ meg}}\right) = 150 \text{ volts}$$

To swing 100 volts/150 volts, or 67 per cent, locates $t/RC = 1.2$ on the time-constant chart. Therefore, $t = 1.2(0.75 \text{ sec}) = 0.9$ sec.

3-21. Find e_o, 5,000 μsec after the 240-volt input step is applied in Fig. 3-73. *Ans.* 223 volts.

Solution. At $t = 0^+$, the capacitor voltage is zero volts. Remember the 0^+ signifies the very instant after the voltage is applied. The maximum possible voltage to which the capacitor can rise is the drop across the 1-meg resistor. Looking back from the capacitor terminals and thevenizing, we find that the capacitor sees an open-circuit voltage, V_{oc}:

$$V_{oc} = 240 \text{ volts} \times \frac{1 \text{ meg}}{1 + 3 \text{ meg}} = 60 \text{ volts}$$

$$Z_{th} = \frac{1 \times 3}{1+3} = \frac{3}{4} \text{ meg}$$

The capacitor-charging time constant is, then,

$$\tau_C = Z_{th}C = 3/4 \text{ meg} \times 0.02\ \mu\text{f} = 0.015 \text{ sec}$$
$$= 15{,}000\ \mu\text{sec}$$

In 5,000 μsec

$$\frac{t}{\tau_C} = \frac{5{,}000\ \mu\text{sec}}{15{,}000\ \mu\text{sec}} = 0.33$$

and this is the value of the exponent x in the equation $e_C/E = 1 - \epsilon^{-x}$. Remember, e_C is the actual change in capacitor voltage that has occurred in an interval t (5,000 μsec in this case), and E is the maximum possible change or swing that can occur in capacitor voltage. Referring to our time-constant chart, which is really

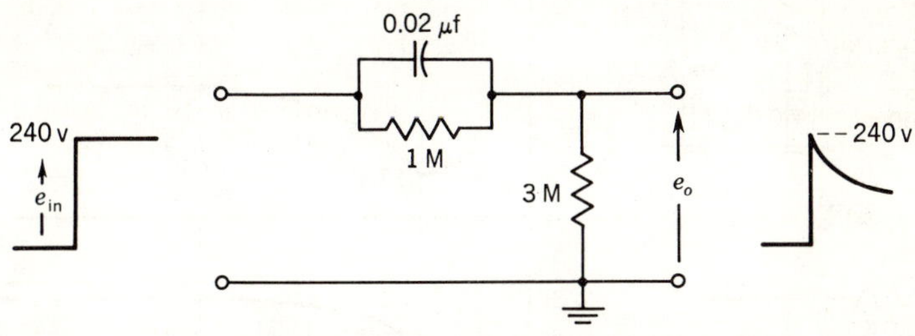

FIG. 3-73. Problem 3-21.

a plot of $1 - \epsilon^{-x}$ for different values of x, we find that for $x = 0.33$, $e_C/E = 0.28$, or 28 per cent. Since the capacitor starts from zero in this problem and swings to a maximum of 60 volts,

$$e_C = 0.28 \times 60 = 16.8 \text{ volts}$$

The output $e_o = 240 - 16.8 = 223$ volts.

In general, we could predict the output waveform of Fig. 3-73. When the input step is applied, C appears as a short circuit, since there is zero volts (no drop) across it at $t = 0^+$. Therefore, e_o equals e_{in}, or 240 volts. As C charges, less voltage is left for the 3-meg resistor, and so e_o decreases. Another way to consider this is that the decrease in capacitor-charging current causes e_o to decrease. The final value of e_o reached is determined solely by the resistive divider, and that is

$$e_{o\,\text{final}} = 240 \text{ volts} \times \frac{3 \text{ meg}}{3 + 1 \text{ meg}} = 180 \text{ volts}$$

3-22. The switch in Fig. 3-74*a* is initially open, and then at t_0 closes for 100 μsec before reopening. Sketch the voltage across R, across C, and between the output terminal and ground.

Solution. With S initially open, $e_o = e_C$, and e_C equals the open-circuit voltage that C looks back into. This is determined by applying our techniques for handling voltage dividers. The reader should prove to himself that $e_C = 34$ volts at t_0. Therefore, at t_0^-, $e_o = 34$ volts. At t_0^+, S closes to ground, and the excess electrons on the ground side of C will flow through R_3 into the upper capacitor terminal. This electron flow produces a drop across R_3 of minus 34 volts, since e_C appears directly across R_3 when S closes. We could then find e_o by adding e_C to e_3. Therefore, $+34$ volts $+ (-34$ volts$) = 0$.

This is surely correct, since closing S short circuits the e_o terminals to ground. The time constant of this capacitor discharge path is $\tau_d = 10$ kilohms $\times$ 0.01 μf = 100 μsec, or 1 time constant. Since the switch is closed for 1 time constant, the capacitor will discharge through 63 per cent of its total swing. This is the same as saying that only 37 per cent of the initial 34 volts will be left in 100 μsec. At this time, $e_C = 12.4$ volts, and $e_R = -12.4$ volts. Of course, e_o is zero with S closed.

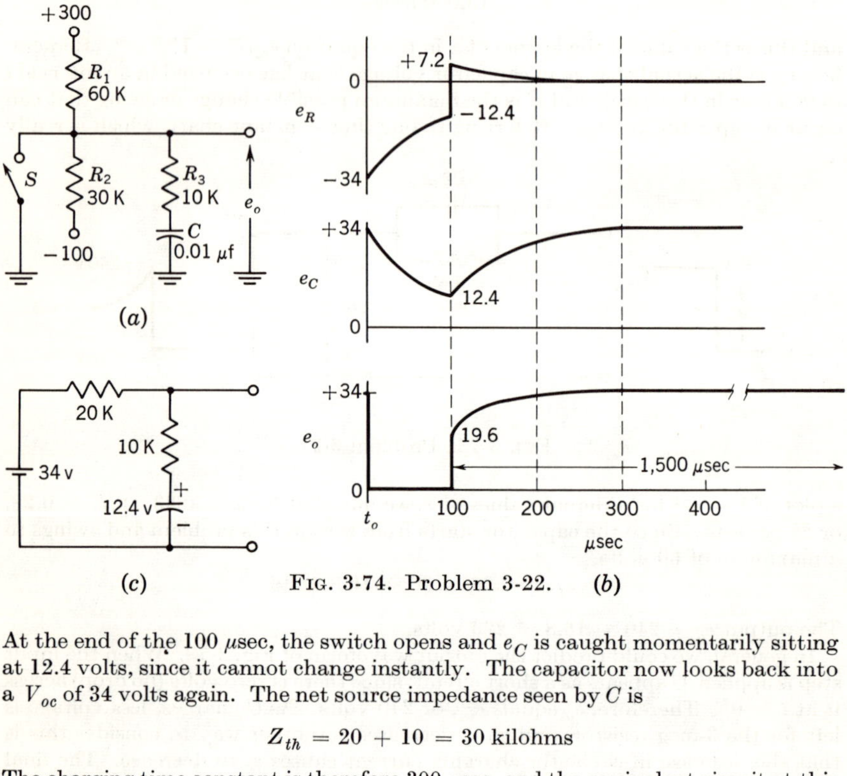

FIG. 3-74. Problem 3-22.

At the end of the 100 μsec, the switch opens and e_C is caught momentarily sitting at 12.4 volts, since it cannot change instantly. The capacitor now looks back into a V_{oc} of 34 volts again. The net source impedance seen by C is

$$Z_{th} = 20 + 10 = 30 \text{ kilohms}$$

The charging time constant is therefore 300 μsec, and the equivalent circuit at this instant is shown in Fig. 3-74*c*. A surge of charging current will flow out of the 34-volt source, to start swinging e_C from 12.4 to 34 volts. At this moment, the charged capacitor looks like a 12.4-volt battery, and the resultant current surge is

$$I_{\text{peak}} = \frac{34 - 12.4}{30} = \frac{21.6 \text{ volts}}{30 \text{ kilohms}} = 0.72 \text{ ma}$$

The drop in voltage across e_R is then 7.2 volts, and we see that e_R has jumped from -12.4 to 7.2 volts (Fig. 3-74*b*):

$$e_o = 12.4 + 7.2 = 19.6 \text{ volts}$$

Since the capacitor is now swinging from 12.4 to 34 volts, with $\tau_C = 300$ μsec, it will take about $5\tau_C$, or 1,500 μsec, from when the switch reopened to recharge to 34 volts.

3-23. Analyze the circuit of Fig. 3-75 in order to determine the output waveform at e_o if S is closed. K is a relay which pulls in at 10 ma and drops out at 5 ma.

The relay coil resistance is 5 kilohms. The relay inductance pull-in and drop-out times may be neglected.

Solution. Closing S will cause a surge of charging current to flow through K and C. Since L is neglected, the peak inrush of current will be limited only by the 5-kilohm coil resistance. The peak current is, then, I_{pk} = 100 volts/5 kilohms = 20 ma, which is more than enough to close the relay. The relay contacts will therefore close, and e_o will jump to +100 volts.

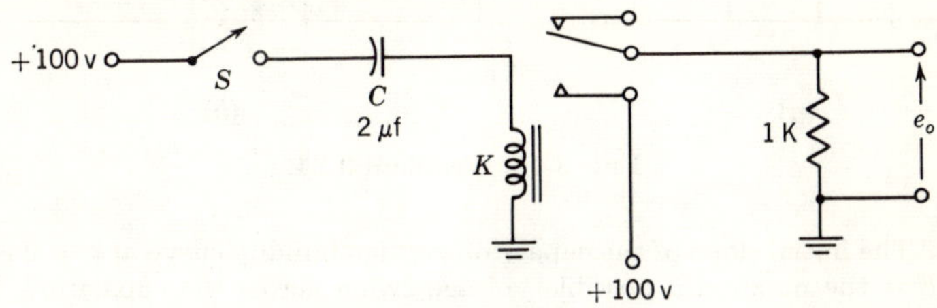

FIG. 3-75. Problem 3-23.

Since the capacitor cannot pass direct current, the current will decay exponentially from 20 ma to 0. The contacts are closed throughout the time it takes the current to decay from 20 ma to 5 ma. The exact equation relating i-versus-t is

$$i = I\epsilon^{-t/RC}$$

where I is the maximum possible current swing, which is I_{pk} in this case. The time constant RC is 10,000 μsec. With the known values plugged into the equation, we have

$$5 \text{ ma} = 20 \text{ ma}\,\epsilon^{-t/10{,}000\,\mu\text{sec}}$$

$$5/20 = 0.25 = \epsilon^{-t/10{,}000\,\mu\text{sec}}$$

We next enter the chart at 0.25, or 25 per cent and intersect the capacitor-current discharge curve, which is of the form ϵ^{-x}. The corresponding value of the exponent is seen to be

$$1.4 = \frac{t}{RC}$$

or

$$t = 1.4 \times 10{,}000\ \mu\text{sec} = 14{,}000\ \mu\text{sec}$$

Therefore, e_o will be a 100-volt rectangular pulse of 14,000-μsec duration.

3-24. The circuit of Fig. 3-76*a* has an output waveshape as shown in Fig. 3-76*b*. What would a d-c voltmeter read when connected across the output terminals?

Ans. 75 volts.

Solution. The area bounded by one cycle is equal to that of a rectangle having a base T and height 60 volts, plus a triangle of base T and height 30 volts:

$$A = 60T + \tfrac{1}{2}T30$$

Remember, the area of a triangle is ½ base × altitude:

$$A = 75T$$

$$\bar{e}_o = \frac{75T}{T} = 75 \text{ volts}$$

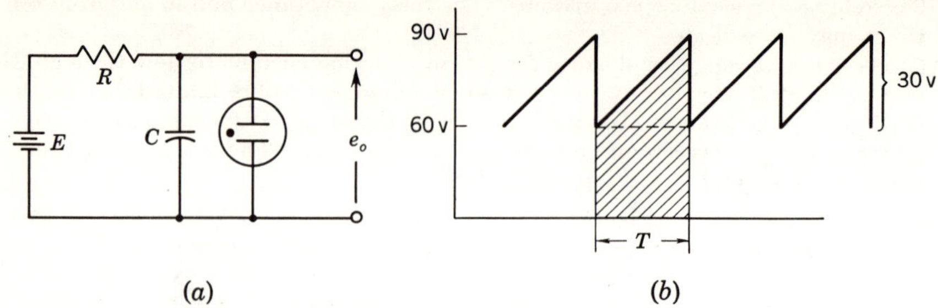

FIG. 3-76. Problem 3-24.

3-25. The initial slope of the capacitor-voltage buildup curve at $t = 0$ is E/RC, where E is the maximum possible voltage swing across the capacitor. Does e_o in Fig. 3-77*a* or Fig. 3-77*b* reach 50 volts at an earlier time? Prove it.

Ans. Fig. 3-77*c*.

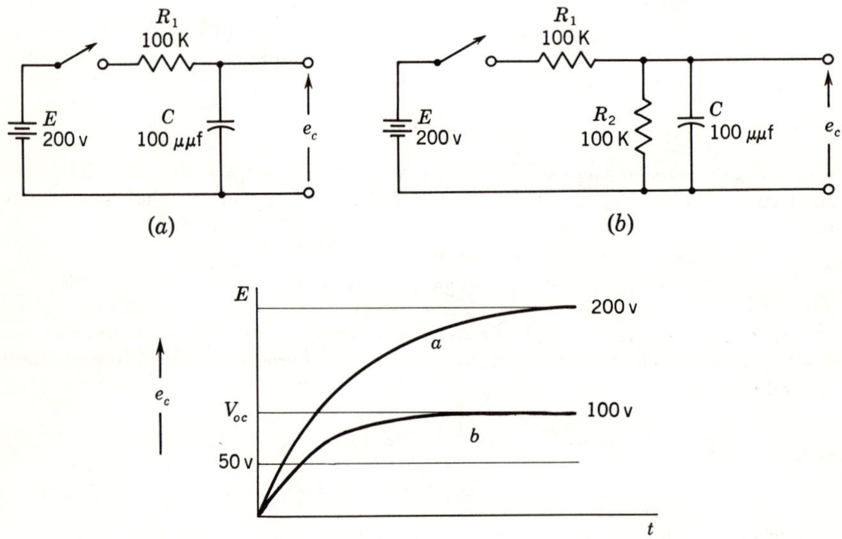

FIG. 3-77. Problem 3-25.

Solution. The initial slope of e_C in Fig. 3-77*a* is

$$\frac{\Delta e_C}{\Delta t} \quad \text{or} \quad \frac{de_C}{dt} = \frac{E}{R_1C} \qquad \text{at } t = 0$$

and in Fig. 3-77*b*, if we thevenize, looking back from C,

$$\frac{de_C}{dt} = \frac{V_{oc}}{Z_{th}C} = \frac{ER_2/(R_1 + R_2)}{R_1R_2C/(R_1 + R_2)} = \frac{E}{R_1C}$$

The initial slope is the same in both cases, but the final value of voltage reached by C in Fig. 3-77a is higher, so e_C will reach 50 volts sooner in the circuit of Fig. 3-77a. This is illustrated in Fig. 3-77c.

3-26. To get to the street from your house, you must go through a long dark alley. You wish to design a circuit which will keep an outdoor light on for 1 min and then go out. This is sufficient time to walk the length of the alley. In the junk box you have available all kinds of resistors and a power supply which delivers 120 volts d-c up to 40 ma. Also available are a SPDT switch and SPDT relay having a 4-kilohm coil with 4-ma pull-in and 2-ma drop-out. Design an RC timing circuit using these components and the smallest possible capacitor. With the light out, the circuit should consume zero standby power.

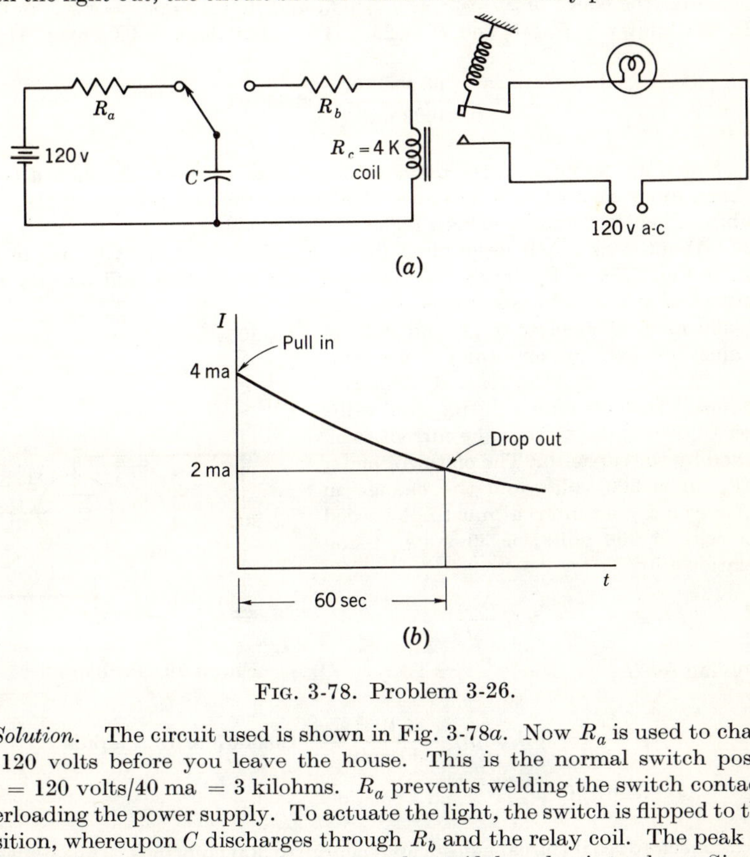

FIG. 3-78. Problem 3-26.

Solution. The circuit used is shown in Fig. 3-78a. Now R_a is used to charge C to 120 volts before you leave the house. This is the normal switch position. R_a = 120 volts/40 ma = 3 kilohms. R_a prevents welding the switch contacts or overloading the power supply. To actuate the light, the switch is flipped to the R_b position, whereupon C discharges through R_b and the relay coil. The peak value of the discharge current must be 4 ma, at least, if the relay is to close. Since C is charged to 120 volts, we can determine R_b from

$$I_{\text{peak}} = 4 \text{ ma} = \frac{120 \text{ volts}}{R_b + 4 \text{ kilohms}}$$

$$(4 \text{ ma})R_b + (4 \text{ ma})4 \text{ kilohms} = 120 \text{ volts}$$

$$R_b = \frac{120 - 16 \text{ volts}}{4 \text{ ma}} = \frac{104 \text{ volts}}{4 \text{ ma}} = 26 \text{ kilohms}$$

Of course a smaller value of R_b could be used if we let the peak current exceed 4 ma. The smaller we make R_b, however, the bigger C must be to keep the discharge of C from occurring too rapidly. It is cheaper to buy resistance than capacitance. (Actually good design practice would permit more than 4 ma of closing current.) The desired current-vs.-time curve is shown in Fig. 3-78*b*. We know the maximum possible swing in current is $I_{max} = 4$ ma. The desired swing is from 4 ma to 2 ma, or,

$$\frac{I}{I_{max}} = \frac{2\text{ ma}}{4\text{ ma}} = 0.5$$

We can enter the curve of $y = \epsilon^{-x}$ at $y = 0.5$, and find $x = 0.7 = t/RC$, or $C = t/0.7R$. We know $t = 60$ sec, and $R = 26 + 4 = 30$ kilohms $= 0.03$ meg. Therefore,

$$C = \frac{60\text{ sec}}{0.7(0.03\text{ meg})} = 2{,}850\ \mu\text{f}$$

Since 2,850 μf at 120 volts represents a substantial investment, the circuit could have been more economically designed if a higher voltage power supply were available. Then R_b could have been higher and C smaller.

3-27. About how much decoupling filter capacitance would you use in the circuit of Fig. 3-79 if V_1 draws 100 ma for 2 μsec when switched on and zero current when cut off? V_1 is switched on once every 500 μsec. The plate-to-ground voltage of V_1 must not vary by more than 10 per cent.

Ans. 6,700 $\mu\mu$f.

Solution. In determining the required value of C, we shall consider all the tube current to be furnished by the capacitor. The plate voltage e_b with V_1 off is 300 volts, and the change in $e_b(\Delta e_b)$ over a 2-μsec interval must not exceed 10 per cent of 300 volts, or 30 volts. Using the relationship

$$i_C = C\frac{de_C}{dt} \approx C\frac{\Delta e_C}{\Delta t}$$

and solving for C,

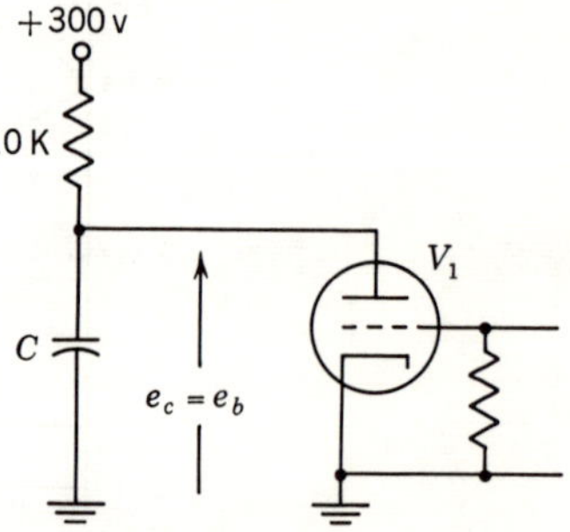

FIG. 3-79. Problem 3-27.

$$C = i_C\frac{\Delta t}{\Delta e_C} = (0.1\text{ amp})\,\frac{2 \times 10^{-6}\text{ sec}}{30\text{ volts}} = 0.0067 \times 10^{-6}\text{ farads}$$

$$= 0.0067\ \mu\text{f}$$

Let us check to see if the capacitor can recharge to 300 volts in the time between the end of the first pulse and the start of the second, which is 488 μsec. With V_1 off, the charging time constant is

$$\tau_C = 10\text{ kilohms} \times 0.0067\ \mu\text{f} = 67\ \mu\text{sec}$$

We know from our curves that in 5 time constants, or $5 \times 67\ \mu\text{sec} = 335\ \mu\text{sec}$, the capacitor will have fully recharged. Since 488 μsec is available, C will fully recover between pulses.

3-28. Determine the output waveform for the circuit of Fig. 3-80*b* if it is driven by the input waveform of Fig. 3-80*a*.

Solution. The natural time constant of the circuit is $RC = 10\ \mu\text{sec}$. Since this is very short as compared with the pulse duration, the circuit will behave as a differentiating network, and will produce an output approximately proportional to the input rate of change:

$$e_o = RC\,\frac{\Delta e_{in}}{\Delta t} = 10\ \mu\text{sec}\,\frac{50\ \text{volts}}{1{,}000\ \mu\text{sec}}$$

$$= 0.5\ \text{volt}$$

When the sawtooth falls, the capacitor voltage momentarily hangs at 0.5 volt, and therefore the output falls to −49.5 volts, as shown in Fig. 3-80*c*.

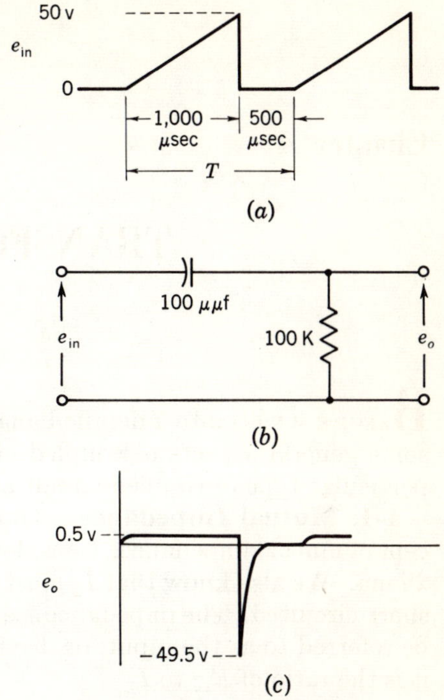

Fig. 3-80. Problem 3-28.

Chapter 4

TRANSFORMERS

Before we go into a detailed analysis of transformers, let us review some general aspects of coupled circuits. A transformer circuit is a particular type of coupled circuit and can be analyzed as such.

4-1. Mutual Impedance. Our first discussion concerns the concept of mutual impedance. Consider Fig. 4-1*a*. Clearly, $I_1 = E_{in}/R_1 =$ 20 ma. We also know that I_2 must equal zero, since points *b* and *d* are short-circuited. The impedance seen looking into terminals *a* and *d* will be referred to as the input, or driving-point, impedance. Specifically, it is the ratio of E_{in} to I_1.

Suppose we were to vary the value of E_{in}. What would be the effect on I_2? There would be none, and I_2 would still equal zero, no matter what value E_{in} assumed. Whenever a circuit (loop) is unaffected by changes in another loop, we say the circuits are not coupled. If variations in one circuit mildly affect another, we say they are loosely coupled. If variations of one strongly affect the other, the coupling is said to be tight. Figure 4-1*a* is a case of extremely loose or zero coupling, while in Fig. 4-1*b* the coupling might be considered moderate. Note that in Fig. 4-1*b* the two loops are coupled through a mutual impedance R_3. Any change in the parameters of loop 1 (E_{in}, R_1, and R_3) will cause a change in I_1. This change in I_1 develops a change in voltage across R_3. Since the voltage across R_3 is a factor in determining I_2, we see that the currents are not independent and the loops are coupled. If R_3 were infinite (an open circuit), then I_1 would be the same current as I_2, and the coupling would be as tight as possible.

In general, loop equations will be of the form

$$E_1 = I_1Z_{11} + I_2Z_{12} \tag{4-1}$$

$$E_2 = I_1Z_{21} + I_2Z_{22} \tag{4-2}$$

That is, for Fig. 4-1*c*, we would write

$$E_A = I_1(R_1 + R_3) + I_2R_3 \qquad E_B = I_1R_3 + I_2(R_2 + R_3)$$

where $R_1 + R_3 = Z_{11}$ $E_A = E_1$

$+R_3 = Z_{12} = Z_{21}$ $E_B = E_2$

$R_2 + R_3 = Z_{22}$

For Fig. 4-1*d*, we would write

$$E_A = I_1(R_1 + R_3) - I_2R_3 \qquad -E_B = -I_1R_3 + I_2(R_2 + R_3)$$

where $R_1 + R_3 = Z_{11}$ $E_A = E_1$

$-R_3 = Z_{12} = Z_{21}$ $-E_B = E_2$

$R_2 + R_3 = Z_{22}$

If we agree that Eqs. (4-1) and (4-2) shall represent the general form of loop equations, the mutual impedance terms Z_{12} and Z_{21} will be

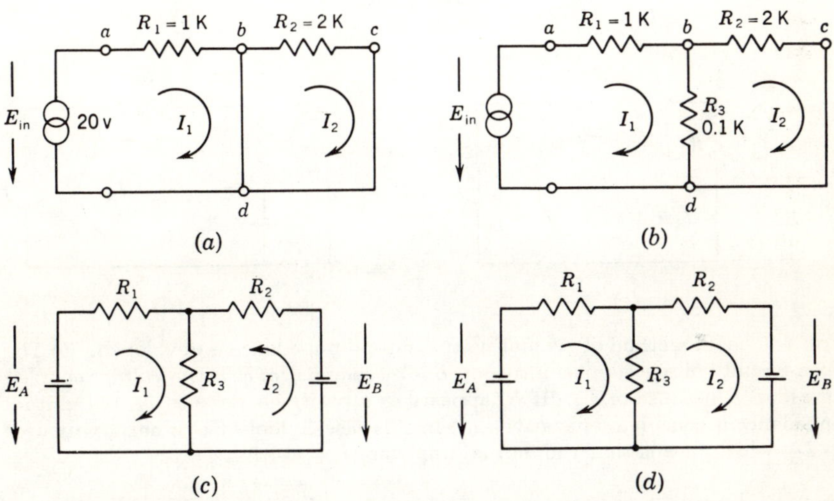

FIG. 4-1. (*a*) The current in loop 1 does not couple any voltage into loop 2, owing to the short circuit between points *b* and *d*. Similarly I_2 has no effect on loop 1. The mutual impedance is zero. (*b*) Loops 1 and 2 coupled through the mutual impedance R_3. (*c*) The mutual impedance R_3 is considered a positive quantity when the loop currents in R_3 are additive. (*d*) The mutual impedance is considered a negative quantity when the loop currents are subtractive in R_3.

written with a plus sign before them. We see, however, that the actual quantity represented by Z_{12} may be positive or negative, depending upon the assumed current directions; for example, in Fig. 4-1*c*, $Z_{12} = +R_3$, while in Fig. 4-1*d*, $Z_{12} = -R_3$. Therefore, it is desirable to define mutual impedance in a manner which is indicative of the current directions assumed.

The mutual impedance Z_{BA} is specifically defined as the ratio of the voltage induced in loop B by a current in loop A, with all loops except A open-circuited. Let us apply this to Fig. 4-2a. The sinusoidal generator E_1 is considered as momentarily frozen in time with the polarity shown. The open-circuit voltage coupled into loop B by I_A can be measured with an infinite-impedance voltmeter inserted in series with loop B, as shown in Fig. 4-2b. The effect of I_A upon loop B can be duplicated by placing a battery having a magnitude of $I_A R_3$ volts into loop B. Loop A can now be thrown away, and only loop B is considered.

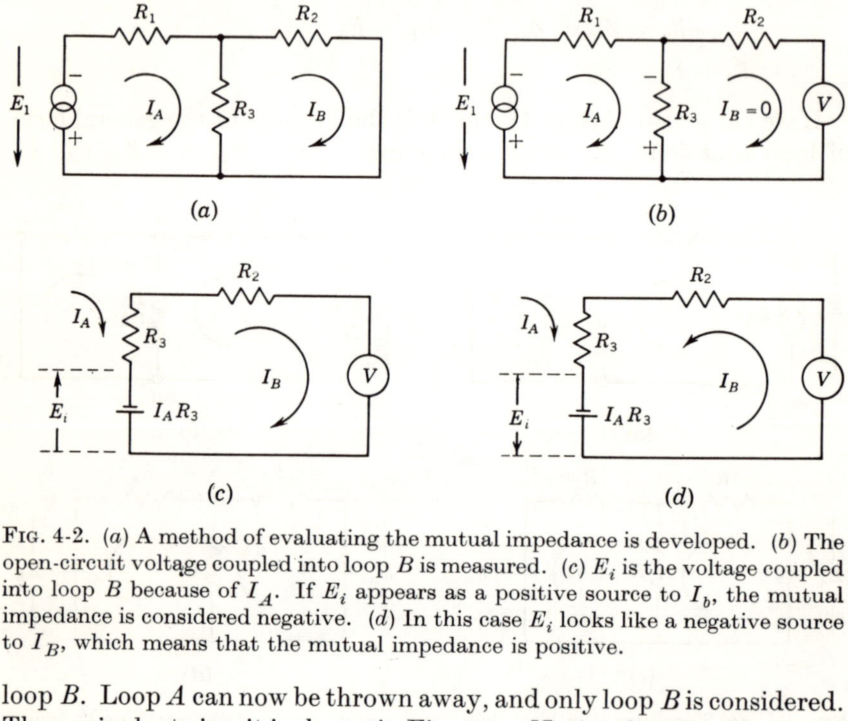

FIG. 4-2. (a) A method of evaluating the mutual impedance is developed. (b) The open-circuit voltage coupled into loop B is measured. (c) E_i is the voltage coupled into loop B because of I_A. If E_i appears as a positive source to I_b, the mutual impedance is considered negative. (d) In this case E_i looks like a negative source to I_B, which means that the mutual impedance is positive.

The equivalent circuit is shown in Fig. 4-2c. Notice that the polarity of the induced voltage is identical with the way I_A polarized R_3. If the positive, or reference, direction for determining the induced voltage E_i is taken in the same direction as I_B, we have

$$Z_{BA} = \frac{\text{voltage induced in loop } B}{\text{current in loop } A} = \frac{E_i}{I_A} = \frac{-I_A R_3}{I_A} = -R_3$$

If I_B had been assumed to flow counterclockwise in Fig. 4-2a, the positive direction of E_i would be as shown in Fig. 4-2d. This time, E_i is a positive quantity equal to $+I_A R_3$, and therefore $Z_{BA} = +R_3$. With this method of defining E_i, that is, always taking the positive direction of E_i in the same direction as I_B, a positive value of mutual impedance indicates that currents are assumed to flow through the mutual impedance in the same direction, while a negative value of mutual impedance indicates that the currents were assumed to flow oppositely through the mutual impedance.

From a source-sink viewpoint, we can say that, if the open-circuit source voltage induced in, say, loop B, by a current in loop A, is such that it aids the assumed direction of I_B, then the mutual impedance is negative. If the induced voltage in loop B due to I_A opposes the assumed direction of I_B, then the mutual impedance is positive.

An exercise left to the reader is to show that in a simple passive T network, as in Fig. 4-2a, and therefore in any network reducible to such a T, $Z_{AB} = Z_{BA}$.

4-2. Self-impedance. We shall frequently refer to the self-impedance of a loop, which, as you will recall from Chap. 1, is the total impedance encountered in traversing that loop. For example, in Fig. 4-3 the self-impedance of loop 1 is $Z_{11} = R_1 + R_3$. The self-impedance of loop 2 is $Z_{22} = R_2 + R_3$. The mutual impedance is $Z_{12} = Z_{21} = +R_3$ and is a positive quantity, since I_1 and I_2 flow through R_3 in the same direction. No particular polarity has been assigned to the generator, since it varies with time. The positive or reference direction of E is, however, definitely stated by the arrow. If we wished, we could assume some instantaneous polarity and freeze the generator in time. This voltage can be adopted as a reference, and the currents with their respective phase relationships evaluated.

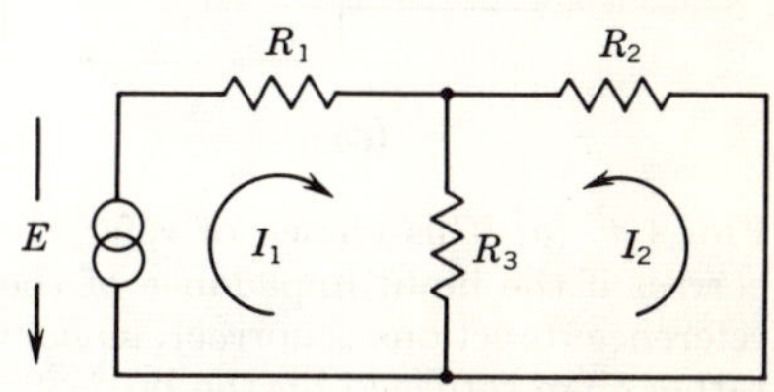

FIG. 4-3. The input impedance seen by the generator is to be determined.

The loop equations for Fig. 4-3 are commonly written as

Loop 1: $$E = I_1Z_{11} + I_2Z_{12} \tag{4-3}$$

Loop 2: $$0 = I_1Z_{12} + I_2Z_{22} \tag{4-4}$$

where $Z_{12} = +R_3$ in this case.

If the assumed direction of I_2 or I_1 were reversed, the general form of the loop equations would be the same, except that Z_{12} would equal a minus R_3.

4-3. Driving-point Impedance. The impedance that E sees looking into the network of Fig. 4-3 is called the driving point, or input, impedance. It is the ratio of E to I_1. Care must be exercised in setting up reference directions when the input impedance Z_{in} is to be evaluated. Much grief can be avoided if the positive input voltage direction is chosen so that the input current will be positive. For example, consider Fig. 4-4a, where we have shown a 2-ohm resistor in a black box. To determine what the input impedance is as seen looking into the box, we would hang a battery across the input terminals and measure the current flowing into the box. Suppose we arbitrarily assigned the reference directions of E and I, as shown in Fig. 4-4a. If a 12-volt battery were hung across the input terminals so that E were a positive quantity, the positive battery terminal would go to the upper input terminal, and we can see that the battery would really be pumping electrons opposite

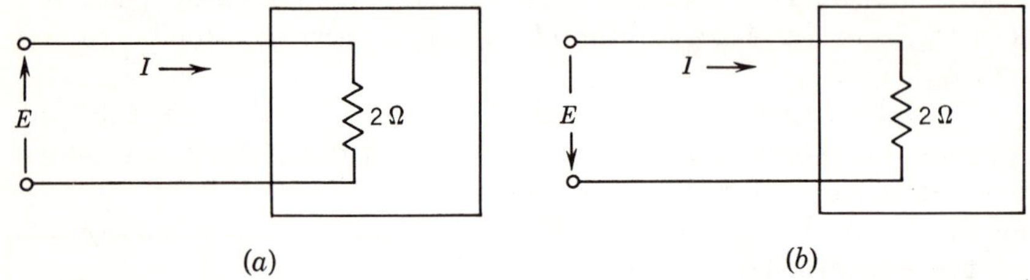

FIG. 4-4. (a) This choice of reference directions for voltage and current is not correct if the input impedance of the box is to be evaluated. (b) This choice of reference directions is correct, since we start by assuming that power is put into rather than returned by the box.

to the positive direction of I. Therefore, I would be a negative quantity. Since R_{in} would be the ratio of E to I, which is a plus E divided by a minus I, we would erroneously conclude (on paper at least) that the box had a negative 2-ohm resistor in it. If we connected the battery so as to make E in Fig. 4-4b a positive quantity, the negative battery terminal would go to the upper input terminal, and I would be a positive quantity. The input resistance in this case would be a $+2$ ohm, which is correct. The positive, or reference, directions assumed for E and I should be chosen so that when E and I are positive quantities, power is put into the box, as in Fig. 4-4b.

To find the driving-point impedance in Fig. 4-3, we can solve the loop 1 and 2 equations by determinants for I_1 and then obtain the ratio of E to I_1, which is Z_{in}:

$$I_1 = \frac{\begin{vmatrix} E & +Z_{12} \\ 0 & Z_{22} \end{vmatrix}}{\begin{vmatrix} Z_{11} & +Z_{12} \\ +Z_{12} & Z_{22} \end{vmatrix}} = \frac{EZ_{22}}{Z_{11}Z_{22} - Z_{12}^2} = \frac{E}{Z_{11} - Z_{12}^2/Z_{22}}$$

The last expression for I_1 was obtained by dividing the numerator and denominator by Z_{22}. Notice that the expression for I_1 is in the form of $I = E/R$, where R is equivalent to the denominator. Therefore, the input impedance that the generator looks into (driving-point impedance) is

$$Z_{in} = Z_{11} - \frac{Z_{12}^2}{Z_{22}}$$

Let us examine this expression for some physical significance. It tells us that the impedance which the generator sees is the self-impedance of loop 1, minus another term made up of the mutual impedance and the self-impedance of loop 2. This additional term is called the reflected impedance. For pure resistances in Fig. 4-3, we would surely expect the driving point impedance to be less than just the self-impedance of loop 1 because R_2 is in parallel with R_3. Let us substitute the R values and see if we get the same result as we would in treating the network as a simple series-parallel circuit. As a simple series-parallel circuit, we can write by inspection

$$R_{in} = R_1 + \frac{R_2R_3}{R_2 + R_3}$$

Now let us see if by some mathematical legerdemain we can prove that

$$Z_{11} - \frac{Z_{12}^2}{Z_{22}} \overset{?}{=} R_1 + \frac{R_2R_3}{R_2 + R_3}$$

Substituting R's for Z's ($Z_{12} = +R_3$),

$$(R_1 + R_3) - \frac{+R_3^2}{R_2 + R_3} = ?$$

Getting all terms over the common denominator $R_2 + R_3$,

$$\frac{(R_1 + R_3)(R_2 + R_3) - R_3^2}{R_2 + R_3} = ?$$

Multiplying through and simplifying,

$$\frac{R_1R_2 + R_1R_3 + R_2R_3}{R_2 + R_3} \overset{?}{=} R_1 + \frac{R_2R_3}{R_2 + R_3}$$

We can see that if we are to prove the two expressions identical, we must isolate an R_1 term. By factoring R_1 we have

$$\frac{R_1(R_2 + R_3) + R_2R_3}{R_2 + R_3} \overset{?}{=} R_1 + \frac{R_2R_3}{R_2 + R_3}$$

Since both terms of the left-hand expression are over the same common denominator, we can write

$$\frac{R_1(R_2 + R_3)}{R_2 + R_3} + \frac{R_2R_3}{R_2 + R_3} = R_1 + \frac{R_2R_3}{R_2 + R_3}$$

and happily we find

$$R_1 + \frac{R_2R_3}{R_2 + R_3} = R_1 + \frac{R_2R_3}{R_2 + R_3}$$

If any of the mathematics is puzzling, the reader should review high school algebra, or study a text such as Cooke.[1]

Had the current directions been chosen so that $Z_{12} = -R_3$, the result would still have been the same, since the Z_{12} term is squared.

4-4. Transfer Impedance. At this point, we might also introduce the concept of a transfer impedance. The transfer impedance between any two loops of a network, say 1 and 2 in our case, is defined as the ratio of a voltage introduced in one loop to the current it produces in the other loop. If there are more than two loops, the other loops are kept closed and any generators in them are replaced by their internal impedances.

In Fig. 4-3, the transfer impedance between the voltage in loop 1 and the current in loop 2 is

$$Z_{T_{12}} = \frac{E}{I_2}$$

Evaluating I_2 by determinants, we have

$$I_2 = \frac{\begin{vmatrix} Z_{11} & E \\ +Z_{12} & 0 \end{vmatrix}}{\begin{vmatrix} Z_{11} & +Z_{12} \\ +Z_{12} & Z_{22} \end{vmatrix}} = -\frac{EZ_{12}}{Z_{11}Z_{22} - Z_{12}^2}$$

Therefore,

$$Z_{T_{12}} = \frac{E}{I_2} = \frac{E}{-EZ_{12}/(Z_{11}Z_{22} - Z_{12}^2)} = -\frac{Z_{11}Z_{22} - Z_{12}^2}{Z_{12}}$$

where $\quad Z_{12} = +R_3$

$$Z_{T_{12}} = -\frac{(R_1 + R_3)(R_2 + R_3) - (+R_3)^2}{+R_3}$$

$$= -\frac{R_1R_2 + R_2R_3 + R_1R_3}{R_3}$$

The reader can prove that $Z_{T_{12}} = Z_{T_{21}}$ by interchanging E and I_2.

Once the transfer impedance between two loops is evaluated, we can easily determine how much current change will be produced in one loop by a voltage change in the other. In a transformer, we might wish to know the secondary current due to the primary voltage.

Let us try a simple problem that illustrates how the concept of transfer impedance can be used.

Problem. In the circuit of Fig. 4-5*a*, the 300-volt power supply is poorly regulated with respect to line-voltage variations. Suppose it were to jump to

[1] Nelson M. Cooke, "Basic Mathematics for Electronics," 2d ed., McGraw-Hill Book Company, Inc., New York, 1960.

360 volts. What would the change in E_o be? Solve the problem by at least two methods, one of which involves the concept of transfer impedance.

Solution 1. One method of evaluating the change in output would be to solve for E_o in the circuit of Fig. 4-5*a* by Thévenin's theorem or Kirchhoff's laws. Then the 300-volt supply is changed to 360 volts, and a new value of E_o is calculated. The change in E_o is the difference between the two calculated values.

Solution 2. Another approach would be to use the superposition theorem. To determine just how effective a 60-volt change in the 300-volt power supply is in determining E_o, we will use the circuit of Fig. 4-5*b*. The 300-volt supply is replaced with a 60-volt supply. Thevenizing to the left of R_4, we have

$$V_{oc} = 60 \text{ volts}\left(\frac{30 \text{ kilohms}}{90 \text{ kilohms}}\right) = 20 \text{ volts}$$

$$Z_{th} = \frac{(60)(30)}{60 + 30} + 10 = 30 \text{ kilohms}$$

The equivalent circuit seen by R_4 is shown in Fig. 4-5*c*, from which we obtain

$$E_o' = 20 \text{ volts}\left(\frac{30 \text{ kilohms}}{60 \text{ kilohms}}\right) = 10 \text{ volts}$$

We see, then, that the output will increase 10 volts for a 60-volt increase in the input. From this information, we could determine the transfer impedance between the loop containing the 60-volt source and the loop containing R_4. It would be

$$Z_T = \frac{60 \text{ volts}}{I_4} = 180 \text{ kilohms}$$

where I_4 is the current through R_4 and, from Fig. 4-5*c*, is $I_4 = 20$ volts/60 kilohms $= 0.33$ ma.

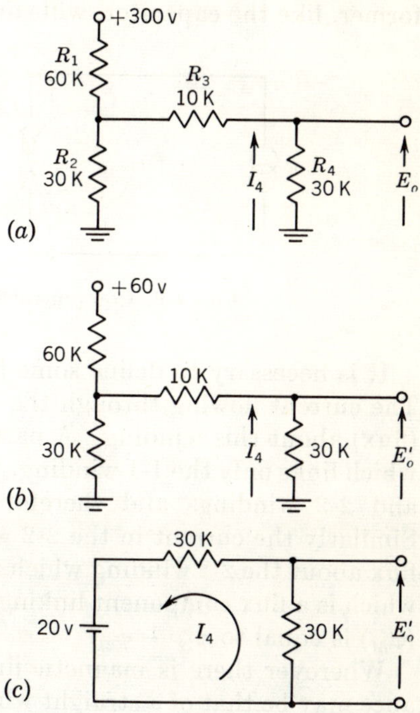

FIG. 4-5. (*a*) The change in E_o due to a 60-volt change in the 300-volt supply is to be determined. (*b*) Since the network is linear, the superposition theorem may be used. (*c*) Thévenin's theorem may be used to simplify the network.

Solution 3. We could also have calculated E_o in Fig. 4-5*a* to be 50 volts. This 50 volts across R_4 means $I_4 = 50$ volts/$R_4 = 1.67$ ma in this figure. The transfer impedance between the loop containing the 300-volt supply and that containing R_4 is, then,

$$Z_T = \frac{300 \text{ volts}}{1.67 \text{ ma}} = 180 \text{ kilohms}$$

The current change in I_4, due to the 60-volt supply change, can then be related through the transfer impedance:

$$\Delta I_4 = \frac{\Delta E_{in}}{Z_T} = \frac{60 \text{ volts}}{180 \text{ kilohms}} = 0.33 \text{ ma}$$

The change in E_o is, then,

$$\Delta E_o = R_4 \Delta I_4 = (30 \text{ kilohms})(0.33 \text{ ma}) = 9.9 \text{ volts} \approx 10 \text{ volts}$$

4-5. The Transformer. We can now tackle the transformer in the general sense of a coupled circuit. The distinguishing feature of a transformer is that the coupling between two loops is due to magnetic flux rather than to a resistance. Furthermore, as we learned in basic electronics, it is only when the current in a winding is changing that a voltage will be induced in another winding. This means that direct currents and voltages cannot get through a transformer. The transformer, like the capacitor, will couple only time-varying signals.

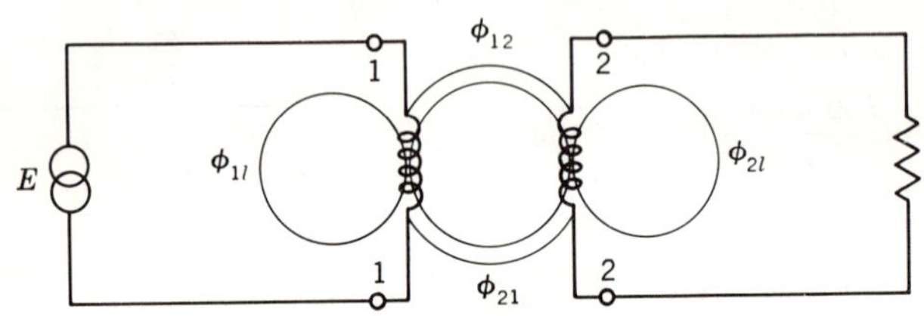

FIG. 4-6. Component fluxes in the transformer.

It is necessary to define some basic parameters. Consider Fig. 4-6. The current flowing through the 1-1 winding sets up a magnetic field (flux) about this winding. A part of this flux, ϕ_{1l}, is the leakage flux which links only the 1-1 winding. The other part, ϕ_{21}, links both the 1-1 and 2-2 windings and therefore is a component of mutual flux. Similarly the current in the 2-2 winding will result in ϕ_{2l}, the leakage flux about the 2-2 winding which does not link the 1-1 winding, and ϕ_{12}, which is a flux component linking both windings. The net mutual flux (ϕ_M) is equal to $\phi_{12} + \phi_{21}$.

Wherever there is magnetic flux, there is inductance. The inductance may be that of a straight wire or of a many-turn coil, but always magnetic flux and inductance go together. We then say that the leakage flux ϕ_{1l} is due to a leakage inductance L_{1l}, which is common only to loop 1. Similarly, ϕ_{2l} is due to the leakage inductance L_{2l}, common only to loop 2. The mutual flux ϕ_M is set up by a mutual inductance M, which is common to both loops. The position and function of these inductances can best be visualized by imagining the transformer inductances in the form of a T, as shown in Fig. 4-7.

Just as the self-inductance of a coil depends upon the number of turns, core material, wire size, and so forth, so the mutual inductance M between two coils depends upon their geometric configuration (proximity and orientation), permeability, and the individual inductance of each coil. The tightness of coupling between the two windings is defined by the coefficient of coupling. A formula relating the coefficient of coupling k to the mutual and self-inductances is $k = M/\sqrt{L_1 L_2}$. When the core material is other than air, M, L_1, and L_2 may not be constant. Instead, these parameters depend upon the coil currents.

In most communication circuits, the value of k is about 0.5 or less. In power work, where high efficiency is important, k is about 0.9 or more. The maximum value of k is 1.

Now we know that the voltage induced in one winding depends upon a current change in another winding. The exact relationship is

$$e_{21} = \pm M \frac{di_1}{dt} \approx \pm M \frac{\Delta i_1}{\Delta t}$$

where e_{21} means the voltage induced in coil 2 by a current change in coil 1. M is the mutual inductance between these coils, and di/dt is the instantaneous rate of change of current with respect to time in coil 1. The meaning of the plus and minus signs will soon be made clear. For the moment just ignore them.

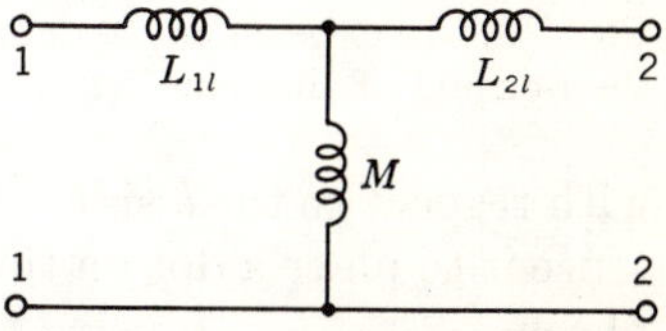

FIG. 4-7. We may imagine the component fluxes as being caused by the leakage and mutual inductances arranged as a T.

Suppose two coils were so coupled that $M = 1$ henry. If the current in one coil changes at a rate of 1 amp/sec, the voltage induced in the other is 1 volt. Increasing the frequency of the current variations raises the induced voltage. For example, were the current to change at a rate of 1 amp/msec, the induced voltage would be 1,000 volts.

When the current changes in a sinusoidal manner, the induced voltage will be a cosine wave. When the sine wave goes through zero, it is *changing* most rapidly. The induced voltage will therefore be a maximum at this time, or 90° out of phase with the current causing it. Similarly, we can write

$$e_{12} = \pm M \frac{di_2}{dt} \approx \pm M \frac{\Delta i_2}{\Delta t}$$

where e_{12} is the voltage induced in coil 1 by a current change in coil 2.

4-6. The Dot Notation. If we could write loop equations for an inductively coupled circuit, the loop currents could then be evaluated. Once the currents are known, all other network characteristics can readily be determined. To write the loop equations, we must know something about the phasing of the windings. This information is necessary to determine whether the voltage induced in loop 1 by a current in loop 2 aids or opposes the assumed direction of current in loop 1. If it aids, e_{12} is considered a positive source in loop 1, and therefore $M\, di_2/dt$, which equals e_{12}, would be written with a plus sign on the source side for loop 1. If e_{12} opposes the assumed direction of i_1, the induced voltage appears as a negative source. Similarly, we must determine the effect of $M\ di_1/dt$ upon the assumed direction of secondary current.

In order to determine the polarity of the induced voltage in one winding due to a current change in another, a simple experiment is performed. The circuit of Fig. 4-8 is set up, and the switch is closed. The voltmeter will kick either up-scale or backwards and then return to zero. The current i_1, in building up through coil 1, must have made the b side of coil 1 negative and the a side positive. If the meter kicks up-scale, we know that making the a side of coil 1 positive drives the a side of coil 2 positive.

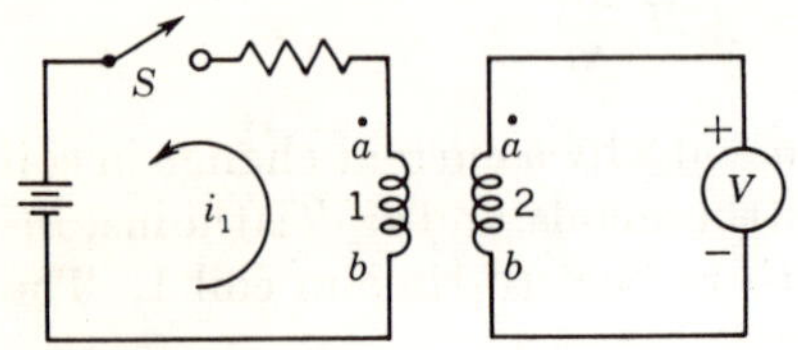

FIG. 4-8. An experiment to determine the phasing of the windings.

Similarly, making the a side of coil 1 negative, which could be due to a decrease in i_1, causes the a side of coil 2 to become negative with respect to the b side. A convenient way of denoting the winding sense is to place a dot on the end of the coils which follow each other in phase.

If in Fig. 4-8 it had been determined that closing caused the meter to read backwards, then the dot on the a side of coil 2 would instead be placed on the b side of coil 2. The dots have nothing to do with the circuit or assumed current directions. The only thing that determines the dots is the manner in which the coil is wound.

The location of the dots must be known before loop equations can be written. This section presents a step-by-step method of writing loop equations for inductively coupled circuits.

Assume that we have the circuit of Fig. 4-9a to analyze. The location of the dots has already been determined as shown. To write the loop equations, use the following procedure:

1. Replace each coil with two generators showing no polarity, as in Fig. 4-9b. One generator in each coil represents the counter emf due to the self-inductance of the winding, and the other represents the induced voltage due to a current in the other winding.

2. Arbitrarily choose a direction of current flow in loop 1, say, counterclockwise. This determines the polarity of the voltage across L_1 as i_1 builds up to its steady-state value. The counter emf develops across L_1 so that it tends to keep the assumed current from increasing. This counter emf ($L_1\ di_1/dt$) will then have the polarity shown in Fig. 4-9c.

3. Observe in Fig. 4-9c that the voltage developed across L_1 because of its self-inductance is such that the dot end of coil 1 is driven positive.

Since the dot voltages follow in phase, the dot end of the $M\,di_1/dt$ generator in loop 2 must be driven positive. Up to this point, everything can be represented by Fig. 4-9*d*.

4. Now arbitrarily assume a current direction for loop 2, and as in loop 1, note how the $L_2\,di_2/dt$ generator is polarized. Let us assume a clockwise current direction in loop 2, as in Fig. 4-9*e*, which polarizes L_2 as shown.

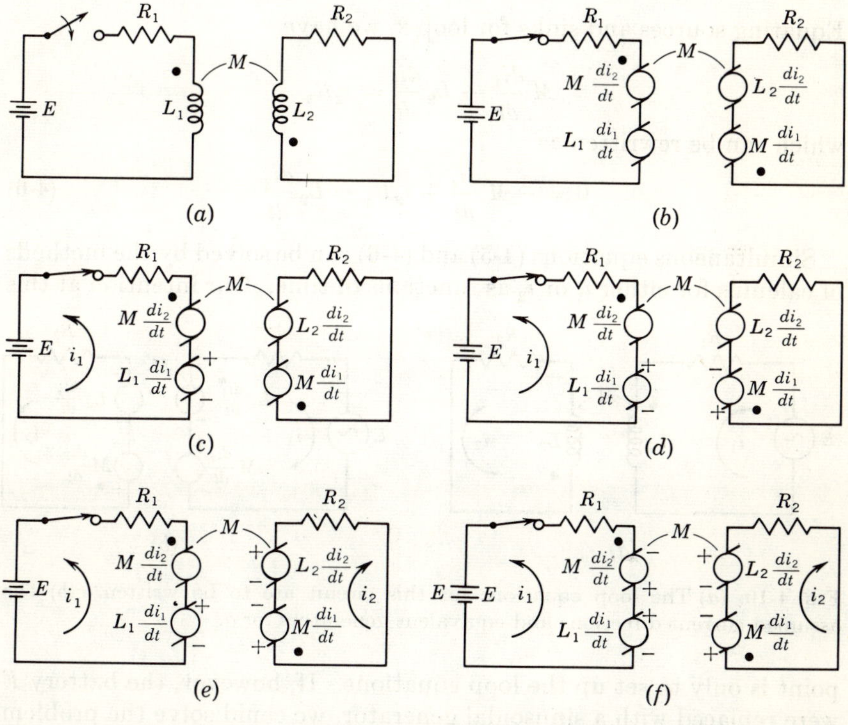

FIG. 4-9. (*a*) A systematic method of writing loop equations for an inductively coupled circuit is to be developed. (*b*) Each coil winding is replaced by two generators, one for the self-induced voltage and the other for the coupled voltage. (*c*) A current direction is arbitrarily assumed for loop 1. This determines the instantaneous polarity of the self-induced voltage. (*d*) Since the dot ends of each coil follow in phase, the voltage induced in loop 2 will have the indicated polarity. (*e*) A current direction for loop 2 is arbitrarily assumed and determines the polarity of the self-induced generator in loop 2. (*f*) Since the dots follow in phase, the voltage induced in loop 1 due to i_2 will be as shown.

5. In Fig. 4-9*e*, we see that the self-induced voltage across L_2 ($L_2\,di_2/dt$) drives the dot end of coil 2 negative. The dot end of the induced generator in coil 1 ($M\,di_2/dt$) must then also go negative. At this point, our analysis can be summed up by Fig. 4-9*f*. We are now in a position

to write our loop equations. Equating sources to sinks for loop 1, we have

$$E - L_1 \frac{di_1}{dt} + M \frac{di_2}{dt} = i_1 R_1$$

which can be rewritten as

$$E = i_1 R_1 + L_1 \frac{di_1}{dt} - M \frac{di_2}{dt} \tag{4-5}$$

Equating sources and sinks for loop 2, we have

$$M \frac{di_1}{dt} - L_2 \frac{di_2}{dt} = i_2 R_2$$

which can be rewritten as

$$0 = -M \frac{di_1}{dt} + i_2 R_2 + L_2 \frac{di_2}{dt} \tag{4-6}$$

Simultaneous equations (4-5) and (4-6) can be solved by the methods of calculus for either i_1 or i_2 as functions of time. Our intention at this

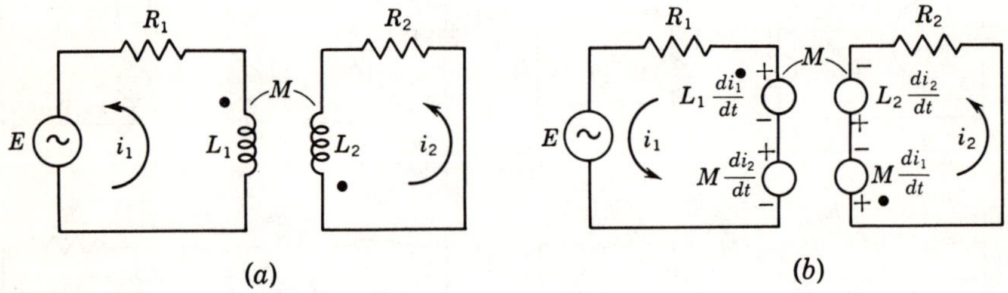

FIG. 4-10. (*a*) The loop equations for this circuit are to be written. (*b*) The assumed current directions and equivalent generators for *a*.

point is only to set up the loop equations. If, however, the battery E were replaced with a sinusoidal generator, we could solve the problem with our present tools. For a sine wave of current, the $L\, di/dt$ generators become $I(jX_L)$ and the $M\, di/dt$ generators become $I(jX_M)$. Equations would then be written as

$$E = I_1(R_1 + jX_{L_1}) - I_2(jX_M)$$

where E and I are the effective values,

$$0 = -I_1(jX_M) + I_2(R_2 + jX_{L_2})$$

Notice that the equations are in the form

$$E = I_1 Z_{11} + I_2 Z_{12}$$
$$0 = I_1 Z_{12} + I_2 Z_{22}$$

where $\quad Z_{12} = -jX_M$

We will solve specific problems shortly but, first, more practice in setting up the equations. Write loop equations for the circuit of Fig. 4-10*a*.

Each coil is replaced with a self-induced and an induced generator, as shown in Fig. 4-10*b*. Since the positive current directions are already specified in Fig. 4-10*a*, the polarity of the self-induced generators will be as shown in Fig. 4-10*b*. The self-induced generator of loop 1 drives the dot end positive as i_1 builds up. With the dot end of L_1 going positive because of i_1, the dot end of the $M\,di_1/dt$ induced generator in loop 2

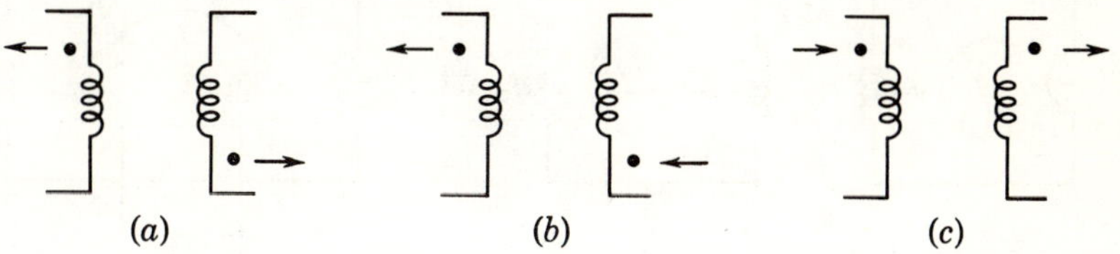

FIG. 4-11. (*a*) When the currents both flow into or out of the dots, the sign of the mutual impedance term is positive. (*b* and *c*) The mutual impedance is negative.

is going positive. Similarly, i_2 is developing a voltage across L_2 which drives the dot end positive in loop 2. The voltage induced in loop 1 by i_2 is then driving the dot end of the $M\,di_2/dt$ generator in loop 1 positive. The loop equations are then

$$\text{Sources} = \text{sinks}$$

$$E - M\frac{di_2}{dt} - L_1\frac{di_1}{dt} = R_1 i_1$$

$$-L_2\frac{di_2}{dt} - M\frac{di_1}{dt} = R_2 i_2$$

If the driving source is sinusoidal, we may in terms of rms values write

$$E = (R_1 + jX_{L_1})I_1 + jX_M I_2$$
$$0 = jX_M I_1 + (R_2 + jX_{L_2})I_2$$

In the future, when writing these equations, we shall transpose the coil generators mentally from the source to the sink side, with the appropriate sign change. A handy rule to remember is that when both assumed current directions either go into or come out of the dot ends of the coils, the sign of the mutual impedance term is the same as the sign of the self-impedance term about that loop. In other words, $Z_{12} = +j\omega M$, as illustrated in Fig. 4-11*a*.

If one of the currents flows into a dot and the other flows out, the sign of the mutual impedance term is opposite to the self-impedance term about the loop (see Fig. 4-11*b* and 4-11*c*). For this case, $Z_{12} = -j\omega M$.

In essence, the dot rule says that if both current arrows flow into or out of the dots, the sign of $j\omega M$ is positive. If one current is directed in and the other out, $j\omega M$ is negative.

4-7. Transformer Input Impedance. When dealing with a transformer circuit in which the coupling is purely inductive, such as Fig. 4-12*a*, an error in the sign of M will not change the magnitude of the currents. The error will alter the phase of I_2 by only 180°.

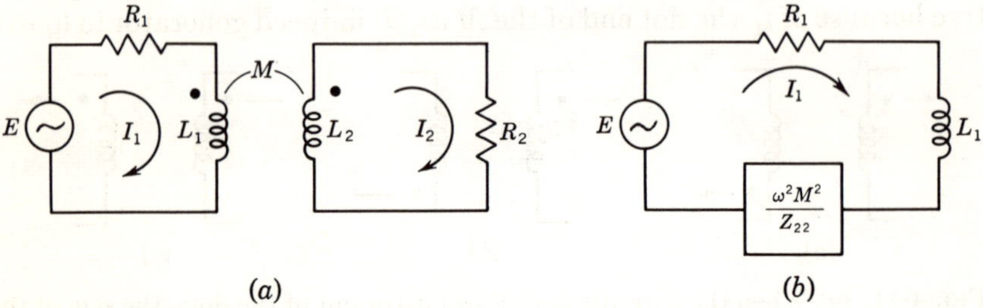

FIG. 4-12. (*a*) An inductively coupled circuit. (*b*) The equivalent circuit that the generator looks into.

If we were to write general loop equations for Fig. 4-12*a*, we should have

$$E = I_1 Z_{11} + I_2 Z_{12} \tag{4-7}$$

$$0 = I_1 Z_{12} + I_2 Z_{22} \tag{4-8}$$

The sign of Z_{12} (the mutual impedance) is negative, since I_1 flows into a dot and I_2 out. (Refer to dot rule.) If we solve for I_1,

$$I_1 = \frac{\begin{vmatrix} E & Z_{12} \\ 0 & Z_{22} \end{vmatrix}}{\begin{vmatrix} Z_{11} & Z_{12} \\ Z_{12} & Z_{22} \end{vmatrix}} = \frac{EZ_{22}}{Z_{11}Z_{22} - Z_{12}^2}$$

By dividing the numerator and denominator of the expression for I_1 by Z_{22}, we arrive at the form

$$I_1 = \frac{E}{Z_{in}} = \frac{E}{Z_{11} - Z_{12}^2/Z_{22}}$$

The input (driving point) impedance that the generator sees is then made up of two terms: Z_{11}, which is the self-impedance of loop 1, and $-Z_{12}^2/Z_{22}$, which is the reflected impedance of what is in loop 2 into loop 1. For an inductively coupled circuit, $Z_{12} = j\omega(\pm M)$. When M is negative, as in Fig. 4-12*a*, $Z_{12} = -j\omega M$. The reflected impedance is $Z_{12}^2/Z_{22} = -[j\omega(-M)]^2/Z_{22}$. It does not really matter if M is plus or minus when we compute Z_{in}, since M^2 is always positive. The coefficient j when squared is $j^2 = -1$. From these considerations, the reflected impedance is therefore

$$-\frac{(-1)\omega^2 M^2}{Z_{22}} = +\frac{\omega^2 M^2}{Z_{22}}$$

The significance of this **result** is that no matter what the sign of M actually is, the input current I_1 can always be determined in a circuit such as Fig. 4-12*a* by computing the driving point impedance as

$$Z_{in} = Z_{11} + \frac{\omega^2 M^2}{Z_{22}}$$

Just *add* the reflected impedance term to the self-impedance of the primary loop, since they appear in series, as shown in Fig. 4-12*b*.

An interesting point revealed by analyzing the expression for Z_{in} further is that inductance in the secondary loop reflects back as capacitance into the primary. Capacitance in the secondary looks like inductance in the primary. This can be seen by imagining Z_{22} to be a pure inductance. Then

$$Z_{22} = X_{L_2}\underline{/90^\circ} \quad \text{and} \quad \omega^2 M^2/X_{L_2}\underline{/90^\circ} = (\omega^2 M^2/X_{L_2})\underline{/-90^\circ}$$

Now $\omega^2 M^2/X_{L_2}$ at an angle of $\underline{/-90^\circ}$ is the same as a *capacitive reactance* whose magnitude is $\omega^2 M^2/X_{L_2}$. In a similar manner, it can be shown that if the net reactance of the secondary is $Z_{22} = X_{C_2}\underline{/-90^\circ}$, the reflected impedance will be inductive at an angle of $\underline{/90^\circ}$. We can also see that when the secondary is open-circuited, or $Z_{22} = \infty$, the reflected impedance is zero. There is no secondary current to set up a flux which will thread the primary.

Let us try an illustrative problem involving these concepts.

Problem. A transformer has the following characteristics:

$L_1 = 3$ henrys $\quad L_2 = 0.05$ henry $\quad k \approx 1$
$R_1 = 20$ ohms $\quad R_2 = 0.08$ ohm $\quad E = 115$ volts, 60 cps

The generator source impedance is negligible, and the secondary load resistance (R_L) is 42 ohms. What is the primary current in Fig. 4-13*a*?

Solution

$$\omega = 2\pi f = 377 \text{ radians/sec}$$

$$M = k\sqrt{L_1 L_2} = \sqrt{0.15} = 0.39 \text{ henry}$$

$$Z_{12} = j\omega M$$

$$Z_{11} = R_1 + j\omega L_1 = 20 + (j377 \text{ ohms})(3 \text{ henrys})$$
$$= 20 + j1{,}131 \text{ ohms}$$

$$Z_{22} = 42.08 + (j377 \text{ ohms})(0.05 \text{ henry}) = 42.08 + j18.85$$
$$= 46.1\underline{/24.1^\circ} \text{ ohms}$$

$$Z_{\text{ref}} = \frac{\omega^2 M^2}{Z_{22}} = \frac{(3.77)^2 \times 10^4 \times (0.39)^2}{46.1\underline{/24.1^\circ}} = \frac{2.14 \times 10^4}{46.1\underline{/24.1^\circ}}$$
$$= 0.0462 \times 10^4\underline{/-24.1^\circ} = 462\underline{/-24.1^\circ} = 421 - j189 \text{ ohms}$$

Note that the inductive impedance of the secondary loop reflects into the primary as a capacitive impedance having a reactive component of $-j189$ ohms. The reflected impedance appears in series with the primary-loop self-impedance. The equivalent circuit that the generator sees is shown in Fig. 4-13*b*. The input impedance is

$$Z_{in} = Z_{11} + Z_{\text{ref}} = 20 + j1{,}131 + 421 - j189$$
$$= 441 + j942 = 1{,}040\underline{/64.9^\circ} \text{ ohms}$$

and the rms input current is

$$I_1 = \frac{115\underline{/0^\circ} \text{ volts}}{1.04\underline{/64.9^\circ} \text{ kilohms}} = 110 \text{ ma}\underline{/-64.9^\circ}$$

Suppose it were desired to bring I_1 into phase with the applied voltage. In other words, we wish to eliminate the reactive component of Z_{in} and bring about

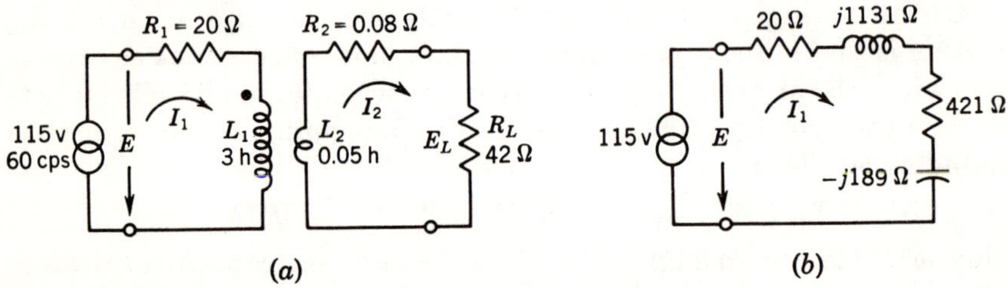

FIG. 4-13. (*a*) The input current (I_1) is to be determined. (*b*) The equivalent circuit that the source looks into.

a condition of primary resonance. An easy way to do this would be to put a capacitor in series with the primary, so that its reactance cancels the inductive $j942$-ohm component of Z_{in}. Therefore,

$$X_C = 942 \text{ ohms} = \frac{1}{\omega C} \quad \text{and} \quad \omega = 377 \text{ radians/sec}$$

Then
$$C = \frac{1}{377 \times 942} = \frac{10^{-5}}{3.77 \times 0.942} = \frac{10^{-5}}{3.55} = 0.281 \times 10^{-5}$$
$$= 2.81\ \mu\text{f}$$

4-8. Transformer Transfer Impedance. We have just completed an analysis which enables us to compute the impedance seen looking into a transformer. Once Z_{in} is evaluated, we can easily determine the input, or primary, current. What about the secondary current in a circuit such as Fig. 4-12*a*? To find I_2, we would set up the determinant based upon Eqs. (4-7) and (4-8):

$$I_2 = \frac{\begin{vmatrix} Z_{11} & E \\ Z_{12} & 0 \end{vmatrix}}{\begin{vmatrix} Z_{11} & Z_{12} \\ Z_{12} & Z_{22} \end{vmatrix}} = -\frac{EZ_{12}}{Z_{11}Z_{22} - Z_{12}^2}$$

The transfer impedance $Z_{T_{12}}$ is therefore

$$Z_{T_{12}} = \frac{E}{I_2} = -\frac{Z_{11}Z_{22} - Z_{12}^2}{Z_{12}}$$

where $Z_{11} = R_1 + j\omega L_1$

$Z_{12} = -j\omega M$ by the dot rule

$Z_{22} = R_2 + j\omega L_2$

By looking at a circuit, we could then evaluate Z_{11}, Z_{12}, and Z_{22}, from which $Z_{T_{12}}$ is calculated. Once $Z_{T_{12}}$ is known, I_2 can be directly determined for any sinusoidal input on the primary side.

We have already shown that the input current is

$$I_1 = \frac{EZ_{22}}{Z_{11}Z_{22} - Z_{12}^2}$$

Therefore,
$$\frac{I_2}{I_1} = \frac{-EZ_{12}/(Z_{11}Z_{22} - Z_{12}^2)}{EZ_{22}/(Z_{11}Z_{22} - Z_{12}^2)} = -\frac{Z_{12}}{Z_{22}}$$

or
$$I_2 = -\frac{Z_{12}}{Z_{22}} I_1$$

The relationship $I_2 = -(Z_{12}/Z_{22})I_1$ is interesting in that it gives us a physical interpretation of the secondary loop. The numerator has the dimensions of ohms times amperes, or volts. The denominator has the dimensions of ohms and the numerator's volts divided by ohms gives I_2, the required dimension of amperes. We see then that the quantity $-Z_{12}I_1$ can be interpreted as the voltage induced in the secondary winding by the current I_1 in the primary. The secondary current I_2 is equal to this induced voltage divided by the secondary loop self-impedance. Let us now evaluate I_2 in Fig. 4-13a:

$$I_2 = -\frac{Z_{12}}{Z_{22}} I_1 = -\frac{(-j\omega M)I_1}{Z_{22}} = \frac{(377)(0.39)\underline{/90^\circ}\text{ ohms} \times 110\text{ ma}\underline{/-64.9^\circ}}{46.1\underline{/24.1^\circ}\text{ ohms}}$$

$$= \frac{16.2\underline{/25.1^\circ}\text{ volts}}{46.1\underline{/24.1^\circ}\text{ ohms}} = 0.35\underline{/1^\circ}\text{ amp}$$

The phase angles of I_1 and I_2 are with respect to the input voltage, which was taken as our zero-degree reference. The load voltage is

$$E_L = I_2R_L = (0.35\underline{/1^\circ}\text{ amp})(42\underline{/0^\circ}\text{ ohms}) = 14.7\underline{/1^\circ}\text{ volts}$$

If the sign of M were positive, I_2 would equal $0.35\underline{/181^\circ}$ amp and the load voltage would also be shifted 180°.

4-9. Thevenizing the Transformer. An alternative approach for computing I_2 would be to thevenize the transformer by looking into the secondary terminals with R_L removed. The open-circuit voltage seen looking into the secondary terminals of Fig. 4-14a is, in general,

$$V_{oc} = -M\frac{di_1}{dt}$$

for any current waveform in the primary. With a sinusoidal primary

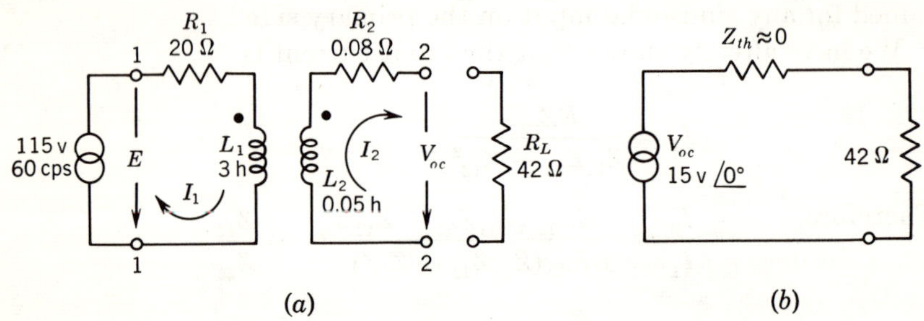

FIG. 4-14. (a) The current I_2 is determined by generating the Thévenin's equivalent circuit that R_L looks into. (b) The Thévenin's equivalent circuit that R_L sees.

current, the general expression becomes $V_{oc} = -j\omega MI_1$, where I_1 is the rms primary current. In this case,

$$I_1 = \frac{E}{Z_{11}} = \frac{115 \text{ volts}}{20 + j1{,}131}$$

$$\approx \frac{115 \text{ volts}}{1{,}131\underline{/90^\circ}} = 102 \text{ ma}\underline{/-90^\circ}$$

The value of I_1 in Fig. 4-13a was different because the secondary winding had a complete circuit and therefore reflected impedance into the primary. With $I_1 = 102 \text{ ma}\underline{/-90^\circ}$, we have

$$V_{oc} = (377)(0.39)\underline{/90^\circ} \times 102\underline{/-90^\circ} \text{ ma} = 15\underline{/0^\circ} \text{ volts}$$

The impedance seen looking into the secondary winding terminals is

$$Z_{th} = Z_{22} - \frac{Z_{12}{}^2}{Z_{11}} = R_2 + j\omega L_2 + \frac{\omega^2M^2}{R_1 + j\omega L_1}$$

We see that Z_{th} is composed of the secondary loop impedance in series with the reflected impedance of the primary loop:

$$
\begin{aligned}
Z_{22} &= R_2 + j\omega L_2 = 0.08 + j377(0.05) = 0.08 + j18.85 \\
&\approx j18.85, \text{ or } 18.85\underline{/90^\circ} \text{ ohms} \\
R_1 + j\omega L_1 &= 20 + j1{,}131 \approx 1{,}131\underline{/90^\circ} \text{ ohms} \\
\omega^2 M^2 &= 2.14 \times 10^4 \\
Z_{th} &= j18.85 + \frac{2.14 \times 10^4}{1.131 \times 10^3\underline{/90^\circ}} = j18.85 + 19\underline{/-90^\circ} \\
&= j18.85 - j19 \approx 0
\end{aligned}
$$

The equivalent circuit that the load looks into is then shown in Fig. 4-14*b*:

$$
I_2 = \frac{V_{oc}}{Z_{th} + R_L} = \frac{15\underline{/0^\circ} \text{ volts}}{42 \text{ ohms}} = 0.35\underline{/0^\circ} \text{ amp}
$$

As you can see, thevenizing the transformer leads to essentially the same result as previously obtained.

4-10. Inductive-Conductive Coupling. In circuits of the type shown in Fig. 4-15*a*, where the coupling is both conductive and inductive, the exact sign of M must be known if serious errors are to be avoided in computations. We shall show that the equivalent inductance of the two coupled coils in Fig. 4-15*a* is *not* $L_1 + L_2$, and that L_{eq} depends on the sign of M. This means, in problems of this nature, that the location of the dots must be known. The best procedure to use in handling circuits of this type is to draw in the equivalent self-induced and externally induced generators in place of each coil. Application of the dot rule is valid, but it can cause confusion in this type of circuit when there are multiple loops with many mutual inductances between loops.

Suppose the dots were placed as in Fig. 4-15*b*. Following the instructions previously outlined, we have Fig. 4-15*c* as the equivalent of Fig. 4-15*b*. The voltage drop across the $M\,di/dt$ generator is of the same polarity as $L_1\,di/dt$. This is the same as if L_1 had been increased to $L_1 + M$. Similarly, L_2 appears as if it had been increased to $L_2 + M$. The equivalent circuit inductance is then

$$
L_{eq} = L_1 + M + L_2 + M = L_1 + L_2 + 2M
$$

If one of the dots is reversed, as in Fig. 4-15*d*, M will be negative. The equivalent circuit in this case is Fig. 4-15*e*. Notice that now the induced voltage in L_1, due to the same current in L_2, tends to buck the self-induced voltage in L_1. This is the same as if L_1 were decreased to

$L_1 - M$. Similarly, L_2 appears as if it had been decreased to $L_2 - M$. The equivalent inductance in this circuit is therefore

$$L_{eq} = L_1 - M + L_2 - M = L_1 + L_2 - 2M$$

If the coupling is tight, making k large, the difference in the equivalent inductance will depend markedly on the sign of M. As an exercise,

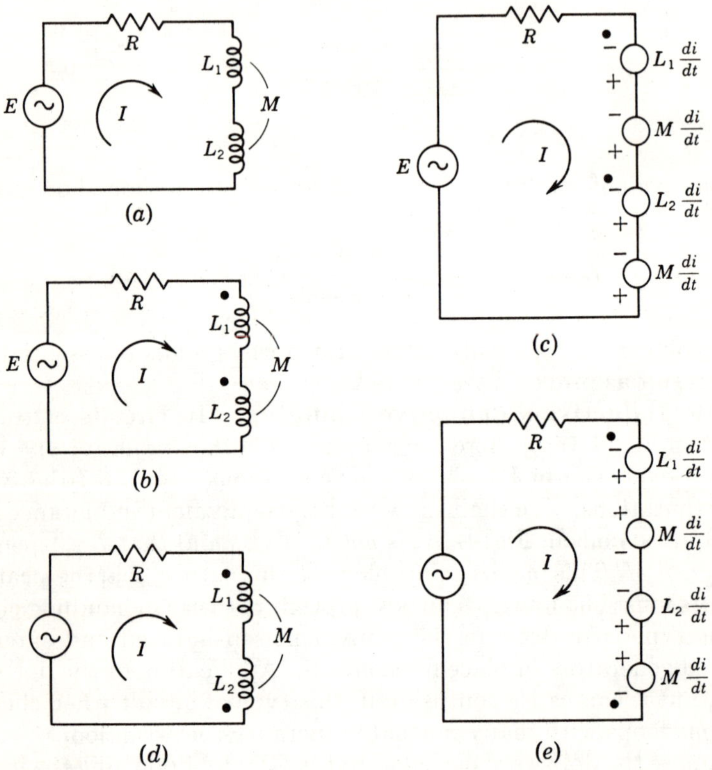

FIG. 4-15. (*a*) The equivalent inductance of the coupled coils is to be determined. (*b*) The location of the dots must be known if I is to be correctly evaluated. (*c*) The equivalent circuit for *b*. (*d*) The circuit of *b* with the phasing of the coils reversed. (*e*) The equivalent circuit of *d*.

the reader might show that if $k = 1$ and $L_1 = L_2$, the equivalent inductance when M is negative is zero and four times as large as either coil when M is positive.

4-11. Measurement of Mutual Inductance. A method of determining the mutual impedance between two coupled coils is the following. Connect two terminals of the transformer of Fig. 4-16*a*, as in Fig. 4-16*b*. Measure the net inductance between terminals A and D with an impedance bridge. Call this equivalent inductance L'_{eq}. Now reverse one of the coils relative to the other, say, as in Fig. 4-16c, and measure the net inductance between terminals A and C. Call this equivalent inductance L''_{eq}. Now either L'_{eq} is greater than L''_{eq} or vice versa. Let us assume L'_{eq} is greater. As an exercise, the reader can then derive the relationships:

$$L'_{eq} = L_1 + L_2 + 2M \tag{4-9}$$

$$L''_{eq} = L_1 + L_2 - 2M \tag{4-10}$$

Solving for M in terms of L'_{eq} and L''_{eq}, we find, from (4-10),

$$L_1 + L_2 = L''_{eq} + 2M$$

Substituting $L''_{eq} + 2M$ for $L_1 + L_2$ in (4-9), we have

$$L'_{eq} = (L''_{eq} + 2M) + 2M \qquad \text{or} \qquad M = \frac{L'_{eq} - L''_{eq}}{4}$$

The self-inductance of each winding may be measured with the impedance bridge by leaving the other winding open. This assumes that the

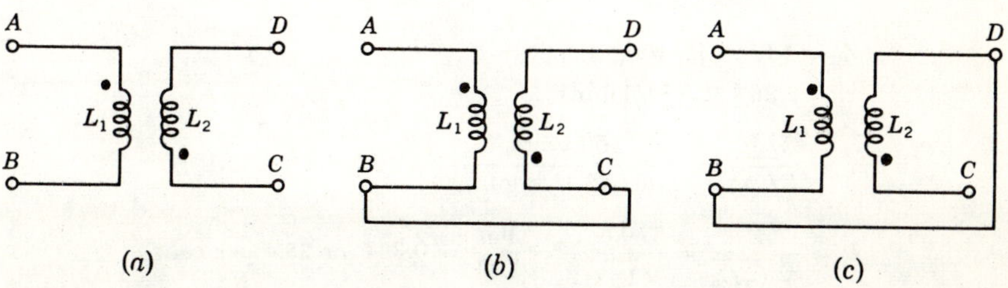

FIG. 4-16. The mutual inductance between two coils may be determined by two measurements.

bridge frequency is low enough to make the effect of winding and interwinding capacitances negligible.

In circuits where both inductive and conductive coupling exist, the location of the dots must be known.

Problem 1. Find the circuit current in Fig. 4-17a, if $R_1 = 1$ kilohm, $R_2 = 1$ kilohm, $L_1 = 1$ henry, $L_2 = 2$ henrys, $M = 0.5$ henry, $E = 100$ volts/100 cps. What is the coefficient of coupling between the two coils?

Solution. Replace L_1 with two generators—one for self-induced voltage due to i_1 in L_1, and another for the voltage induced in L_1 by the same current i_1 in coil L_2. Similarly, L_2 is replaced with two generators. By observing the manner in which the self-inductances are polarized, we can determine the polarity of the induced generators via the dots. This leads to Fig. 4-17b. The loop equation is then

$$\begin{aligned} E &= i_1R_2 + L_2\frac{di_1}{dt} - M\frac{di_1}{dt} - M\frac{di_1}{dt} + L_1\frac{di_1}{dt} + i_1R_1 \\ &= i_1(R_1 + R_2) + (L_1 + L_2 - 2M)\frac{di_1}{dt} \end{aligned}$$

Applying the dot rule, we see that the current goes into one dot and out the other. Therefore, the self and mutual terms will have opposite signs, and M will be negative.

Since E is a sinusoidal generator, we can write

$$E = (R_1 + R_2)I_1 + j\omega(L_1 + L_2 - 2M)I_1$$
$$= [R_1 + R_2 + j\omega(L_1 + L_2 - 2M)]I_1 = ZI_1$$

where E and I_1 are effective values.

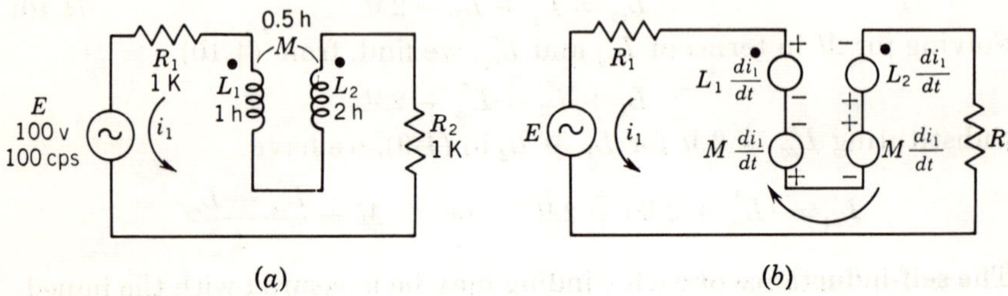

FIG. 4-17. (a) The current is to be determined. (b) Equivalent circuit.

$$Z = 2\text{ K} + j2\pi100(2\text{ henrys})$$
$$= 2.36\underline{/\ 32.1^\circ}\text{ kilohms}$$

$$I = \frac{E\underline{/\ 0^\circ}}{Z\underline{/\ \phi^\circ}} = \frac{100\text{ volts}}{2.36\underline{/\ 32.1^\circ}\text{ kilohms}} = 42.3\text{ ma}\underline{/\ -32.1^\circ}$$

$$k = \frac{M}{\sqrt{L_1L_2}} = \frac{0.5}{\sqrt{1 \times 2}} = \frac{0.5}{1.41} = 0.354\text{, or 35.4 per cent}$$

Problem 2. Find the output voltage (E_o) in the circuit of Fig. 4-18a. The coefficient of coupling between the two coils is 50 per cent.

Solution. If $k = 0.5 = M/\sqrt{L_1L_2}$,

$$M = 0.5\sqrt{0.1 \times 0.01} = 1.59 \times 10^{-2}\text{ henry}$$

Both i_1 and i_2 flow through L_1. Therefore, there will be two self-induced generators in L_1, and i_2 in flowing through L_2 also induces a voltage in L_1 because of the mutual coupling M. We then replace L_1 with a total of three generators. Only i_2 flows through L_2, giving rise to a self-induced generator in L_2. Since i_1 and i_2 flow through L_1, which is coupled to L_2, both these currents will induce voltages in L_2. We then replace L_2 with three generators. The equivalent circuit is shown in Fig. 4-18b. Coil polarities are determined in the usual manner by observing the assumed current directions in relation to the dots. Under steady-state a-c conditions, the loop equations around loop 1 would be

$$E = I_1\left(R_1 + j\omega L_1 - j\frac{1}{\omega C}\right) + I_2(-j\omega M - j\omega L_1)$$

$$= +I_1\left[R_1 + j\left(\omega L_1 - \frac{1}{\omega C}\right)\right] - I_2[j\omega(L_1 + M)]$$

and around loop 2, we have

$$0 = -I_1(j\omega L_1 + j\omega M) + I_2(j\omega L_1 + j\omega M + j\omega M + j\omega L_2) + I_2 R_2$$
$$= -I_1[j\omega(L_1 + M)] + I_2[j\omega(L_1 + L_2 + 2M)] + I_2 R_2$$

From this point on, the problem is straightforward, and the reader should have no trouble evaluating E_o.

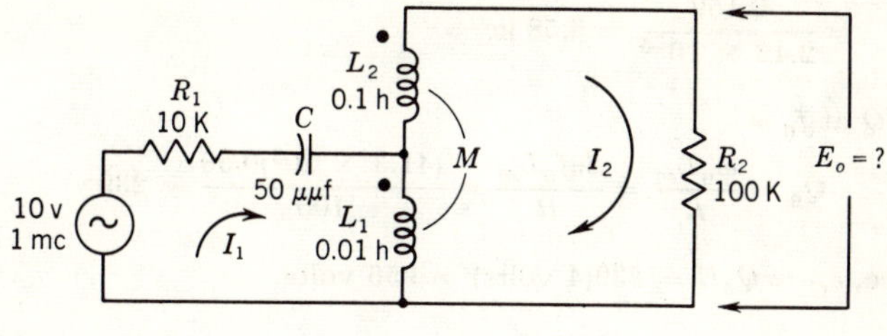

(a)

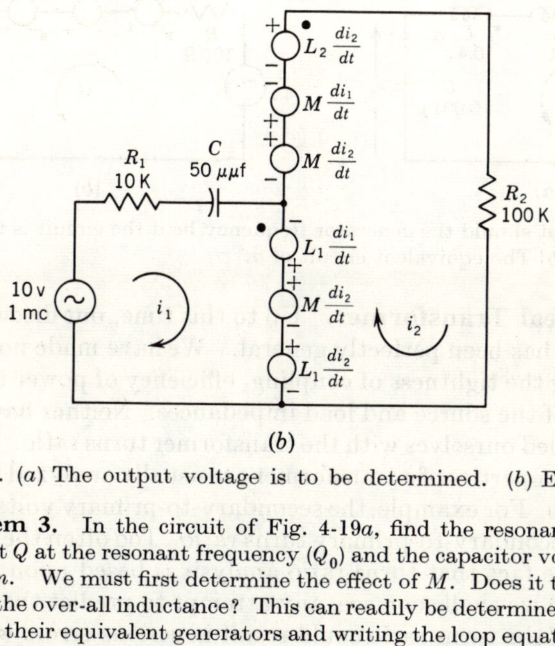

(b)

FIG. 4-18. (a) The output voltage is to be determined. (b) Equivalent circuit.

Problem 3. In the circuit of Fig. 4-19a, find the resonant frequency (f_o), the circuit Q at the resonant frequency (Q_0) and the capacitor voltage (e_C) at f_0.

Solution. We must first determine the effect of M. Does it tend to increase or decrease the over-all inductance? This can readily be determined by replacing the coils with their equivalent generators and writing the loop equation for Fig. 4-19b. Now I in flowing through L_1 makes the dot end of the equivalent self-induced generator ($L_1\, di_1/dt$) negative. This means that the induced voltage ($M\, di/dt$) in coil L_2 will make the dot end of the coil negative. And I_1 in flowing through L_2 makes the dot end of the equivalent self-induced generator ($L_2\, di_1/dt$) negative. The voltage induced in L_1, which is ($M\, di_1/dt$), due to the current in L_2 will then drive the dot end of L_1 negative.

The loop equation for Fig. 4-19b with a sinusoidal current is

$$E = I_1 R + I_1 jX_{L_1} + I_1 jX_M + I_1 jX_{L_2} + I_1 jX_M + I_1(-jX_c)$$
$$= I_1[R + j(X_{L_1} + X_{L_2} + 2X_M) - jX_c]$$

The total inductive reactance is, clearly, $j(X_{L_1} + X_{L_2} + 2X_M)$. This means that the total inductance is $L_1 + L_2 + 2M$. The resonant frequency (f_0) is then

$$f_0 = \frac{1}{2\pi\sqrt{(L_1 + L_2 + 2M)C}} = \frac{0.159}{\sqrt{0.58 \text{ henry} \times 10^{-9} \text{ farad}}}$$

$$= \frac{0.159}{2.42 \times 10^{-5}} = 6.58 \text{ kc}$$

The circuit Q at f_0 is

$$Q_0 = \frac{\omega_0 L_{eq}}{R} = \frac{2\pi f_0 L_{eq}}{R} = \frac{(41.3 \times 10^3)0.58}{100} = 239$$

At resonance, $e_C = Q_0 E = 239(4 \text{ volts}) = 956$ volts.

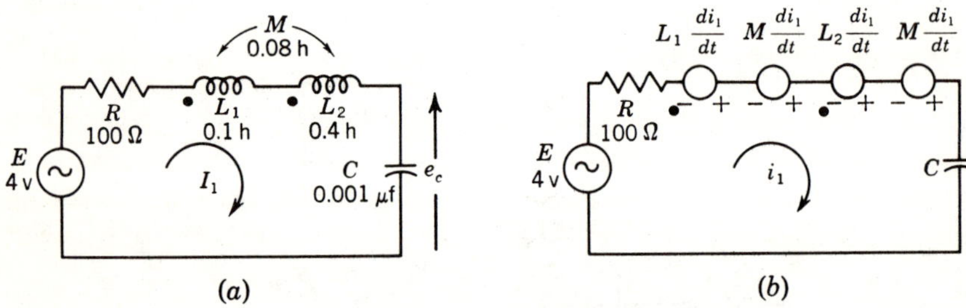

FIG. 4-19. (*a*) What should the generator frequency be if the circuit is to appear series-resonant? (*b*) The equivalent circuit of *a*.

4-12. The Ideal Transformer. Up to this time, our discussion of the transformer has been perfectly general. We have made no restrictions concerning the tightness of coupling, efficiency of power transfer, or magnitudes of the source and load impedances. Neither have we in any way concerned ourselves with the transformer turns ratio. In basic electronics the properties of a transformer are usually analyzed in terms of the turns ratio. For example, the secondary-to-primary voltage ratio is equal to the secondary-to-primary turns ratio. Too often the student loses sight of the fact that turns-ratio analysis is based upon an *ideal* transformer. As we shall soon see, any attempt to predict the voltage ratio in terms of the turns ratio could result in serious errors when the transformer is not ideal. Since many of the iron-core type of transformers used in power and audio work approximate ideal transformers, a discussion of this subject is in order.

First of all, what do we mean by an ideal transformer? An ideal transformer has 100 per cent coupling. In other words, $k = 1$, and there is negligible leakage inductance and leakage flux. An ideal transformer has zero losses. This means negligible d-c winding resistance. It also means negligible a-c winding resistance and core losses at the frequencies of interest. In an ideal transformer either the primary winding reactance ($j\omega L_1$) and/or the secondary winding reactance ($j\omega L_2$) is very large relative to any impedance in the secondary loop and therefore $Z_{22} \approx j\omega L_2$.

An important relationship in the *ideal* transformer is that the inductance ratio of the windings is equal to the square of the turns ratio. That is,

$$\frac{L_1}{L_2} = \left(\frac{N_1}{N_2}\right)^2$$

If the transformer is ideal and has zero losses, the power put into the primary winding will equal the power dissipated in the load, as shown in Fig. 4-20.

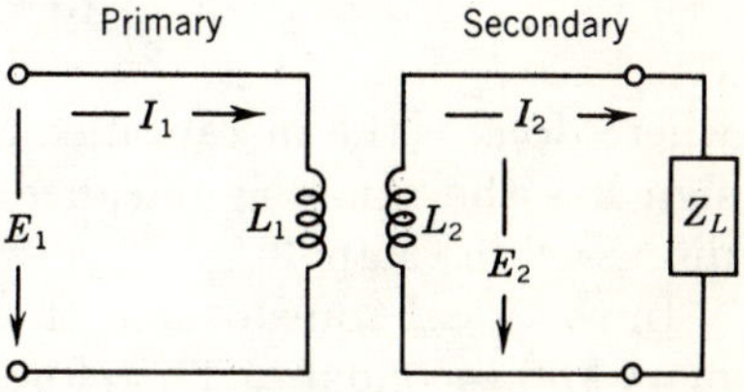

FIG. 4-20. If the transformer is ideal the power put into the primary winding will equal that developed in the load.

4-13. Current and Voltage Relationships. The first circuit characteristic of an ideal transformer that we shall investigate is the relationship between the secondary and primary currents. In general,

$$\frac{I_2}{I_1} = \frac{Z_{12}}{Z_{22}} = \frac{j\omega M}{R_2 + j\omega L_2 + Z_L}$$

Since $k = 1$ in the ideal transformer, $M = \sqrt{L_1 L_2}$, and therefore

$$\frac{I_2}{I_1} = \frac{j\omega\sqrt{L_1 L_2}}{R_2 + j\omega L_2 + Z_L}$$

Also, $j\omega L_2 \gg R_2 + Z_L$, which simplifies the expression to

$$\frac{I_2}{I_1} = j\,\frac{\omega\sqrt{L_1 L_2}}{j\omega L_2} = \sqrt{\frac{L_1}{L_2}}$$

But $L_1/L_2 = N_1^2/N_2^2$, and therefore

$$\frac{I_2}{I_1} = \frac{N_1}{N_2}$$

We may also express the primary and secondary winding voltages in terms of the turns ratio. Since the transformer is ideal, and $P_p = P_L$, we may write

$$E_1 I_1 = E_2 I_2 \qquad \text{or} \qquad \frac{E_1}{E_2} = \frac{I_2}{I_1}$$

But $I_2/I_1 = N_1/N_2$; therefore,

$$\frac{E_2}{E_1} = \frac{N_2}{N_1}$$

Actually this derivation assumes that the phase angle between E_1 and I_1 is the same as that between E_2 and I_2. This is true in an ideal transformer.

We may also prove that $E_2/E_1 = N_2/N_1$ by Faraday's law, which states that the voltage induced in a coil of N turns by a changing flux is

$$e_L = -N\frac{d\phi}{dt}10^{-8}\text{ volts} \approx -N\frac{\Delta\phi}{\Delta t}10^{-8}\text{ volts}$$

where $d\phi/dt$ is the instantaneous rate of change of flux. The negative sign has the same significance as in the relationship $e_L = -L\,di/dt$, discussed in Chap. 3.

In the ideal transformer, the same flux (no leakage) links both primary and secondary. Therefore, the time rate of change of flux ($d\phi/dt$) must be the same in both windings. Therefore, in the primary,

$$e_1 = -N_1\frac{d\phi}{dt}10^{-8}$$

and in the secondary,

$$e_2 = -N_2\frac{d\phi}{dt}10^{-8}$$

and hence

$$\frac{e_2}{e_1} = \frac{N_2}{N_1}$$

The voltages may be instantaneous or effective values.

This derivation involved no assumption about the phase angle between E_1 and I_1, or between E_2 and I_2. We can now show that the phase angle in both windings is identical, since

$$P_p = P_L \qquad \text{or} \qquad E_1I_1\cos\phi_1 = E_2I_2\cos\phi_2$$

we have

$$\frac{\cos\phi_1}{\cos\phi_2} = \frac{E_2I_2}{E_1I_1} = \frac{N_2}{N_1}\frac{N_1}{N_2} = 1 \qquad \text{or} \qquad \cos\phi_1 = \cos\phi_2$$

4-14. Impedance Matching. Now let us investigate the problem of impedance matching with transformers. First of all, what do we mean by impedance matching? Impedance matching is usually practiced to achieve maximum power transfer. In basic electronics, we were told that a generator will deliver maximum power to a load when the load impedance is equal to the generator or source impedance. Actually, there is a little more to this. The generator will deliver maximum power to a load when the load impedance is the conjugate of the source impedance. In other words, if the source impedance is $R_g + jX_{L_g}$, the load should be equal to $R_L - jX_{C_L}$, where $R_L = R_g$, and $X_{C_L} = X_{L_g}$. The idea is to achieve a resonant circuit in which the reactive components cancel, leaving the load resistance equal to the source resistance. Similarly, if the source were capacitive in nature, the load would present an inductive component in the primary circuit. If the circuit is such that a conjugate impedance match cannot be achieved, the next best thing to do is to match the magnitudes of the source and load impedances.

Occasionally there are instances in which a change of impedance levels is desired—but not necessarily that change which is required for maximum power transfer. For example, a vacuum-tube amplifier having a certain internal resistance will deliver maximum power to a load when the load resistance is equal to the amplifier resistance. This may, however, result in the amplifier operating with excessive distortion. If the design criterion is minimum distortion, then a value of load resistance that fulfills this requirement must be chosen.

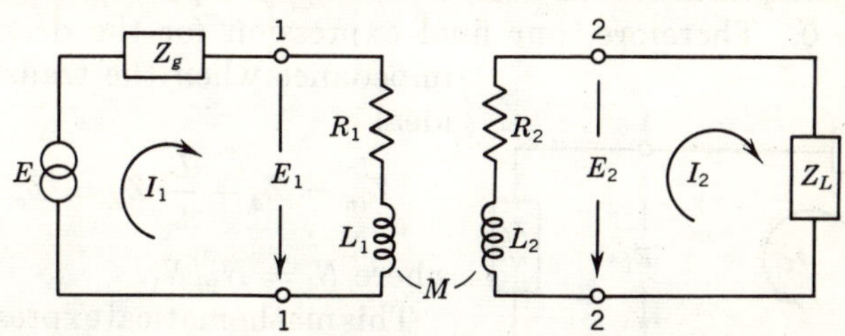

FIG. 4-21. The transformer is assumed to be ideal.

The general problem involved in impedance transformation is to have a source look into a pair of terminals and see a particular value of impedance, and at the same time to have the load look into another pair of terminals and see some other desired impedance. In this section we shall restrict ourselves to impedance transformation with ideal transformers. Impedance transformation can also be accomplished with nonideal transformers, but the mathematics gets cumbersome. The basic technique for this type of problem has been developed, however, and the reader should have no trouble in evaluating the impedance seen looking into a pair of nonideal transformer terminals.

Consider the transformer circuit of Fig. 4-21. In our study of coupled circuits we showed that the driving point impedance of loop 1 is

$$Z_{in} = Z_{11} - \frac{{Z_{12}}^2}{Z_{22}}$$

and that for an inductively coupled circuit this becomes

$$Z_{in} = Z_{11} + \frac{\omega^2 M^2}{Z_{22}} = Z_g + R_1 + j\omega L_1 + \frac{\omega^2 M^2}{Z_L + R_2 + j\omega L_2}$$

Putting the expression for Z_{in} over a common denominator, we have

$$Z_{in} = \frac{Z_gZ_L + Z_gR_2 + Z_gj\omega L_2 + R_1Z_L + R_1R_2 + R_1j\omega L_2 + j\omega L_1Z_L + j\omega L_1R_2 - \omega^2L_1L_2 + \omega^2M^2}{Z_L + R_2 + j\omega L_2}$$

Now if the transformer is assumed ideal, $M = \sqrt{L_1L_2}$, $R_1 \approx 0$, $R_2 \approx 0$, $j\omega L_1 \gg Z_g$, and $j\omega L_2 \gg Z_L$ and $L_1/L_2 = N_1^2/N_2^2 = 1/N^2$. With these assumptions, the terms multiplied by R_1 and R_2 in the numerator can be neglected, and the denominator essentially becomes $j\omega L_2$. If we also substitute L_1L_2 for M^2, we now have

$$Z_{in} = \frac{Z_gZ_L + Z_gj\omega L_2 + j\omega L_1Z_L}{j\omega L_2} = \frac{Z_gZ_L}{j\omega L_2} + Z_g + \frac{L_1}{L_2}Z_L$$

The $Z_gZ_L/j\omega L_2$ term can be neglected, since $j\omega L_2 \gg Z_L$, which means $Z_L/j\omega L_2 \approx 0$. Therefore, our final expression for the driving-point impedance when the transformer is ideal is

$$Z_{in} = Z_g + \frac{L_1}{L_2}Z_L = Z_g + \frac{Z_L}{N^2}$$

where $N = N_2/N_1$.

This mathematical expression may be physically interpreted to mean that the generator looks into the source impedance Z_g, acting in series with the load impedance divided by the square of the turns ratio. The equivalent circuit that E looks into is shown in Fig. 4-22.

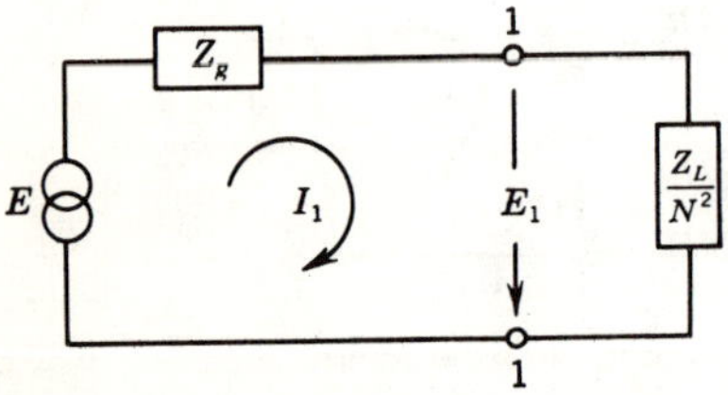

FIG. 4-22. In an ideal transformer the impedance seen looking into the primary winding will equal the load impedance divided by the square of the turns ratio.

If the reader finds this derivation for Z_{in} difficult, a simpler though not so fundamental approach is herewith presented. In Fig. 4-22, the impedance that E looks into must be equal to Z_g plus the impedance between the 1-1 terminals of the transformer ($Z_{1\text{-}1}$). Note that $Z_{1\text{-}1}$ is not the same as the loop impedance Z_{11}. Now the impedance $Z_{1\text{-}1}$ must surely be equal to the voltage across the 1-1 terminals divided by the current flowing into them. We can then write

$$Z_{in} = Z_g + Z_{1\text{-}1}$$

But

$$Z_{1\text{-}1} = \frac{E_1}{I_1}$$

Since $E_1 = E_2N_1/N_2$ and $I_1 = I_2N_2/N_1$, we have

$$Z_{1\text{-}1} = \frac{E_2N_1/N_2}{I_2N_2/N_1} = \frac{E_2}{I_2}\frac{N_1^2}{N_2^2}$$

But $E_2/I_2 = Z_L$, since Fig. 4-21 plainly shows that $E_2 = I_2 Z_L$. Therefore,

$$Z_{1\text{-}1} = Z_L \frac{N_1^2}{N_2^2} = \frac{Z_L}{N^2}$$

Substituting in the first expression, we have

$$Z_{in} = Z_g + \frac{Z_L}{N^2}$$

which again is physically interpreted by Fig. 4-22.

A third approach to developing the impedance relationships of an ideal transformer is based upon its being 100 per cent efficient:

Power into primary winding = power into load

$$P_1 = P_2$$

Therefore
$$\frac{E_1^2}{Z_{1\text{-}1}} = \frac{E_2^2}{Z_L}$$

which can be rewritten as

$$\frac{E_1^2}{E_2^2} = \frac{Z_{1\text{-}1}}{Z_L}$$

But
$$\frac{E_1}{E_2} = \frac{N_1}{N_2} = \frac{1}{N}$$

Therefore
$$\left(\frac{N_1}{N_2}\right)^2 = \frac{1}{N^2} = \frac{Z_{1\text{-}1}}{Z_L}$$

Now let us see how Fig. 4-22 can aid us in analyzing circuits involving ideal transformers.

Problem 1. Suppose you had a 4-ohm loudspeaker available and an amplifier. If the power output tube requires a 16-kilohm load for specified operation, what turns ratio would you specify for the output transformer?

Solution. If the power output tube requires a 16-kilohm load, this means that the impedance seen looking into the 1-1 transformer terminals must be 16 kilohms. In an ideal transformer, the impedance between the 1-1 terminals is the reflected impedance, Z_L/N^2. We can then write

$$\frac{Z_L}{N^2} = 16 \text{ kilohms}$$

where $Z_L = 4$ ohms

Therefore,
$$N^2 = \frac{4}{16{,}000} = \frac{1}{4{,}000}$$

$$N = \frac{1}{\sqrt{4{,}000}} = \frac{1}{63.2}$$

For every turn on the secondary there should be 63.5 turns on the primary. The transformer would not, however, be built with just one turn on the secondary and 63.5 on the primary. Many turns would be used on each winding, to make the winding reactance much greater than any other loop impedance. The proper turns ratio must, however, be preserved.

Problem 2. The turns ratio of the transformer in Fig. 4-23 has been chosen for maximum power transfer. Find N, Z_{in}, I_1, E_1, I_2, E_2, and the power developed in the loudspeaker (P_L).

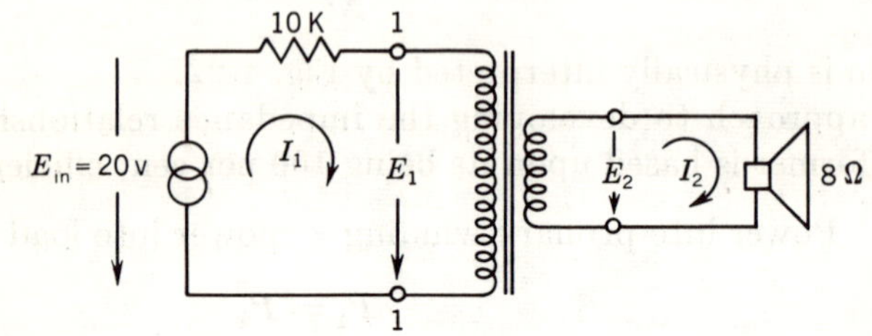

FIG. 4-23. The power in the loudspeaker is to be determined.

Solution. To find N, we will make use of the fact that the turns ratio has been chosen for maximum power transfer. This means that the reflected impedance of the load as it appears between the 1-1 terminals must match (equal) the 10-kilohm impedance. Therefore,

$$\frac{Z_L}{N^2} = 10 \text{ kilohms}$$

or $$N^2 = \frac{8}{10{,}000}$$

Therefore $$N = \frac{N_2}{N_1} = \frac{\sqrt{8}}{\sqrt{10{,}000}} = \frac{2.84}{100} = 2.84 \times 10^{-2}$$

$$Z_{in} = 10 + \frac{Z_L}{N^2} = 10 + 10 = 20 \text{ kilohms}$$

$$I_1 = \frac{20 \text{ volts}}{20 \text{ kilohms}} = 1 \text{ ma}$$

$$E_1 = I_1\frac{Z_L}{N^2} = 1 \text{ ma} \times 10 \text{ kilohms} = 10 \text{ volts}$$

$$I_2 = \frac{I_1}{N} = \frac{1}{2.84 \times 10^{-2}} = 0.352 \times 10^2 = 35.2 \text{ ma}$$

$$E_2 = E_1N = 10 \times 2.84 \times 10^{-2} = 0.284 \text{ volts}$$

$$P_L = E_2I_2 = 0.284 \text{ volts} \times 35.2 \text{ ma} = 10 \text{ mw} = E_1I_1$$

$$= 10 \text{ volts} \times 1 \text{ ma} = 10 \text{ mw}$$

Problem 3. If the load resistance on the secondary side of an ideal transformer is reduced to zero, what impedance is seen looking into the primary terminals?

Solution. In an ideal transformer, the impedance seen looking into the primary terminals is equal to the secondary load resistance divided by the time ratio squared. If the load is zero ohms, the reflected resistance is also zero. The primary current is then limited by the primary winding resistance only. This is why shorting the secondary of a power transformer will cause fuses to blow in the primary side.

Problem 4. What if the transformer of the previous problem were not ideal and the load were shorted?

Solution. The impedance seen looking into the primary would now depend upon transformer characteristics other than the turns ratio. The reflected impedance would depend upon the value of M and the secondary winding resistance R_2 and inductance L_2. The self-impedance of the primary winding would then also appear in series with the reflected resistance. The exact relationship is

$$Z_{1\text{-}1} = R_1 + j\omega L_1 + \frac{\omega^2 M^2}{R_2 + j\omega L_2}$$

Frequently we encounter circuits in which an ideal transformer drives multiple loads, as in Fig. 4-24*a*. How do we analyze a circuit of this

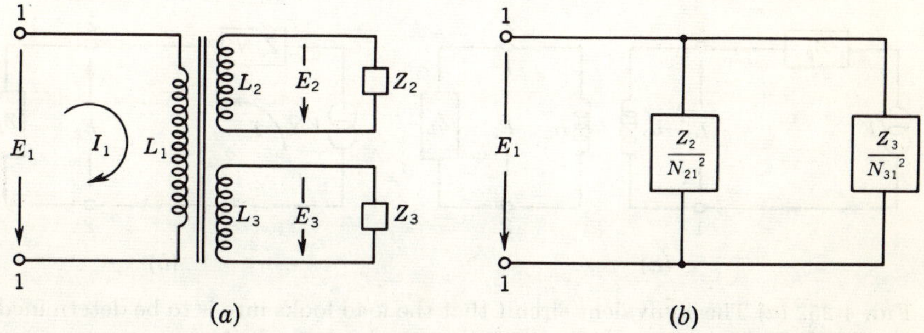

FIG. 4-24. (*a*) An ideal transformer driving multiple loads. (*b*) The equivalent circuit seen looking into the 1-1 terminals.

nature? Intuitively, we might sense that impedance reflected by Z_3 into the primary would appear in parallel with the reflected impedance of Z_2. The reason we sense this is that the more loads you hang on the transformer, the greater the primary current becomes, which implies that the total reflected impedance is decreasing. Intuition, however, can at times lead us astray. Therefore, let us develop a rigorous analysis of the circuit.

The power put into the transformer primary is $E_1^2/Z_{1\text{-}1}$. Since the transformer is ideal, the primary power must equal the sum of the load powers. We can then write

$$\frac{E_1^2}{Z_{1\text{-}1}} = \frac{E_2^2}{Z_2} + \frac{E_3^2}{Z_3} \tag{4-11}$$

We can relate E_2 and E_3 to E_1 in terms of the turns ratios. Let $N_{21} = N_2/N_1$, and $N_{31} = N_3/N_1$; then

$$\frac{E_1^2}{Z_{1\text{-}1}} = \frac{(E_1N_2/N_1)^2}{Z_2} + \frac{(E_1N_3/N_1)^2}{Z_3} \tag{4-12}$$

$$\frac{E_1^2}{Z_{1\text{-}1}} = \frac{E_1^2N_{21}^2}{Z_2} + \frac{E_1^2N_{31}^2}{Z_3}$$

$$\frac{1}{Z_{1\text{-}1}} = \frac{1}{Z_2/N_{21}^2} + \frac{1}{Z_3/N_{31}^2} \tag{4-13}$$

Equation (4-13) is in the same form as that for two resistors in parallel:

$$\frac{1}{R_{eq}} = \frac{1}{R_1} + \frac{1}{R_2}$$

where R_1 is analogous to Z_2/N_{21}^2 and R_2 to Z_3/N_{31}^2. The equivalent circuit for Fig. 4-24*a* is, then, that of Fig. 4-24*b*.

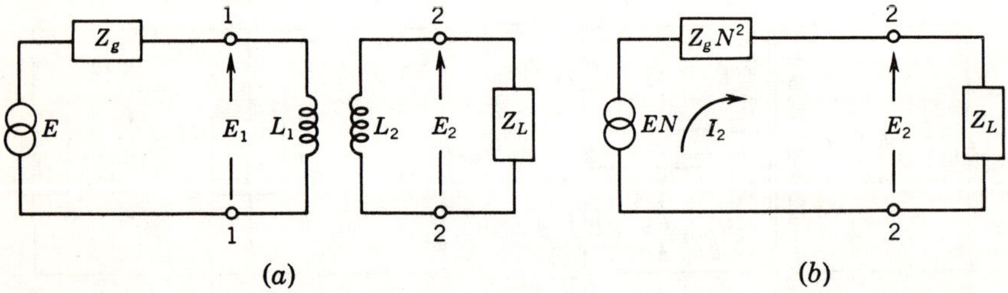

FIG. 4-25. (*a*) The equivalent circuit that the load looks into is to be determined. (*b*) The Thévenin's equivalent circuit that the load of *a* looks into.

Up to this time, we have been analyzing the ideal transformer by considering the reflected impedance of the secondary in the primary loop. There might be occasions when we would prefer to work with the secondary loop. Therefore, we would be concerned with the reflected impedance of the primary into the secondary loop. This work can be easily done by thevenizing the transformer. Consider the ideal transformer circuit of Fig. 4-25*a*. To develop the Thévenin's equivalent looking into the 2-2 terminals, we must first disconnect Z_L and then determine V_{oc}. V_{oc} will be equal to E_1N_2/N_1. With Z_L disconnected, the self-impedance of loop 1 is simply $Z_{11} = Z_g + j\omega L_1$. In an ideal transformer, $j\omega L_1 \gg Z_g$, and therefore the applied voltage is $E \approx E_1$. V_{oc} is then equal to $E(N_2/N_1)$. To find Z_{th} looking back into terminals 2-2, we shall use the relationship $Z_{th} = V_{oc}/I_{sc}$, where I_{sc} is the current that will flow through a short circuit across terminals 2-2. When 2-2 is shorted, $Z_L = 0$ and the reflected impedance into the primary is 0 ohms. The primary current (I_1) is $I_1 = E/Z_g + Z_{\text{ref}} = E/Z_g$. The secondary short-circuit current is, then,

$$I_{sc} = I_1 \frac{N_1}{N_2} = \frac{E}{Z_g}\frac{N_1}{N_2}$$

Therefore $$Z_{th} = \frac{V_{oc}}{I_{sc}} = \frac{EN_2/N_1}{(E/Z_g)(N_1/N_2)} = Z_g\left(\frac{N_2}{N_1}\right)^2 = Z_gN^2$$

The equivalent circuit that Z_L looks into is then that of Fig. 4-25*b*.

An alternative method of deriving the equivalent circuit of Fig. 4-25*b* is to start with the expression for I_1,

$$I_1 = \frac{E}{Z_g + Z_L/N^2}$$

$$I_2 = \frac{I_1}{N} = \frac{E}{Z_g + Z_L/N^2}\frac{1}{N} = \frac{E}{(N^2Z_g + Z_L)/N} = \frac{EN}{Z_L + N^2Z_g}$$

This last expression for I_2 is of the form $I = E'/Z$, where E' is analogous to EN and Z, the loop impedance, is analogous to $Z_L + N^2Z_g$. The equivalent circuit is again that of Fig. 4-25*b*.

4-15. Equivalent Circuits. The response of a transformer-coupled circuit to nonsinusoidal waveforms is difficult to determine and generally beyond the scope of this text. We shall, however, develop equivalent circuits for the transformer which can be used to study the effect of transformer parameters with nonsinusoidal excitation. The usual transformer circuit with its separate primary and secondary loops is not as familiar as a conductively coupled network. We will utilize the principle that any four-terminal network can, at a specific frequency, be reduced to an equivalent T network. We shall therefore try to develop an equivalent T network for a transformer in which the elements (arms) are conductively rather than inductively coupled. The concept of an equivalent circuit and the technique used in developing one might seem strange at first. For this reason we shall practice a bit with equivalent circuits of more familiar networks.

Suppose we had the very simple two-terminal network of Fig. 4-26*a* and wished to develop the equivalent circuit of Fig. 4-26*b*. What we really want to know is what relationship there must be between R_1, R_2 and R_{eq} so that the two networks are indistinguishable when looking into their respective input terminals. We can very obviously see that R_{eq} must equal $R_1 + R_2$. Usually the equivalent circuit is not quite so obvious, and therefore we wish to develop a rigorous procedure for finding one. A method that we shall use is to apply a voltage to the input terminals of the network. This causes a current to flow which depends upon the input impedance that the generator sees looking into the input terminals. Now if the network of Fig. 4-26*a* is to be equivalent to that of Fig. 4-26*b*, the *same* voltage, when applied to the input terminals of each, should result in the *same* input current. In effect, we are saying that the input impedance of each must be the same. Figures 4-26*c* and *d* show the network of Figs. 4-26*a* and *b* with the same voltage source connected to the input terminals. Be sure to choose the same

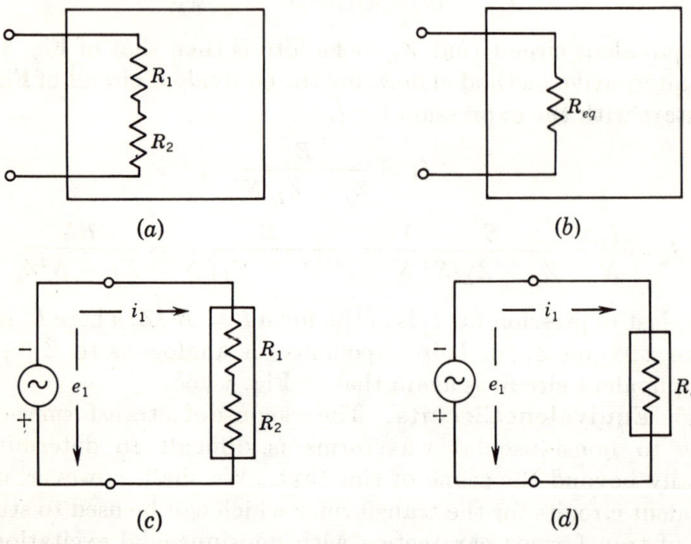

FIG. 4-26. (*a*) A simple network whose equivalent is to be determined. (*b*) By inspection $R_{eq} = R_1 + R_2$. (*c*) If *d* is to contain an equivalent network, the ratio of e_1 to i_1 must be the same as in *c*. (*d*) The same voltage and current are assumed to drive the equivalent network.

voltage and current directions in both cases. Also, the input-current direction chosen should be that in which the generator is actually trying to force current flow. Otherwise, the input resistance would erroneously appear negative. There are networks in which the input impedance is negative, as we shall later see. We shall, however, always begin an analysis of this nature with the current direction such that the source is positive (aiding).

We are now in a position to write and compare loop equations for Figs. 4-26*c* and *d*:

For Fig. 4–26*c*, $$e_1 = i_1(R_1 + R_2)$$

For Fig. 4–26*d*, $$e_1 = i_1(R_{eq})$$

By comparing the coefficients of i_1 in each equation, we can see that if e_1 and i_1 are to be the same in both cases, $R_1 + R_2$ must equal R_{eq}.

Let us try this technique on Fig. 4-27*a* and generate the equivalent circuits of Fig. 4-27*b*. Because of the parallel branches, this circuit is best analyzed by the nodal method: For Fig. 4-27*a*,

$$i_1 = i_a + i_b$$

$$i_1 = \frac{0 - (-e_1)}{R_A} + \frac{0 - (-e_1)}{R_B} = \frac{e_1}{R_A} + \frac{e_1}{R_B} = e_1\left(\frac{1}{R_A} + \frac{1}{R_B}\right)$$

Rewriting this equation so that we can compare it with that of Fig. 4-27*b*, we have

$$e_1 = i_1 \frac{1}{1/R_A + 1/R_B} = i_1 \frac{R_A R_B}{R_A + R_B}$$

For Fig. 4-27*b*,

$$e_1 = i_1 R_{eq}$$

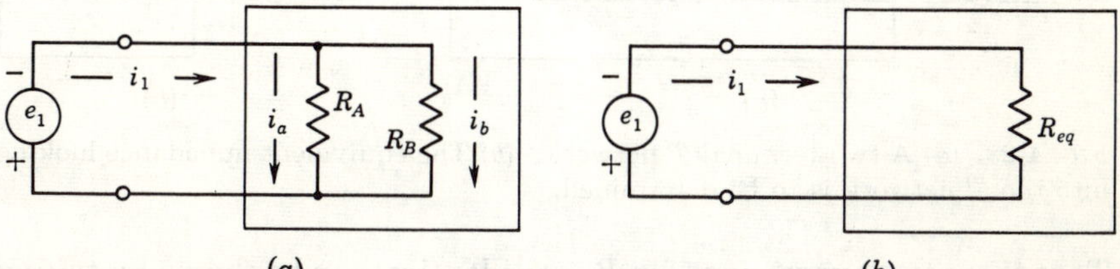

FIG. 4-27. (*a*) The original network whose equivalent is to be derived. (*b*) The equivalent network.

Comparing coefficients of i_1, we see that

$$\frac{R_A R_B}{R_A + R_B} = R_{eq}$$

This is surely in agreement with what we have learned about parallel resistors in basic electronics.

Let us try a more difficult problem. Convert the circuit of Fig. 4-28*a* to Fig. 4-28*b*.

$$e_1 = i_1(R_1 + R_2) - i_2 R_2$$
$$0 = -i_1 R_2 + i_2(R_2 + R_3)$$

$$i_1 = \frac{\begin{vmatrix} e_1 & -R_2 \\ 0 & R_2 + R_3 \end{vmatrix}}{\begin{vmatrix} R_1 + R_2 & -R_2 \\ -R_2 & R_2 + R_3 \end{vmatrix}} = \frac{e_1(R_2 + R_3)}{(R_1 + R_2)(R_2 + R_3) - R_2^2}$$

$$e_1 = i_1 \frac{(R_1 + R_2)(R_2 + R_3) - R_2^2}{R_2 + R_3} = i_1 \frac{R_1 R_2 + R_2 R_3 + R_1 R_3}{R_2 + R_3}$$

Comparing coefficients of i_1 with $e_1 = i_1 R_{eq}$, we see that

$$R_{eq} = \frac{R_1R_2 + R_2R_3 + R_1R_3}{R_2 + R_3}$$

From basic electronics we can tell by inspection that R_{eq} must equal

$$R_{eq} = R_1 + \frac{R_2R_3}{R_2 + R_3}$$

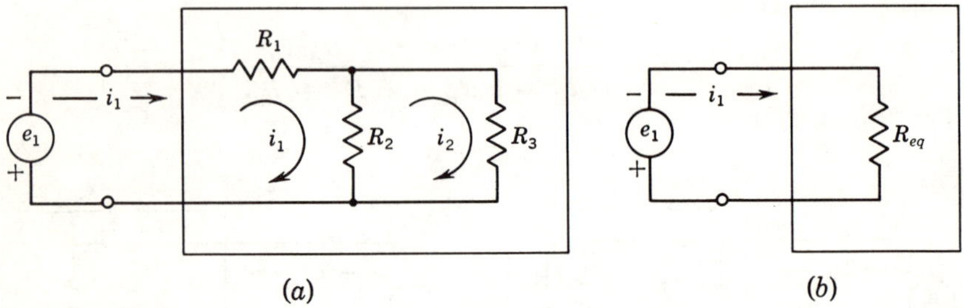

FIG. 4-28. (*a*) A two-terminal T network. (*b*) The equivalent impedance looking into the T network is to be determined.

That these two expressions for R_{eq} are identical can be shown by taking the derived expression for R_{eq} and wrestling it into the more familiar form:

$$R_{eq} = \frac{R_1R_2 + R_2R_3 + R_1R_3}{R_2 + R_3} = \frac{R_1R_2 + R_1R_3}{R_2 + R_3} + \frac{R_2R_3}{R_2 + R_3}$$

$$= R_1\frac{R_2 + R_3}{R_2 + R_3} + \frac{R_2R_3}{R_2 + R_3} = R_1 + \frac{R_2R_3}{R_2 + R_3}$$

Up to this point we have been dabbling with rather simple two-terminal networks. Many networks, including the transformer, are four-terminal networks. These networks consist of a pair of input terminals, to which power is usually applied, and a pair of output terminals, to which a load is connected. The output current and voltage may be entirely different from the input voltage and current.

A very simple four-terminal network is that shown in Fig. 4-29. In this case, $i_1 = i_2$, since there are no shunt branches within the network. The output voltage developed across the load R_L is not, however, the same as e_1, owing to voltage drop within the network. Another simple four-terminal network, in which e_1 and e_2 are equal but i_1 and i_2 are not, is shown in Fig. 4-30.

The T network of Fig. 4-31 is a most common four-terminal network. The impedance seen looking into the input terminal naturally depends upon the value of R_L. When writing the loop equations, however, we do not want R_L to appear. It is what is inside the box that we are interested in. By writing the equations in terms of e_1, i_1 and e_2, i_2, the effect of R_L is accounted for, since $R_L = e_2/i_2$. Similarly, the impedance seen looking into the output terminals will depend upon the generator impedance at the input end. We have been considering generators with

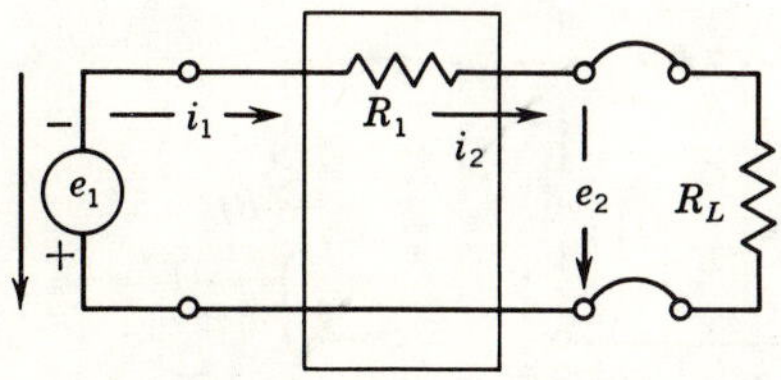

FIG. 4-29. An elementary four-terminal network in which the input and output currents are identical.

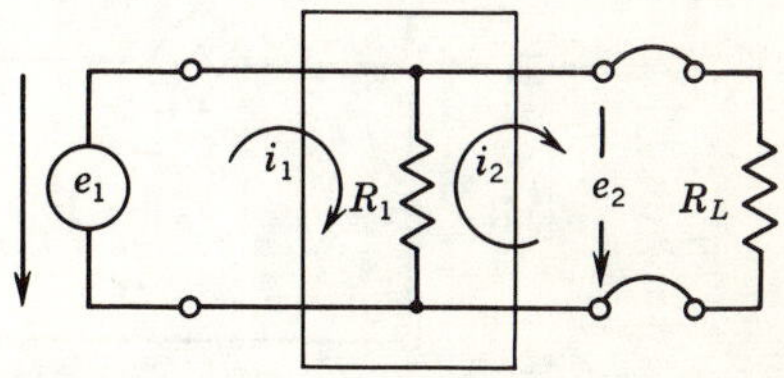

FIG. 4-30. The input and output voltages are identical in this four-terminal network.

zero internal impedance connected to the input terminals. If the generator impedance is not zero, the voltage e_1 is to be considered as the actual voltage applied to the input terminals.

It is not really necessary to connect our zero impedance generator to the input terminals. We could just as well have said that when a current i_1 is forced into the input terminals, a voltage e_1 will appear between

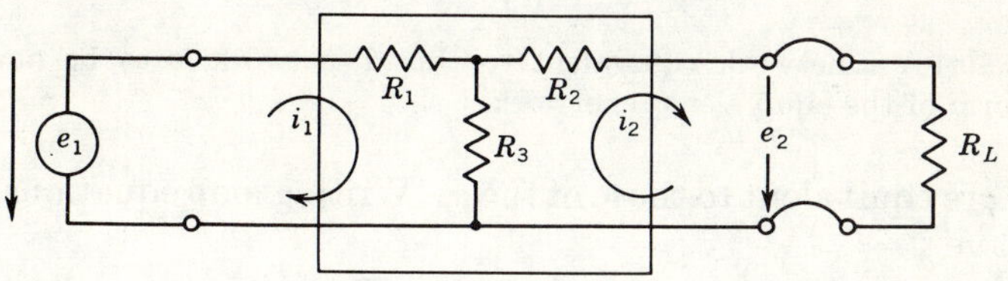

FIG. 4-31. A four-terminal T network.

them. Similarly when a current i_2 flows out of the output terminals, a voltage e_2 appears between them. It is immaterial whether we consider the voltage as producing the current or vice versa.

Now let us see if we can use our method of generating circuits to convert the T network of Fig. 4-32*a* into the π network of Fig. 4-32*b*. The reader may recognize this as the delta-wye transformation generally covered in basic electronics.

The current flowing into the T network of Fig. 4-32*a* develops a voltage e_1 across the input terminals 1-1. This current is supplied by a generator whose presence need not be indicated. A load is connected across the output (2-2) terminals, and a voltage e_2 is developed across

R_L. The same is true for the π network of Fig. 4-32*b*. What we want to know is what relationship must exist between the elements of the π and the T networks for them to be equivalent.

The conditions for equivalency will be developed by examining the terminals voltages and currents. With the same load on both networks, they will surely be identical if the voltages and currents at the terminals

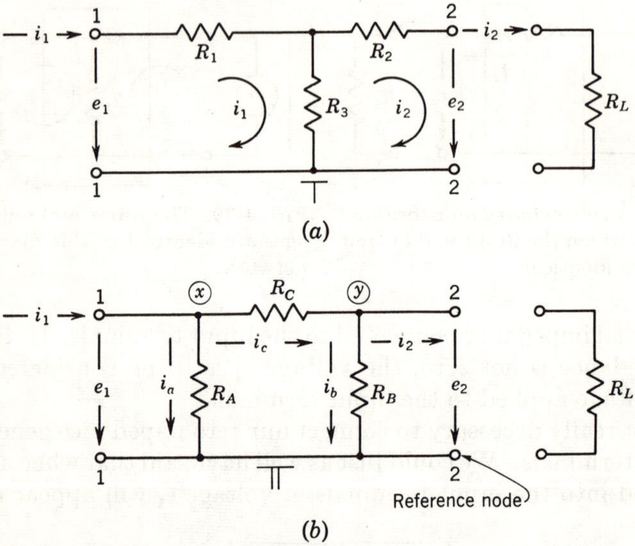

FIG. 4-32. (*a*) A π network equivalent to this T network is to be developed. (*b*) The form of the equivalent π network.

of the T are equivalent to those of the π. Writing loop equations for the T, we have

$$e_1 = (R_1 + R_3)i_1 - R_3 i_2 \tag{4-14}$$

$$-e_2 = -R_3 i_1 + (R_2 + R_3)i_2 \tag{4-15}$$

which in general form may be written

$$e_1 = Z_{11}i_1 + Z_{12}i_2 \tag{4-16}$$

$$-e_2 = Z_{12}i_1 + Z_{22}i_2 \tag{4-17}$$

where $\quad Z_{11} = R_1 + R_3 \qquad (4\text{-}18)$

$$Z_{12} = -R_3 \quad (Z_{12} = Z_{21} \text{ in passive networks}) \tag{4-19}$$

$$Z_{22} = R_2 + R_3 \tag{4-20}$$

Solving for i_1 and i_2, we have

$$i_1 = \frac{\begin{vmatrix} e_1 & Z_{12} \\ -e_2 & Z_{22} \end{vmatrix}}{\begin{vmatrix} Z_{11} & Z_{12} \\ Z_{12} & Z_{22} \end{vmatrix}} = \frac{e_1 Z_{22} + e_2 Z_{12}}{Z_{11}Z_{22} - Z_{12}^2} \tag{4-21}$$

$$i_2 = \frac{\begin{vmatrix} Z_{11} & e_1 \\ Z_{12} & -e_2 \end{vmatrix}}{\begin{vmatrix} Z_{11} & Z_{12} \\ Z_{12} & Z_{22} \end{vmatrix}} = \frac{-e_2 Z_{11} - e_1 Z_{12}}{Z_{11}Z_{22} - Z_{12}^2} \tag{4-22}$$

Let

$$\begin{aligned} \Delta &= Z_{11}Z_{22} - Z_{12}^2 \\ &= (R_1 + R_3)(R_2 + R_3) - R_3^2 \\ &= R_1R_2 + R_1R_3 + R_2R_3 \end{aligned}$$

We may now write (remember $Z_{12} = -R_3$)

$$i_1 = e_1 \frac{Z_{22}}{\Delta} + e_2 \frac{Z_{12}}{\Delta} = e_1 \frac{R_2 + R_3}{\Delta} - e_2 \frac{R_3}{\Delta} \tag{4-23}$$

$$i_2 = -e_1 \frac{Z_{12}}{\Delta} - e_2 \frac{Z_{11}}{\Delta} = e_1 \frac{R_3}{\Delta} - e_2 \frac{R_1 + R_3}{\Delta} \tag{4-24}$$

Now let us write some nodal equations for the π network. At node x,

$$i_1 = i_A + i_C \tag{4-25}$$

where $i_A = \dfrac{e_1}{R_A}$

i_C = voltage at y with respect to reference node, minus the voltage at x with respect to reference node, all divided by R_C. The voltage at y with respect to reference node is minus e_2, since the head of the e_2 arrow lies on the reference node. Similarly the voltage at x is minus e_1. Therefore,

$$i_C = \frac{e_y - e_x}{R_C} = \frac{-e_2 - (-e_1)}{R_C} = \frac{e_1 - e_2}{R_C}$$

$$i_1 = e_1\left(\frac{1}{R_A} + \frac{1}{R_C}\right) - \frac{e_2}{R_C} \tag{4-26}$$

At node y we have

$$i_2 = i_C - i_B = \frac{e_1 - e_2}{R_C} - \frac{e_2}{R_B} = \frac{e_1}{R_C} - e_2\left(\frac{1}{R_B} + \frac{1}{R_C}\right) \tag{4-27}$$

Comparing coefficients of (4-26) and (4-27) with (4-23) and (4-24), we see that

$$\frac{R_2 + R_3}{\Delta} = \frac{1}{R_A} + \frac{1}{R_C} \tag{4-28}$$

$$\frac{R_3}{\Delta} = \frac{1}{R_C} \tag{4-29}$$

$$\frac{R_1 + R_3}{\Delta} = \frac{1}{R_B} + \frac{1}{R_C} \tag{4-30}$$

Substituting (4-29) in (4-28), we obtain

$$\frac{R_2 + R_3}{\Delta} = \frac{1}{R_A} + \frac{R_3}{\Delta}$$

$$\frac{1}{R_A} = \frac{R_2}{\Delta}$$

and therefore

$$R_A = \frac{\Delta}{R_2}$$

Similarly, using (4-29) and (4-30), we obtain

$$R_B = \frac{\Delta}{R_1}$$

and from (4-29),

$$R_C = \frac{\Delta}{R_3}$$

which is the complete solution.

4-16. The T-Equivalent Circuit. We are given the inductively coupled circuit of Fig. 4-33*a* and wish to construct a T-equivalent circuit as in Fig. 4-33*b*. If the currents i_1 and i_2 and the terminal voltages e_1 and e_2 for both configurations are the same, then the two circuits must be equivalent. Our problem then is to express Z_A, Z_B, and Z_C in terms of known transformer parameters. The technique we shall use is to write and compare loop equations for both configurations and see what values Z_A, Z_B, and Z_C must have in order to keep the currents and voltages identical.

In Fig. 4-33*a*, if we follow the current i_1 around its loop, we have $e_1 = i_1Z_{11} + i_2Z_{12}$. We can see that $Z_{12} = -j\omega M$, since i_1 flows into a dot and i_2 flows out of a dot. (Refer to dot rule.) Going around loop 2 in Fig. 4-33*a*, we have $-e_2 = i_1Z_{12} + i_2Z_{22}$, where

$$Z_{11} = R_1 + j\omega L_1$$

$$Z_{12} = -j\omega M$$

$$Z_{22} = R_2 + j\omega L_2$$

Doing the same for the T network, we have

$$e_1 = i_1 Z_{11_T} + i_2 Z_{12_T}$$
$$-e_2 = i_1 Z_{12_T} + i_2 Z_{22_T}$$

where
$$Z_{11_T} = Z_A + Z_C$$
$$Z_{12_T} = -Z_C$$
$$Z_{22_T} = Z_B + Z_C$$

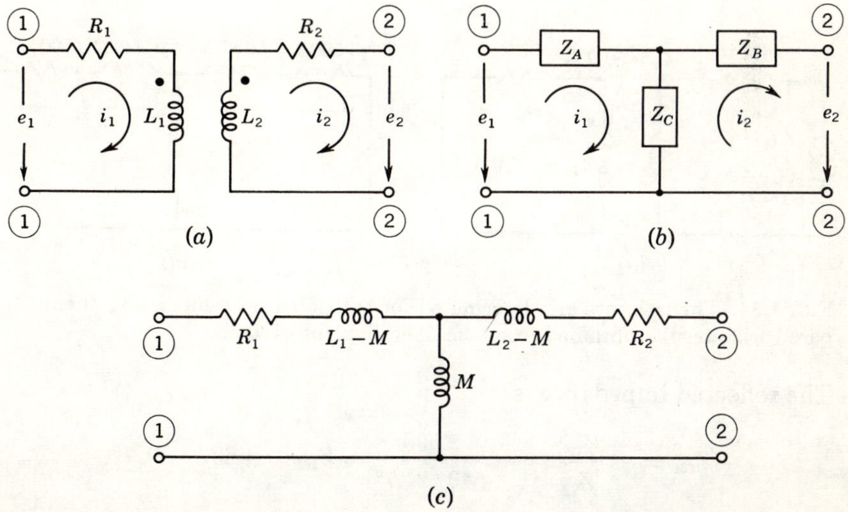

FIG. 4-33. (*a*) An inductively coupled circuit. (*b*) A T network whose arms are to be equivalent to the circuit of *a*. (*c*) The arms of the equivalent T network.

For the networks to be identical, we must have

$$Z_{11} = Z_{11_T}$$
$$Z_{12} = Z_{12_T}$$
$$Z_{22} = Z_{22_T}$$

or

$$R_1 + j\omega L_1 = Z_A + Z_C \tag{4-31}$$
$$j\omega M = Z_C \tag{4-32}$$
$$R_2 + j\omega L_2 = Z_B + Z_C \tag{4-33}$$

Equations (4-31), (4-32), and (4-33) may be solved for

$$Z_C = j\omega M, \quad Z_A = R_1 + j\omega(L_1 - M), \quad \text{and} \quad Z_B = R_2 + j\omega(L_2 - M).$$

These results may be summarized in Fig. 4-33*c*.

As an exercise in using the T-equivalent circuit, solve for I_2 in Figs. 4-34a and b. Compare the results:

$$Z_{11} = 500 + j1{,}000 = 1{,}120\underline{/63.4^\circ}$$

$$-Z_{12} = -j50 = 50\underline{/-90^\circ}$$

or

$$Z_{12} = j50$$

$$Z_{22} = 500 + j200 = 540\underline{/21.7^\circ}$$

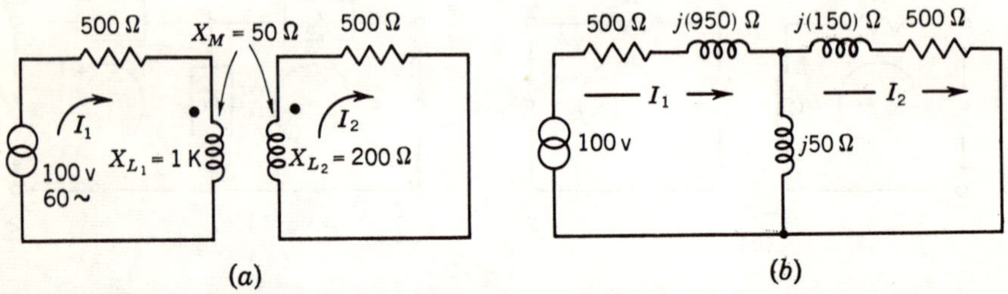

FIG. 4-34. The currents are determined for transformer circuit a and then compared with results obtained from the T equivalent of b.

The reflected impedance is

$$Z_{\text{ref}} = -\frac{Z_{12}^{\,2}}{Z_{22}} = +\frac{2{,}500}{540\underline{/21.7^\circ}} = Z_{\text{ref}} = 4.63\underline{/-21.7^\circ}$$

$$= 4.3 - j1.7 \text{ ohms}$$

The driving-point impedance is then

$$Z_{in} = Z_{11} + Z_{\text{ref}} = 504.3 + j998.3 \approx 500 + j1{,}000$$

Apparently the secondary is very loosely coupled to the primary.[1] Therefore,

$$I_1 = \frac{E}{Z_{in}} = \frac{100 \text{ volts}}{1{,}120\underline{/63.4^\circ}} = 89.2\underline{/-63.4^\circ} \text{ ma}$$

$$I_2 = +j\frac{\omega M}{Z_{22}} I_1 = \frac{(50\underline{/90^\circ})(89.2\underline{/-63.4^\circ})}{540\underline{/21.7^\circ}} = 8.25\underline{/4.9^\circ} \text{ ma}$$

[1] Not that we are interested at the moment, but k could be evaluated as follows:

$$k = \frac{M}{\sqrt{L_1 L_2}} = \frac{\omega M}{\omega\sqrt{L_1 L_2}} = \frac{\omega M}{\sqrt{\omega^2 L_1 L_2}} = \frac{\omega M}{\sqrt{(\omega L_1)(\omega L_2)}}$$

$$= \frac{50}{\sqrt{1{,}000 \times 200}} = 0.11, \text{ or } 11 \text{ per cent}$$

Using Fig. 4-34*b*,

$$100 \text{ volts} = I_1Z_{11} + I_2Z_{12}$$
$$0 = +I_1Z_{12} + I_2Z_{22}$$

where $Z_{11} = 500 + j1{,}000 = 1{,}120\underline{/63.4°}$

$Z_{12} = -j50$

$Z_{22} = 500 + j200 = 540\underline{/21.7°}$

$$I_2 = \frac{\begin{vmatrix} Z_{11} & 100 \\ Z_{12} & 0 \end{vmatrix}}{\begin{vmatrix} Z_{11} & +Z_{12} \\ +Z_{12} & Z_{22} \end{vmatrix}} = \frac{-100Z_{12}}{Z_{11}Z_{22} - Z_{12}^2} = \frac{-100(50\underline{/-90°})}{(1{,}120\underline{/63.4°})(540\underline{/21.7°}) + 2{,}500}$$

$$= \frac{5 \times 10^3\underline{/90°}}{605{,}000\underline{/85.1°} + 2{,}500} = \frac{5 \times 10^3\underline{/90°}}{51{,}600 + j603{,}000 + 2{,}500}$$

$$\approx \frac{5 \times 10^3\underline{/90°}}{54{,}100 + j603{,}000} \approx \frac{5 \times 10^3\underline{/90°}}{606 \times 10^3\underline{/85.1°}} = 8.25 \text{ ma}\underline{/4.6°}$$

We see that I_2, as computed from the equivalent circuit, checks with the computed value of I_2 for the actual circuit of Fig. 4-34*a*.

One unfortunate point about this equivalent circuit is that M may be greater than L_1 or L_2 when the coupling is tight. This means that one of the arms of the T will consist of a negative inductance. While this may not be physically realizable with passive elements, this equivalent circuit can still be used for analysis. The analysis is restricted to a single frequency, since a negative inductor is not quite the same as a capacitor, as X_L varies directly with frequency and X_C varies inversely. At a *single* frequency, however, a negative inductor can be replaced with a capacitor.

Transient waveforms contain many frequency components. This means that the equivalent circuit of Fig. 4-33*c* is not very convenient to work with when one of the arms is negative. We shall therefore investigate an equivalent circuit in which all the arms can consist of conventional positive inductances.

4-17. The Primary-side Equivalent Circuit. You will recall that for an inductively coupled circuit, such as in Fig. 4-35*a*, we showed that

$$Z_{in} = Z_{11} - \frac{Z_{12}^{\ 2}}{Z_{22}}$$

where $Z_{11} = R_1 + j\omega L_1$

$Z_{12} = \pm j\omega M$

$Z_{22} = R_2 + j\omega L_2 + Z_L$

Suppose Z_{12} were multiplied by an arbitrary number a, and Z_{22} by a^2. We would then have

$$Z_{in} = Z_{11} - \frac{(aZ_{12})^2}{a^2Z_{22}}$$

The reflected impedance part of Z_{in} has an a^2 in the numerator and an a^2 in the denominator, so that the actual value of Z_{in} is unchanged. This means that, looking into the primary terminals, the value of Z_{in} is unchanged, and the circuit looks no different as far as the generator is concerned.

Now let us look at conditions on the load side. The secondary current, you will recall, is

$$I_2 = -\frac{Z_{12}I_1}{Z_{22}}$$

We have just shown that operating with the number a in the described manner did not alter Z_{in}. Therefore, I_1 will surely remain the same. However, Z_{12} and Z_{22} were modified by a and a^2, respectively. Therefore, the new value of I_2, called I_2', is

$$I_2' = -\frac{aZ_{12}}{a^2Z_{22}}I_1 = -\frac{1}{a}\left(\frac{Z_{12}}{Z_{22}}I_1\right) = \frac{I_2}{a}$$

Since $(Z_{12}/Z_{22})I_1$ is the original value of I_2 before operating with a, we see that the new value of I_2 is equal to the original value divided by a.

We can now draw the circuit of Fig. 4-35*b*, which is equivalent to Fig. 4-35*a* *only* from the input or primary side. The value of Z_{11} remains unchanged. For Z_{12} to increase by the factor a, we must multiply M by a so that $aZ_{12} = j\omega(aM)$. Similarly, since

$$Z_{22} = R_2 + j\omega L_2 + Z_L$$
$$a^2Z_{22} = a^2(R_2 + j\omega L_2 + Z_L) = a^2R_2 + j\omega a^2L_2 + a^2Z_L$$

If Z_L should consist of, say, an R, an L, and a C, the R and L components would be multiplied by a^2, but C would now look like C/a^2, since

$$a^2X_C = a^2\frac{1}{j\omega C} = \frac{1}{j\omega C/a^2}$$

In Fig. 4-35*a*, the load voltage $E_2 = I_2Z_L$. In the modified circuit of Fig. 4-35*b*, the new load voltage (E_2') is $E_2' = (I_2/a)a^2Z_L = aI_2Z_L = aE_2$.

The reader may be wondering what point there is in generating a one-sided equivalent circuit to a transformer. Furthermore, we are no better off than before, since we still do not have an equivalent T that we can use for transient analysis. But we can take the one-sided equivalent circuit of Fig. 4-35*b* and convert it to an equivalent T. Then we can see if there is any choice in the arbitrary value of a, which will avoid negative inductances. If we succeed in avoiding negative inductances, the remaining step is to correct the secondary current and terminal voltage,

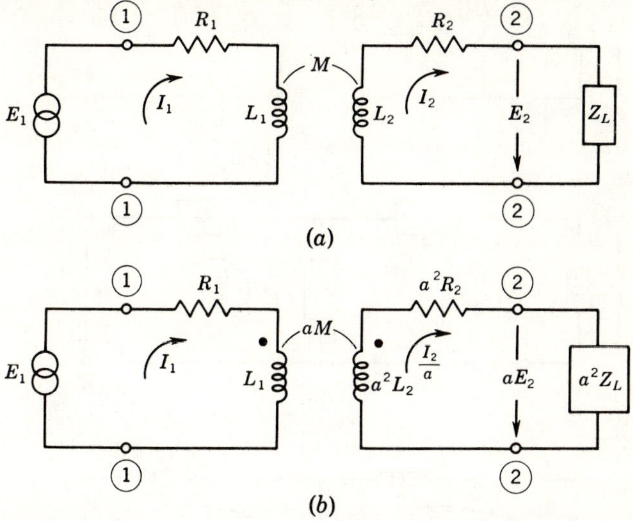

FIG. 4-35. (a) An inductively coupled circuit. (b) The primary side equivalent circuit.

since I_2 was divided by the factor a, and E_2 was multiplied by a. This correction can easily be made by connecting a perfect transformer between the actual load Z_L and the output terminals of the one-sided

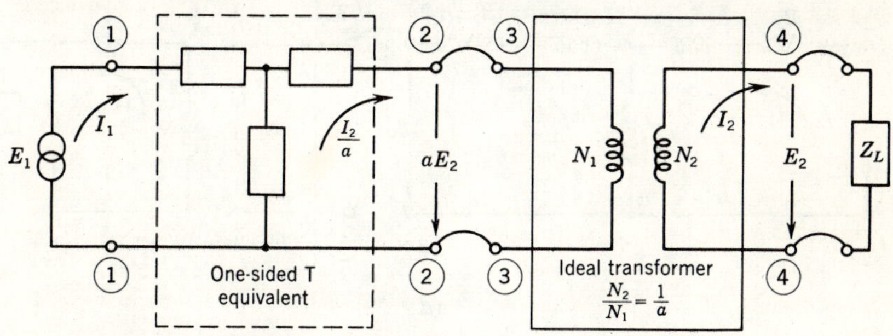

FIG. 4-36. Adding an ideal transformer to the 2-2 terminals converts the one-sided T equivalent to a fully equivalent circuit.

T equivalent, as in Fig. 4-36. If we choose the turns ratio of our ideal transformer so that $N_2/N_1 = 1/a$, the primary voltage aE_2 which appears at the 3-3 terminals will be transformed back to E_2 volts at the 4-4 terminals. Likewise, the current into the 3-3 terminals, which is I_2/a,

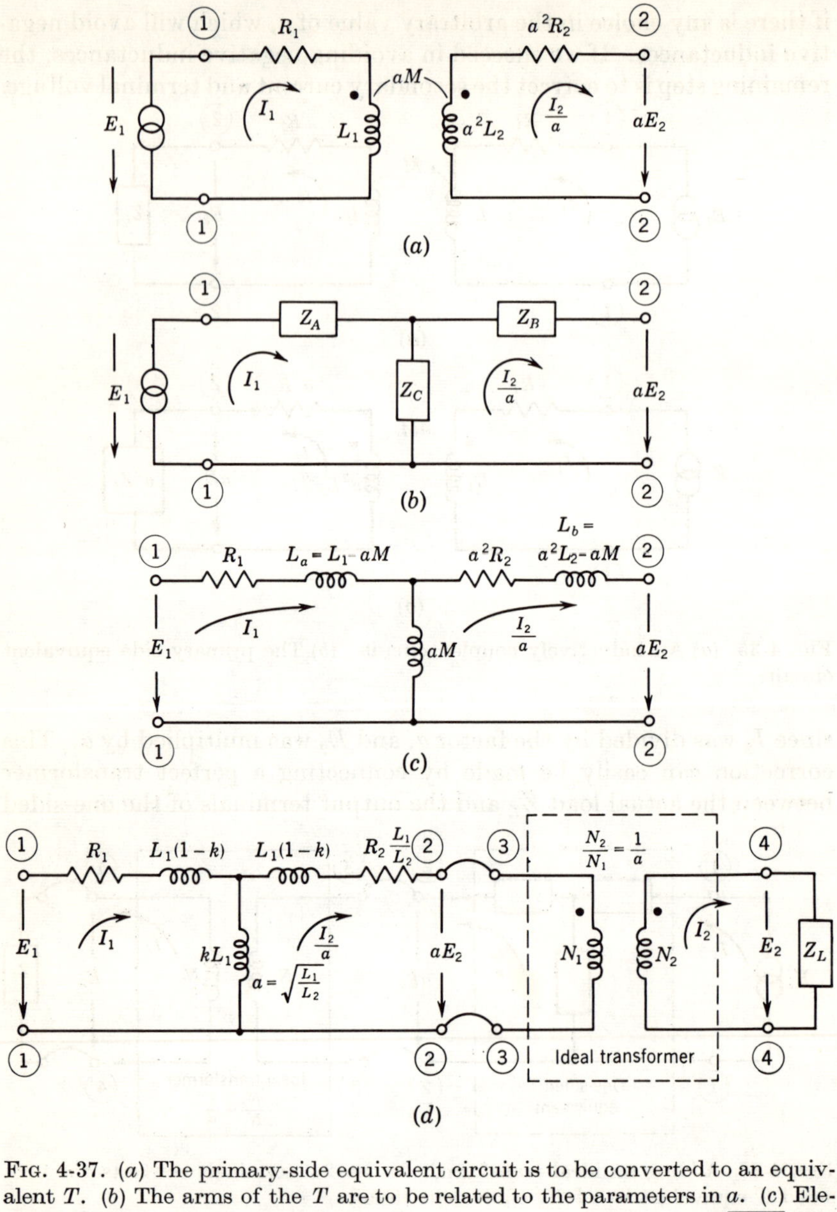

FIG. 4-37. (*a*) The primary-side equivalent circuit is to be converted to an equivalent T. (*b*) The arms of the T are to be related to the parameters in a. (*c*) Elements of the primary side equivalent T. (*d*) If we choose $a = \sqrt{L_1/L_2}$, the series inductances L_a and L_b will be equal and the parameters will have the indicated values.

will be transformed back to I_2 at the 4-4 terminals. We see then, that by the addition of an ideal transformer to the one-sided T equivalent, we have actually generated between the 1-1 and 4-4 terminals a true equivalent circuit. The ideal transformer introduces no frequency or phase distortion, is 100 per cent efficient and changes only the voltage, current, and impedance levels. The load Z_L at the 4-4 terminals actually appears at the 3-3 terminals as $Z_L/(N_2/N_1)^2 = a^2 Z_L$.

Now that we see what we are trying to do, let us again resume the problem of generating a one-sided T equivalent. The one-sided inductively coupled equivalent circuit of Fig. 4-35*b* is again repeated for convenience as Fig. 4-37*a*. The T equivalent is shown in Fig. 4-37*b*. Now let us see how Fig. 4-37*b* was obtained from Fig. 4-37*a*. Writing and comparing loop equations, we have, for Fig. 4-37*a*,

$$E_1 = I_1(R_1 + j\omega L_1) - \frac{I_2}{a} j\omega a M$$

$$-aE_2 = -I_1(j\omega a M) + \frac{I_2}{a}(a^2 R_2 + j\omega a^2 L_2)$$

For Fig. 4-37*b*,

$$E_1 = I_1(Z_A + Z_C) - \frac{I_2}{a} Z_C$$

$$-aE_2 = -I_1 Z_C + \frac{I_2}{a}(Z_B + Z_C)$$

Comparing coefficients, we see that

$$R_1 + j\omega L_1 = Z_A + Z_C$$

$$j\omega a M = Z_C$$

$$a^2 R_2 + j\omega a^2 L_2 = Z_B + Z_C$$

Therefore,

$$Z_A = R_1 + j\omega(L_1 - aM)$$

$$Z_B = a^2 R_2 + j\omega(a^2 L_2 - aM)$$

Our one-sided T-equivalent circuit is then as shown in Fig. 4-37*c*.

Inspection of Fig. 4-37*c* shows that if we arbitrarily choose $a = 1$, Fig. 4-37*c* reduces to the previously developed exact T-equivalent circuit of Fig. 4-33*c*. This, however, was the one that could produce negative inductances.

We could also choose a value for a which is such that the inductors in the series arms of Fig. 4-37*c* are equal. That is, we will let

$$L_a = L_b$$

$$L_1 - aM = a^2 L_2 - aM$$

and solve for the value of a which will cause this equality. If a is evaluated in this expression, we find that $a = \sqrt{L_1/L_2}$. Now let us see what $L_a = L_1 - aM$ is equal to when $a = \sqrt{L_1/L_2}$:

$$L_a = L_1 - aM$$

Substituting for a and M,

$$L_a = L_1 - \sqrt{\frac{L_1}{L_2}}\,(k\sqrt{L_1L_2}) = L_1 - k\sqrt{\frac{L_1}{L_2}\,L_1L_2}$$
$$= L_1 - kL_1 = L_1(1-k)$$

Since k is always less than 1, L_a will surely be a positive inductance. Similarly, L_b will be an equal positive inductance because we have chosen a so that $L_a = L_b$.

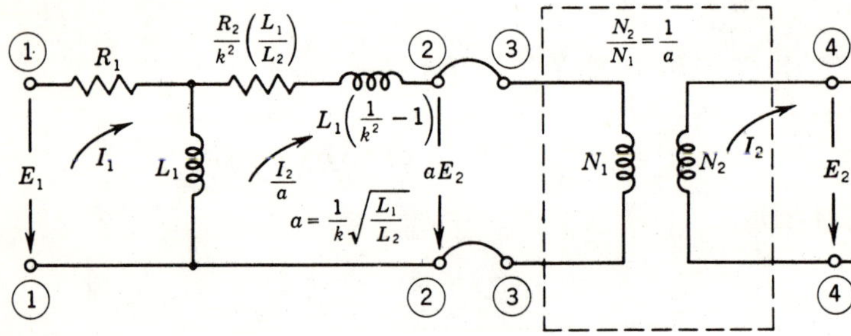

FIG. 4-38. If we choose $a = 1/K(\sqrt{L_1/L_2})$, the series inductance L_a is eliminated.

The shunt arm (aM) can also be expressed in terms of k:

$$aM = \sqrt{\frac{L_1}{L_2}}\,(k\sqrt{L_1L_2}) = kL_1$$

The complete equivalent circuit for $a = \sqrt{L_1/L_2}$ is shown in Fig. 4-37d. Again an ideal transformer has been added to generate a true equivalent circuit between the 1-1 and the 4-4 terminals.

We could devise another equivalent circuit based upon Fig. 4-37c by choosing a value for a such that the inductance $L_a = 0$. That is, set $L_1 - aM = 0$ and, therefore,

$$a = \frac{L_1}{M} = L_1/\,k\sqrt{L_1L_2} = \frac{1}{k}\sqrt{\frac{L_1}{L_2}}$$

This value of a will make L_b equal to

$$L_b = a^2L_2 - aM = \frac{L_1{}^2L_2}{M^2} - \frac{L_1M}{M}$$
$$= L_1\left(\frac{L_1L_2}{k^2L_1L_2} - 1\right) = L_1\left(\frac{1}{k^2} - 1\right)$$

Since k is less than 1, the quantity $1/k^2$ is surely greater than 1. This means that L_b will always be a positive inductor for $a = L_1/M = 1/k\sqrt{L_1/L_2}$. The shunt arm of the T equivalent is

$$aM = \frac{L_1}{M} M = L_1$$

The equivalent circuit for $a = L_1/M$ is then as shown in Fig. 4-38.

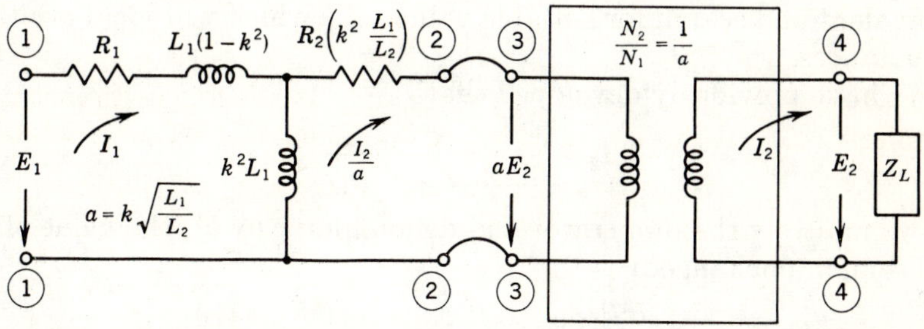

FIG. 4-39. If we choose $a = K\sqrt{L_1/L_2}$, the series inductance L_b is eliminated.

If we wish, we can find the value for a which will make $L_b = 0$:

$$a^2L_2 - aM = L_b = 0$$
$$a^2L^2 = aM$$
$$aL_2 = M$$
$$a = \frac{M}{L_2} = \frac{k\sqrt{L_1L_2}}{L_2} = k\sqrt{\frac{L_1}{L_2}}$$

For this choice of a, we have

$$La = L_1 - aM = L_1 - \frac{M}{L_2} M$$
$$= L_1 - \frac{k^2L_1L_2}{L_2} = L_1(1 - k^2)$$

which must be a positive inductor also. The shunt arm will be

$$aM = \frac{M}{L_2} M = k^2L$$

The equivalent circuit for $a = M/L_2 = k\sqrt{L_1/L_2}$ is then as shown in Fig. 4-39.

All these equivalent circuits may be used for either air- or iron-core transformers. In core-type transformers, however, where the windings are on a common core and k is close to unity, we can substitute N_2/N_1 for $\sqrt{L_2/L_1}$, or N_1/N_2 for $\sqrt{L_1/L_2}$. The same substitutions may also be made in the next group of equivalent circuits discussed.

4-18. The Secondary-side Equivalent Circuit. It is also possible to generate equivalent transformer circuits which are identical with an actual transformer only when looking into the secondary side. The technique employed is essentially the same as was used in generating circuits equivalent on the primary side. First we shall examine the expression for I_2. Then we shall introduce a scale factor b, in such a manner that the secondary current I_2 is unchanged. After that we can convert the inductively coupled circuit to a conductively coupled T equivalent and search for suitable values of b which will yield positive inductances.

We have previously developed that

$$I_2 = -\frac{EZ_{12}}{Z_{11}Z_{22} - Z_{12}^2}$$

If we multiply the numerator and denominator by b^2, the value of I_2 will remain unchanged:

$$I_2 = -\frac{b^2E_1Z_{12}}{b^2(Z_{11}Z_{22} - Z_{12}^2)} = -\frac{(bE_1)(bZ_{12})}{Z_{22}(b^2Z_{11}) - (bZ_{12})^2}$$

We see that I_2 is unchanged if:

1. The driving voltage E_1 is multiplied by b.
2. The mutual impedance is multiplied by b.
3. The primary loop impedance is multiplied by b^2.
4. The secondary loop impedance remains unchanged.

Now let us see what has happened to I_1 as a result of this:

$$\frac{I_2}{I_1} = -\frac{Z_{12}}{Z_{22}}$$

$$I_1 = -\frac{I_2Z_{22}}{Z_{12}}$$

Since I_2 and Z_{22} are unchanged, but Z_{12} is increased b times, it follows that I_1 will be decreased b times, or $I_1' = I_1/b$, where I_1' is the new value of I_1.

Figure 4-40a is the original transformer circuit, and Fig. 4-40b is the equivalent circuit, looking into the 4-4, or secondary, terminals. We can always add an ideal transformer on the primary side to convert bE_1 back to E_1, and I_1/b to I_1. The circuit is then a true equivalent of Fig. 4-40a.

We shall now convert Fig. 4-40b to its T equivalent, Fig. 4-40c. Writing and comparing loop equations, we have, for Fig. 4-40b,

$$bE_1 = \frac{I_1}{b}(b^2R_1 + b^2j\omega L_1) - I_2(j\omega bM)$$

$$-E_2 = -\frac{I_1}{b}j\omega bM + I_2(R_2 + j\omega L_2)$$

and for Fig. 4-40*c*,

$$bE_1 = \frac{I_1}{b}(Z_A + Z_C) - I_2 Z_C$$

$$-E_2 = -\frac{I_1}{b} Z_C + I_2(Z_B + Z_C)$$

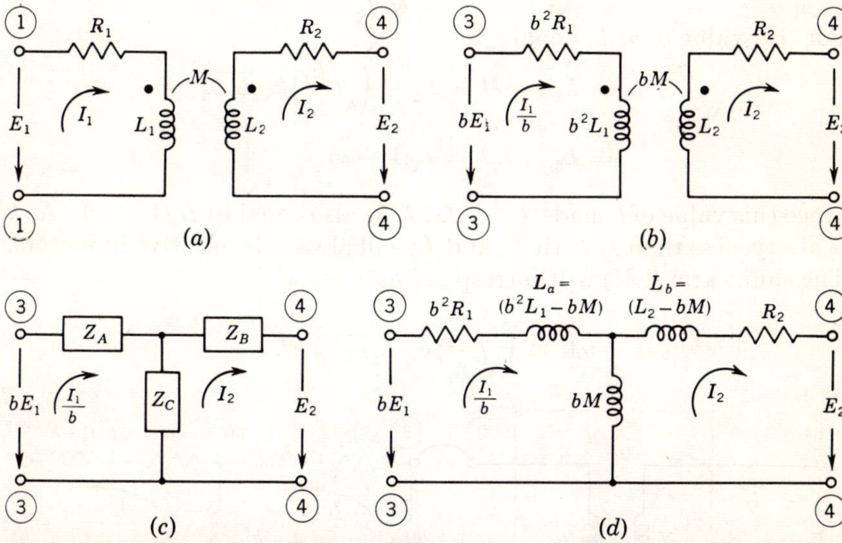

FIG. 4-40. (*a*) We wish to generate a secondary side equivalent circuit. (*b*) This circuit is equivalent to *a* on the secondary side only. (*c*) The *T* equivalent of *b* is to be generated. (*d*) The secondary-side *T* equivalent of *a*.

For Fig. 4-40*c* to be equivalent to Fig. 4-40*b*, the following relationships must hold true:

$$b^2R_1 + b^2 j\omega L_1 = Z_A + Z_C$$

$$j\omega bM = Z_C$$

$$R_2 + j\omega L_2 = Z_B + Z_C$$

Therefore,

$$Z_A = b^2R_1 + j\omega(b^2L_1 - bM)$$

$$Z_B = R_2 + j\omega(L_2 - bM)$$

$$Z_C = j\omega bM$$

These results can be summarized in Fig. 4-40*d*.

We can now experiment with different values of b. If we choose $b = 1$, the circuit of Fig. 4-40*d* reduces to that of Fig. 4-33*a*. This, however, may again lead to negative inductors when M is large. Suppose we try a value of b which is such that $L_a = L_b$. Then

$$b^2L_1 - bM = L_2 - bM$$
$$b^2L_1 = L_2$$
$$b = \sqrt{\frac{L_2}{L_1}}$$

For this value of b, L_b equals

$$L_b = L_2 - bM = L_2 - \left(\sqrt{\frac{L_2}{L_1}}\right)k\sqrt{L_1L_2}$$
$$= L_2 - L_2k = L_2(1 - k)$$

Since this value of b made $L_a = L_b$, L_a is also equal to $L_2(1 - k)$. As k is always less than 1, both L_a and L_b will always be positive inductors. The shunt arm (bM) will correspondingly equal

$$bM = \left(\sqrt{\frac{L_2}{L_1}}\right)k\sqrt{L_1L_2} = kL_2$$

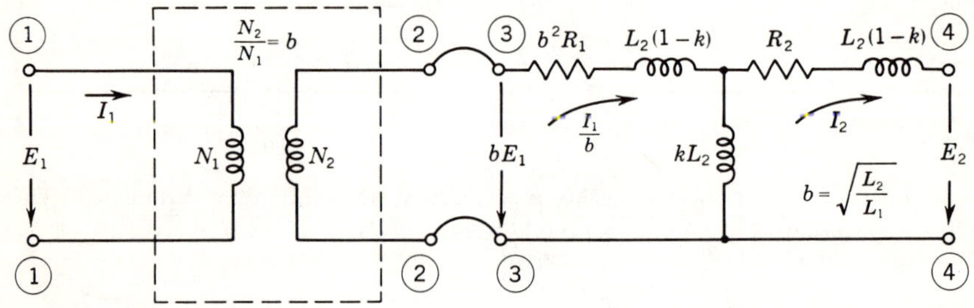

FIG. 4-41. The series inductances L_a and L_b are equal if we choose $b = \sqrt{L_2/L_1}$.

The complete T equivalent for $b = \sqrt{L_2/L_1}$ is shown in Fig. 4-41. It is understood that an ideal transformer is to be added to the left of the 3-3 terminals in order to make the one-sided T equivalent truly equivalent from either end. The turns ratio is $N_2/N_1 = b$ in the ideal transformer. This assures that the voltage bE_1 and the current I_1/b at the 3-3 terminals become E_1 and I_1 at the 1-1 terminals.

We could choose a value of b such that $L_a = 0$:

$$b^2L_1 - bM = 0$$
$$bL_1 = M$$
$$b = \frac{M}{L_1} = \frac{k\sqrt{L_1L_2}}{L_1} = k\sqrt{\frac{L_2}{L_1}}$$

For this value of b, L_b is equal to

$$L_b = L_2 - bM = L_2 - \left(k\sqrt{\frac{L_2}{L_1}}\right)(k\sqrt{L_1 L_2})$$

$$= L_2 - k^2 L_2 = L_2(1 - k^2)$$

The shunt arm is

$$bM = \frac{M}{L_1} M = \frac{k^2 L_1 L_2}{L_1} = k^2 L_2$$

The T equivalent for $b = k\sqrt{L_2/L_1}$ is then as shown in Fig. 4-42.

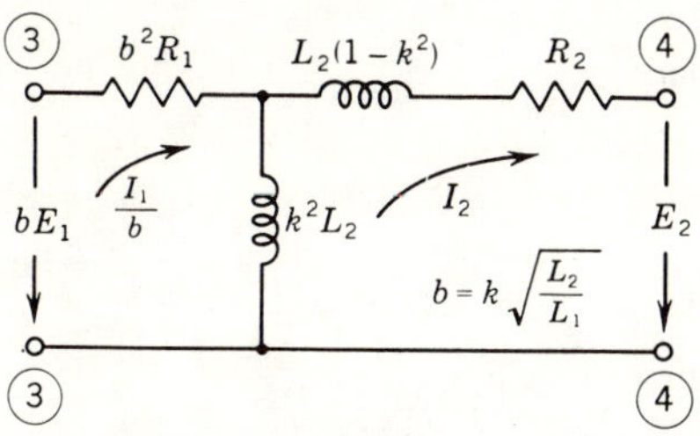

FIG. 4-42. Choosing $b = k\sqrt{L_2/L_1}$ eliminates the series inductance L_a.

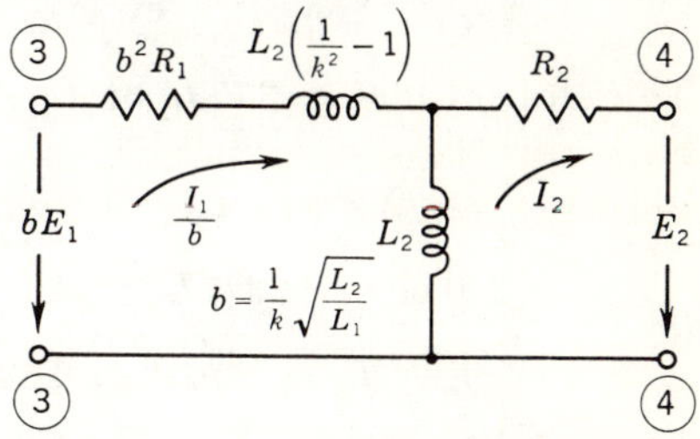

FIG. 4-43. Series inductance L_b will equal zero if we choose $b = (1/k)\sqrt{L_2/L_1}$.

Another choice of b is to make $L_b = 0$:

$$L_2 - bM = 0$$

$$b = \frac{L_2}{M} = \frac{L_2}{k\sqrt{L_1 L_2}} = \frac{1}{k}\sqrt{\frac{L_2}{L_1}}$$

For this choice of b, L_a is equal to

$$L_a = b^2 L_1 - bM$$

$$= \left(\frac{1}{k^2}\frac{L_2}{L_1}\right)L_1 - \left(\frac{1}{k}\sqrt{\frac{L_2}{L_1}}\right)(k\sqrt{L_1 L_2})$$

$$= \frac{L_2}{k^2} - L_2 = L_2\left(\frac{1}{k^2} - 1\right)$$

L_a in this case is also a positive inductor, since k is less than one, and k^2 is therefore less than k. Therefore, 1 divided by a number less than 1 is surely greater than 1.

The shunt arm (bM) is

$$bM = \frac{L_2}{M} M = L_2$$

The T-equivalent circuit for $b = 1/k\sqrt{L_2/L_1}$ is shown in Fig. 4-43.

An infinite number of equivalent circuits could be generated by choosing various values of a or b. The ones developed here will, however, suffice for our purposes.

Previously, we solved the circuit of Fig. 4-44a for I_2. Let us try to solve for I_2, using the one-sided equivalent circuit of Fig. 4-44b, which is equivalent on the secondary side only. We will use $b = k\sqrt{L_2/L_1}$. First, we shall evaluate the parameters L_1, L_2, and M from $X_{L_1} = 2\pi f L_1 = \omega L_1$:

$$\omega = 2\pi \frac{\text{radians}}{\text{cycle}} (60 \text{ cps}) = 377 \frac{\text{radians}}{\text{sec}}$$

Therefore, $1{,}000 = 377L_1$ or $L_1 = 1{,}000/377 = 2.65$ henrys.

$$X_{L_2} = \omega L_2$$

$$200 \text{ ohms} = 377L_2$$

$$L_2 = 200/377 = 0.530 \text{ henry}$$

$$X_M = \omega M$$

$$50 = 377M, \text{ or } M = 50/377 = 0.133 \text{ henry}$$

$$k = \frac{M}{\sqrt{L_1 L_2}} = \frac{0.133}{\sqrt{2.65 \times 0.53}}$$

$$= \frac{0.133}{\sqrt{1.40}} = \frac{0.133}{1.19} = 0.112$$

$$b = k\sqrt{\frac{L_2}{L_1}} = 0.112\sqrt{\frac{0.530}{2.65}} = 0.112\sqrt{0.2}$$

$$= 0.112(0.45) = 0.0504$$

$$b^2 = 0.00253 = 2.53(10^{-3})$$

$$b^2 R_1 = 2.53(10^{-3})5(10^2) = 1.265 \text{ ohms}$$

$$k^2 = 0.0124 = 1.24(10^{-2})$$

$$k^2 L_2 = 1.24(10^{-2})(0.530) = 0.657(10^{-2})$$

$$L_2(1 - k^2) = 0.530(0.988) = 0.524$$

$$bE_1 = 0.0504(100) = 5.04 \text{ volts}$$

$$j\omega k^2 L_2 = j377(0.657)10^{-2} = j2.48$$

$$j\omega L_2(1 - k^2) = j377(0.524) = j193$$

Solving for I_2 by thevenizing to the left of the x-x terminals in Fig. 4-44c, we have

$$V_{oc} = \frac{(5.04)2.48\underline{/90^\circ}}{127 + j2.48} = \frac{12.5\underline{/90^\circ}}{2.78\underline{/62.9^\circ}} = 4.5\underline{/27^\circ}$$

$$Z_{th} = \frac{(1.27\underline{/0^\circ})(2.48\underline{/90^\circ})}{1.27 + j2.48} = 1.13\underline{/27^\circ} = 1.01 + j0.513$$

$$I_2 = \frac{4.5\underline{/27^\circ}}{1.01 + j0.513 + 500 + j193} \approx \frac{4.5\underline{/27^\circ}}{501 + j194}$$

$$= \frac{4.5\underline{/27^\circ}}{549\underline{/21.1^\circ}} = 8.23\underline{/5.9^\circ}$$

The solution obtained with the one-sided equivalent circuit of Fig. 4-44b agrees closely with previously derived values for I_2.

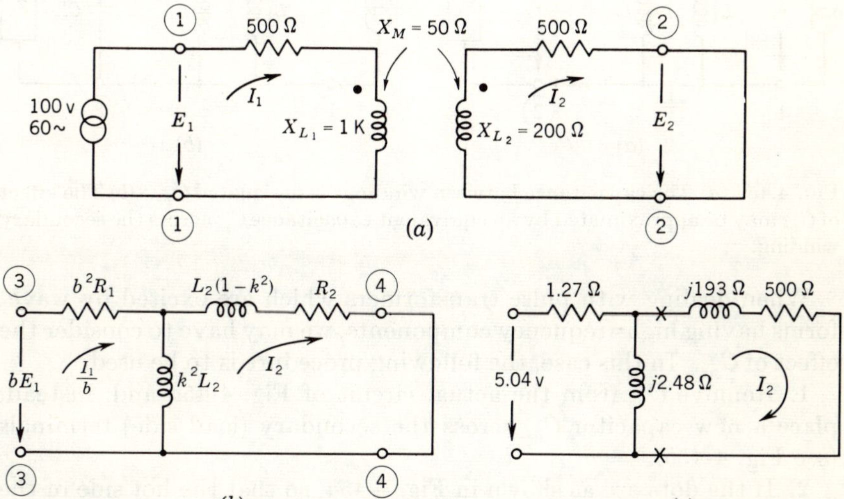

FIG. 4-44. Circuit a is to be solved with the aid of the T equivalent shown in b. (c) The current I_2 may easily be determined in this equivalent circuit.

There is no particular advantage in solving a single-frequency problem of this type with such an equivalent circuit. These one-sided equivalent circuits find their greatest utility in analyzing transformer circuits, where the driving waveform is nonsinusoidal and has a broad frequency spectrum.

4-19. Capacitance Effects in the Transformer. Up to now, we have been neglecting the effect of distributed capacitance within a transformer. The primary winding will have capacitance between the primary turns. The effect of this distributed capacitance can be approximated by considering a lumped capacitor (C_p) placed across the primary terminals. Similarly, the effect of secondary winding capacitance can be approximated with a capacitor (C_s) placed across the secondary.

We seldom have to contend with the capacitance between the primary and secondary windings (C_{ps}), since this capacitance is usually negligible in well-designed audio-transformers. Frequently, an electrostatic shield is placed between primary and secondary windings. This increases C_p and C_s somewhat, but renders C_{ps} negligible.

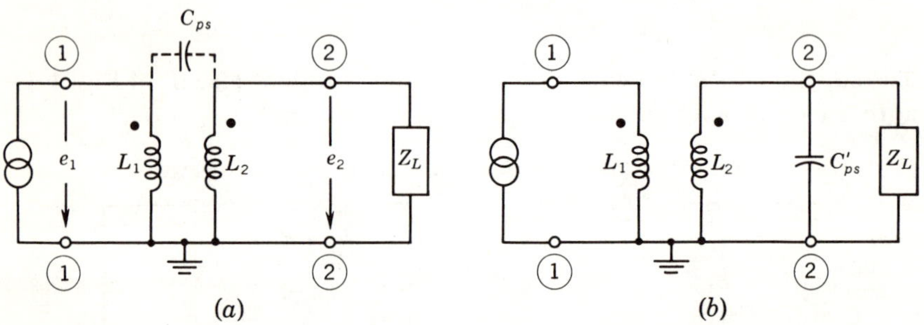

FIG. 4.45. (a) The capacitance between windings is designated C_{ps}. (b) The effect of C_{ps} may be approximated by an equivalent capacitance C'_{ps} across the secondary winding.

When dealing with pulse transformers which are excited by waveforms having high-frequency components, we may have to consider the effect of C_{ps}. In this case, the following procedure is to be used.

1. Remove C_{ps} from the actual circuit of Fig. 4-45*a* and, instead, place a new capacitor C'_{ps} across the secondary (load side) terminals (see Fig. 4-45*b*).

2. If the dots are as shown in Fig. 4-45*a*, so that the hot side of the secondary is following the hot side of the primary in polarity, C'_{ps} will be related to C_{ps} by

$$C'_{ps} = \left(\frac{e_2}{e_1} - 1\right)^2 \frac{C_{ps}}{3} \approx \left(\frac{N_2}{N_1} - 1\right)^2 \frac{C_{ps}}{3}$$

In a fairly good pulse transformer e_2/e_1 is essentially equal to the turns ratio N_2/N_1.

3. If either one of the dots in Fig. 4-45*a* is reversed, so that the hot side of the secondary is 180° out of phase with the hot side of the primary,

$$C'_{ps} = \left[\left(\frac{e_2}{e_1}\right)^2 - \frac{e_2}{e_1} + 1\right]\frac{C_{ps}}{3} \approx \left[\left(\frac{N_2}{N_1}\right)^2 - \frac{N_2}{N_1} + 1\right]\frac{C_{ps}}{3}$$

C'_{ps} appears larger when the voltages are out of phase because a larger potential difference is developed across C_{ps}. That is, when one side of C_{ps} is swinging plus and the other side minus, the magnitude of the voltage across C_{ps} is greater than when both sides are swinging plus. With the larger voltage, more charge is forced into C_{ps}. This is equivalent to having a voltage equal to the inphase case across C_{ps}, but with a greater capacitance, since $q = Ce$.

All these relationships are approximate but adequate for practical purposes. For an advanced theoretical discussion of pulse transformers, the reader is referred to Vol. 5 of the M.I.T. Radiation Laboratory Series, "Pulse Generators," by Glasoe and Lebacqz.[1] An exceptionally lucid treatment of pulse transformers with considerable design data is found in "Electronic Transformers and Circuits" by Reuben Lee.[2]

4-20. Transformer Transient Response. The transient analysis of multiple loop circuits containing more than one reactive element is very complex. A thorough background in calculus is required, and once again the reader is urged to master this subject and thereby enormously broaden his horizons. Recommended texts are:

J. E. Thompson, "Calculus for the Practical Man," D. Van Nostrand Company, Inc., Princeton, N.J., 1946.

A. E. Richmond, "Calculus for Electronics," McGraw-Hill Book Company, Inc., New York, 1958.

Henry M. Nodelman and Frederick W. Smith, "Mathematics for Electronics," McGraw-Hill Book Company, Inc., New York, 1956.

Essentially, there are two basic approaches, the Fourier technique and straightforward transient analysis. The Fourier technique breaks the existing waveform down into its harmonic components. The circuit response is then determined for each component frequency, using conventional a-c analysis methods. If the network is linear, the superposition principle applies, and all the individual responses are totaled to obtain the over-all response waveform. Straightforward transient analysis involves writing a number of loop equations containing not only the unknown quantities but terms involving their time rate of change. Both techniques are beyond the scope of this text, and therefore we must restrict ourselves to the simplest cases and make many approximations. The only driving voltages we shall consider are either steps or rectangular pulses. If the transformer will reproduce a square wave of a given PRF, it will reproduce any other waveform of the same PRF, which justifies the choice of driving waveform.

[1] G. N. Glasoe and J. V. Lebacqz, "Pulse Generators," McGraw-Hill Book Company, Inc., New York, 1948.

[2] Reuben Lee, "Electronic Transformers and Circuits," John Wiley & Sons, Inc., New York, 1955.

Some familiarity with transformer transient response will be gained by the detailed analysis of several illustrative problems. The reader should work his way through each step of every problem. It is in this way that a feeling for the analytic techniques involved is obtained.

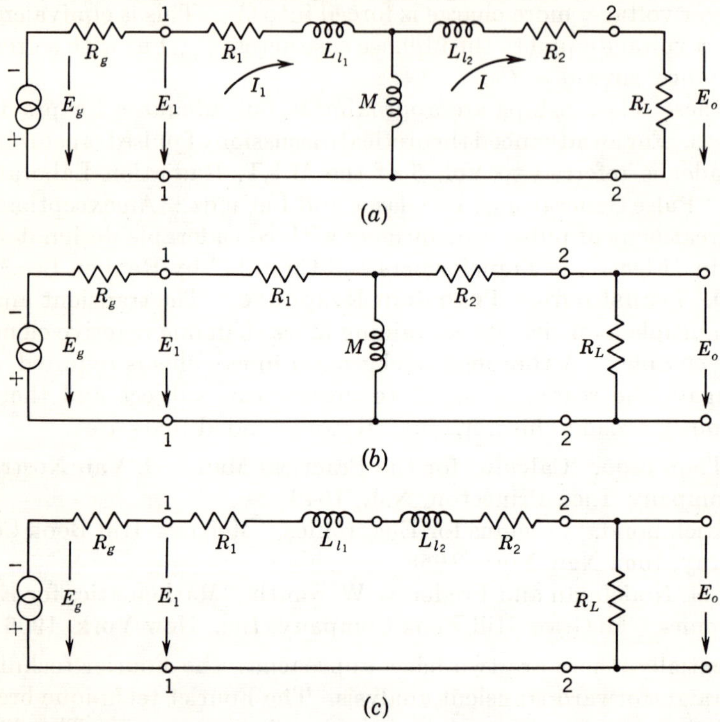

FIG. 4-46. (a) In a good pulse transformer the mutual inductance (M) is much larger than the leakage inductances (L_{l_1} and L_{l_2}). (b) The equivalent circuit at low frequencies. (c) The equivalent circuit at high frequencies.

In Sec. 4-5, we showed that the three transformer fluxes may be considered as being due to the currents in the three inductances of Fig. 4-7, which is repeated here as Fig. 4-46a. Resistors R_1 and R_2 are the winding resistances of the 1-1 and 2-2 windings. For simplicity, the transformer is considered as having a one-to-one turns ratio. If the turns ratio is not unity, then an ideal transformer having the required turns ratio may be inserted between the 2-2 terminals and the load (R_L). Recall that L_{l_1} is the leakage inductance of winding 1, corresponding to the flux about winding 1, which does not cut winding 2. Similarly L_{l_2}, the leakage inductance of winding 2, corresponds to the flux about winding 2, which does not cut winding 1. The mutual inductance M corresponds to the mutual flux which threads both windings and is responsible for coupling between the two windings.

Since a pulsed waveform contains many frequency components, let us see how the output in Fig. 4-46*a* varies with the frequency of E_g. At very low frequencies, all the inductive reactances will be small. A small value of X_{l_1} and X_{l_2} is conducive to keeping E_o from being attenuated severely. On the other hand, a small value of X_M tends to shunt the signal away from the load, which reduces E_o. For good low-frequency response, we want X_M as large as possible. At high frequencies, the situation is no different. To keep E_o from falling off in magnitude, the shunt branch X_M should have a high impedance, while the leakage reactances and winding resistances should be as low as possible. Apparently, the transformer must have large mutual inductance and small leakage inductances if it is to reproduce a broad frequency spectrum without excessive distortion.

To make X_M large at low frequencies means that the two windings should have large inductances (L_1 and L_2) and should be tightly coupled. This is plainly so, since $M = k\sqrt{L_1L_2}$. Unfortunately, the more turns that are placed on a winding, the greater the leakage flux becomes, and this is detrimental to the high-frequency output.

In a good audio- or pulse transformer, the leakage inductances are much less than the mutual inductance. Therefore, a low-frequency equivalent circuit for Fig. 4-46*a* might be that of 4-46*b*. Frequently, R_1 and R_2 may be neglected in comparison with R_g and R_L, which further simplifies the equivalent circuit. At high frequencies, X_M is presumably so large that it may be considered an open circuit. The high-frequency equivalent circuit is then as shown in Fig. 4-46*c*.

Now suppose E_g is a pulse rather than a sinusoidal generator. At the leading edge of the pulse with its high-frequency components, the equivalent circuit of Fig. 4-46*c* determines the response. The leakage inductances prevent the instantaneous buildup of current. Since E_o is proportional to the current, the output will build up just as fast as the current does. If the time constant in Fig. 4-46*c* is short relative to the time constant of Fig. 4-46*b*, we are justified in using the two equivalent circuits. The buildup of current through the mutual inductance determines the output response during the flat part of the input pulse. The leakage inductances affect E_o in the same manner that stray capacitance across R_L would. The mutual inductance affects E_o in the same manner that a coupling capacitor does in an *RC*-coupled circuit. Fig. 4-46*a* is a general description of the transformer equivalent circuit. Specifically, any of the one-sided equivalent circuits of Secs. 4-17 or 4-18 could be used. Any inductances in the series arms are leakage inductances, while the shunt arm is the mutual inductance.

Let us try some illustrative examples to get some feeling for the thinking involved.

Problem 1. Sketch e_o versus time after the switch is closed in Fig. 4-47*a*. $e = -10$ volts; $R_1 = 2$ kilohms, $L_1 = 1$ henry, $L_2 = 2$ henrys, and $M = 0.5$ henry.

Solution. Since the secondary winding (L_2) is open-circuited, there will be no secondary current to react back into the primary loop. The only current to be considered is therefore i_1. The open-circuit output voltage (e_o) is related to i_1 by

$$e_o = M \frac{di_1}{dt}$$

Our problem is then to evaluate di_1/dt, where di_1/dt is the instantaneous rate of change of i_1 with respect to time. With the secondary open-circuited, the primary loop consists of a simple RL circuit. At the instant of closure, i_1 will be zero and then gradually build up exponentially to its final value of e/R_1. The output will jump to its peak positive value when the switch closes, since the primary current is changing most rapidly at this time. It will be positive because the dot end of L_1 is driven negative upon closure, and the dot end of L_2 follows in phase. After i_1 reaches its final value, di_1/dt must equal zero, and therefore e_o equals zero.

Now for a detailed solution. The primary circuit time constant is $L_1/R_1 = 0.5(10^{-3})$ sec $= 500\ \mu$sec. The maximum primary current is

$$I_M = \frac{10 \text{ volts}}{2 \text{ kilohms}} = 5 \text{ ma}$$

In one time constant, or 500 μsec,

$$i_1 = 63 \text{ per cent of } 5 \text{ ma} = 3.14 \text{ ma}$$

In two time constants (1,000 μsec) the current is

$$87 \text{ per cent of } 5 \text{ ma} = 4.35 \text{ ma}$$

In 1,500 μsec and 2,000 μsec, we have

$$i_1 = 4.75 \text{ and } 4.9 \text{ ma}$$

In 2,500 μsec, the steady-state value of 5 ma may be assumed. We may now sketch i_1 as a function of time, as shown in Fig. 4-47*b*.

The instantaneous rate of change of i_1 with respect to time is the slope of the current-vs.-time curve. The slope at different points may be determined by constructing straight-line tangents to the point in question. In Fig. 4-47*b*, tangents were constructed at $t = 0$, 500, and 1,000 μsec. At t_0, the tangent line seems to rise at a rate of 7 ma in 500 μsec. The voltage induced in the secondary winding is therefore,

$$e_o = M \frac{di}{dt} = 0.5 \text{ henry} \left(\frac{7 \text{ ma}}{500\ \mu\text{sec}} \right) = \frac{0.5(7)(10^{-3})}{5(10^{-4})} = 7 \text{ volts}$$

At 500 μsec, the slope (t_1) is 3.1 ma in 1,000 μsec. Therefore,

$$e_o = 0.5 \text{ henry} \left(\frac{3.1 \text{ ma}}{1{,}000\ \mu\text{sec}} \right) = 1.5 \text{ volts}$$

At 1,000 μsec, the slope is 4.34 ma in 3,000 μsec. Therefore,

$$e_o = 0.5 \text{ henry} \left(\frac{4.34 \text{ ma}}{3{,}000\ \mu\text{sec}} \right) = 0.72 \text{ volt}$$

These points are plotted as the solid curve in Fig. 4-47c, which gives us some picture of the output. We shall soon see that there is quite a bit of error involved in this method. It does, however, put us in the ball park and indicates a general method of deriving an output waveform where the output is proportional to the input rate of change.

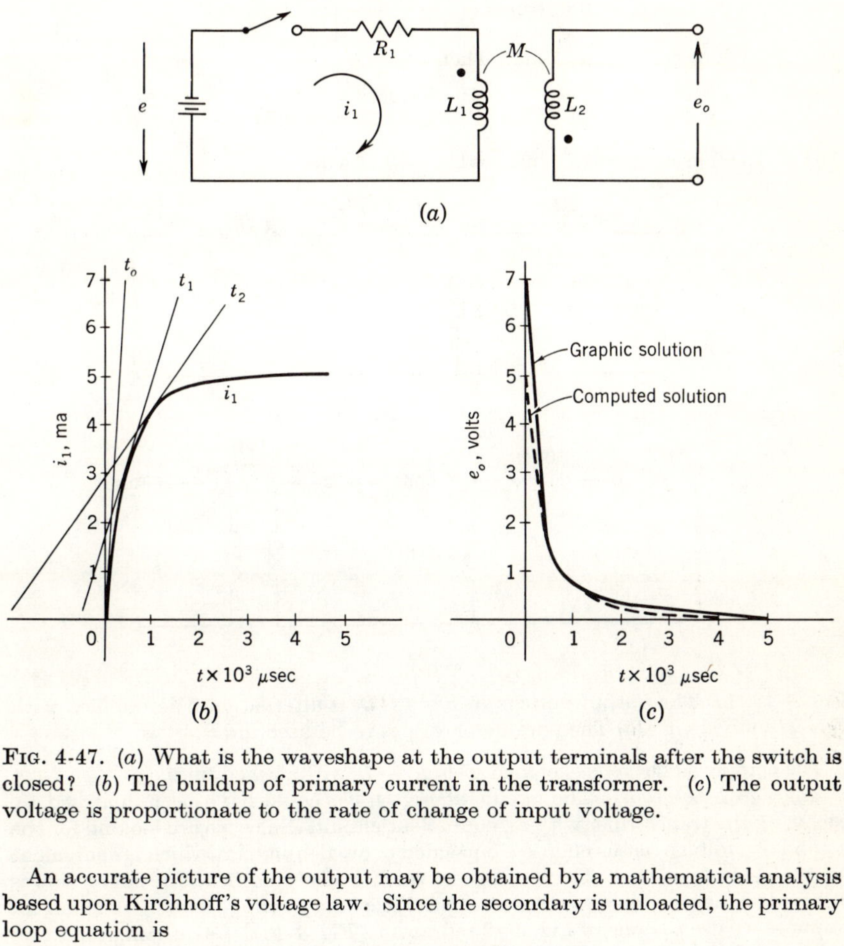

FIG. 4-47. (a) What is the waveshape at the output terminals after the switch is closed? (b) The buildup of primary current in the transformer. (c) The output voltage is proportionate to the rate of change of input voltage.

An accurate picture of the output may be obtained by a mathematical analysis based upon Kirchhoff's voltage law. Since the secondary is unloaded, the primary loop equation is

$$e = i_1R_1 + L_1\frac{di_1}{dt}$$

Solving for di_1/dt,

$$\frac{di_1}{dt} = \frac{e - i_1R_1}{L_1}$$

Normally, we could not solve this differential equation without some background in calculus. Owing to the simplicity of the circuit, however, we were able to evaluate i_1 as a function of time, as shown in Fig. 4-47b. Therefore, di_1/dt and then e_o may be accurately determined. For example, at t_0, $i_1 = 0$. Then

$$e_o = M\frac{di_1}{dt} = M\frac{e - i_1R_1}{L} = 0.5\text{ henry}\left(\frac{10 - 0}{1\text{ henry}}\right) = 5.0\text{ volts}$$

which is a good deal less than 7 volts. At $t = 500$ μsec, $i_1 = 3.14$ ma. Then,

$$e_o = 0.5\text{ henry}\left[\frac{10 - 3.14(2)}{1\text{ henry}}\right] = 1.8\text{ volts}$$

At $t = 1{,}000$ μsec, $i_1 = 4.35$ ma. Then

$$e_o = 0.5\text{ henry}\left[\frac{10 - 4.35(2)}{1\text{ henry}}\right] = 0.65\text{ volt}$$

At $t = 1{,}500$ μsec, $i_1 = 4.75$ ma, and $e_o = 0.25$ volt.

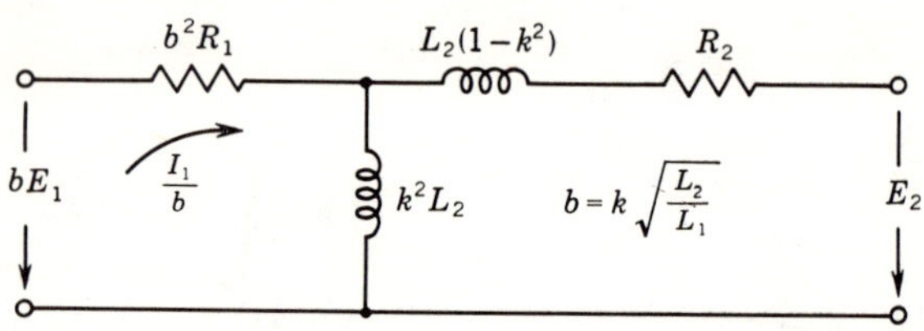

(a)

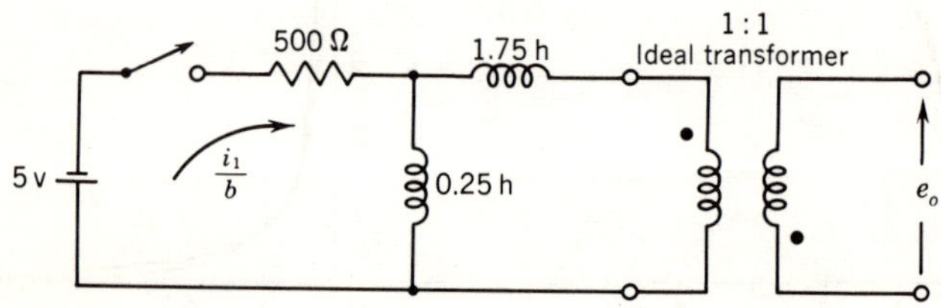

(b)

FIG. 4-48. (a) The output voltage in Fig. 4-47a is alternatively determined with this T equivalent. (b) The parameter values to be used in (a).

The computed curve is sketched in Fig. 4-47c with dashed lines.

Alternative Solution. The circuit of Fig. 4-47a could have been analyzed by replacing the transformer with an equivalent circuit. Since we are looking for the secondary voltage, let us choose a one-sided equivalent circuit which is equivalent on the secondary side. That is one in which everything in the primary side is referred to the secondary, so that the actual secondary voltage remains unchanged. Suppose we use the one of Fig. 4-42 redrawn as Fig. 4-48a. Calculating the exact values of the parameters for the equivalent circuit, we have

$$b = k\sqrt{\frac{L_2}{L_1}} = \frac{M}{\sqrt{L_1L_2}}\sqrt{\frac{L_2}{L_1}} = \frac{M}{L_1} = \frac{0.5\text{ henry}}{1\text{ henry}} = 0.5$$

$$bE_1 = 0.5(10) = 5\text{ volts}$$

$$b^2R_1 = 0.25(2{,}000) = 500\text{ ohms}$$

$$k^2L_2 = bM = 0.5(0.5) = 0.25\text{ henry}$$

$$L_2(1 - k^2) = L_2 - bM = 2 - 0.25 = 1.75\text{ henrys}$$

$$R_2 = 0$$

The equivalent circuit with the actual parameter values is shown in Fig. 4-48*b*. The 1 : 1 ideal transformer on the right side of the figure was added to phase the output properly, since the simple one-sided T does not allow for the dot markings in Fig. 4-47*a*.

Analysis of Fig. 4-48*b* indicates that at the instant the switch closes, e_o will jump to 5 volts. This is concluded in the following manner. The open-circuited secondary of the 1 : 1 ideal transformer will reflect an infinite impedance in series with the 1.75-henry inductor, and therefore, negligible current will flow in this loop. The drop across the 1.75-henry inductor is therefore negligible, and the voltage impressed across the input terminals of the ideal transformer will be the same as that across the 0.25-henry coil. At the instant the switch closes, the current through the 0.25-henry coil will be zero, and the voltage across it 5 volts. The dot end of the input winding on the ideal transformer is therefore driven 5 volts negative. The output winding dot follows in phase, which means that e_o will be a positive 5 volts in the arrow direction. The time constant for the buildup of current through the 0.25-henry inductor is L/R = 0.25 henry/0.5 kilohm = 500 μsec, which is the same as before.

Problem 2. Suppose we wished to determine the manner in which I_1 varies with time in the circuit of Fig. 4-49*a*, if the switch is closed at t_0.

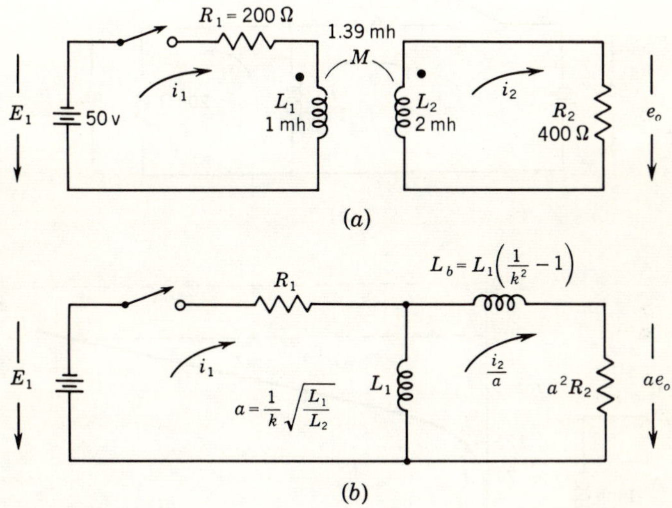

FIG. 4-49. (*a*) How does i_1 vary with time after the switch closes? (*b*) T equivalent.

Solution. To facilitate the analysis of i_1, let us use the primary-side equivalent circuit of Fig. 4-38, which is repeated as Fig. 4-49*b* for convenience and with some modification. Since our only interest at the moment is in i_1, there is no need to draw the ideal transformer to correct the secondary current and voltage. We can consider the output terminals of the ideal transformer as being connected to R_2 (see Fig. 4-38), which means that R_2 will appear across the 2-2 terminals as a^2R_2.

This is shown in Fig. 4-49*b*. The actual circuit parameters in Fig. 4-49*b* will be

$$R_1 = 200 \text{ ohms}$$
$$L_1 = 1 \text{ mh}$$
$$a = \frac{1}{k}\sqrt{\frac{L_1}{L_2}} = \frac{\sqrt{L_1 L_2}}{M}\sqrt{\frac{L_1}{L_2}} = \frac{L_1}{M} = \frac{1 \text{ mh}}{1.39 \text{ mh}} = 0.72$$
$$a^2 = 0.52$$
$$L_b = L_1\left(\frac{1}{k^2} - 1\right) = a^2 L_2 - aM = 0.52(2) - 0.72(1.39) = 0.04 \text{ mh}$$
$$a^2 R_2 = 0.52(400) = 208 \text{ ohms}$$

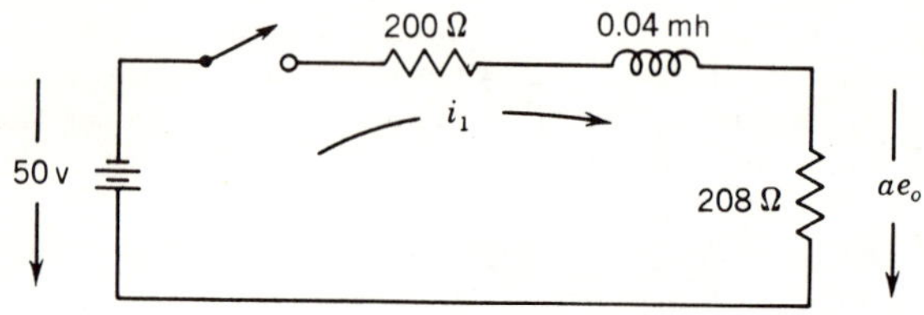

(*a*)

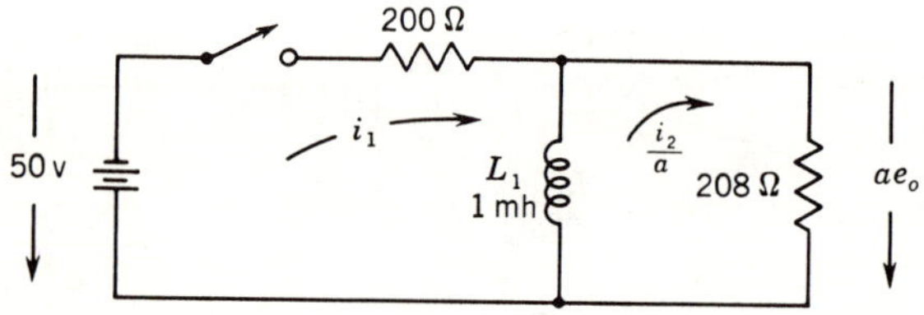

(*b*)

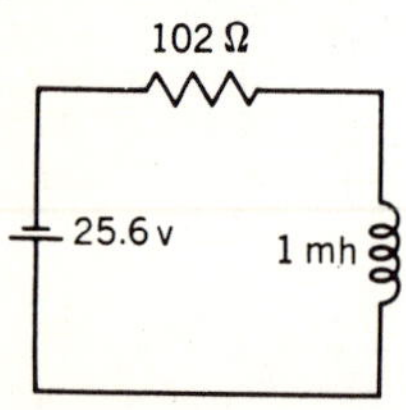

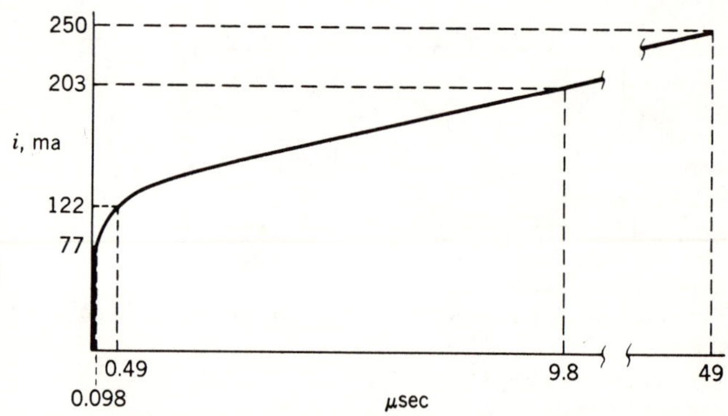

(*c*) (*d*)

FIG. 4-50. (*a*) The pulse buildup time of Fig. 4-49*a* is determined from this equivalent circuit. (*b*) The tilt during the flat portion of the input is determined from this equivalent circuit. (*c*) The current through the shunt inductance (L_1) may be determined from the Thévenin's equivalent circuit that it looks into. (*d*) Primary current following closure of the switch.

At the instant the switch is closed (t_0), the shunt inductance L_1, sometimes called magnetizing inductance, is so much greater than the leakage inductance L_b that the shunt arm looks like an open circuit. The equivalent circuit for the buildup of i_1 is therefore that of Fig. 4-50*a*. The time constant for the equivalent circuit at t_0^+ is therefore L/R = 0.04 mh/408 ohms = 0.098 μsec. The final value of i_1 in the equivalent circuit of Fig. 4-50*a* is 50 volts/408 ohms = 122 ma. Therefore, in one time constant, or 0.098 μsec, i_1 will equal 0.63(122 ma) = 77 ma. In 5 time constants or 0.49 μsec, i_1 will have reached its steady-state value of 122 ma. Actually i_1 will be slightly more, owing to the buildup of current in the shunt arm L_1. It is assumed to be negligible, however, since L_1 is so much larger that L_b.

As time goes by, the current through the shunt arm builds up, causing an additional increase in i_1, but at a slower rate because of the longer time constant. The flat-top pulse response is essentially determined by the equivalent circuit of Fig. 4-50*b*. Thevenizing, looking back from the inductor in Fig. 4-50*b*, we obtain Fig. 4-50*c*. The final (steady-state) current through L_1 will be 25.6 volts/102 ohms = 250 ma. This is also the maximum possible change in current through L_1, since the initial current in L_1 was zero. The time constant is L/R = 1 mh/102 ohms = 9.8 μsec. Therefore, in 5 time constants (49 μsec) all of i_1 will flow through L_1, and i_1 will equal 250 ma. In one time constant (9.8 μsec), the inductor current will be 0.63(250 ma) = 157 ma. With 157 ma flowing through the 102-ohm resistor in Fig. 4-50*c* there will be 25.6 − (157 ma)(102 ohms) = 9.6 volts across the inductor. If there are 9.6 volts across the inductor in 9.8 μsec, there will also be 9.6 volts across the 208-ohm resistor in Fig. 4-50*b*; and the resistor current is therefore 9.6 volts/208 ohms = 46.1 ma. The current i_1 at this time is the sum of the currents in L_1 and the 208-ohm resistor.

A plot of i_1 versus time is shown in Fig. 4-50*d*. Since the current buildup time in Fig. 4-50*a* is so small, it was neglected when the equivalent circuit of Fig. 4-50*b* was used.

Problem 3. Determine the voltage waveform across R_2 in Fig. 4–49*a*, if the switch is closed for 100 μsec and then reopened.

Solution. From the equivalent circuit of Fig. 4-49*b* in the previous problem, we see that the secondary loop current i_2/a must be zero at t_0 because of leakage inductance L_b. Therefore, e_o must be zero at t_0. In 0.098 μsec we found i_1 in Fig. 4-50*a* to be 77 ma. Therefore, i_1, which equals i_2/a in this figure, develops 77 ma (208 ohms) = 16 volts for ae_o. Note, this is ae_o, and not e_o. Since $a = 0.72$, the actual value of e_o must be ae_o/a = 16 volts/0.72 = 22.2 volts. The same result would be obtained if we replaced the 208-ohm resistor with an ideal transformer having a turns ratio of $N_2/N_1 = 1/a$ and then placed a 400-ohm resistor across the secondary. At t = 0.49 μsec, i_1 was determined as 122 ma in Fig. 4-50*a*. Therefore, the voltage across the 208-ohm resistor is ae_o = 122 ma (208 ohms) = 25.4 volts, and $e_o = {}^{25.4}/_{0.72}$ = 35.2 volts.

At the end of 0.49 μsec, we switched to the equivalent circuit of Fig. 4-50*b*. The voltage across L_1 at 9.8 μsec was previously determined to be 9.6 volts, and this is ae_o. Therefore, e_o at this time is 9.6 volts/0.72 = 13.3 volts. At 49 μsec, the voltage across L_1 was zero, and therefore $e_o = 0$.

A plot of e_o versus time is shown in Fig. 4-51. An interesting check upon the peak value of the e_o waveform is based upon the fact that voltages are transformed according to the turns ratio in an ideal transformer. Since the leakage inductance was very small in Fig. 4-50*a*, we could consider the transformer as being approximately ideal. The fact that $k = M/\sqrt{L_1L_2} = 1.39/\sqrt{(1\text{ mh})(2\text{ mh})} = 0.98$ substantiates that the windings are tightly coupled. Furthermore, Fig. 4-49*a* does not indicate any winding resistances. The turns ratio N_2/N_1 is

$$\frac{N_2}{N_1} = \sqrt{\frac{L_2}{L_1}} = \frac{2\text{ mh}}{1\text{ mh}} = 2$$

The 400-ohm load (R_2) will be reflected into the primary as $400/(N_2/N_1)^2 = 100$ ohms. This is the impedance seen looking into the primary terminals as shown in

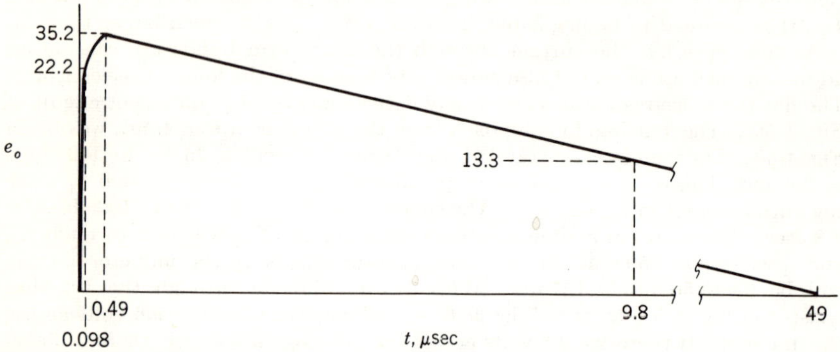

FIG. 4-51. The output voltage response when the switch is closed in Fig. 4-49*a*.

Fig. 4-52. The voltage across the 1-1, or primary, terminals at the instant the switch is closed is then

$$E_{1\text{-}1} = 50 \times \frac{100}{100 + 200} = 16.7 \text{ volts}$$

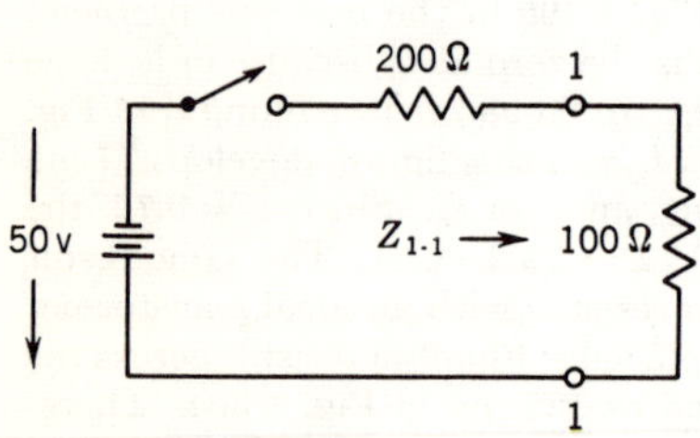

FIG. 4-52. If the transformer in Fig. 4-49*a* were an ideal 2 : 1 transformer, this is the equivalent circuit that the battery would see.

Since the turns ratio is 2 : 1, the peak secondary output for this ideal transformer would be 2(16.7 volts), or 33.4 volts, as compared with 35.2 volts for the nonideal transformer.

Problem 4. Assume the switch in Fig. 4-49*a* has been closed for a long time and then suddenly opened at t_0^+. Determine the output voltage (e_o) waveform.

Solution. Here we must proceed with caution, since we are opening an inductive load on the switch. Looking at Fig. 4-49*a*, we might be tempted to say that an arc will appear across the switch contacts because of the collapsing magnetic field about L_1. A much more informative interpretation of what happens may be gleaned from Fig. 4-49*b*, repeated here as Fig. 4-53*a*.

With the switch closed, we previously showed that the steady-state value of i_1 was $E_1/R_1 = 250$ ma, and $i_2 = 0$. If the switch suddenly opened, the current in L_1 would tend to remain constant, and L_1 would therefore look like a generator which tries to keep i_1 flowing in the same direction. This is shown in Fig. 4-53*b*, except that i_1 is now i_2/a. At this point, we are faced with a serious dilemma. In

Chap. 3, we stated that the current through an inductor cannot change instantaneously. Then how can i_2/a at t_0^+ be equal to 250 ma to satisfy this requirement in L_b? In Chap. 3 we restricted ourselves to problems involving a single inductor of constant value. Here we have a two-inductor problem in which L_b is effectively switched in series with L_1.

To solve this problem, we shall make use of the theorem of constant flux linkages, which states, "The sum of the flux linkages about a closed loop cannot change instantaneously." A flux linkage is one line of flux linking one turn of wire. The total flux linkages (ψ) about a coil is, then, the summation of the turns times the flux linking these turns. The total flux linkages can more conveniently be expressed in terms of the current through the coil and its inductance:

$$\psi = Li$$

As an illustration of this theorem, consider this problem. At t_0, the switch across L_2 in Fig. 4-53c is suddenly opened. How does the current behave? Prior to the switch opening, the current in L_1 is $I_C = E/R$, while that in L_2 is zero. The flux linkages about the loop are therefore

$$\psi_C = L_1 I_C = L_1 \frac{E}{R}$$

When the switch opens, the instantaneous flux linkages about the loop must be

$$\psi_O = L_1 I_O + L_2 I_O$$

where I_O is the current immediately after the switch opens. According to the theorem of constant flux linkages, we can say that instantaneously

$$\psi_C = \psi_O$$

$$L_1 \frac{E}{R} = (L_1 + L_2) I_O$$

and therefore

$$I_O = \frac{L_1}{L_1 + L_2} \frac{E}{R}$$

If we assume that the switch opens instantaneously, I_C must have jumped abruptly to its new value I_O.

The discerning reader might well ask, "Granting that the current can jump instantaneously, would not infinite voltages be produced in the circuit, since the coil voltage is $e_L = L\,di/dt$? A jump in current surely implies an infinite di/dt." This reasoning is quite correct, and high voltages may be produced. They will not, however, be infinitely high, because the arc which forms across the switch contacts, owing to the high self-induced voltages, lasts for a small but nevertheless finite time. It is during this time that the current through L_1 decreases and the current through L_2 increases to the new value I_O. The jump in current indicated by the pure mathematics of the theorem of constant flux linkages is not truly instantaneous in a physical system. This does not mean that the theorem is useless. Even though we cannot exactly specify the behavior of the current during the small jump interval, we can specify it right afterward, which is what the theorem really does tell us. Furthermore, if the switch opens rapidly, so that the arc is quickly extinguished relative to the recovery time constant of the circuit, the current change may essentially be considered as a jump. For example, assume in Fig. 4-53c that $E = 10$ volts, $R = 10$ ohms, $L_1 = 1$ henry, $L_2 = 2$ henrys, and that the switch, arc included, opens in 10 μsec. Sketch the current waveform.

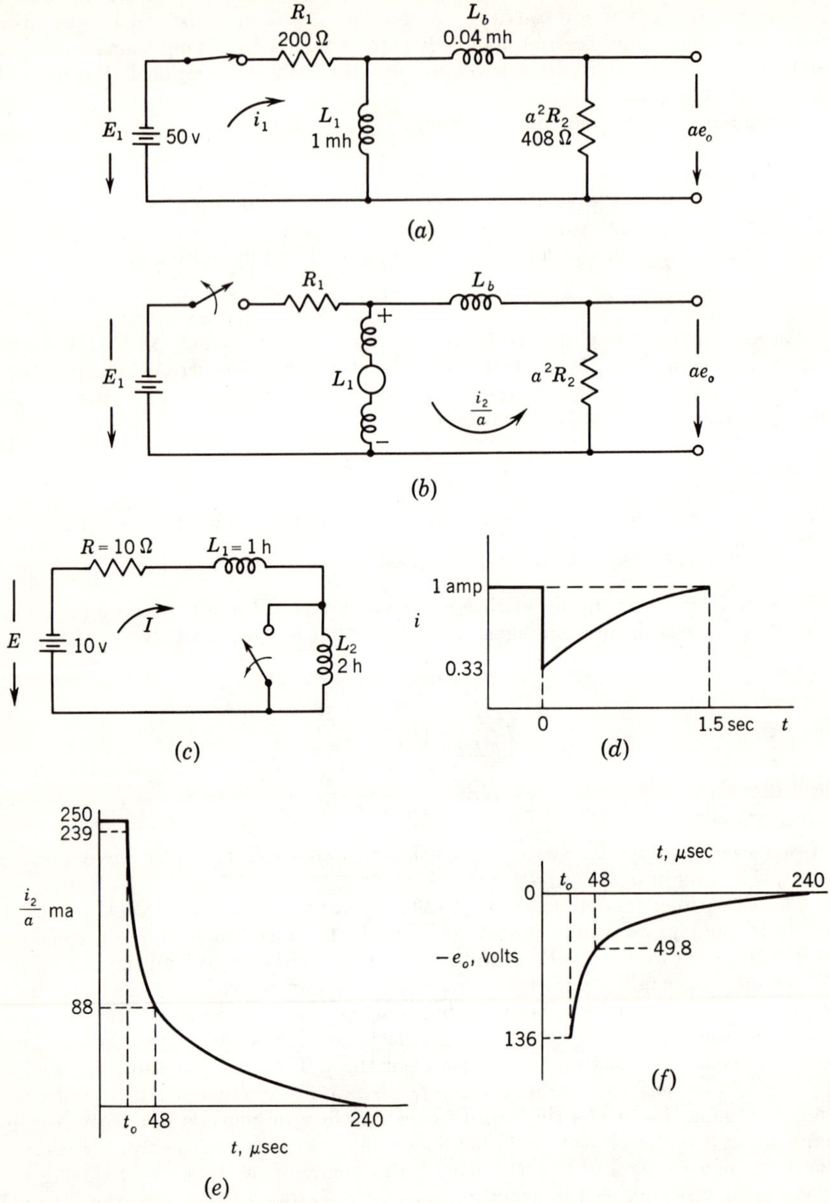

FIG. 4-53. (*a*) Prior to opening the switch the current in L_1 is i_1 and in L_b it is zero. (*b*) When the switch is suddenly opened, i_1 becomes identical with i_2/a. (*c*) The theorem of constant flux linkages enables us to analyze this circuit. (*d*) The current waveform when the switch in *c* opens. (*e*) The current waveform after the switch opens in *a*. (*f*) The output voltage waveform for 3*a*.

With the switch closed, $I_C = 10$ volts/10 ohms $= 1$ amp. The instant the switch fully opens, we have

$$I_O = \frac{L_1}{L_1 + L_2}\frac{E}{R} = \frac{10\text{ volts}}{30\text{ ohms}} = 0.33\text{ amp}$$

The final value of current with the switch open is again $E/R = 1$ amp. The recovery time of the system is now, however, dependent upon the time constant $(L_1 + L_2)/R$, or 300,000 μsec. The current waveform would therefore appear as shown in Fig. 4-53*d*.

Returning to Prob. 4 and Fig. 4-53*a* and *b* with our new assortment of weapons, we may reason as follows. Before the switch opens, the only flux linkages present are those about L_1. Therefore,

$$\psi_C = i_1 L_1 = \frac{50\text{ volts}}{200\text{ ohms}}(1\text{ mh}) = 0.25(10^{-3})\text{ amp-henry}$$

At this time, ae_o equals zero. The moment the switch opens (arc extinguished), the flux linkages about the loop containing L_1 and L_b equal

$$\psi_O = \frac{i_2}{a}(L_1 + L_b) = \frac{i_2}{a}(1.04\text{ mh})$$

Since, instantaneously,

$$\psi_O = \psi_C$$

$$\frac{i_2}{a} = \frac{0.25(10^{-3})\text{ amp-henry}}{1.04(10^{-3})\text{ henry}} = 0.239\text{ amp}$$

The corresponding value of e_o is

$$ae_o = -\frac{i_2}{a}a^2R_L = -0.239(408) = -97.5\text{ volts}$$

Therefore

$$e_o = \frac{ae_o}{a} = \frac{-97.5}{0.72} = -136$$

The current waveform will have the shape indicated in Fig. 4-53*e*. With the switch open, the time constant of the system is

$$\frac{L}{R} = \frac{L_1 + L_b}{a^2R^2} = \frac{1.04\text{ mh}}{(0.72)^2(408\text{ ohms})} = 48\ \mu\text{sec}$$

Therefore, in 48 μsec, the current at the end of the jump (239 ma) will swing through 0.63(239 ma) $=$ 151 ma. The actual current is, therefore, $239 - 151 =$ 88 ma, and the corresponding output is

$$e_o = -\frac{88\text{ ma }(0.408\text{ kilohms})}{0.72} = -49.8\text{ volts}$$

In 5 time constants, 240 μsec, e_o will equal zero. The output voltage waveform is illustrated in Fig. 4-53*f*.

Note that if the leakage inductance were zero, all of i_1, 250 ma, would flow through a^2R_2 when the switch opens. The corresponding peak output voltage is $ae_o = -102$, or $e_o = -142$ volts.

The effect of winding and stray capacitances which we have been ignoring would be to round off the abrupt changes in voltage. Depending upon the circuit parameters, the output voltage may be a single rounded pulse or damped sinusoid with e_o actually swinging positive. Usually, shock excitation, such as opening the switch, will cause oscillations (ringing). An analysis which includes the effects of the various capacitances is quite complicated, and beyond the scope of this book. The reader is referred to "Electronic Transformers and Circuits" by Reuben Lee[1] and "Pulse and Digital Circuits" by Millman and Taub.[2]

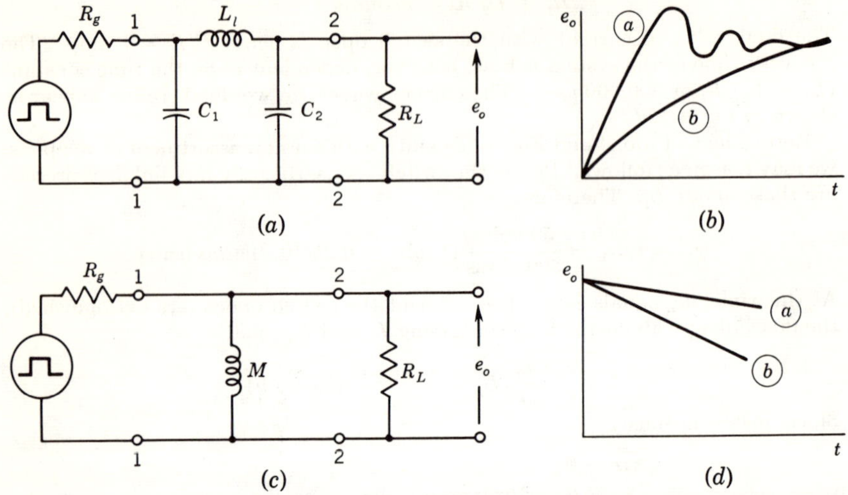

FIG. 4-54. (*a*) The effects of winding capacitance upon the leading edge of the pulse may be determined from this equivalent circuit. (*b*) The output voltage may be underdamped as in curve *a* or overdamped as in curve *b*. (*c*) The output during the flat top part of the pulse depends upon the shunt inductance *M*. (*d*) The larger *M* is, the smaller the tilt in the output waveform will be.

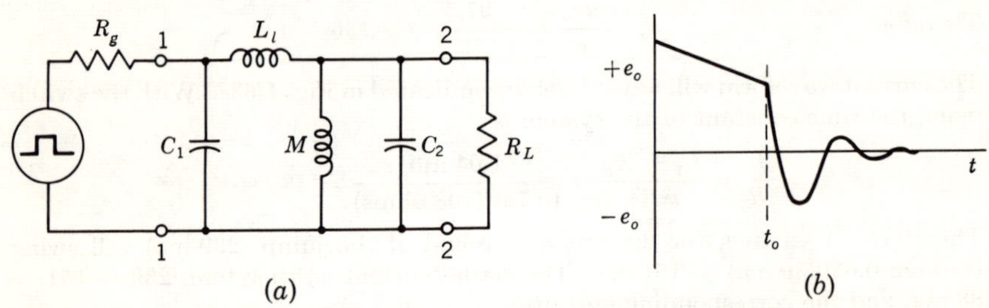

FIG. 4-55. (*a*) This circuit is useful in determining the backswing at the trailing edge of the pulse. (*b*) Backswing which may occur when the input falls to zero.

The effect of winding capacitances upon the leading edge of the pulse may be determined from the equivalent circuit of Fig. 4-54*a*. The winding resistances have been neglected, since they are usually much less than the source (R_g) and load (R_L) resistances. Capacitors C_1 and C_2 represent the lumped winding capacitances, plus the effect of capacitance between windings. Load capacitance may be lumped with C_2, while source capacitance can be lumped with C_1. Depending upon the

[1] *Ibid.*

[2] Jacob Millman and Herbert Taub, "Pulse and Digital Circuits," McGraw-Hill Book Company, Inc., New York, 1956.

parameters, e_o will have the shape of curve a or b in Fig. 4-54b. In the oscillatory case (curve a), the peak output voltage may exceed the applied input.

During the flat-top portion of the input pulse, the capacitive effects may be neglected. The equivalent circuit (assuming the leakage inductance is small compared to the mutual inductance) is shown in Fig. 4-54c. The output waveform appears in Fig. 4-54d. The droop in the output is determined by the time constant made up of M and the equivalent resistance it looks back into. Curve a in Fig. 4-54d corresponds to the case in which the time constant is greater than that of curve b.

The trailing edge response will depend upon the equivalent circuit of Fig. 4-55a. Assuming the pulse is turned off at t_0, the response will more or less appear as shown in Fig. 4-55b. Because of L_l and C_1, it is possible to have some high-frequency oscillation superimposed upon the back-swing oscillation of Fig. 4-55b.

SUMMARY

In this chapter, we have established methods of analyzing inductively coupled circuits. Concepts such as mutual, self-, and driving-point impedance were introduced, since they are basic in the analysis of any type of coupled circuit. With the aid of the dot notation, we learned how to write loop equations which were subsequently solved for currents and voltages in various parts of an inductively coupled network. By properly interpreting these equations, we were able to develop some equivalent circuits. For example, the concept of reflected impedance was developed, which enabled us to determine the impedance seen looking into a transformer with a given load. We also thevenized the transformer in order to obtain an equivalent circuit looking into the output terminals.

By postulating certain characteristics for the general transformer, the concept of an ideal transformer was developed. Since many power, audio-, and pulse transformers approximate the ideal transformer, the latter was studied extensively. The turns ratio was shown to be a convenient way of analyzing the current, voltage, and impedance relationships between the primary and secondary windings.

We then developed a conductively coupled T-equivalent circuit for the transformer, which could be used only at a single frequency if one of the arms was a negative inductance. In an effort to circumvent this difficulty, a number of one-sided equivalent circuits were developed in which everything was referred to the primary or secondary winding. By a suitable choice of parameter, the problem of negative inductances was avoided. These one-sided equivalent circuits proved very useful in analyzing the response of a transformer over a broad frequency spectrum. Since pulsed waveforms contain many frequency components, these equivalent circuits were used to determine the transient response of a transformer.

It is hoped that the material in this chapter will furnish the reader with the background and stimulus to pursue the subject further.

PROBLEMS

4-1. Determine the resonant frequency looking into the input terminals of the circuit in Fig. 4-56.

Solution

$$L_1 + L_2 + 2M = 0.58 \text{ henry}$$

$$f_r = \frac{1}{2\pi\sqrt{L_{eq}C}} = \frac{1}{2\pi\sqrt{0.58(10^{-9})}}$$

$$\approx 6{,}600 \text{ cps}$$

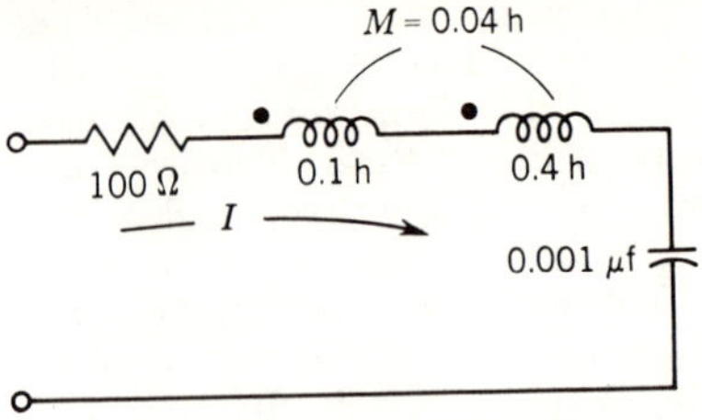

FIG. 4-56. Problem 4-1.

4-2. What is the current in Fig. 4-57?

Solution

$$E = I[R_1 + R_2 + j\omega(L_1 + L_2 - 2M)] = IZ_{in}$$

$$Z_{in} = 2{,}000 + j2\pi(100 \text{ cps})(2 \text{ henrys}) = 2{,}360\underline{/32.1^\circ}$$

$$I = \frac{E}{Z_{in}} = 42.3\underline{/-32.1^\circ} \text{ ma}$$

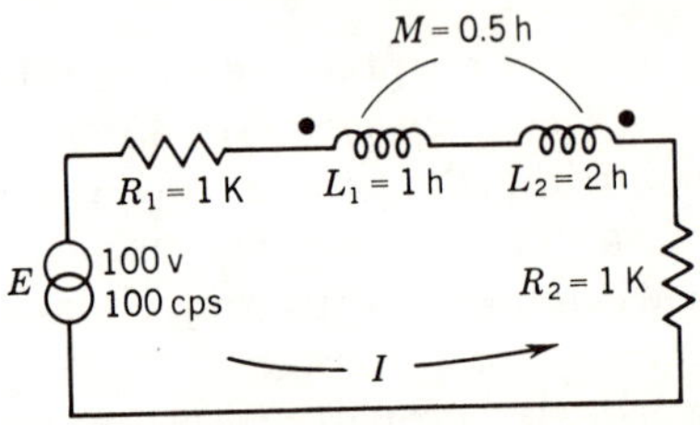

FIG. 4-57. Problem 4-2.

4-3. What sort of load should Z_x be for maximum power transfer in Fig. 4-58?

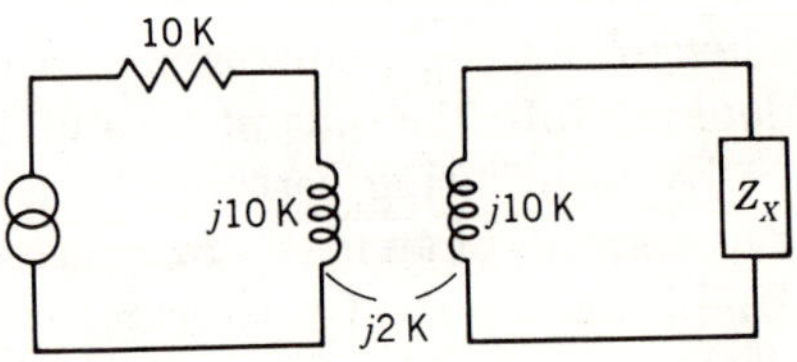

FIG. 4-58. Problem 4-3.

4-4. An ideal transformer is used to match an amplifier to a pair of 200-ohm earphones for maximum power transfer (Fig. 4-59). What is the required turns ratio? What is the power developed in the earphones if the generator has an open circuit voltage of 1 volt and an internal impedance of 12 kilohms?

Solution

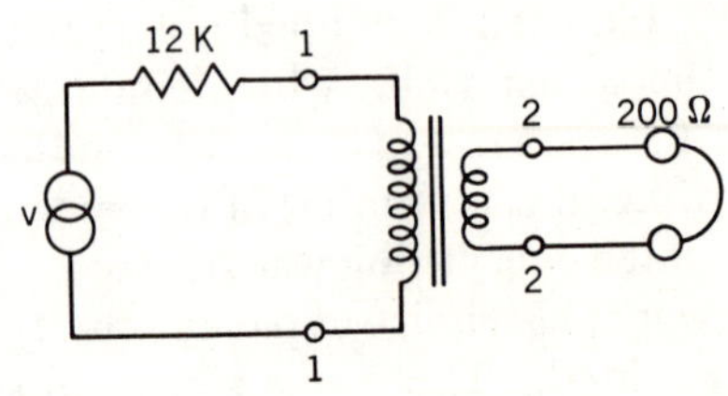

FIG. 4-59. Problem 4-4.

$$\frac{N_1}{N_2} = \sqrt{\frac{Z_{1\text{-}1}}{Z_{2\text{-}2}}} = \sqrt{\frac{12{,}000}{200}} = \sqrt{60} = 7.8$$

The power put into the primary winding, which is (0.5 volts)^2/12 kilohms = 0.02 Mw, is all transferred to the load.

4-5. What turns ratio is required for maximum power transfer in the circuit of Fig. 4-60? Assume that the transformer is ideal.

Solution

$$\frac{N_1}{N_2} = \sqrt{\frac{20{,}000}{4}} = 70.5$$

$$\frac{N_1}{N_3} = \sqrt{\frac{20{,}000}{8}} = 50$$

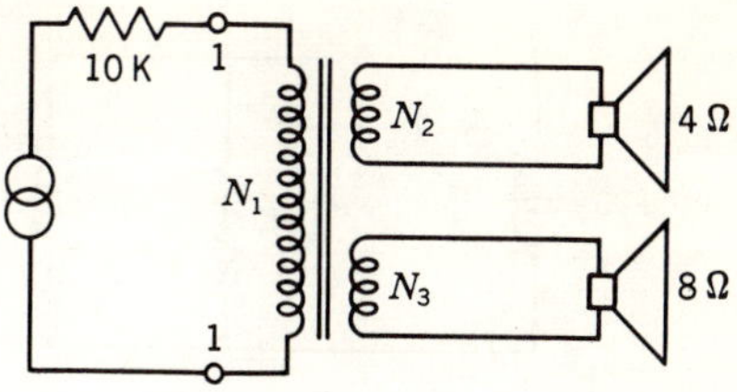

FIG. 4-60. Problem 4-5.

4-6. Determine I_2 in Fig. 4-61.

Solution

$$M = 0.1\sqrt{1 \times 4} = 0.2 \text{ henry}$$

$$\omega = 2\pi f = 6.28 \times 159 = 1{,}000 \text{ radians/sec}$$

$$Z_{22} = 1{,}000 + j4{,}000 = 4{,}140\underline{/76^\circ}$$

$$Z_{11} = 1{,}000 + j1{,}000 = 1{,}410\underline{/45^\circ}$$

$$Z_{12}{}^2 = \omega^2 M^2 = 10^6(0.04) = 4 \times 10^4 \text{ ohms}$$

$$\begin{aligned} I_1 &= \frac{1.00\underline{/0^\circ}}{Z_{11} + Z_{12}{}^2/Z_{22}} \\ &= \frac{100\underline{/0^\circ}}{1{,}000 + j1{,}000 + 4 \times 10^4/(4.14 \times 10^3\underline{/76^\circ})} \\ &= \frac{100\underline{/0^\circ}}{1{,}000 + j1{,}000 + 9.65\underline{/-76^\circ}} \\ &\approx \frac{100\underline{/0^\circ}}{1{,}000 + j1{,}000} = \frac{100}{1{,}410\underline{/45^\circ}} = 70.7\underline{/-45^\circ} \text{ ma} \end{aligned}$$

$$\begin{aligned} I_2 &= -\frac{j\omega M I_1}{Z_{22}} = \frac{-j(10^3)(0.2)}{4{,}140\underline{/76^\circ}}(70.7\underline{/-45^\circ}) \\ &= -j3.41\underline{/-121^\circ} = 3.41\underline{/-121^\circ + 270^\circ} = 3.41\underline{/149^\circ} \text{ ma} \end{aligned}$$

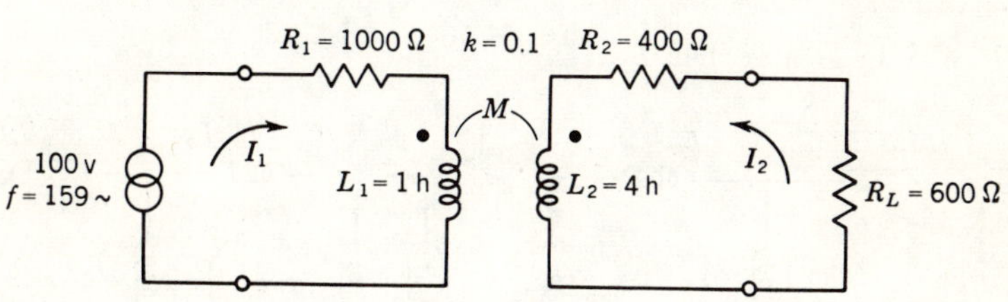

FIG. 4-61. Problem 4-6.

4-7. Compute I_{in} in Fig. 4-62*a*. HINT: Use the series-parallel conversions of Chap. 2 in order to obtain the simplified circuit of Fig. 4-62*b*.

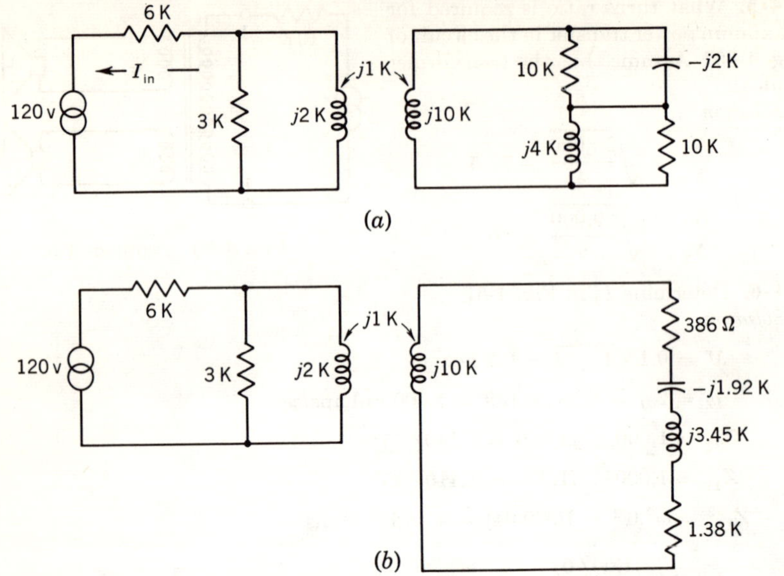

FIG. 4-62. Problem 4-7.

4-8. Prove that E_2 can be greater than E_1 in the T-equivalent circuit of Fig. 4-63*a* if $k = 1$.

Solution. If $k = 1$, $M = \sqrt{L_1L_2}$, and Fig. 4-63*a* is equivalent to Fig. 4-63*b*.

$$E_2 = I_1 j\omega M = I_1 j\omega \sqrt{L_1L_2}$$

$$I_1 = \frac{E}{j\omega(L_1 - \sqrt{L_1L_2} + \sqrt{L_1L_2})} = \frac{E}{j\omega L_1}$$

Therefore,
$$E_2 = \frac{E}{j\omega L_1} j\omega \sqrt{L_1L_2} = E\sqrt{\frac{L_2}{L_1}}$$

and L_2 may be greater than L_1.

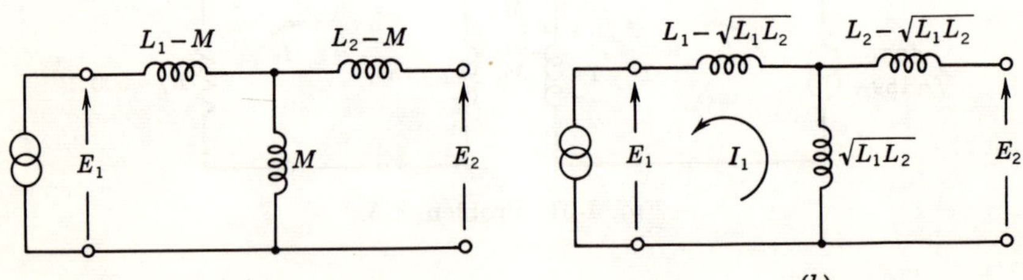

FIG. 4-63. Problem 4-8.

4-9. (*a*) Draw the primary-side equivalent circuit for the transformer circuit of Fig. 4-64*a*. (*b*) Solve for I_1. (*c*) Solve for I_2, using the primary-side equivalent circuit. Let $E_1 = 120$ volts at 1,000 radians/sec

$$R_1 = 5 \text{ ohms} \qquad R_2 = 5 \text{ ohms}$$
$$L_1 = 1 \text{ henry} \qquad R_L = 1 \text{ kilohm}$$
$$L_2 = 4 \text{ henry} \qquad C = 0.25\ \mu\text{f}$$
$$M = 2 \text{ henry} \qquad a = \sqrt{\frac{L_1}{L_2}}$$

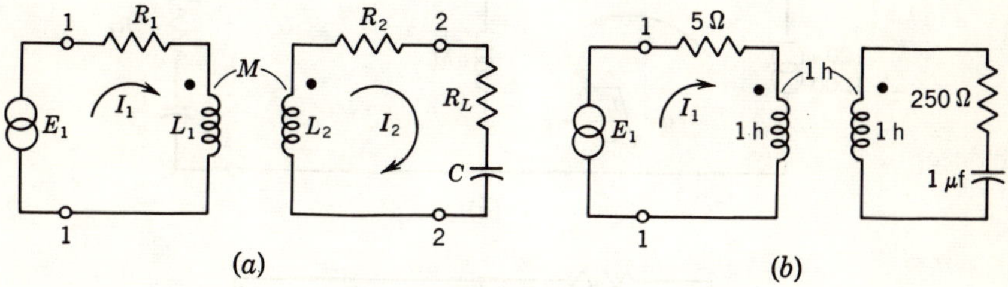

FIG. 4-64. Problem 4-9.

Solution

$$a = \sqrt{\frac{L_1}{L_2}} = \sqrt{\frac{1}{4}} = \frac{1}{2}$$

$$a^2 = 1/4$$

$$a^2L_2 = 1/4(4) = 1 \text{ henry and } X'_{L_2} = 1 \text{ kilohm}$$

$$aM = 1/2(2) = 1 \text{ henry and } X'_M = 1 \text{ kilohm}$$

a^2R_2 is negligible

$$a^2R_L = \frac{1 \text{ kilohm}}{4} = 250 \text{ ohms}$$

$$\frac{C}{a^2} = 1\ \mu\text{f and } X'_C = 1 \text{ kilohm}$$

The equivalent circuit is shown in Fig. 4-64*b*.

$$Z_{22} = 250 + j(1{,}000 - 1{,}000) = 250 \text{ ohms}$$

$$Z_{\text{ref}} = \frac{aZ_{12}{}^2}{a^2Z_{22}} = \frac{10^6}{250} = 4 \text{ kilohms}$$

$$Z_{11} = 5 + j1{,}000 \approx j1{,}000$$

$$I_1 = \frac{120}{Z_{11} + Z_{\text{ref}}} = \frac{120}{4{,}000 + j1{,}000} = \frac{120}{4{,}120\underline{/14^\circ}} = 29\underline{/-14^\circ}$$

$$\frac{I_2}{a} = -(j\omega M')\frac{I'_1}{Z'_{22}} = \frac{1{,}000(29\underline{/-14^\circ})\underline{/-90^\circ}}{250}$$

$$= 116\underline{/-104^\circ} \text{ ma}$$

$$I_2 = \frac{I_2 a}{a} = 1/2(116\underline{/-104^\circ}) = 58\underline{/-104^\circ} \text{ ma}$$

4-10. Construct a primary-side T-equivalent circuit for Fig. 4-64a with equal-leakage inductances, and solve for I_1.

Solution. The circuit used is that of Fig. 4-37c, which is in essence repeated here as Fig. 4-65a. Using the data in the previous problem, we find $L_a = L_b = 0$. This must be so, since

$$k = \frac{M}{\sqrt{L_1 L_2}} = \frac{2}{\sqrt{1(4)}} = 1$$

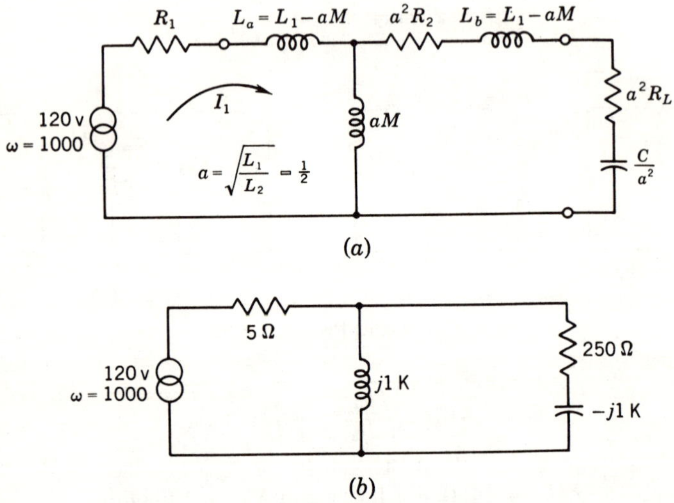

FIG. 4-65. Problem 4-10.

and therefore the leakage inductances are zero. The circuit therefore simplifies to Fig. 4-65b, in which

$$Z_{in} \approx Z_{eq} = \frac{(1{,}000\underline{/90^\circ})(250 - j1{,}000)}{250 + j(1{,}000 - j1{,}000)} = 4{,}100\underline{/14^\circ}$$

Therefore,

$$I_1 = \frac{120\underline{/0^\circ}}{4{,}100\underline{/14^\circ}} = 29\underline{/-14^\circ} \text{ ma}$$

which checks the solution in Prob. 4-9.

Chapter 5

GRAPHIC ANALYSIS OF VACUUM-TUBE CIRCUITS

THE DESIGNING OF a new electronic circuit is often a simpler task than the analysis of an existing one. The designer will usually choose a circuit to function in some desired manner. Then by calculation, experience, and intuition, he proceeds to evaluate parameters. Somewhere along the line he must choose suitable quiescent operating points for the tubes or transistors employed. The often arbitrary choice of operating points is another factor in determining circuit parameters and performance.

The technician, troubleshooter, or field engineer, however, is frequently faced with the problem of correcting some difficulty or of evaluating performance. Part of his problem is to determine the quiescent operating points so that the direct voltages and currents that he measures can be intelligently compared with predicted values. Furthermore, knowing the quiescent operating points can enable him to evaluate circuit performance under actual operating conditions. For example, given an input signal and a schematic diagram, he should be able to say whether a circuit will behave as a linear amplifier or as a clipper. If signal levels are low, graphic analysis is impracticable and the techniques of linear small-signal analysis are employed. But the quiescent operating points must still be determined if the small signal parameters such as μ, r_p, and g_m are to be correctly evaluated. It is for these reasons that this entire chapter is devoted to graphic analysis.

5-1. The Vacuum Tube. In basic electronics the reader no doubt became acquainted with the fundamental theory of the vacuum tube. In brief, most tube types contain two or three basic elements. These are:

1. A directly or indirectly heated cathode, whose function is to furnish free electrons.

2. One or more electrodes, usually in the form of a grid, to control and shape the electron stream within the tube. If there are no grids at all, the tube is known as a diode.

3. A plate to collect the electron stream.

When the filament temperature is raised by passing a current through it, free electrons are boiled out of the surrounding cathode. The amount of heat energy that has to be put into the cathode, to get some average number of electrons out of it, depends upon the chemical composition of the cathode surface. Other forms of energy, such as light, can also be used to release free electrons, as in photoelectric cells.

The electrons form a cloud about the cathode, which is known as the space charge. It is negative and tends to repel electrons coming out of the cathode surface. Therefore, actual current through the tube depends upon (1) the rate of cathode emission, (2) the space charge density, and (3) the electrostatic field distribution within the tube. Item 3 in turn depends upon the geometry of the tube and the voltages applied to the various electrodes.

From our point of view, the physics of the vacuum tube is best summarized by characteristic curves relating the currents and voltages of the various tube elements. The first set of curves we shall discuss comprises those of the diode.

5-2. The Diode. The diode consists of a cathode and plate. This is symbolized in Fig. 5-1. The cathode current in the diode is identical with the plate current, whereas in the more complicated tubes this is not necessarily so. For a given tube type, operating with a given cathode temperature, the plate current will be determined by the actual voltage existing between plate and cathode (e_b). Figure 5-2 is a typical diode characteristic curve.

Notice that in region one there is still some plate current flowing, even though the plate is negative with respect to the cathode. This current is due to the fact that some small percentage of the emitted electrons has sufficient kinetic energy to overcome the repelling field of the negative plate. For typical receiver-type diodes, the plate current will completely cut off with about 0.2-volt negative bias on the plate. With zero plate voltage, the plate current is in the neighborhood of 30 μa, which is virtually zero for most practical purposes.

In region two, for the curve E_{f_1}, we see that the plate current is quite dependent upon e_b. Here an increase in plate voltage draws more electrons out of the space charge. This means that the electrons trying to emerge from the cathode surface will not be as severely repelled. Hence, more emerge and i_b increases. If we continue to increase e_b, a point will be reached where all the emitted electrons are collected and the space charge is negligible.

This is the condition in region three and is referred to as emission-limited saturation. E_{f_1} is some filament voltage which raises the cathode to a temperature T_1. To get more plate current at a value of e_b, which biases the diode into region five, the filament voltage must be increased to E_{f_2}. The tube manufacturer generally recommends specific values of filament voltage at which the tube should be operated.

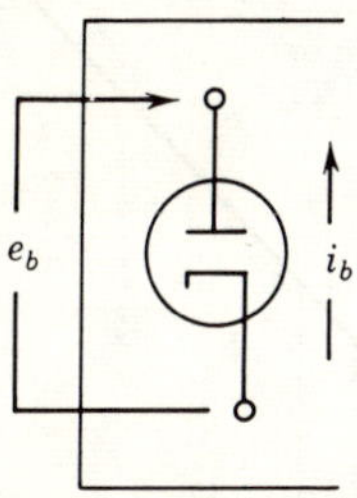

FIG. 5-1. Voltage e_b is the voltage at the plate with respect to the cathode. Current i_b is the positive direction of electron flow from cathode to plate.

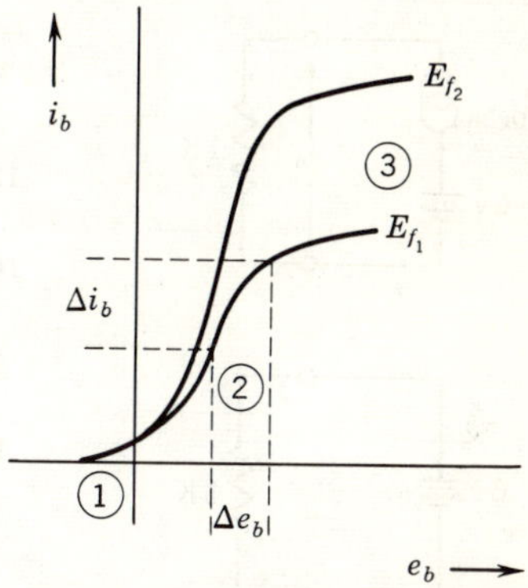

FIG. 5-2. The diode characteristics may be roughly divided into three regions.

5-3. Analysis of Characteristic Curves. Inspection of Fig. 5-2 indicates that the diode is a nonlinear device. That is, equal changes in voltage do not produce equal changes in current over the whole operating range of voltages. For example, a change of Δe_b volts in region two produces a much greater change in i_b, that is, Δi_b, than it would in region three, where the curve is flatter. Physically, this could be interpreted to mean that the resistance presented by the diode in region two to a variation in voltage is less than in region three. This resistance to variational changes is known as the dynamic plate resistance (r_p) and is different from the static plate resistance (R_p). We shall now investigate the meaning of static (R_p) and dynamic (r_p) plate resistances further.

Let us sketch the volt-ampere characteristic of a linear 1-kilohm resistor, as shown in Fig. 5-3*d*. To do this, we use Ohm's law. If we assume a voltage, say, 10 volts, the corresponding current is then $i_b =$ 10 volts/1 kilohm $=$ 10 ma. This gives us one point on the curve. With $e_b = 6$ volts, $i_b = 6$ ma; and with $e_b = 0$ volts, $i_b = 0$. These points may then be connected, and the result is a straight line. The result will always be a straight line if the resistor is purely linear. The ratio of voltage to current at any point on this curve (line) is 1 kilohm. Conversely the ratio of current to voltage at any point is 0.001 mho, which is the conductance or reciprocal of 1 kilohm. The ratio of a fixed voltage at any point to the corresponding fixed current at that point is the static resistance of the device. For the linear 1-kilohm resistor, the static resistance (R_p) is 1 kilohm at every point along the curve. Now

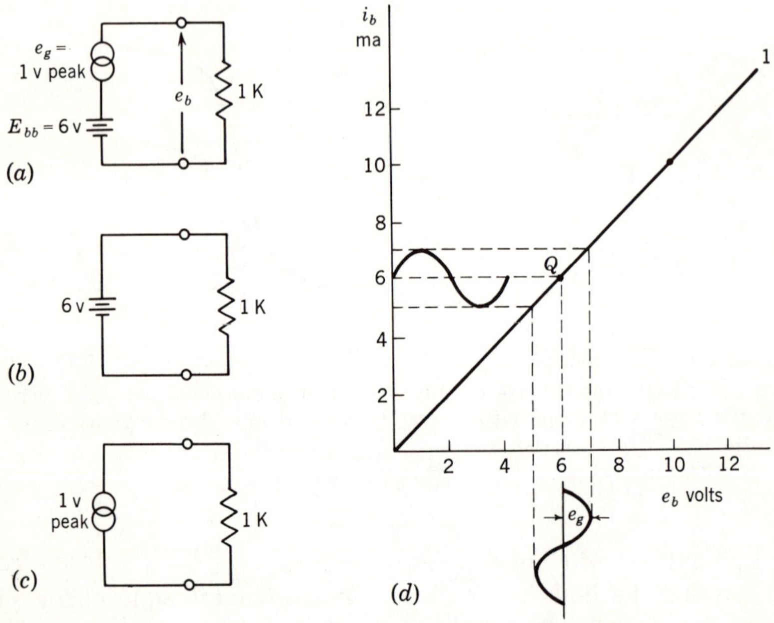

FIG. 5-3. (a) Static and variational currents which may be analyzed by the superposition theorem. (b) The static or quiescent current is determined. (c) The variational current is determined. (d) The ratio of the current variation due to a voltage variation about some quiescent operating point defines the dynamic plate resistance.

suppose we put the 1-kilohm resistor into the circuit of Fig. 5-3a. Let the a-c source have a peak value of 1 volt. The resultant current which flows through the 1-kilohm resistor will be due to the effect of E_{bb} and e_g combined. By the superposition theorem we can break up Fig. 5-3a into Fig. 5-3b and c. From Fig. 5-3b we can compute the steady, or d-c, value of the current, which is $I_b = 6$ ma. Graphically, we obtain the same result by entering the graph at 6 volts and reading off the corresponding current, which is 6 ma. This point, established by the d-c, or bias, source, is known as the quiescent operating point (Q). From Fig. 5-3c we see that the peak value of alternating current will be 1 ma. The ratio of the alternating current in Fig. 5-3c to the a-c voltage is the dynamic resistance r_p. For this case of a simple linear resistor, both the dynamic and static resistances are the same.

The volt-ampere characteristic of Fig. 5-3*d* indicates nothing more than a linear 1-kilohm resistor. Now suppose we had a device with the volt-ampere characteristic of Fig. 5-4*f*. The first thing we notice is that the device is nonlinear, since the slope of the volt-ampere characteristic is different in regions one and two. In region one, where the curve is a straight line and passes through the origin, the static and dynamic resistances are the same. For example, with $e_b = 2$ volts, $i_b = 4$ ma. Therefore, $R_p = 2$ volts/4 ma $= 0.5$ kilohm. Similarly, we see that a 2-volt peak-peak variation in e_b (same as Δe_b) about the quiescent value of $e_b = 2$ volts produces a 4 = ma peak-peak variation in i_b (same as Δi_b) about the quiescent value of 4 ma. The dynamic resistance about the operating point 2 volts, 4 ma is

$$r_p = \frac{\Delta e_b}{\Delta i_b} = \frac{2 \text{ volts peak-peak}}{4 \text{ ma peak-peak}} = 0.5 \text{ kilohm}$$

In region two the static resistance (R_p) will differ from point to point, even though the line is straight. This is because the curve of region two, if extended into region one, would not go through the origin but would instead intercept the current axis at 6 ma. For example, at point *a*, $R_p = 4$ volts/8 ma $= 500$ ohms. At point *b*, $R_p = 8$ volts/10 ma $=$ 800 ohms, and at point *c*, $R_p = 12$ volts/12 ma $= 1{,}000$ ohms. The fact that the curve in region two is a straight line does, however, denote that the dynamic resistance is constant, since equal increments of voltage variation will produce equal increments of current variation. To determine the dynamic resistance in this linear region, we can vary the plate voltage by some amount, say, $\Delta e_b = 2$ volts peak-peak, and read off the corresponding current variation, which is $\Delta i_b = 1$ ma peak-peak. The dynamic resistance about the operating point of 8 volts, 10 ma is then $r_p = 2$ volts/1 ma $= 2$ kilohms. Notice that r_p and R_p may differ widely.

5-4. The Diode Equivalent Circuit. What sort of an equivalent circuit can we construct for the device having the volt-ampere characteristic of Fig. 5-4*f*. In region one the equivalent circuit is simply that of a 0.5-kilohm resistor. As the voltage across the device increases, the current through it increases just as if the device were a 0.5 kilohm resistor. When the voltage across the device reaches 4 volts, however, its characteristics suddenly change. The equivalent circuit must now fit a device with the volt-ampere characteristic of Fig. 5-4*g*. The line *ac* has been extended to intercept both axes.

Here the device is one which, with 0 volts between its end terminals has 6 ma flowing through it or with -12 volts across it has zero ma

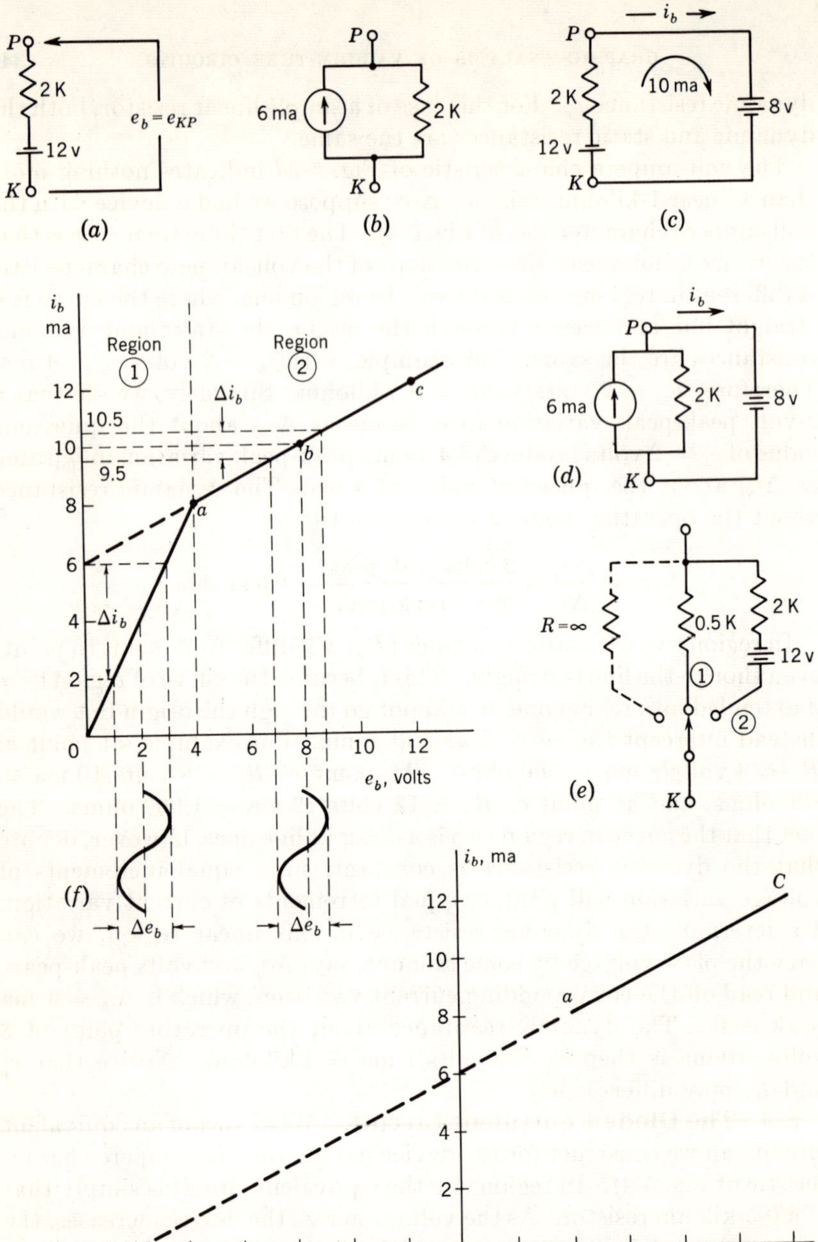

FIG. 5-4. (*a*) The diode equivalent circuit in region 2. (*b*) The current source equivalent circuit in region 2. (*c*) Using the equivalent circuit to determine the diode current when 8 volts is applied to it. (*d*) Using the current source equivalent circuit to determine the plate current. (*e*) An equivalent circuit to characterize operation in any of the three regions. (*f*) The variational resistance in region 2 is greater than in region 1. (*g*) The characteristics of region 2 analyzed in detail.

flowing through it. The dynamic resistance of the device is 2 kilohms. The device could conceivably be a vacuum diode, in which case we would plot plate current (i_b) versus plate-to-cathode voltage (e_b). A possible equivalent circuit that gives $i_p = 0$ with $e_{kp} = -12$ volts and an internal resistance of 2 kilohms is shown in Fig. 5-4*a*. The open-circuit voltage is -12 volts, and $Z_{th} = 2$ kilohms. Remember that this circuit does not become effective until e_b reaches 4 volts. We measure e_b with respect to the cathode, that is, $e_b = e_{kp}$.

The voltage-source equivalent circuit of Fig. 5-4*a* could be converted to a current-source equivalent by applying Norton's theorem. The result is shown in Fig. 5-4*b*. The short-circuit current is $I_{sc} = 12$ volts/2 kilohms $= 6$ ma. To verify the accuracy of these equivalent circuits, let us say that 8 volts is applied between terminals *p* and *k*, as shown in Fig. 5-4*c*. The resultant current is $i_b = 10$ ma.

Checking with the volt-ampere characteristic of Fig. 5-4*g*, we see that with 8 volts across the device the current is truly 10 ma. This verifies Fig. 5-4*a*. To verify the current-source equivalent, put 8 volts across Fig. 5-4*b*, as shown in Fig. 5-4*d*, and solve for the current through the 8-volt source. This can be done in several ways. We can convert the current source and 2-kilohm resistor back into Fig. 5-4*c*, which has already been solved, or evaluate i_b by inspection of Fig. 5-4*d*. If we consider just the effect of the current source, and replace the 8-volt source by its zero internal impedance, the current source will see 2 kilohms shunted by a short circuit. Hence, all 6 ma of the current source will flow through the 8-volt source.

Now if we replace the current source by its internal impedance, which is infinite, we have 8 volts across a 2-kilohm resistor, which gives 4 ma. Since the 6- and 4-ma currents are flowing in the same direction through the 8-volt battery, the total current is $i_b = 10$ ma. The complete equivalent circuit for our device is then as shown in Fig. 5-4*e*. The switch is in position one for $0 < e_b < 4$ volts, and in position two for $e_b > 4$ volts. We could have added a third switch position for $e_b < 0$. That is, when the *p* end is negative with respect to the *k* end, the current is zero and the internal resistance is infinite.

In Fig. 5-5*b*, the heavy curve is the volt-ampere characteristic of a single section of a typical dual diode. The curve was judiciously broken up into three straight line segments. We have included the part of the curve where $i_b = 0$, for all negative values of e_b, as the first segment. Then the characteristics in each region are investigated. In region one with 0 volts, there is zero current. In fact, for any value of negative voltage between the end terminals there is zero current. The device in region one can then be represented by an infinite resistance, or open circuit.

In region two with zero volts, the device passes zero current. With 1 volt (the right end of this segment) the device passes 2.5 ma. Since we started from zero, $\Delta e_b = 1$ volt and $\Delta i_b = 2.5$ ma. Therefore, $r_p = 400$ ohms. In region three, we see that a half-volt change in e_b produces a 2-ma change in i_b. Therefore, r_p in region three equals 250 ohms. If we extend the third segment so that it intercepts the horizontal axis, we see that with $i_p = 0$ the equivalent circuit must show an open-circuit voltage of 0.5 volt with the plate end positive, since the axis is intercepted in the

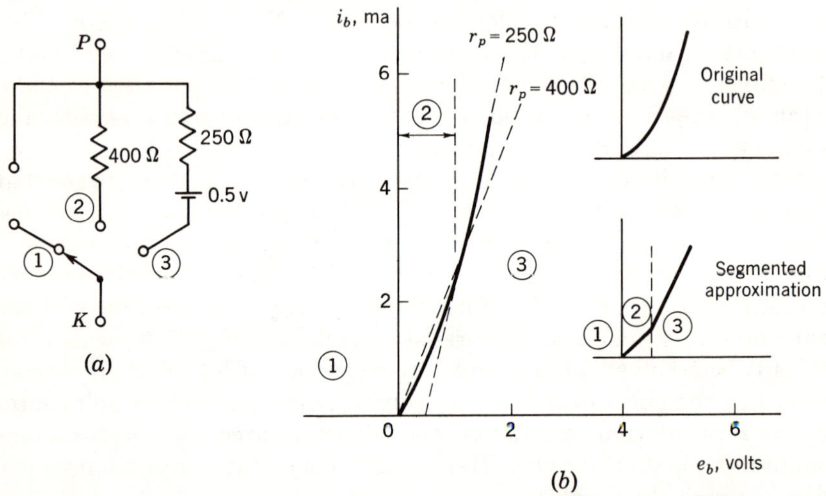

FIG. 5-5. (a) An equivalent circuit based upon the straight-line approximations of b. (b) The diode characteristics may be approximated by straight-line segments.

forward-bias region. The final equivalent is shown in Fig. 5-5a. These equivalent circuits afford us a compact means of expressing the device characteristics and in addition may be used to evaluate circuit behavior.

5-5. Q-point Evaluation. Generally the diode is operated in conjunction with a load resistor R, as shown in Fig. 5-6a. The applied voltage E_{in} may be an a-c, a d-c, or a combined source. For the moment, let us assume E_{in} is a d-c source of 2 volts. E_{in} could also represent the instantaneous value of an a-c source as it goes through 2 volts. If $R =$ 500 ohms and the diode characteristic is that of Fig. 5-6b, what will the current and voltages in the circuit be? Figure 5-6b is the same as Fig. 5-5a. There are two approaches to this problem at our level, the first of which is a graphic analysis.

Plot the volt-ampere curve of the 500-ohm resistor by assuming a voltage across the 500-ohm resistor. For example, with 2.5 volts a 500-ohm resistor passes 5 ma. This gives us a point to plot. Since this resistor is linear, we can draw a straight line from the origin.

The resistor and diode are in series, so that the same current must flow through each. Assume a value of current, say 1 ma. The drop across the tube with 1 ma through it is 0.5 volt. The drop across the resistor is also 0.5 volt. The drop across both tube and resistor can be graphically added with dividers and is about 1 volt.

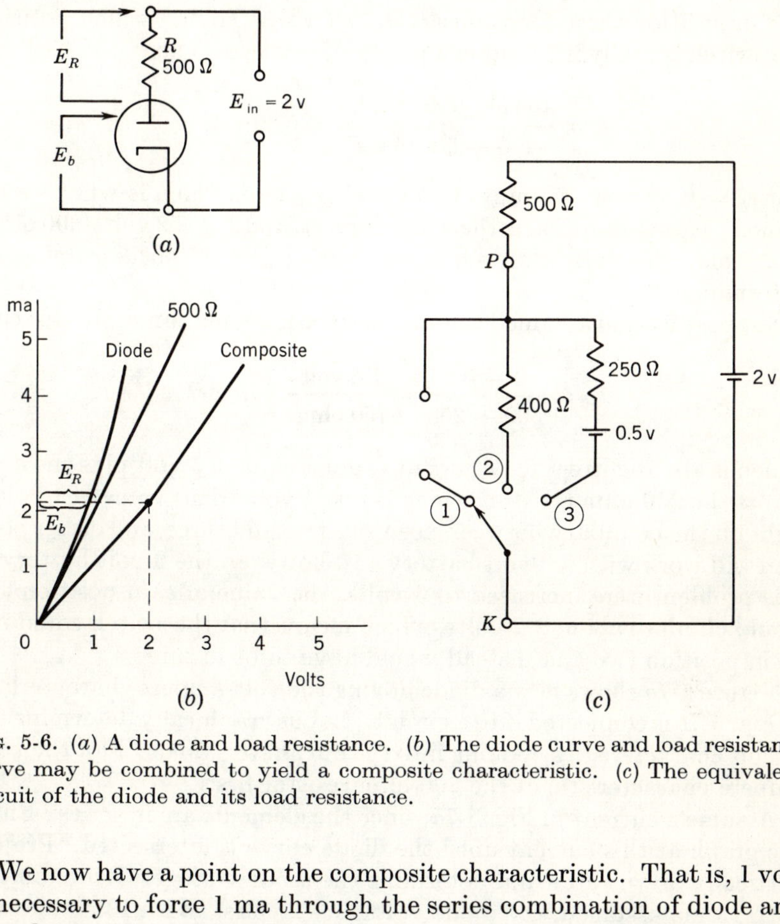

FIG. 5-6. (a) A diode and load resistance. (b) The diode curve and load resistance curve may be combined to yield a composite characteristic. (c) The equivalent circuit of the diode and its load resistance.

We now have a point on the composite characteristic. That is, 1 volt is necessary to force 1 ma through the series combination of diode and 500-ohm resistor. Repeat for the other points. For example, with 4.5 ma assumed, the diode drop is 1.5 volts and the resistor drop 2.25 volts. The total applied voltage must then be 3.75 volts. Sufficient points are taken so that a smooth composite curve can be drawn. We see that when 2 volts is impressed across the composite device, about 2.1 ma of quiescent current (I_{b0}) flows. From the curves, we see that the quiescent resistor voltage $E_{R0} \approx 1.1$ volts, while the quiescent tube voltage E_{b0} is about 0.75 volt.

Let us see how these results compare with an analysis using the equivalent circuit developed in Fig. 5-5*a*. The complete circuit is now that of Fig. 5-6*c*. The switch will be in position one, two, or three. It will be in position one if point p is negative with respect to point k. Because of the polarity of the 2-volt battery, it obviously is not, in this case. The switch, you will recall, is in position two for values of $0 < e_b \leq 1$ volt, and in position three for values of $e_b \geq 1$ volt. To find which position the switch is really in, assume a position, say two. Then

$$e_b = \frac{400 \text{ ohms} \times 2 \text{ volts}}{400 + 500 \text{ ohms}} = 0.89 \text{ volt}$$

For $e_b = 0.89$ volt, we must be in position two, which is what we assumed. The assumption is therefore correct, and $I_{b0} = 2$ volts/900 ohms $= 2.2$ ma, which is in close agreement with $I_{b0} = 2.1$ ma as graphically determined.

Suppose we had assumed the switch to be in region three. In this case

$$I_{b0} = \frac{2 - 0.5}{500 + 250} = \frac{1.5 \text{ volts}}{750 \text{ ohms}} = 2 \text{ ma}$$

which is about correct for I_{b0}. But e_b must equal 0.5 volt plus the drop across the 250-ohm resistor, which is just 1 volt. This value of 1 volt is right on the boundary between positions two and three, and either position will work with a 2-volt battery. If, however, the 2-volt battery in this problem were increased to 3 volts, the value of e_b in position two would climb to about 1.2 volts, which means that the switch could not be in position two and instead would have to be in three.

Figure 5-7*a* shows how a diode having the volt-ampere characteristic of Fig. 5-7*b* is connected into a circuit. Let us graphically determine the circuit characteristics, looking into the output terminals. Plot the volt-ampere characteristic of the 200-ohm resistor first.

Assume a current in Fig. 5-7*a*, since the elements are in series. Enter the graph at this current until the diode curve is intersected. Project vertically downwards and determine the diode voltage. For example, with 13 ma assumed, the diode voltage is 2 volts. With 13 ma flowing through the 200-ohm resistor, 2.6 volts appear across it. With a pair of dividers, add the diode voltage to the resistor voltage along $i = 13$ ma. The combined voltage (4.6 volts) must be the voltage applied to force 13 ma through the circuit. Since the diode characteristic is a straight line in this region, we need not plot further points, but can draw a line (dashed) from the origin to the 13-ma, 4.6-volt point. This line is the combined characteristic of the diode and 200-ohm resistor in series. Since 5 volts is applied to the combination, we enter the chart at 5 volts and proceed vertically until the combined characteristic is intersected. The resultant current is about 14 ma. Since the output voltage is taken off the diode, the diode voltage is the V_{oc} seen looking

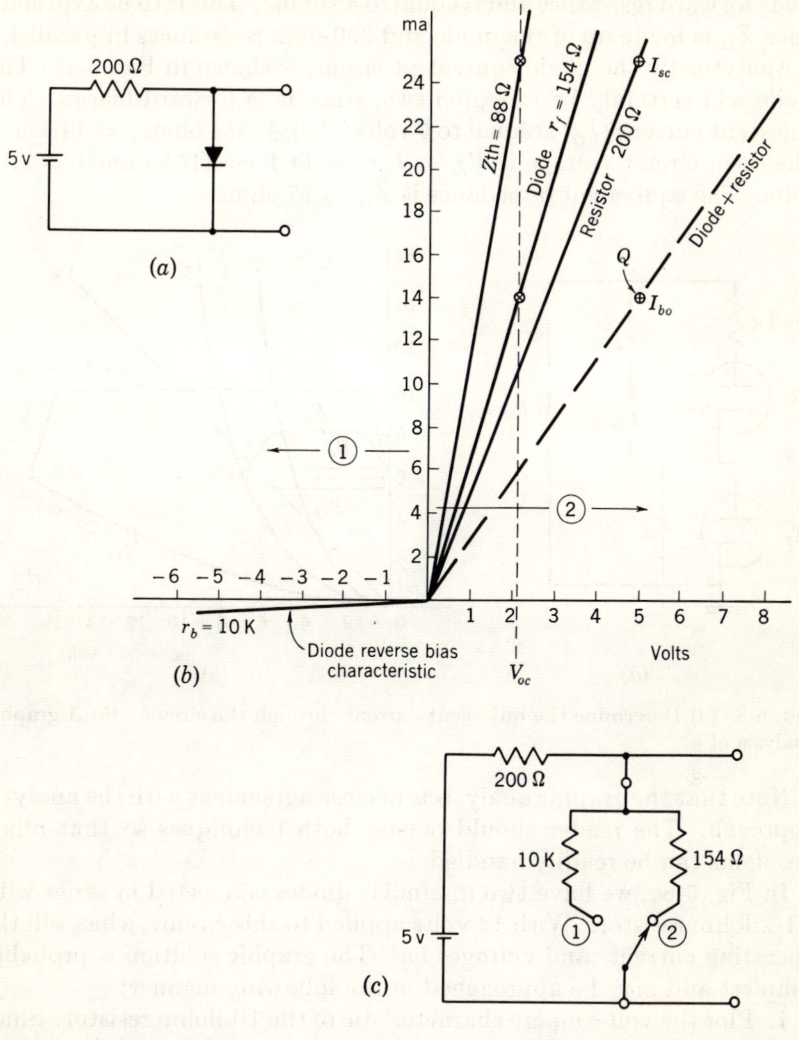

FIG. 5-7. (*a*) A semiconductor diode circuit. (*b*) A graphic analysis of *a*. (*c*) An equivalent circuit for *a*.

into the output terminals. The Z_{th} seen looking into the output terminals can be found as follows. First, find the short-circuit current (I_{sc}) that would flow through the output terminals if they were shorted. This is simply the intersection of the 5-volt projection with the 200-ohm characteristic. In Chap. 1, we learned that $Z_{th} = V_{oc}/I_{sc}$. We therefore have a point on the Z_{th} characteristic, since we know V_{oc} and I_{sc}. Notice that the slope of the Z_{th} characteristic is steeper than the slope of the diode forward resistance and is equal to 88 ohms. This is to be expected, since Z_{th} is made up of the diode and 200-ohm resistances in parallel.

Analytically the diode equivalent circuit is shown in Fig. 5-7*c*. The diode will certainly be in region two, since it is forward-biased. The quiescent current (I_Q) is equal to 5 volts/(200 + 154 ohms) = 14.1 ma. The open-circuit voltage is $V_{oc} = I_Q r_f = 14.1$ ma (154 ohms) = 2.17 volts. The equivalent impedance is $Z_{th} = 87$ ohms.

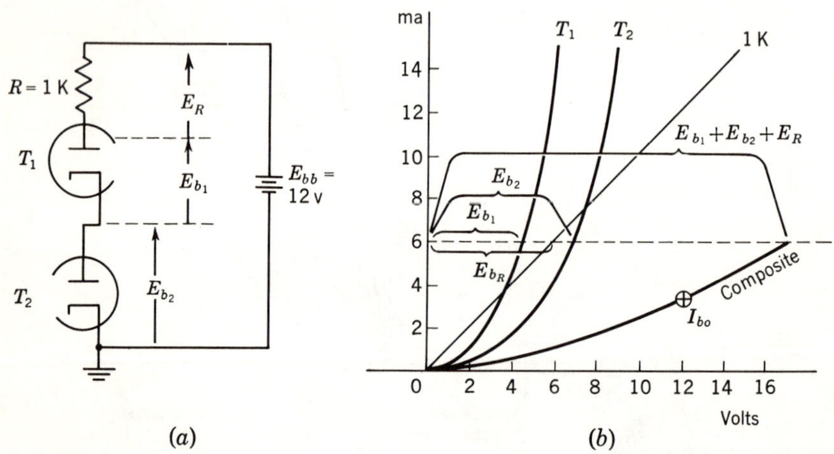

FIG. 5-8. (*a*) Determine the quiescent current through the circuit. (*b*) A graphic analysis of *a*.

Note that the graphic analysis is in close agreement with the analytic approach. The reader should master both techniques so that allied problems can be readily handled.

In Fig. 5-8*a*, we have two dissimilar diodes connected in series with a 1-kilohm resistor. With 12 volts applied to this circuit, what will the operating currents and voltages be? The graphic solution is probably simplest and may be approached in the following manner:

1. Plot the volt-ampere characteristic of the 1-kilohm resistor. Since 10 volts across 1 kilohm gives 10 ma by Ohm's law, 10 volts, 10 ma is a point on the 1-kilohm characteristic. Since it is a linear element, a straight line connecting this point to the origin can be drawn at once.

2. Since the elements are all in series, the same current must flow through each element. Therefore, assume a current, determine the voltage across each element for this assumed current, and then add all the voltages to give the voltage that must be applied to force this assumed

current through the composite circuit. This operation will give a volt-ampere point on the composite characteristic. For example, assume 6 ma in Fig. 5-8*b*. With a pair of dividers, add the respective voltages, E_{b_1}, E_{b_2}, and E_R, to get a point on the composite curve. Since the composite curve is partly made up of nonlinear elements, we cannot draw a straight line to the origin, but must also plot additional points.

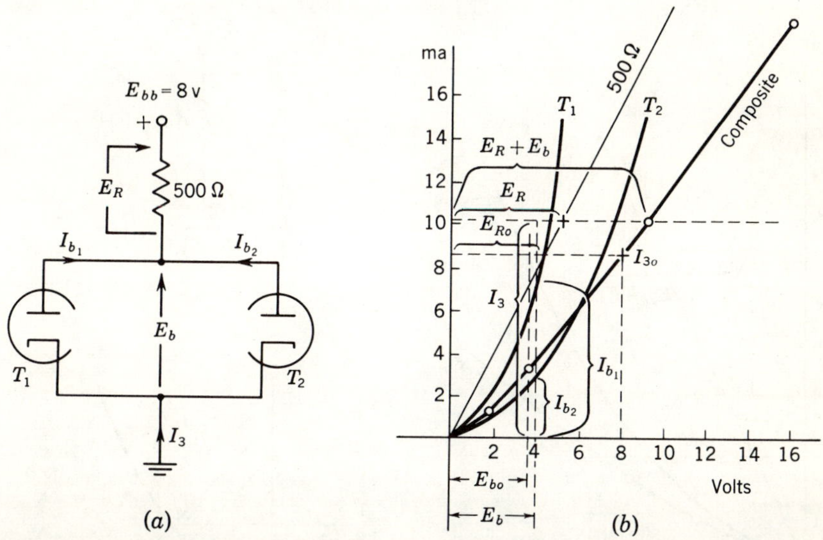

FIG. 5-9. (*a*) Two dissimilar diodes in parallel feeding a common load resistor. (*b*) A graphic analysis of *a*.

3. After the composite curve is drawn, enter the graph at 12 volts until the composite curve is intersected. The corresponding value of quiescent current is I_{b0} and is about 3.4 ma. The tube voltages E_{b1} and E_{b2} are about 3.2 and 5.2 volts, respectively.

In Figure 5-9*a*, we have two dissimilar diodes in parallel. Find the various currents and voltages. To solve this problem we shall do the following:

1. Plot the volt-ampere characteristic of the 500-ohm resistor as in Fig. 5-9*b*.
2. Assume a voltage E_b across the diodes, since they are in parallel. Let us assume 4 volts.
3. With $E_b = 4$ volts, we can find the corresponding currents I_{b_1} and I_{b_2}, which are about 7.2 and 3.2 ma, respectively.
4. The currents I_{b_1} and I_{b_2} will give a combined current I_3 of about 10.2 ma, which can be obtained by graphically adding I_{b_1} and I_{b_2} with dividers.

5. Now I_3 flows through a 500-ohm resistor. Projecting I_3 horizontally into the 500-ohm resistor curve gives the voltage across the 500-ohm resistor E_R, which is about 5.1 volts.

6. If we add E_R to the assumed value of $E_b = 4$ volts, the resultant voltage must be that value of E_{bb} which would cause I_3 to equal 10.2 ma. This is then a point on the composite characteristic.

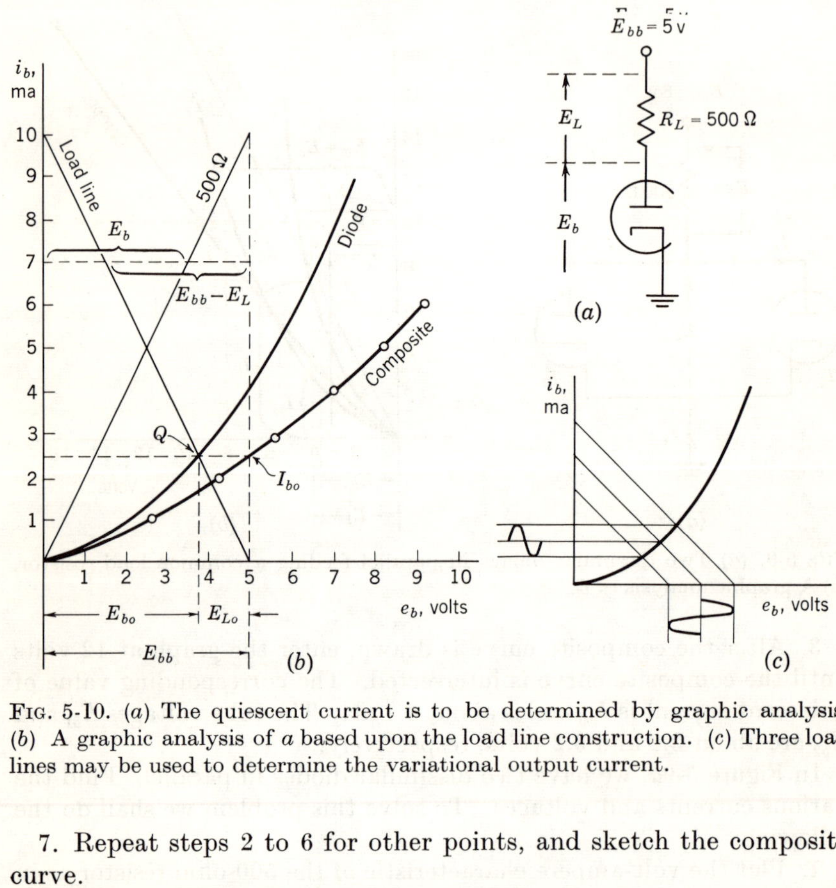

FIG. 5-10. (*a*) The quiescent current is to be determined by graphic analysis. (*b*) A graphic analysis of *a* based upon the load line construction. (*c*) Three load lines may be used to determine the variational output current.

7. Repeat steps 2 to 6 for other points, and sketch the composite curve.

8. Enter the composite curve at the actual value of $E_{bb} = 8$ volts, and read I_3 actual, which is called I_{30} and is about 8.6 ma.

9. The quiescent drop across the 500-ohm resistor E_{R0} is about 4.2 volts from the graph, or 4.3 volts if calculated by computing the IR drop across the 500-ohm resistor with $I_{30} = 8.6$ ma.

10. The difference between $E_{bb} = 8$ volts and E_{R0} is E_{b0}. This may be calculated or graphically determined. To avoid cluttering Fig. 5-9*b* further, it will be calculated as $E_{b0} = 8 - 4.3 = 3.7$ volts.

11. Now enter the chart at 3.7 volts, and intersect the T_2 curve at about $I_{b20} = 2.8$ ma and I_{b10} at about 6.2 ma.

12. Check to see how closely I_{b20} and I_{b10} equal I_{30}. We have, respectively, 2.8 ma + 6.2 ma = 9.0 ma, as compared with 8.6 obtained in step 8. Considering all the graphic work and accumulation of errors, this discrepancy is understandable.

There is an alternative approach to determining the quiescent operating point in the diode circuit of Fig. 5-10*a*. Suppose the diode has the characteristics shown in Fig. 5-10*b*. As in the previous problems, we could assume values of i_b, determine E_L and E_b for each assumed i_b, and then add E_L and E_b to get points on the composite characteristic. After the composite curve is obtained, the value $E_{bb} = 5$ volts is projected into it and I_{b0} is determined. The composite curve of Fig. 5-10*b* was obtained in this manner.

The alternative approach is to assume values of i_b and determine the voltage across R_L for this assumed value of i_b. For example, with $i_b =$ 7 ma, $E_L = 7 \text{ ma} \times 0.5 \text{ kilohm} = 3.5$ volts if computed or if graphically determined by observing that voltage which intersects the 500-ohm curve at 7 ma. To keep the diagrams from getting too cluttered, these values will usually be computed. Now if we subtract this value of E_L from the known E_{bb}, the remainder must be E_b. This too may either be done graphically or computed. If it is done graphically, lay off the distance E_L with a pair of dividers, as shown in Fig. 5-10*b* and then subtract this distance from E_{bb}. This practice can be repeated for different values of i_b. The line will be straight, since E_L is directly proportional to the current through R_L and is subtracted from a constant quantity E_{bb}. This line is commonly known as a load line. It can very easily be plotted in the following manner.

When $i_b = 0$, the drop across R_L is zero and therefore the full value of E_{bb} appears across the diode. This is the point at which the load line intersects the voltage axis. The maximum possible value that i_b can assume in Fig. 5-10*a* will occur if the diode has been shorted ($e_b = 0$). This current must be 5 volts/500 ohms = 10 ma and is the point at which the load line intersects the current axis. The quiescent current (I_{b0}) is determined by the intersection of this load line with the diode characteristic. The same value of $I_{b0} = 2.5$ ma was obtained when the composite characteristic was used. The quiescent tube drop E_{b0} and load drop E_{L0} can be directly read from this quiescent operating point Q.

If E_{bb} were changed to some other value, a new load line could be calculated. For example, let E_{bb} have a sinusoidal signal impressed upon it, as shown in Fig. 5-10*c*. Load lines could be plotted for the maximum and minimum values, in addition to the quiescent value of E_{bb}. The intersection of these load lines with the diode characteristic will give the quiescent, maximum, and minimum values of plate current. The load lines will be parallel to each other and have a slope of $-1/R_L$.

5-6. The Triode. Now let us turn our attention to the triode. The essential difference between the triode and the diode is the presence of a control grid. The control grid (termed "grid" for short) is much closer to the cathode than to the plate. It will therefore exert a strong effect on the space charge. As a rule, the control grid is maintained at some negative potential with respect to the cathode. Under this condition, the number of electrons landing on the grid will be extremely small in most tubes. If the grid is permitted to become positive with respect to the cathode, it will collect a significant number of electrons and will cause the plate current to be less than the cathode current. Because of the grid's meshlike structure, its physical presence does not impede the plate current. The voltage between grid and cathode, e_c, does, however, exert a profound influence on the plate current. Even though the grid is negative with respect to the cathode, plate current can still flow because the positive plate field overrides the negative grid field. But the grid can be made sufficiently negative to completely mask the plate field and thereby completely cut off the plate current. Figure 5-11*a* illustrates a triode circuit and standard symbols.

The following symbols will usually be employed in our discussion of vacuum-tube circuitry:

E_L = d-c voltage across load
E_{cc} = d-c grid bias
E_{bb} = d-c plate supply
E_g = rms value of a-c input excitation voltage if this excitation is sinusoidal
E_p = rms value of a-c output voltage if output is sinusoidal
e_g = instantaneous a-c grid voltage, measured at grid with respect to cathode terminal
e_c = total (a-c and d-c) instantaneous signal between grid and cathode
e_p = instantaneous a-c signal between plate and cathode
e_b = e_{kp} = total (a-c and d-c) instantaneous voltage between plate and cathode of tube
i_p = instantaneous signal component of plate current, positive in the direction from cathode to plate through the tube
i_b = total (a-c and d-c) instantaneous value of plate current, positive in the direction from cathode to plate through the tube
I_b = average, or direct, current in the plate circuit
E_b = average, or d-c, potential between plate and cathode

Generally we use lower case for instantaneous values, *b* and *p* subscripts for plate circuits, c and g subscripts for grid circuits, and the subscript 0 for quiescent values.

5-7. Plate Dissipation. The electrons which are boiled out of the cathode surface have a low emission velocity. As they travel towards

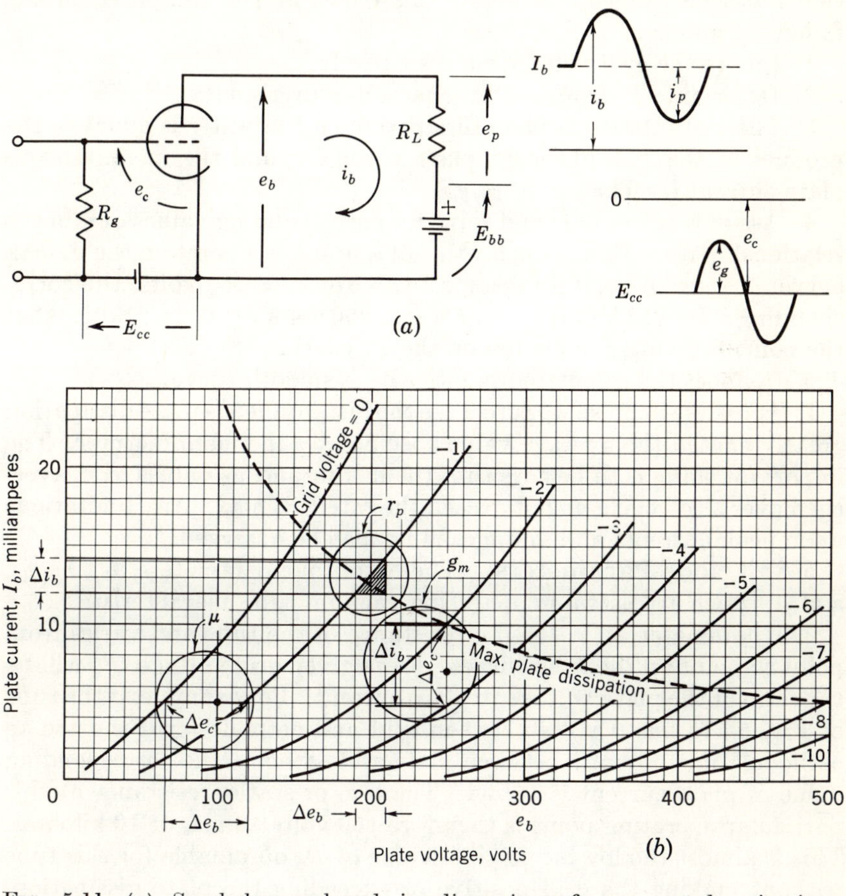

FIG. 5-11. (a) Symbols employed in discussion of vacuum-tube circuitry. (b) Graphic constructions for the small-signal parameters.

the plate, they are continuously accelerating and arrive at that electrode with considerable kinetic energy. Upon striking the plate, this kinetic energy is converted to heat, which, added to any heat picked up from the cathode, raises the plate temperature. This heat must not become excessive, or gases trapped on the plate material will be released with harmful effects. Therefore, the tube manufacturer will specify a maximum plate dissipation. For small receiving-tube voltage amplifiers, 2.5 watts of plate dissipation (P_d) is typical. Figure 5-11b is a plot of the plate characteristics for a half-section of a 12AT7 dual triode. The P_d of this tube is rated at 2.5 watts per section. This is the average plate dissipation. In Fig. 5-11*b* you will notice a dashed curve labeled "Max Plate Dissipation." This curve tells us that all tube operation must be carried out below this curve and is computed in the following manner:

1. Ignore the grid voltage curves entirely.
2. Determine P_d from the tube manufacturer's data.
3. The instantaneous plate dissipation (p_d) in watts is equal to the product of the instantaneous plate voltage e_b and the instantaneous plate current i_b. That is, $p_d = e_b i_b$.
4. Assume values of e_b and solve for corresponding values of i_b in the relationship $i_b = P_d/e_b$. Each value of e_b and i_b is a point on the p_d max curve. For example, if we assume a value of $e_b = 250$ volts, the corresponding i_b for a 12AT7 is $i_b = 2.5$ watts/250 volts $= 10$ ma. Notice that the point 250 volts, 10 ma lies on the P_d curve.
5. Connect the calculated points with a smooth curve.

It is possible in pulse work to have the instantaneous plate dissipation greatly exceed the average value specified by the manufacturer. The important thing is to be sure that the instantaneous values of p_d averaged over a full cycle do not exceed the rated P_d max. Any limitations upon peak currents and voltages must also be observed.

5-8. Triode Parameters. There are four parameters, R_p, r_p, g_m, and μ, which are useful in analyzing vacuum-tube characteristics.

The parameter R_p is the d-c resistance of the tube and varies from point to point on the tube curves. For a given grid voltage, R_p will be the ratio of the plate voltage to plate current. For example, if the tube of Fig. 5-11*b* has 0 volts between grid and cathode (grid shorted to cathode) and 150 volts between plate and cathode, the corresponding value of plate current is 15 ma. The d-c, or static, resistance at this particular operating point is therefore 150 volts/15 ma, or 10 kilohms. This is almost the lowest possible value of R_p obtainable for this tube without making the grid positive or exceeding the plate dissipation. When the tube is cut off, R_p is infinite.

The parameter r_p is the dynamic or variational resistance of the tube. It is the ratio of a small change in plate voltage to a corresponding change in plate current, with the grid voltage held constant. In other words, it is a measure of how effective a variation in plate voltage is in producing a change in plate current. Symbolically, this is written

$$r_p \approx \frac{\Delta e_b}{\Delta i_b}\bigg/ e_c \text{ constant} \qquad \text{or} \qquad r_p = \frac{de_b}{di_b}\bigg/ e_c$$

where the d symbol implies an infinitesimally small change.

The dynamic plate resistance will also vary from point to point because of the nonlinear nature of the tube. In general, we can see that r_p will be smallest in the region where grid voltage is zero and plate voltage is highest, that is, where the plate current is largest. The curve is fairly steep here. On the other hand, the curves flatten out as the grid voltage (bias) is increased and the plate voltage decreases, which indicates a high r_p. Suppose we wish to determine r_p at a specified operating point, say, with the grid bias (E_{cc}) fixed at -1 volt and 200 volts on the plate. To do this, construct a small triangle as shown in Fig. 5-11*b*. The base of this triangle is $\Delta e_b = 20$ volts. The corresponding Δi_b along the curve of -1 grid volt is $\Delta i_b = 2.1$ ma. Therefore, $r_p = 20$ volts/2.1 ma, or 9.5 kilohms. This is at an operating or quiescent point of 200 volts of plate voltage, 13 ma of plate current, and -1 volt of grid voltage. The static resistance at this point is 200 volts/13 ma or 15.4 kilohms. These results tell us that if we had a circuit such as that of Fig. 5-12*a*, the reading of the d-c ammeter would be 200 volts/R_p, or 13 ma, and the a-c ammeter would read 2 volts/$r_p = 0.21$ ma.

If a large a-c source causes variations about the quiescent point to become large enough, the current variation for the series-aiding case will be different from the bucking case. This is distortion and is due to the nonlinear characteristics of the tube. The value of r_p for the series-aiding half cycle is different than for the series-opposing half cycle and therefore does not mean much. For this reason, r_p is called a small-signal parameter. At times, particularly in pulse work, we use an average value of r_p derived from very large variations in plate voltage.

The next parameter to discuss is the transconductance g_m. We know that changing the grid voltage is going to vary the plate current. To determine just how effective the grid is in controlling plate current, we must hold the plate voltage constant, since it too determines plate current. This therefore defines g_m for us. It is the ratio of a small change in plate current to a small change in grid voltage, with the plate voltage held constant. This is written

$$g_m \approx \left.\frac{\Delta i_b}{\Delta e_c}\right/ e_b \text{ constant} \qquad \text{or} \qquad g_m = \left.\frac{di_b}{de_c}\right/ e_b$$

The value of g_m also varies from point to point and is usually highest when the plate current is high, as with zero grid bias. To determine the value of g_m at some specific operating point, proceed as follows: Observe the variation in current produced by a variation in grid voltage along a line of constant plate voltage passing through the operating point. For example, assume one half of a 12AT7, as in Fig. 5-12*a*, with a quiescent operating point at 250 volts and 7 ma. Since e_b is fixed at 250 volts, we will slide above and below the Q point along the 250-volt line and read off the plate currents corresponding to the closest grid voltage intersections. For example, with $e_c = -2$ volts, $i_b = 10$ ma and with $e_c = -3$ volts, $i_b = 4.8$ ma. Therefore, g_m in this vicinity is about

$$g_m = \frac{\Delta i_b}{\Delta e_c}\bigg/ e_b = \frac{10 - 4.8}{3 - 2} = \frac{5.2 \text{ ma}}{1 \text{ volt}}$$
$$= 5.2 \times 10^{-3} \text{ mho, or } 5{,}200 \ \mu\text{mhos}$$

Usually g_m is expressed in micromhos as shown.

The manufacturer lists g_m as 4,500 μmhos for a 12AT7 with 7 ma of quiescent current (Fig. 5-12*b*). The error in our calculation is due to curvature of the tube's characteristics. Later we shall discuss more accurate ways of evaluating the small-signal parameters.

A sample problem might illustrate the physical significance of g_m. Suppose a tube has a relay coil in its plate circuit that has negligible resistance, so that the plate voltage does not vary much with variations in plate current. Let us say the tube is sitting at a Q point, at which $i_b = 5$ ma and $g_m = 5{,}000$ μmhos. If the relay requires 10 ma to close, about how much change in grid voltage will be needed to close the relay? We see that the relay current must rise from 5 to 10 ma, or $\Delta i_b = 5$ ma. Since $g_m = \Delta i_b / \Delta e_c$, we can solve for Δe_c and get

$$\Delta e_c = \frac{\Delta i_b}{g_m} = \frac{5 \text{ ma}}{5 \times 10^3 \times 10^{-6} \text{ mho}} = 1 \text{ volt}$$

The last triode parameter we shall discuss is μ (pronounced mu). The μ is a measure of how much more effective the control grid field is in controlling plate current than the plate's electrostatic field. Specifically μ is defined as the ratio of a small change in plate voltage to a small change in grid voltage with the plate current held constant. This is written

$$\mu \approx \frac{\Delta e_b}{\Delta e_c}\bigg/ i_b \text{ constant} \qquad \text{or} \qquad \mu = \frac{de_b}{de_c}\bigg/ i_b$$

The μ can also be determined graphically in the neighborhood of any specific operating point, as follows. Let us say we had a Q point of $e_b = 100$ volts and $i_b = 5$ ma. Slide along the line $i_b = 5$ ma until the Q point is bracketed by two values of e_c. Then note the required change in e_b to keep i_b constant. In this example, sliding to the *left* of the Q point along the 5-ma line, we intersect the point $e_c = 0$, $e_b = 68$ volts. Sliding to the *right* of the Q point, we have $e_c = -1$, $e_b = 121$ volts. The approximate value of μ is then

$$\mu = \frac{\Delta e_b}{\Delta e_c} = \frac{121 - 68 \text{ volts}}{1 - 0 \text{ volt}} = 53$$

Note that μ is a dimensionless parameter, since it is the ratio of two voltages. Comparison of this calculated value with the manufacturer's listed value of $\mu = 54$ (Table 5-1) when $i_b = 3.7$ ma shows that we are in the right ball park.

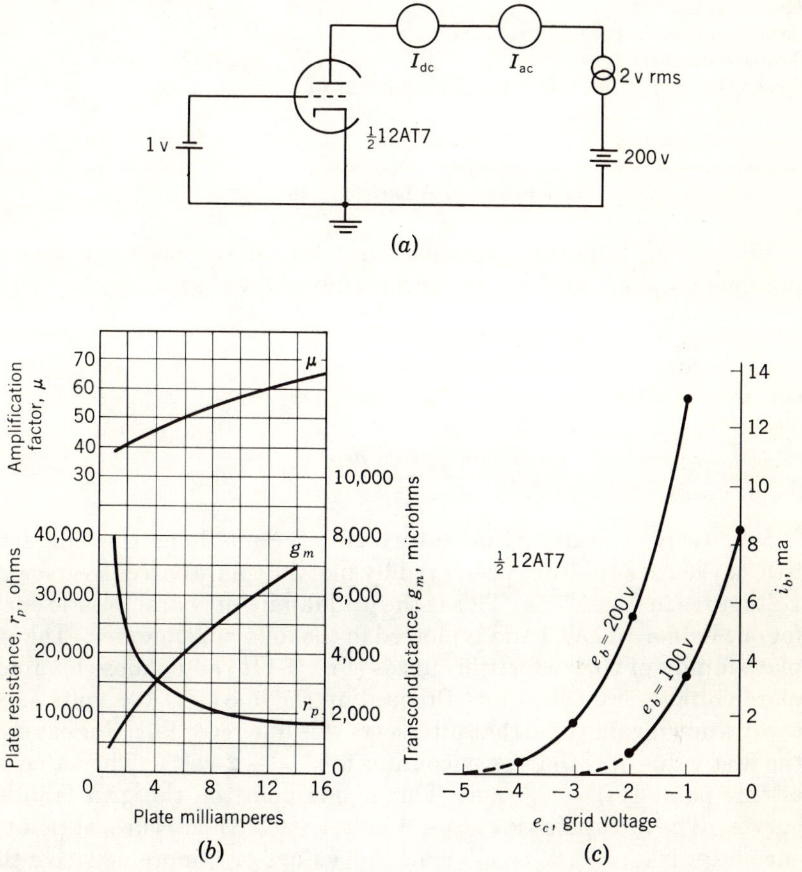

FIG. 5-12. (a) The readings of the d-c and a-c meters will depend upon the static and dynamic plate resistance of the tube. (b) Average characteristics for each unit of a type 12AT7 at 250 plate volts. (*RCA*.) (c) The plate current may be plotted as a function of the grid voltage.

For many tube types there are curves available which indicate the value of the small-signal parameters versus plate current or grid voltage. These curves, however, are usually plotted for a specific value of plate voltage. At some other value of plate voltage, even though the current is still the same, r_p, g_m, and μ might be quite different. Figure 5-12*b* and Table 5-1 show how this information is displayed.

TABLE 5-1

CHARACTERISTICS OF 12AT7 HIGH-μ TWIN TRIODE*

Plate voltage	100	180	250
Grid voltage	−1	−1	−2
Amplification factor	54	62	55
Transconductance, μmhos	4,000	6,600	5,500
Grid voltage (approximate) for plate current of 10 ma	−6	−8	−12
Plate current, ma	3.7	11	10

* SOURCE: Radio Corporation of America.

There is an important relationship which exists between the three parameters μ, g_m, and r_p. We can readily show that

$$\mu = g_m r_p$$

since

$$g_m = \frac{di_b}{de_c} \qquad \text{and} \qquad r_p = \frac{de_b}{di_b}$$

$$g_m r_p = \frac{di_b}{de_c}\frac{de_b}{di_b} = \frac{de_b}{de_c} = \mu$$

5-9. Grid Family Characteristics. Sometimes the characteristic curves of a tube are presented by plotting plate current versus grid voltage, as in Fig. 5-12*c*. This is the grid family of characteristic curves for one half of a 12AT7 and is plotted in the following manner. Take the plate family of characteristic curves (Fig. 5-11*b*) and choose a value of plate voltage, say 100 volts. Proceed up the line $e_b = 100$ volts and at every known value of e_c that intersects this line record i_b. For example, the first value of e_c that we encounter is $e_c = -2$ volts. The value of i_b at this point is $i_b = 0.7$ ma. This is one point on the grid family of curves. The next point is $e_c = -1$ volt, $i_b = 3.5$ ma. The last point we encounter is $e_c = 0$, $i_b = 8.5$ ma. For values of e_c more negative than -2 volts, the shape of the curve can be approximated. The whole process is now repeated for another value of constant e_b, say, 200 volts as shown. The curve for $e_b = 200$ volts was not plotted for values more positive than $e_c = -1$ volt, since the maximum plate dissipation was being exceeded. We shall primarily concern ourselves with the plate family of characteristic curves.

Although they are seldom used, we could have plotted constant-current curves instead of constant-grid-voltage curves as in the plate family or constant plate voltage as in the grid family. In this case, we will start with the plate family, choose a value of i_b, and plot all the values of e_b versus e_c that give that value of i_b.

5-10. The Tetrode. A serious disadvantage of the triode in many applications is the rather large value of grid-to-plate capacitance C_{gp}. In certain circuits, feedback through this interelectrode capacitance will cause undesired oscillations to occur. In other circuits C_{gp} can cause excessive high-frequency attenuation, as we shall see when the Miller effect is discussed.

In an effort to reduce C_{gp}, another grid was inserted between the control grid and the plate. The purpose of this grid was to act as an electrostatic shield between the control grid and plate. This grid was therefore

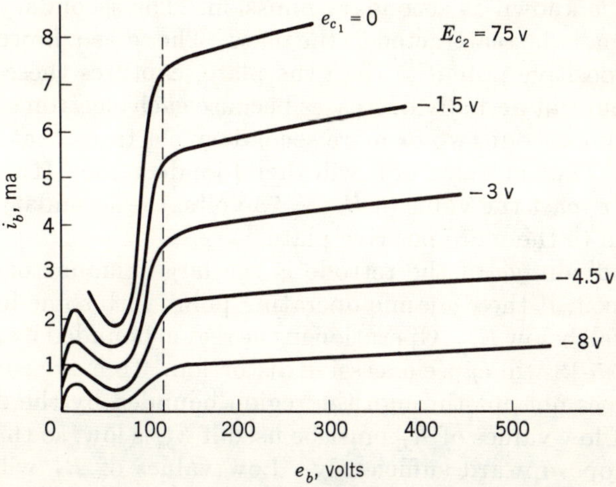

FIG. 5-13. Tetrode characteristics.

called the screen grid, and since it is the second grid away from the cathode, it is designated G_2. The control grid is called G_1. The screen grid is generally operated at some positive voltage equal to or less than the E_{bb} supply. The screen grid captures a small percentage of electrons emitted by the cathode, which results in screen current (i_{c_2}) and screen power dissipation.

Since the screen tends to shield the plate field from the control grid, it also reduces the ability of the plate to control the space charge. This means that a change in e_b cannot produce as much of a change in i_b as it would in the same tube without a screen grid. The r_p of the tube must therefore be greater, since $r_p = \Delta e_b/\Delta i_b$ with e_c constant. Furthermore, the μ must be higher, since the effectiveness of the plate in controlling the space charge has been reduced in comparison with the control grid.

The plate characteristics of a tetrode are shown in Fig. 5-13. The slope of the characteristic curves can be explained in the following way. Starting at the extreme left, we see that with E_{c_2} held constant at 75 volts, increasing e_b increases i_b for values of e_b up to 12 volts approximately. We then see that further increases in e_b in the range of 12 to 75 volts actually cause i_b to decrease. This is most unusual, since increasing the voltage across a positive resistance element normally increases the current. If an increase of voltage decreases the current, the element exhibits negative dynamic resistance. The d-c, or static, resistance in this region is positive. The a-c, or dynamic, resistance is, however, negative.

Physically, what is happening is that the electrons striking the plate now have sufficient kinetic energy to knock out secondary electrons. This effect is known as secondary emission. The secondary electrons would normally be reattracted to the plate. The screen, however, being at a more positive potential than the plate, captures these electrons. The plate current actually decreases because each electron striking the plate may knock out two or more secondary electrons that are lost to the screen. The net value of i_b will therefore decrease. If we continue to increase e_b past the value of $E_{c_2} = 75$ volts, the secondary electrons are redrawn to the more positive plate.

The disadvantage of the tetrode is the large amount of distortion that can occur if the dynamic operating point with some load should cause e_b to fall below E_{c_2}. Operation in the region bounded by the dotted line in Fig. 5-13 will cause excessive distortion. To make sure that the load line does not cut through the region bounded by the dotted line means that low values of R_L must be used if E_{bb} is low, so that the load line will slope upward sufficiently. Low values of R_L will result in reduced gain and poor efficiency, however, since the load does not match the internal resistance of the tube. If high values of R_L are used, excessive and dangerous values of E_{bb} must be used if the distortion region is to be cleared. For this reason the tetrode has become virtually obsolete.

5-11. The Pentode. In an effort to overcome the secondary emission effect of the tetrode, the pentode was developed. A third grid, called the suppressor grid (G_3), is inserted in the region between the screen and the plate. This grid is usually operated at cathode potential and is therefore negative with respect to the plate. Secondary electrons are therefore repelled back into the plate.

An additional benefit derived from the addition of G_3 is that C_{gp} is reduced even further than in tetrodes. Since the plate's electric field is more effectively shielded from the space charge about the cathode, μ and r_p are even higher than in tetrodes. The g_m is about the same as for triodes. Typical values are $\mu = 100$ to 4,000, $r_p = 100$ kilohms to 2 meg, $g_m = 1{,}500$ to 5,000 μmhos. Grid-to-plate capacitance is in the order of 0.004 $\mu\mu$f for typical receiving pentodes, as compared with 1.6 $\mu\mu$f for typical triodes. The characteristic curves of a typical pentode are

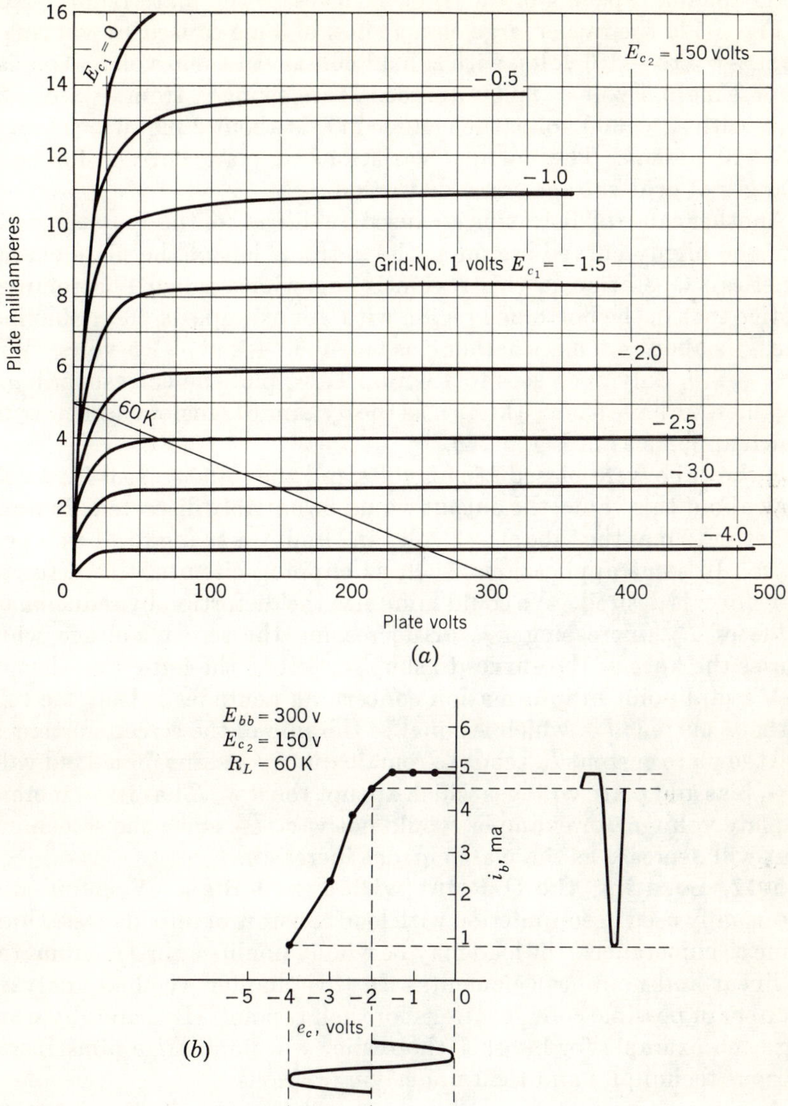

FIG. 5-14. (*a*) A 60-K load line drawn upon the average plate characteristics of a 6AU6 pentode. (*RCA*.) (*b*) Clipping will result with excessive grid drive.

shown in Fig. 5-14*a*. In general, the g_m increases if either the grid bias decreases or the screen or plate voltage increases. Even though the pentode gives higher gain than a triode, it tends to produce more distortion than a triode for the same swing in grid voltage, partly because of the unequal spacing of the G_1 bias curves on the plate family. Note in Fig. 5-14*a* the rather large change in i_b of 3 ma caused by varying e_c from -0.5 to -1.0 volts, with e_b held constant at 300 volts. This is a g_m of 3 ma/0.5 volt $= 6{,}000\ \mu$mhos. If we swing e_c from -2 to -2.5 volts with $e_b = 300$ volts, the change in i_b is about 2 ma, which is a g_m of 4,000 μmhos. The unequal variations of plate current for equal changes of grid voltage cause distortion.

Another cause of distortion is caused by shifting of the operating point into the highly curved region at the extreme left of the plate characteristics. Operation in this region is sometimes called "bottoming." Notice that in the bottomed region with, for example, a 60-kilohm load line, i_b is about 4.9 ma whether e_c is 0, -0.5, -1, or -1.5 volts. With $e_c = -2$, i_b only decreases to 4.6 ma. If we plot the corresponding i_b, e_c values, we have a curve known as the dynamic transfer characteristic, which appears as in Fig. 5-14*b*.

If the tube were biased at -2 volts and the input varied by 2 volts peak about this value, the output would be horribly distorted. It would be better to bias the tube at -3 volts and limit the swing to a 1-volt peak value. In some applications, such as clipping circuits, this distorted waveform is desired. We could minimize the distortion by reducing the value of R_L, increasing E_{bb}, or decreasing the screen voltage which causes the knee of the curves to shift further to the left.

A useful point of information concerning pentodes is that the total cathode current I_K, which is equal to the sum of the screen current I_{c_2} and the plate current I_b, tends to remain quite constant for a fixed value of G_1 bias and plate voltages which are not too low. That is, an increase in plate voltage, for example, would not vary I_K, since the screen current will decrease as the plate current increases.

5-12. Locating the Q Point with Fixed Bias. Vacuum tubes are usually used in conjunction with load resistors or impedances. Under some circumstances, the load may be wholly nonlinear or a combination of linear and nonlinear elements. In this chapter we shall analyze a number of possible configurations for their Q point. By carefully working each example for himself, the reader will familiarize himself with various techniques and their underlying reasons.

As a starting point let us endeavor to determine the Q point of the tube in Fig. 5-15*a*. Since the grid is tied directly to the cathode, $e_c = 0$. Furthermore, this value of e_c is fixed and in no way dependent upon the tube current. If the short between grid and cathode were removed and a bias battery E_{cc} put in its place, the grid-to-cathode voltage would be clamped or fixed to the value $e_c = E_{cc}$. This type of bias is therefore known as fixed bias.

To determine the Q point, we could use one of two approaches. The first is to get the composite characteristic by plotting the 6-kilohm volt-ampere characteristic on the tube curves, as shown in Fig. 5-15*b*. This, of course, is a straight line. To plot this line, assume a convenient voltage across the resistor and determine the corresponding current. Say

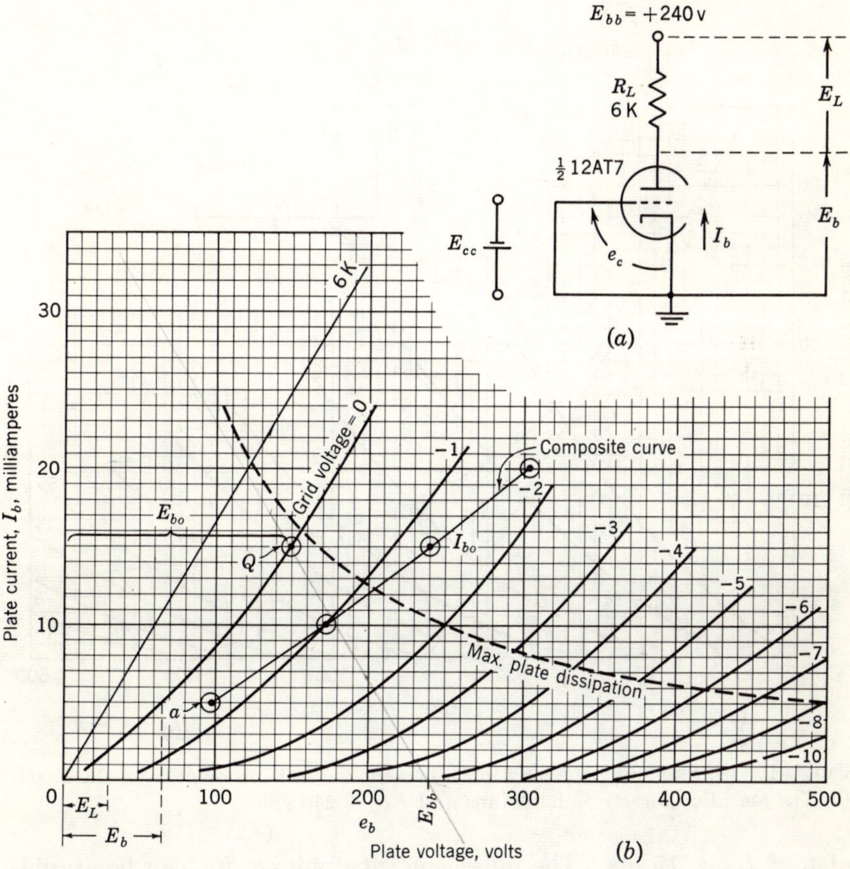

FIG. 5-15. (*a*) The Q point is to be determined. (*b*) A composite characteristic may be made for the tube and load resistor.

60 volts are assumed: then the current must be 60 volts/6 kilohms = 10 ma. Draw a straight line from the origin to this point. Now assume values of I_b and determine the corresponding values of E_b and E_L for a point on the composite curve. For example, if $I_b = 5$ ma, $E_B = 68$ volts and $E_L = 30$ volts. The total applied voltage to pump this current through the circuit is then $E_b + E_L = 98$ volts, which can be graphically added with dividers for convenience. Plot the point 98

volts and 5 ma as shown by point a on Fig. 5-15b. Now repeat the process, choosing other values of I_b until the composite curve can be drawn.

The quiescent current I_{b0} that will flow into the composite circuit is determined by entering the composite curve at the actual applied voltage of 240 volts. The intersection of E_{bb} and the composite curve occurs

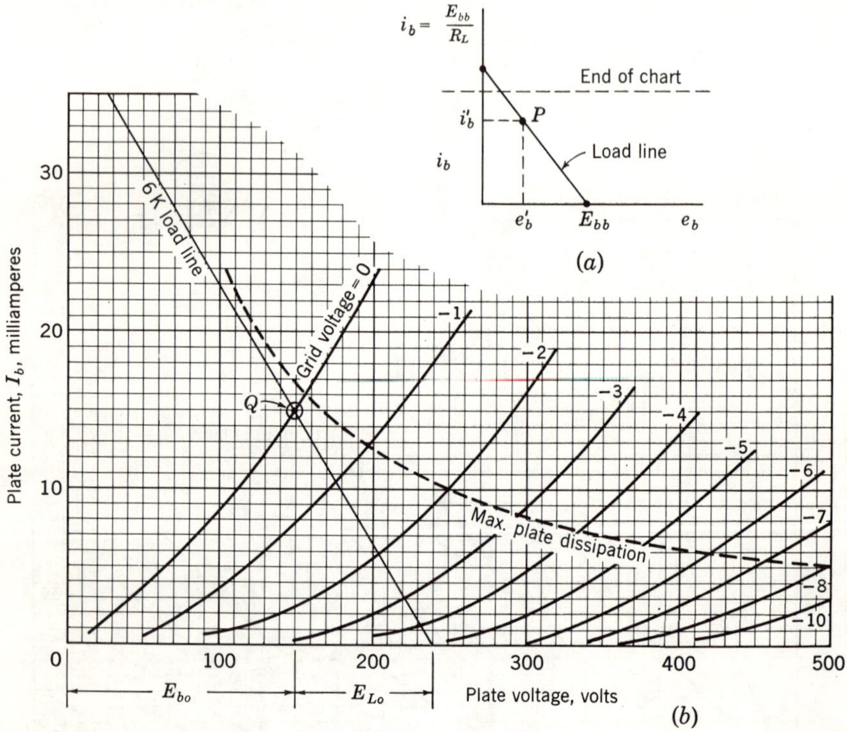

FIG. 5-16. (a) Sketching the load line when the current intercept is off the chart. (b) The load line for $R_L = 6$ kilohms and $E_{bb} = 240$ volts.

at $i_b = I_{b0} = 15$ ma. The quiescent tube voltage E_{b0} can be graphically determined by noting what voltage must exist across the tube with I_{b0} flowing through it. From Fig. 5-15b we see that $E_{b0} = 150$ volts. The quiescent tube voltage could also have been determined by subtracting $E_L = (I_{b0})6{,}000$ from E_{bb}.

The other approach to the problem of finding the Q point in Fig. 5-15a is to draw the 6-kilohm load line. One point of the load line terminates on $E_{bb} = 240$ volts. This corresponds to the case where the tube is an open circuit, so that $E_b = E_{bb}$. The other end of the load line terminates on a value of $I_b = 240$ volts/6 kilohms $= 40$ ma. This corresponds to considering the tube a short circuit. Since 40 ma is beyond the limits of the chart, some other point along the load line must be determined. We can do this by using the relationships of similar triangles. Refer to Fig. 5-16*a*, which illustrates the problem. The coordinates of point P must be determined, since this point is on the chart. Using similar triangles, we can write the following proportions. The height of the large triangle E_{bb}/R_L is to the height of the small triangle i_b' as the base of the large triangle E_{bb} is to the base of the small triangle $(E_{bb} - e_b')$. That is,

$$\frac{E_{bb}/R_L}{i_b'} = \frac{E_{bb}}{E_{bb} - e_b'}$$

We know that $E_{bb} = 240$ volts and $E_{bb}/R_L = 40$ ma. Simply assume a value of i_b' that is surely on the chart, say, 20 ma, and solve for e_b' to get the coordinates of the point P:

$$e_b' = E_{bb} - i_b'R_L = 240 - 20 \text{ ma}(6 \text{ kilohms}) = 120 \text{ volts}$$

The intersection of the 6-kilohm load line with the $e_c = 0$ curve in Fig. 5-16*b* gives the same operating point that we obtained by generating the composite characteristic.

If the grid voltage were fixed to some other level, say, -2 volts, the Q point would move to $E_{b0} = 205$ volts, $I_{b0} = 6$ ma. If the grid bias were increased to -6 volts, the tube would be completely cut off, with $E_{b0} = 240$ volts and $I_{b0} = 0$ ma.

In solving for the Q point by the load-line technique, we are in reality performing a graphic solution to a group of simultaneous equations. We can surely write, for Fig. 5-15*a*,

$$E_b = E_{bb} - I_bR_L = 240 - I_b(6{,}000) \tag{5-1}$$

$$E_b = f(I_b, e_c) \tag{5-2}$$

$$e_c = E_{cc} = 0 \tag{5-3}$$

Equation (5-1) is the equation for the load line and tells us what E_b must be for any I_b. Furthermore, it tells us that when I_b equals zero, E_b must equal E_{bb}. Also, when E_b equals zero, I_b must equal E_{bb}/R_L. These two extreme cases are the points used in the load-line construction. Equation (5-2) is represented by the characteristic curves for the tube. It states that E_b is a function of I_b and e_c. The exact expression for this function is unknown to us. The tube curves, however, plot it. Equations (5-1) and (5-2) in themselves cannot be solved simultaneously, because e_c is a third unknown. Equation (5-3), however, fixes e_c at zero. When the load line is drawn on the tube curves, the intersection of the load line with every e_c curve is a possible solution to (5-1) and (5-2). When we fix e_c at zero, the third equation is used to locate the specific operating point. In the case of a diode, E_b is dependent on I_b only and therefore two equations suffice.

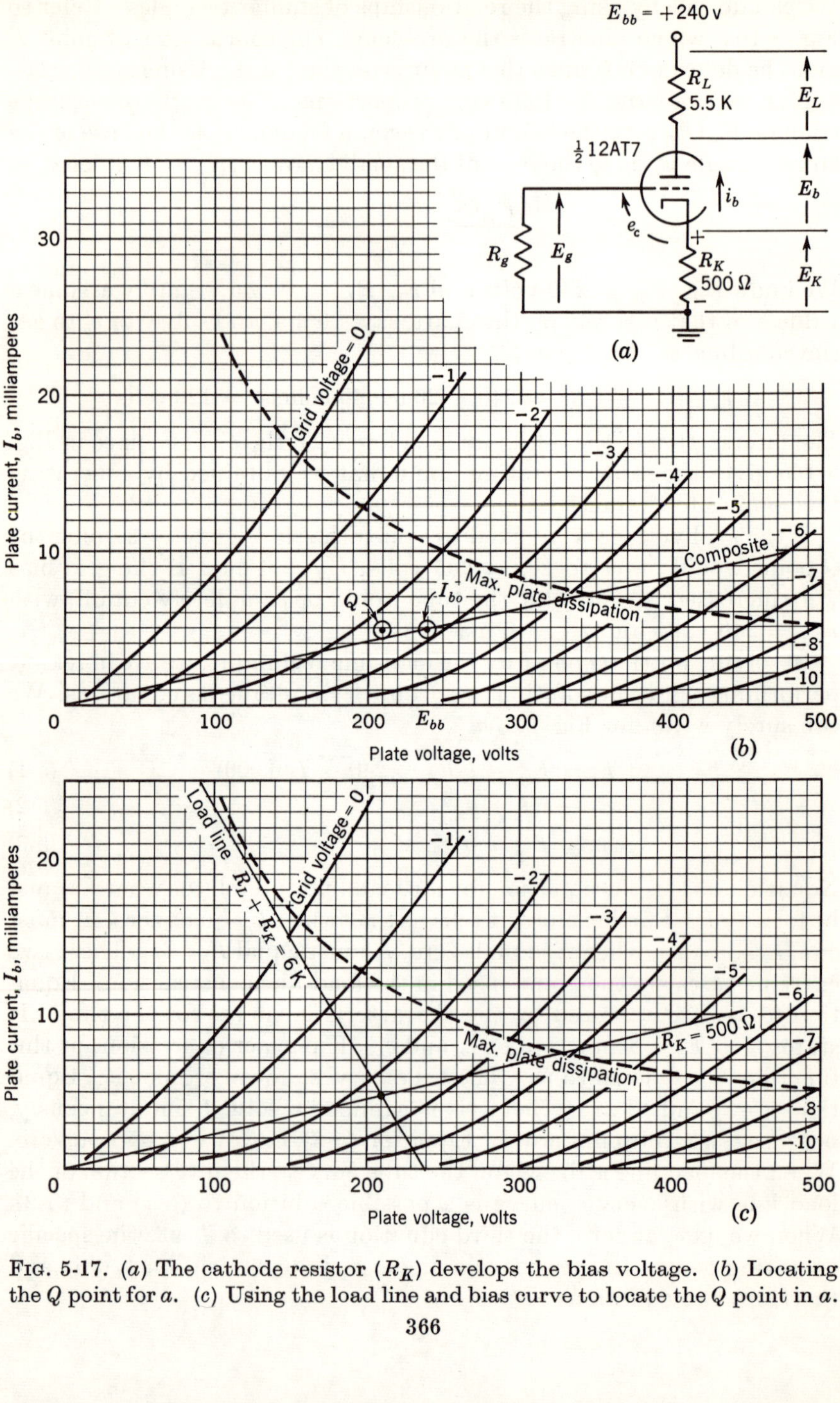

FIG. 5-17. (*a*) The cathode resistor (R_K) develops the bias voltage. (*b*) Locating the Q point for *a*. (*c*) Using the load line and bias curve to locate the Q point in *a*.

5-13. Locating the Q Point with Self-bias. Now suppose we were faced with the circuit of **Fig.** 5-17a. Here the grid bias is undetermined as yet and depends upon the voltage developed across R_K, which is E_K. As i_b (or I_b for d-c considerations) flows through R_K, it develops a voltage E_K with the polarity shown. The voltage drop across R_K can be seen to bias the grid negative with respect to the cathode. Furthermore, the grid current will be zero under these conditions, and so $e_c = E_K$ and $E_g = 0$.

How do we find the Q point for the tube? If we could determine the quiescent current, we would know everything about the circuit. Let us start out by using the techniques we have developed and make suitable modifications if we run into problems.

The first thought that might occur to us is to assume values of i_b, since all the elements are in series. With this assumed i_b, we could determine E_L and E_K without any difficulty by applying Ohm's law. Of course, we could also plot the volt-ampere characteristics of R_L and R_K and graphically determine E_L and E_K for the assumed current. Our graphic solutions would become cluttered with a lot of detail, however, and, therefore, calculated values will be used whenever possible. Once i_b is assumed, say, $i_b = 5$ ma, we have $E_L = (5 \text{ ma})5.5$ kilohms $= 27.5$ volts, $E_K = 5 \text{ ma}(0.5 \text{ kilohm}) = 2.5$ volts. To find E_b for this assumed value of i_b, enter the plate characteristics at 5 ma and locate where 5 ma intersects with $e_c = -E_K = -2.5$ volts. Projecting vertically downward from this point, we could obtain E_b. Then adding E_b to E_K and E_L would give us a point on the composite characteristic. But in doing this, we encounter our first problem. There is no curve plotted for $e_c = -2.5$ volts. We might interpolate the curve, but at best this would be a sketchy procedure. Our problem is that the assumed value of i_b when multiplied by R_K gives a value of $e_c = -E_K$, which is unpublished. To circumvent this difficulty, let us instead assume a value of e_c which is published and compute the corresponding i_b. Then we can pretend we assumed this i_b originally and proceed as before. By steps we then have:

1. Assuming values of e_c (same as $-E_K$ in Fig. 5-17a) and computing corresponding values of i_b, we have

e_c, volts $= -E_K$	0	-1	-2	-3	-4	-5
i_b, ma $= E_K/R_K$	0	2	4	6	8	10

2. Using the tube plate characteristics, fill in the values of E_b corresponding to the i_b, e_c points of step 1. For example, with $e_c = 0$, $i_b = 0$,

$E_b = 0$. With $e_c = -1$, $i_b = 2$ ma, we read $E_b = 79$ volts. With $e_c = -2$, $i_b = 4$ ma, the value of E_b is read as 179 volts, and so on:

e_c	0	−1	−2	−3	−4	−5
i_b	0	2	4	6	8	10
E_b	0	79	179	265	342	420

3. To get the composite characteristics of the tube and associated resistors, we must determine the voltages E_L and E_K for the assumed currents and add them to E_b to get that value of E_{bb} which would have to be applied to pump this assumed current through this circuit. The easiest way to do this is to add R_L and R_K and multiply the sum by the assumed values of i_b. Tabulating this calculation, we have

i_b	0	2	4	6	8	10
$E_L + E_K = i_b(6{,}000)$	0	12	24	36	48	60

4. Now add E_b to $E_L + E_K$ for the E_{bb} required to give the i_b assumed and retabulate all the information prior to plotting the composite curve (Table 5-2).

Table 5-2

$e_c = -E_K$	0	−1	−2	−3	−4	−5
$i_b = I_b$	0	2	4	6	8	10
$e_b = E_b$	0	79	179	265	342	420
$E_L + E_K$	0	12	24	36	48	60
E_{bb}	0	91	203	301	390	480

5. Plot the E_{bb}, i_b values of any column in step 4. For example, we would mark a point on the composite curve having the coordinates $i_b = 4$, $E_{bb} = 203$. Connect these points to determine the shape of the composite curve as shown in Fig. 5-17*b*. The composite curve will virtually be a straight line in most cases and is due to the negative feedback voltage developed across R_K. We will learn more of this later.

6. Determine the exact Q point for the tube by entering the composite curve at the value of E_{bb} actually applied (240 volts) and reading $I_{b0} = 4.9$ ma. The quiescent tube voltage E_{b0} is then determined by subtracting $E_{L0} + E_{K0}$ from E_{bb}. That is, $E_{b0} = 240 - (4.9 \text{ ma})6$ kilohms $= 210.6$ volts. Mark the point I_{b0}, E_{b0} as the Q point for the tube. The value of bias (e_c) the tube will have is $E_{K0} = I_{b0}R_K = (4.9 \text{ ma})0.5$ kilohm $= 2.45$ volts. E_{L0} is equal to $I_{L0}R_L$.

There is an alternative and simpler approach to finding the Q point of the tube in a circuit such as that of Fig. 5-17*a*. It is based upon the fact that E_{bb} is known, and, therefore, if we assume values of i_b, the tube voltage will be $E_b = E_{bb} - i_b(R_L + R_K)$. This is a linear equation, and can be plotted as a 6-kilohm load line. One end of this load line (the voltage-axis intercept) will lie on the point $E_b = E_{bb}$ for i_b equal to zero. The other end of the load line (the current-axis intercept) will lie on the point $i_b = E_{bb}/(R_L + R_K)$. The distance from E_{bb} to any point on this load line is $E_L + E_K$. The distance from the origin to any point on this load line is E_b. Therefore, as the first step, we have:

1. Plot the load line for $R_L + R_K$, as shown in Fig. 5-17*c*.
2. Assume values of e_c and compute the corresponding values of i_b as in step 1 of the alternative method:

e_c, volts	0	−1	−2	−3	−4	−5
i_b, ma	0	2	4	6	8	10

3. In step 2 we computed the exact value of i_b that must flow through R_K to give the assumed values of e_c. Plot these values of e_c, i_b, and draw a smooth curve through them, which will almost be a straight line. For example, with $e_c = -2$, $i_b = 4$ ma. These points are plotted on the line labeled $R_K = 500$ ohms in Fig. 5-17*c*.
4. Now there must be a value of i_b which satisfies the load-line equation $E_b = E_{bb} - i_b(R_L + R_K)$ and develops just the right bias across R_K actually to cause this i_b to flow. The curve plotted in step 3 relates bias to i_b. The quiescent current I_{b0} is that current that satisfies both conditions and is the point where the R_K curve intersects the load line. Note that the Q point of Fig. 5-17*c* is the same as that determined in Fig. 5-17*b*.

The simultaneous equations solved by this procedure are

$$E_b = E_{bb} - i_b(R_L + R_K) \qquad \text{load-line construction}$$

$$E_b = f(i_b, e_c) \qquad \text{tube curves}$$

$$e_c = -i_bR_K = -E_K \qquad \text{bias-curve construction}$$

Problem. Find the Q point for the circuit of Fig. 5-18*a*. What is E_K in this circuit?

Solution. 1. Plot the load line as shown in Fig. 5-18*b*.

2. Plot the cathode load line by assuming values of e_c. For example, with $e_c = -5$ volts, $i_b = 1$ ma.

3. Intersection of these lines is the Q point with $E_{b0} \approx 295$ volts, $I_{b0} \approx 1.1$ ma, and $E_{c0} \approx -5.3$ volts if read graphically. Since I_{b0} is most accurately determined from the curves, use it to calculate other voltages. Therefore, $E_{K0} = 5.5$ volts, $E_{b0} = 294.5$ volts.

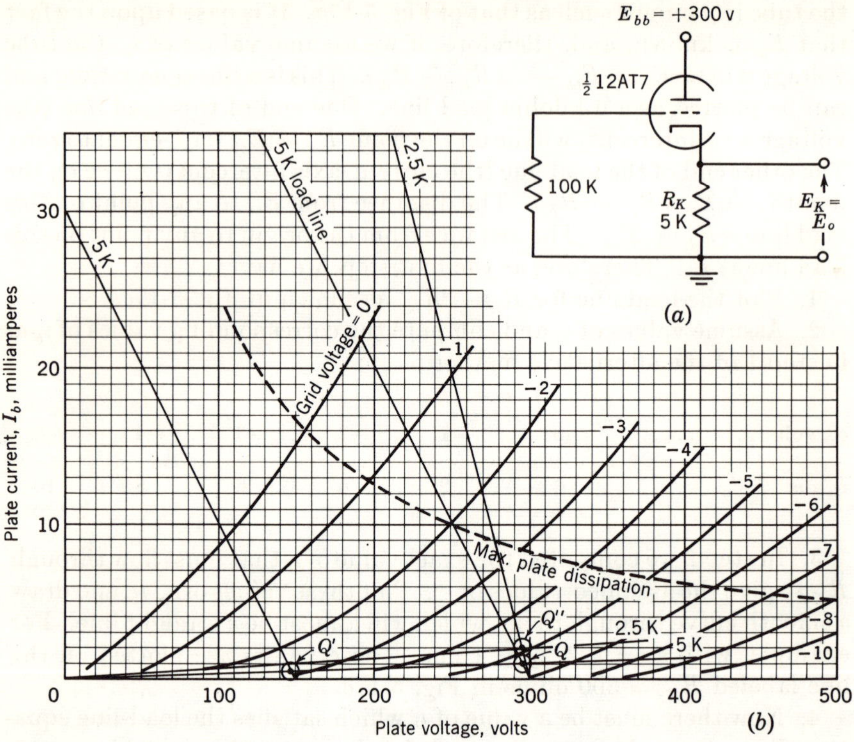

FIG. 5-18. (*a*) A cathode follower. (*b*) Verifying the stability of a cathode follower graphically.

Notice that, as R_K increases, the bias point is shifted closer and closer to cutoff. The tube can never be fully cut off, however, because some current must flow through R_K to develop bias.

The reader may recognize this circuit as that of the highly stable cathode follower. The stability of the circuit to variations in E_{bb} can be determined by plotting a new value of E_{bb}, say, 150 volts, and noting the corresponding change in plate current. If this is done, the new Q point, Q', is at $E'_{b0} = 148$ volts, $I'_{b0} = 0.6$ ma. The plate current has changed no more than $(1.1 - 0.6)$ ma $= 0.5$ ma. If R_K were zero, making $e_c = 0$, and a 5-kilohm resistor were put in the plate lead, the change in plate current would be from $i_b = 21.7$ ma at the intersection of the $e_c = 0$ point on the load line to $i_b = 9$ ma on the $E_c = 0$ curve and 150-volt, 5-kilohm load line. This is a change of $\Delta i_b = 12.7$ ma, as compared with 0.5 ma. The apparent dynamic resistance that the E_{bb} supply looks into when R_K is present is much greater than when it is absent.

Specifically, with R_K in the circuit, the dynamic resistance of the over-all circuit is

$$r_{in} = \frac{\Delta e_b}{\Delta i_b} = \frac{300 - 150}{1.1 - 0.6} = \frac{150 \text{ volts}}{0.5 \text{ ma}} = 300 \text{ kilohms}$$

With $R_K = 0$,

$$r'_{in} = \frac{\Delta e_b}{\Delta i_b} = \frac{300 - 150}{21.7 - 9} = \frac{150 \text{ volts}}{12.7 \text{ ma}} = 11.8 \text{ kilohms}$$

Notice the tremendous difference between r_{in} and r'_{in}. It is much greater than that caused by the mere addition of a 5-kilohm resistor. This apparent increase in dynamic resistance is due to the plate current developing bias across R_K, which tends to oppose variations in the E_{bb} supply. For example, an increase in E_{bb} tends to increase the current. The current increase, however, increases the tube bias, and this increases the tube resistance, which tends to minimize the possible increase in current. Do not forget that a change in e_c can exert a greater effect on plate current than can a change in e_b.

It might also be interesting to investigate the impedance seen looking back into the terminals across the 5-kilohm resistor. The open-circuit voltage V_{oc} in Fig. 5-18*a* is $E_{K0} = 5.5$ volts. If we place a short circuit across these terminals, we obtain the short-circuit current I_{sc}. Now I_{sc} will be the current that would flow through the tube with $e_c = 0$ and $E_{bb} = 300$ volts since R_K is shorted. We can get this information from the tube curves by noting that if the $e_c = 0$ curve were extended so that it intersected the $E_{bb} = 300$-volt line, the value of i_b would be about 39 ma. Do not try this on a circuit, since the plate dissipation of the tube is greatly exceeded. The apparent dynamic output resistance is then equal to the change in output voltage divided by the change in output current. Therefore,

$$Z_{th} = r_o = \frac{V_{oc}}{I_{sc}} = \frac{5.5 \text{ volts}}{39 \text{ ma}} = 140 \text{ ohms}$$

Notice that the apparent dynamic output resistance is much less than R_K, perhaps because the d-c resistance of the tube is in parallel with R_K. The R_p with $e_b = 300$ volts and $e_c = 0$ is $R_P = 300$ volts/39 ma $= 7.7$ kilohms. However, 7.7 kilohms in parallel with 5 kilohms is 3 kilohms and not 140 ohms. Another unusual aspect of this circuit! The dynamic output resistance is quite low. Physically, this means that drawing a little current through a load placed across the output terminals (in parallel with R_K) should not change the output voltage E_K very much. Let us see if this is so by a graphic analysis. Suppose we assume a 5-kilohm load hung in parallel with R_K so that the total R_K is 2.5 kilohms. What will the new E_K, E''_K be?

To solve for E''_K, simply draw a 2.5-kilohm load line as shown in Fig. 5-18*b*, and then draw a 2.5-kilohm cathode load line. We see that I''_{b0} is 1.95 ma and $E''_K = (1.95 \text{ ma})2.5 \text{ kilohms} = 4.9$ volts. The change in E_K is only 0.6 volt.

5-14. Locating the Q Point with Combination Bias. In Fig. 5-19*a* the grid voltage (e_c, that is) is developed by the fixed source, E_{cc}

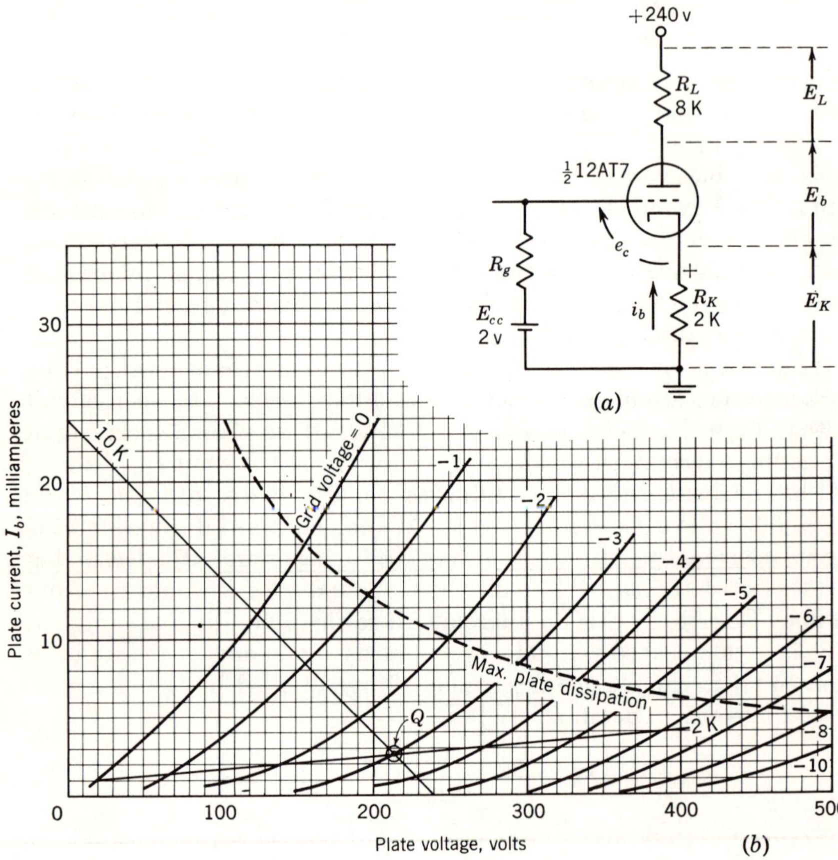

FIG. 5-19. (*a*) Determine the Q point. (*b*) Evaluating the Q point.

and the self-bias voltage across R_K. The voltage on the grid with respect to the cathode ($e_{kg} = e_c$) is readily determined by starting at the cathode, going down in algebraic potential by an amount $i_b R_K$ as we go through R_K and then up by E_{cc}. We shall assume zero grid current:

$$e_c = -i_b R_K + E_{cc} = -E_K + E_{cc}$$

A simple method of determining the Q point in this problem is:

1. Plot the load line for the total load that the tube sees. That is,

$E_b = E_{bb} - i_b(R_L + R_K) = 240$ volts $- i_b(10$ kilohms). This will be a straight line having a slope equal to $-1/10^4$ and having e_b and i_b intercepts of 240 volts and 24 ma, as shown in Fig. 5-19*b*.

2. Assume values of E_K which will result preferably in published values of e_c when plugged into the equation $e_c = -E_K + E_{cc}$. To do this, let us solve for E_K in $E_K = E_{cc} - e_c$, and determine the corresponding values of i_b:

e_c	0	−2	−4	−6
E_K	2	4	6	8
$E_K/R_K = i_b$, ma	1	2	3	4

3. Plot the e_c, i_b points, and connect them with a smooth curve to get the cathode load line.

4. Locate the Q point at the intersection of these lines. Note that $I_{b0} = 2.5$ ma, $E_{c0} = -3$ volts, and $E_{b0} = 213$ volts.

In Fig. 5-20*a*, determine the tube's Q point for a few values of I_{b_2} in the range of 0-2 ma. The current I_{b_2} might be flowing into another tube whose cathode is also tied to R_K.

1. Plot the total load line; that is, use $R_L + R_K = 6$ kilohms, as in Fig. 5-20*b*.

2. Plot the bias curve $e_c = -(I_{b_1} + I_{b_2})R_K$. To be sure of getting published values of e_c, let us instead solve for i_{b_1} values necessary to give published e_c values. We shall also assume values of I_{b_2} in the range 0-2 ma:

$$e_c = -(I_{b_1} + I_{b_2})R_K$$
$$-e_c = I_{b_1}R_K + I_{b_2}R_K$$
$$I_{b_1}R_K = -(e_c + I_{b_2}R_K)$$
$$I_{b_1} = \frac{e_c + I_{b_2}R_K}{R_K}$$

(The values of e_c plugged into this equation are negative in themselves.) If $I_{b_2} = 0$,

e_c	0	−1	−2	−3	−4
I_{b_1}	0	2	4	6	8

The Q point above is at $I_{b0} = 4.9$ ma. $E_{K0} = 2.45$ volts. (I_{b0} is the quiescent value of I_{b_1}.) If $I_{b_2} = 1$ ma,

e_c	0	−1	−2	...	−4
I_{b_1}	−1	1	3	...	7

The Q point above is at $I_{b0} = 4.3$ ma, $E_{K0} = R_K(4.3 + 1)$ ma $= 2.6$ volts. To get $e_c = 0$ with $I_{b_2} = 1$ ma, the current I_{b_1} would have to be equal and opposite to I_{b_2}. This of course cannot occur, owing to the unilateral conductivity of the tube. If $I_{b_2} = 2$ ma,

e_c	0	−1	−2	−4
I_{b_1}	−2	...	2	6

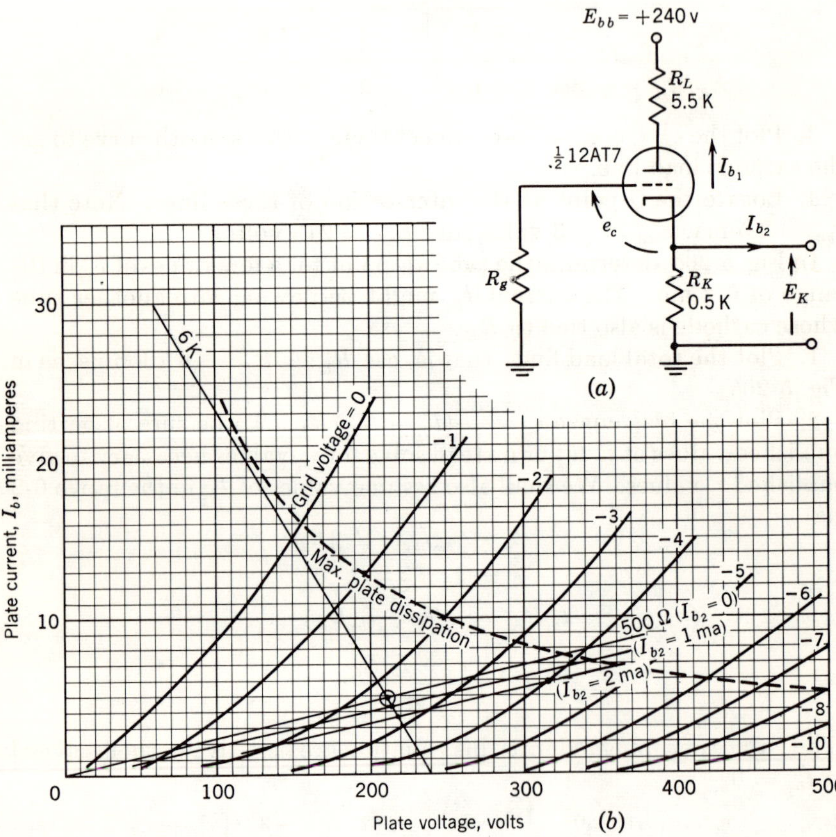

FIG. 5-20. (a) The Q point will shift as I_{b_2} varies. (b) Analyzing the shift in the Q point for different values of I_{b_2}.

The Q point above is at $I_{b0} = 3.7$ ma and $E_{K0} = R_K(3.7 + 2)$ ma $=$ 2.8 volts. From the tube curves we see that if we wished to cut I_{b_1} off, the bias must equal -5 volts. For I_{b_2} to cut off I_{b_1}, it must equal $I_{b_2} =$ 5 volts/0.5 kilohm $= 10$ ma.

Actually, we have neglected a minor consideration in finding the Q

points for different values of I_{b_2}. The current I_{b_2}, in flowing through the 0.5-kilohm resistor, actually reduces the value of E_{bb} that the tube looks back into, and therefore the 6-kilohm load lines should be redrawn and shifted slightly to the left. For example, suppose $I_{b_2} = 10$ ma, which cuts the tube off. The tube voltage, when cut off in this manner, is 240 volts − (10 ma)0.5 kilohm = 235 volts. Since the change in the plate-voltage intercept is only 5 volts, the increased accuracy is hardly justified by the labor involved.

Suppose we wished to determine E_K as a function of I_2 in Fig. 5-21*a* Circuits of this type with R_L equal to zero are frequently found in digital computers and are called cathode-follower clamps. The bias voltage E_{cc} is usually a tap on a voltage divider.

We can write the following equations for Fig. 5-21*a*:

$$E_b = f(I_1, e_c) \tag{5-4}$$

$$e_c = -E_K + E_{cc} \tag{5-5}$$

$$E_K = (I_1 + I_2)R_K \tag{5-6}$$

$$E_b = E_{bb} - E_K - I_1R_L \tag{5-7}$$

Substituting (5-6) into (5-7), we have

$$\begin{aligned} E_b &= E_{bb} - (I_1 + I_2)R_K - I_1R_L \\ &= E_{bb} - I_2R_K - I_1(R_K + R_L) \end{aligned} \tag{5-7a}$$

Equation (5-7*a*) tells us that if there is an I_2 flowing through R_K, the load line must be drawn from a point on the plate voltage axis equal to $E_{bb} - I_2R_K$ instead of E_{bb}. In other words, one effect of I_2 is to reduce the E_{bb} supply so far as the tube is concerned. The other effect is to modify the bias, as shown by Eqs. (5-5) and (5-6). Figure 5-21*a* is therefore equivalent to Fig. 5-21*b*. With the aid of Fig. 5-21*b*, we can readily determine the Q point and value of E_K for any assumed I_2 as shown in Fig. 5-21*c*.

For example, with $I_2 = 0$, the Q point is found by drawing the 4-kilohm load line with 238 volts as the e_b-axis intercept. A bias curve is then constructed from

$$e_c = -I_1R_K - I_2R_K + E_{cc} \qquad \text{or} \qquad I_1 = \frac{E_{cc} - I_2R_K - e_c}{R_K} \tag{5-5a}$$

and with $I_2 = 0$,

$$I_1 = \frac{11 - e_c}{1K}$$

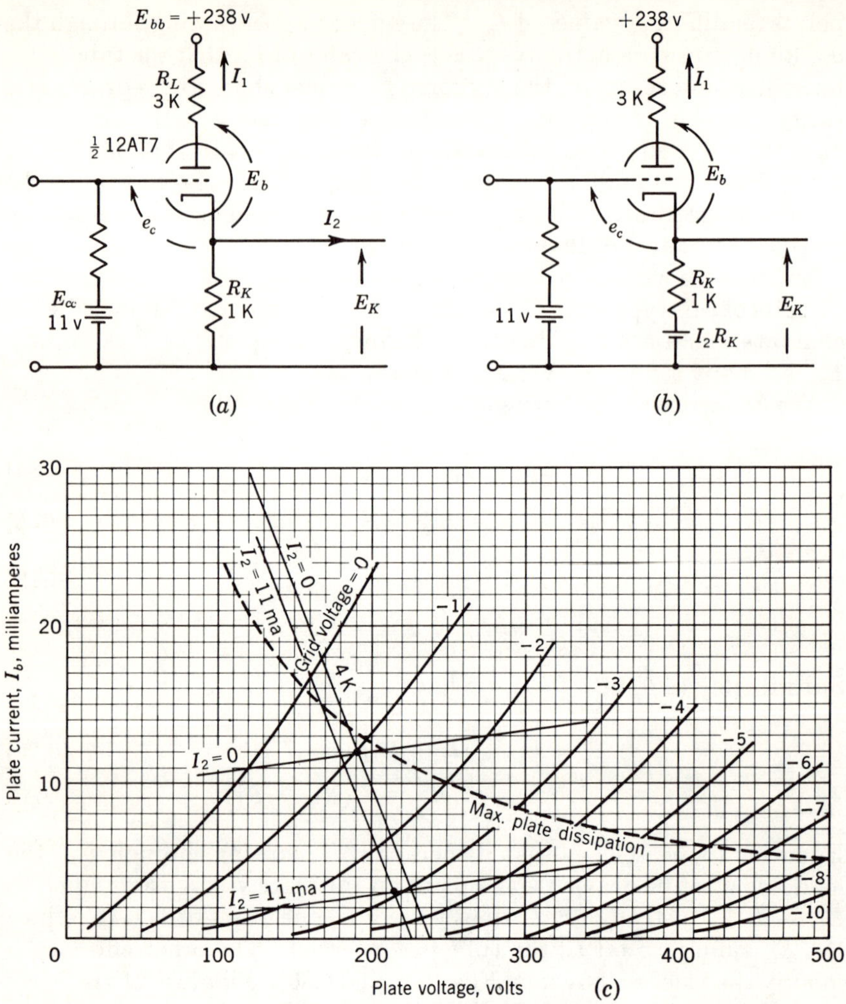

FIG. 5-21. (a) Determine the Q point. (b) The effect of I_2 is equivalent to inserting a battery of I_2R_K volts in series with R_K. (c) Determining the Q point as I_2 varies.

Assuming values of e_c, we can tabulate corresponding I_1 values and draw a bias curve:

e_c	0	−2	−4
I_1, ma ...	11	13	15

The intersection of the bias curve for $I_2 = 0$ with the $I_2 = 0$ load line is the Q point for $I_2 = 0$. At this Q point, $I_1 = I_{10} = 12$ ma, and from Eq. (5-6), $E_K = 12$ volts.

If we assume $I_2 = 11$ ma, the corresponding load line will intersect the e_b axis at 227 volts and of course lie parallel to the load line for $I_2 = 0$. Using Eq. (5-5*a*), the bias curve for $I_2 = 11$ ma can be constructed by assuming values of e_c again:

e_c	0	−2	−4
I_1, ma ...	0	2	4

The Q point in this case lies on $I_1 = I_{10} = 3$ ma. The value of E_K is then, by Eq. (5-6), equal to 14 volts. Note that a change in I_2 of 11 ma has caused E_K to vary by only 2 volts. The apparent source impedance looking into the E_K terminals is therefore 2 volts/11 ma = 182 ohms.

5-15. The Dynamic Q Point. Sometimes the dynamic Q point with an input signal is different from the static Q point without a signal. Let us call the Q point with no signal Q_0 and that with signal Q_d. As an extreme illustration of the Q point shift, consider the following problem.

Figure 5-22*a* is to be analyzed for the values Q_0, Q_d, and the readings of the d-c voltmeters and milliammeters. Also, determine P_d for both conditions. The switch is closed for 20 μsec and opened for 80 μsec during each cycle of operation.

First, draw the 5-kilohm load-line curve, as in Fig. 5-22*b*. The current-axis intercept is 300 volts/5 kilohms = 60 ma, which is off the chart. By similar triangles we deduce, however, that the load line will pass through the point 150 volts/30 ma. With the input signal off, the bias is −6 volts. From the curves we see that this is just enough bias to keep the tube cut off. Therefore, Q_0 is at $I_{b0} = 0$, $E_{b0} = 300$ volts, $E_{L0} = 0$ volts.

When the input signal jumps to its peak value of 6 volts, the grid-to-cathode voltage is zero. The Q point jumps to the values $i_b = 21.8$ ma, $e_c = 0$, and $e_b = 191$ volts. The drop across the load resistor is 109 volts at this time. This instantaneous Q point is not Q_d, however. To get Q_d, we must determine the average, or d-c, values of E_b and I_b. The instantaneous waveforms for this circuit are displayed in Fig. 5-22*c*.

The average value of i_b is

$$I_b = 21.8 \text{ ma}\left(\frac{20\ \mu\text{sec}}{100\ \mu\text{sec}}\right) = 4.4 \text{ ma}$$

This is what the d-c milliammeter would read. (Recall that the ratio of pulse presence to the pulse period is the duty cycle.) The average value of e_b which the E_b voltmeter reads is

$$E_b = \frac{191 \text{ volts}(20\ \mu\text{sec}) + 300 \text{ volts}(80\ \mu\text{sec})}{100\ \mu\text{sec}} = 278.2 \text{ volts}$$

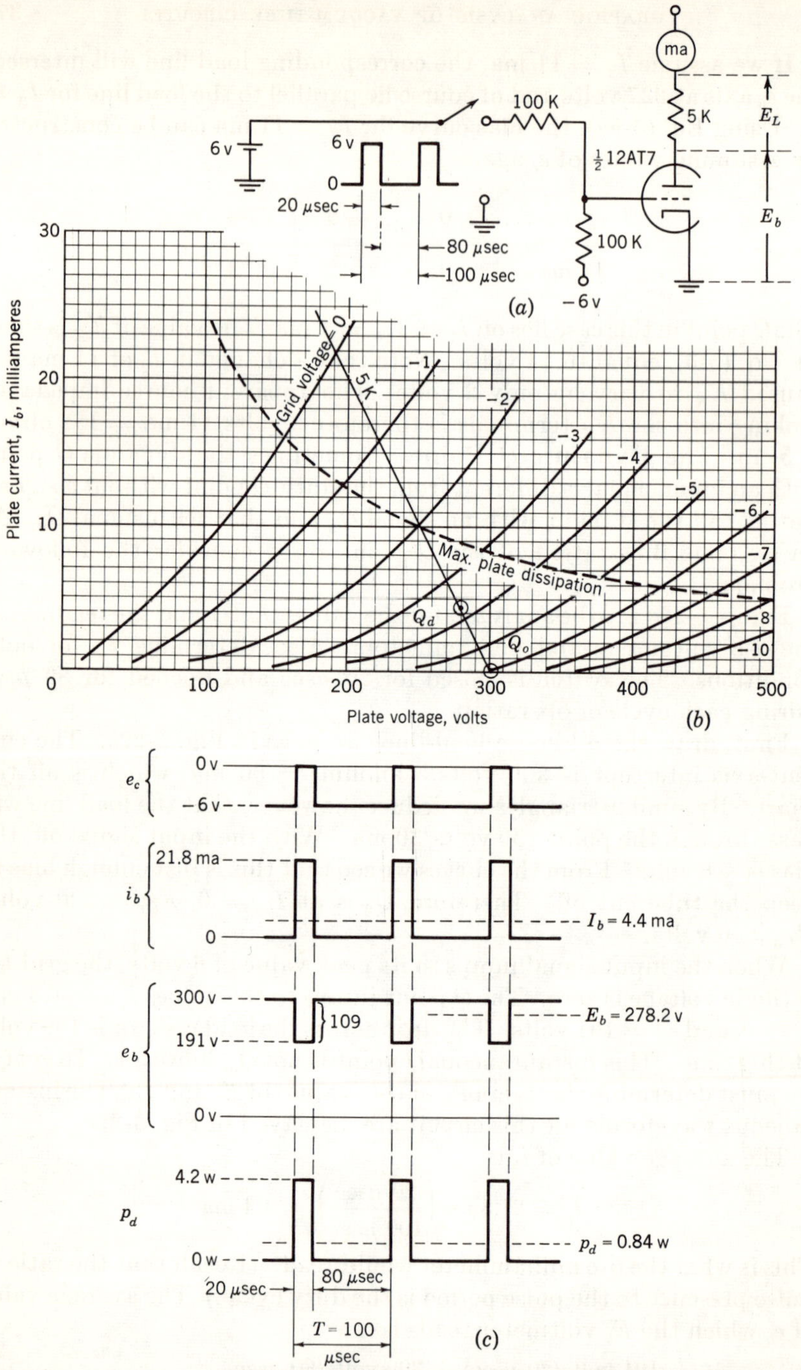

FIG. 5-22. (*a*) The average plate voltage, current, and dissipation are to be determined. (*b*) The Q point with the signal is not always the same as without the signal. (*c*) Waveforms for *a*.

The average value of e_L is

$$E_L = 109 \text{ volts}\left(\frac{20\ \mu\text{sec}}{100\ \mu\text{sec}}\right) = 21.8 \text{ volts}$$

which is also the difference between E_{bb} and E_b.

The peak power will occur when the tube is passing current and is equal to $P_{d_{\max}} = 21.8$ ma(191 volts) $= 4.2$ watts. The average power is

$$P_d = 4.2 \text{ watts}\left(\frac{20\ \mu\text{sec}}{100\ \mu\text{sec}}\right) = 0.84 \text{ watt}$$

Note that the average power is well below the maximum rated value of 2.5 watts.

The coordinates of Q_d are $e_b = 278.2$ volts, $i_b = 4.4$ ma. Be sure to compute the power by taking the average value of the power versus time curve and *not* the product of the average (d-c) voltage (E_b) and the average current (I_b). The average value of the product of two functions is not the same as the product of the average values.

5-16. The Dynamic Load Line. In the circuit of Fig. 5-22*a*, the value of plate current for any value of plate voltage is a point on the load line. Another way of expressing this is to say that the load line is the locus of all the possible operating points for that circuit. This load line is called the static load line.

Now suppose we had the circuit of Fig. 5-23*a* and for the moment imagine there is no capacitor in the circuit. The static load line will be as shown in Fig. 5-23*b*. At any value of e_c there will be an i_b, e_b point on this particular load line. For example, when the input goes to 5 volts, the net bias (e_c) is essentially zero, and $i_b = 9.2$ ma, $e_b = 108$ volts.

Now let us put the capacitor back into the circuit and see what possible values the e_b, i_b points might have. Assume that the tube is initially cut off. The capacitor is charged to 200 volts. When the input pulse comes in, the grid voltage instantaneously jumps to zero. The capacitor, however, cannot change its voltage instantaneously, and the capacitor voltage is the tube plate voltage. With 200 volts on the plate and zero volts on the grid, the operating point must have jumped straight up from point 1 to point 2. The capacitor is discharging through the tube, and the peak tube current is about 23 ma. The grid voltage stays at zero for 20 μsec. The final value of voltage to which the capacitor discharges will be determined by point three on the load line, since this is the static voltage across the tube with $e_c = 0$ and a 10-kilohm load. Therefore, the operating point of the tube must slide down the $e_c = 0$ curve from point 2 to point 3. The apparent source resistance that the capacitor looks back into will be that presented by the a-c resistance of the tube in parallel with the 10-kilohm load resistor. We can determine this a-c resistance by noting that the corresponding change in i_b produced the change in e_b between points 2 and 3 with e_c held constant at

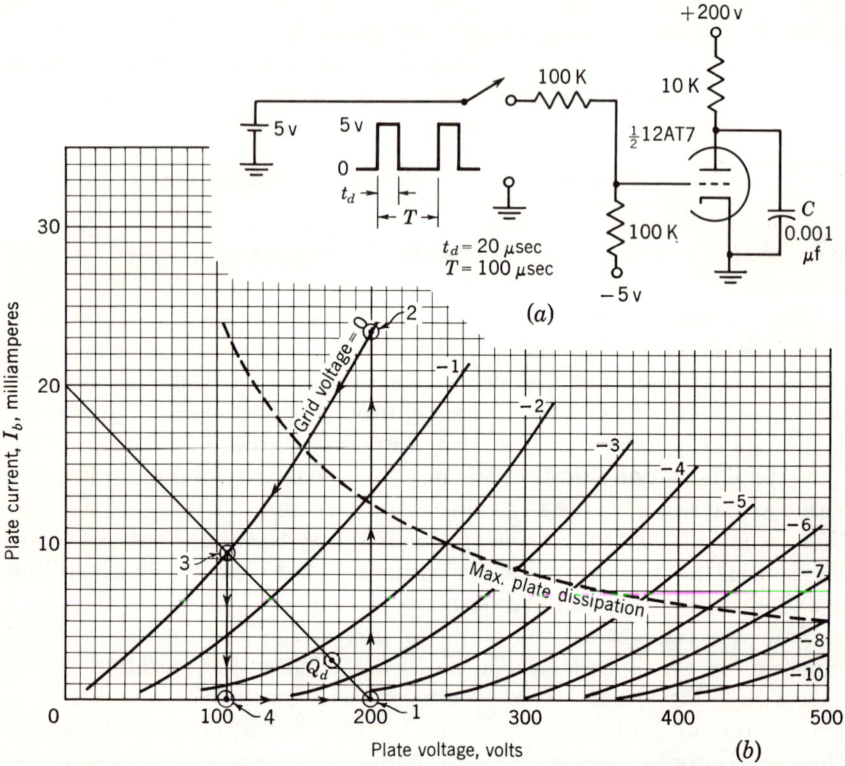

FIG. 5-23. (a) The locus of operating points with a reactive load is not restricted to the load line. (b) A graphic description of the operating point locus in a.

zero. Notice how this fits in with our basic definition of r_p:

$$r_p = \frac{\Delta e_b}{\Delta i_b}\bigg/ e_c \text{ constant} = \frac{(200 - 108) \text{ volts}}{(23 - 9.2) \text{ ma}} = \frac{92}{13.8} = 6.7 \text{ kilohms}$$

The apparent source resistance that the capacitor discharges through is then

$$R_{eq} = \frac{r_p R_L}{r_p + R_L} = \frac{(6.7)(10)}{16.7} = 4 \text{ kilohms}$$

The discharge time constant is $\tau_d = R_{eq}C = 4$ μsec. Since τ_d is about one-fifth as large as the pulse duration of 20 μsec, we can be sure that the capacitor will actually discharge down to point 3.

When the pulse ends, the tube is suddenly switched off again, and appears as an open circuit. The capacitor now charges toward 200 volts again with the charging time constant, $\tau_c = 10$ kilohms(0.001 μf) = 10 μsec. The 80-μsec "off" interval is then adequate for the capacitor to recharge to 200 volts.

The instant the tube switches off at point 3, the current ceases instantaneously, since the bias is now more than enough to cause cutoff for lesser plate voltage of point 3. The operating point therefore jumps from point 3 to point 4. The tube is held off while e_b builds up from the value at point 4 to that of point 1 again. Even though the voltage is changing, the current is still unvarying, which means that r_p is infinite. The arrows of Fig. 5-23*b* trace out the dynamic load line, which is the actual locus of operating points with a signal present. The dynamic load line, as we have just seen, can be radically different from the static load line. We shall pursue this matter of dynamic load line further.

At this point, the reader may be wondering why we did not thevenize the circuit, looking back from the capacitor by considering the d-c, or static, tube resistance (R_p) in parallel with R_L. That is, use the R_p value at point 3 as the tube resistance when it switches on, since this value of R_p will be the correct value to give a final capacitor voltage of 108 volts.

For example, R_p at point 3 is R_p = 108 volts/9.2 ma = 11.8 kilohms. We might then be tempted to say that the equivalent circuit due to the tube switching on is that of Fig. 5-24*a*. The final capacitor voltage will be

$$V_{oc} = 200 \text{ volts}\left(\frac{11.8 \text{ kilohms}}{10 + 11.8 \text{ kilohms}}\right) = 108 \text{ volts}$$

It is true that this equivalent circuit represents the final conditions at point 3, but it does not account for the fact that the operating point must have jumped from point 1 to point 2, where R_p = 200 volts/23 ma = 8.7 kilohms. The equivalent circuit of Fig. 5-24*a* is perfectly correct for determining final conditions, but not transitional conditions. This is because the tube is a nonlinear device, and Thévenin's theorem applies to linear devices. There will be times when we can apply Thévenin's theorem to nonlinear circuits, but with discretion. In the circuit we have just discussed, the R_p of point 3 would be used in many cases, although it is not quite correct. Experience in the field will give the reader a feeling for what approximations are valid in various applications.

While we are on the subject, there is one more approach we could consider. The plate characteristic for $e_c = 0$ could be nicely approximated by two straight-line segments. One would be from the origin to point 3, and the other from point 3 to point 2, as shown in Fig. 5-24*b*.

The equivalent circuit for the two regions that the segments appear in would be as shown in Fig. 5-25*a*. In region I we see a device which has no open-circuit voltage, since the segment goes through the origin. The internal resistance of the device is determined by the change in current

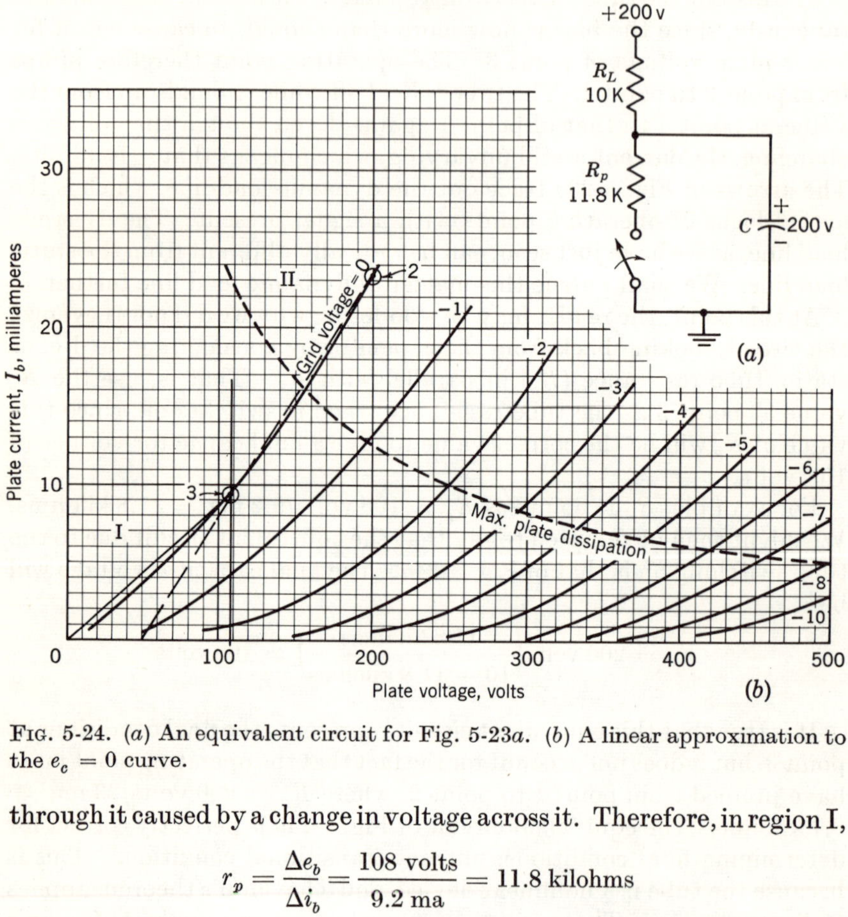

FIG. 5-24. (*a*) An equivalent circuit for Fig. 5-23*a*. (*b*) A linear approximation to the $e_c = 0$ curve.

through it caused by a change in voltage across it. Therefore, in region I,

$$r_p = \frac{\Delta e_b}{\Delta i_b} = \frac{108 \text{ volts}}{9.2 \text{ ma}} = 11.8 \text{ kilohms}$$

In region II we see that by extending the straight-line segment until it intersects the e_b axis, we have a device which exhibits an open-circuit voltage of 50 volts with zero current through it. The internal impedance of the device is the reciprocal of the slope of the segment in region II and is

$$r_p = \frac{\Delta e_b}{\Delta i_b} = \frac{200 - 108 \text{ volts}}{23 - 9.2 \text{ ma}} = 6.7 \text{ kilohms}$$

We need not concern ourselves with region I, since the dynamic operating point does not move into it.

Now let us see how well this equivalent circuit works out. Assume that the tube has just switched on, so that the capacitor is still charged to 200 volts, as shown in Fig. 5-25*b*. The peak current through the tube will be

$$i_{b_{pk}} = \frac{200 - 50 \text{ volts}}{6.7 \text{ kilohms}} = 22.4 \text{ ma}$$

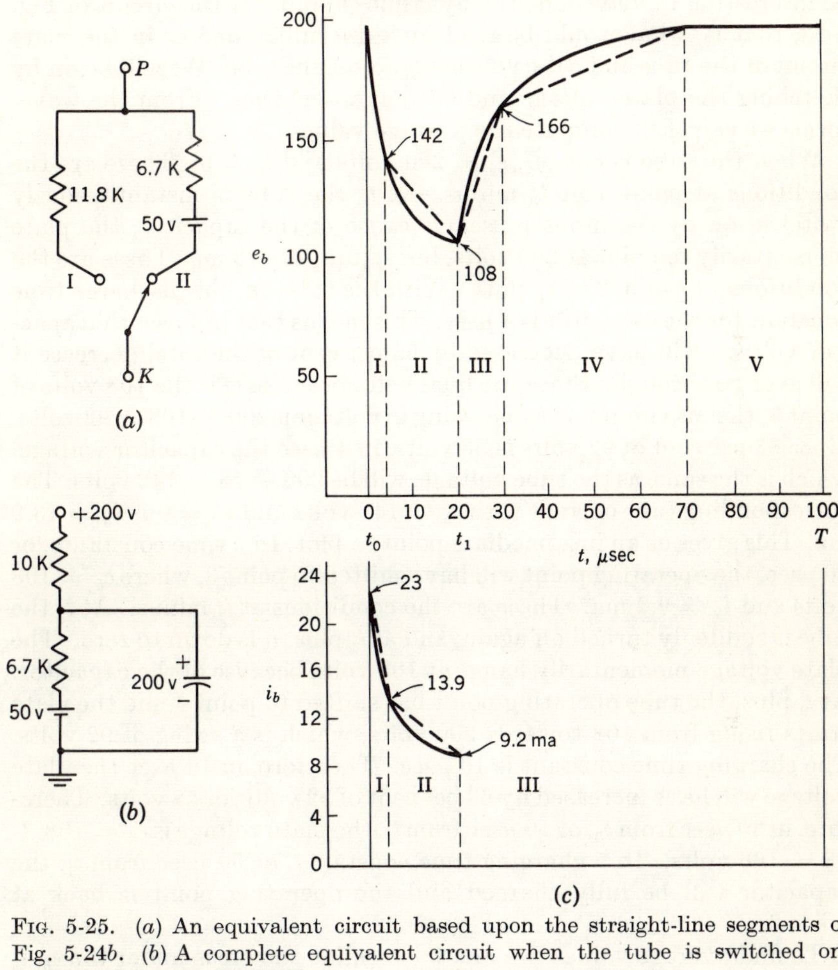

FIG. 5-25. (*a*) An equivalent circuit based upon the straight-line segments of Fig. 5-24*b*. (*b*) A complete equivalent circuit when the tube is switched on. (*c*) Wave-forms for Fig. 5-23*a*.

which compares very closely with 23 ma as read from the plate characteristics. The final capacitor voltage (same as E_b) will be the static drop across the 6.7-kilohm resistor added to the 50-volt source.

$$E_b = 50 + (200 - 50)\frac{6.7 \text{ kilohms}}{16.7 \text{ kilohms}} = 110 \text{ volts}$$

This is quite close to the 108 volts previously determined for E_b at point 3.

We have delved rather deeply into the detailed mechanics of a particular circuit, in order to show that the dynamic load line or path of operating points can be very different from the static load line. It might be interesting to determine the dynamic Q point for the circuit of Fig. 5-23, that is, what would be read by a d-c milliammeter in the plate circuit of the tube and a d-c voltmeter across the tube. We will begin by sketching the plate voltage and current waveforms. From the waveforms we can then compute the average values.

When the tube is cut off, $e_b = 200$ volts and $i_b = 0$. These are the conditions at point 1 at t_0 minus. At t_0 the tube is instantaneously switched on by the input pulse. Because of the capacitor, the plate momentarily remains at 200 volts, but i_b jumps to 23 ma. These are the conditions at point 2 at t_0 plus. With the tube on, the discharge time constant for the capacitor is 4 μsec. This means that in 4 μsec the capacitor voltage will have decreased by 63 per cent of the total decrease it will ever go through. Since the final voltage across C is the 108 volts of point 3, the maximum possible swing in voltage is $200 - 108 = 92$ volts. Since 63 per cent of 92 volts is 58 volts, in 4 μsec the capacitor voltage, which is the same as the tube voltage, will be $200 - 58 = 142$ volts. The corresponding tube current with $e_b = 142$ volts and $e_c = 0$ is $i_b = 13.9$ ma. This gives us an intermediate point to plot. In 5 time constants, or 20 μsec, the operating point will have shifted to point 3, where $e_b = 108$ volts and $i_b = 9.2$ ma. These are the conditions at t_1 minus. At t_1 the tube is suddenly turned off again, and at t_1 plus, i_b is down to zero. The plate voltage momentarily hangs at 108 volts because of the capacitor. At t_1 plus, the tube operating point has shifted to point 4 and the plate starts rising from 108 towards 200 volts, which is a swing of 92 volts. The charging time constant is 10 μsec. Therefore, in 10 μsec the plate voltage will have increased by 63 per cent of 92 volts, or 58 volts. Therefore, in 30 μsec from t_0, or 10 μsec from t_1, the plate voltage is $e_b = 108 + 58 = 166$ volts. In 5 charging time constants, or 50 μsec from t_1, the capacitor will be fully charged and the operating point is back at point 1.

To find average values, we must determine the areas under the e_b, i_b curves. By methods of calculus this could be done exactly. We can, however, approximate the area by drawing the curves on graph paper and counting squares or by dividing the area bounded by the curve into thin rectangular strips and summing the areas of the strips as we have previously done. We could also approximate the curve by straight-line segments, which is probably the easiest way. The straight-line approximations are shown by dashed lines in Fig. 5-25c.

The area under the e_b curve in section I is

$$
\begin{aligned}
A_{\text{I}} &= (142 \text{ volts})(4\ \mu\text{sec}) + \tfrac{1}{2}(4\ \mu\text{sec})(58) = 568 + 116 \\
&= 684 \text{ volt-}\mu\text{sec} \\
A_{\text{II}} &= (108)(16\ \mu\text{sec}) + \tfrac{1}{2}(16\ \mu\text{sec})(34) = 1{,}730 + 272 \\
&= 2{,}002 \text{ volt-}\mu\text{sec} \\
A_{\text{III}} &= (108 \text{ volts})(10\ \mu\text{sec}) + \tfrac{1}{2}(10\ \mu\text{sec})(58 \text{ volts}) \\
&= 1{,}080 + 290 = 1{,}370 \text{ volt-}\mu\text{sec} \\
A_{\text{IV}} &= (166 \text{ volt})(40\ \mu\text{sec}) + \tfrac{1}{2}(40\ \mu\text{sec})(34 \text{ volt}) \\
&= 6{,}640 + 680 = 7{,}320 \text{ volt-}\mu\text{sec} \\
A_{\text{V}} &= (200 \text{ volt})(30\ \mu\text{sec}) = 6{,}000 \text{ volt-}\mu\text{sec}
\end{aligned}
$$

The summation of the areas from A_{I} to and including A_{V} is 17,376 volt-μsec. The dynamic average value of e_b is the area under the curve divided by T:

$$E_{bd} = \frac{17{,}376 \text{ volt-}\mu\text{sec}}{100\ \mu\text{sec}} = 173.8 \text{ volts}$$

The area under the i_b curve is

$$
\begin{aligned}
A_{\text{I}} &= (13.9 \text{ ma})(4\ \mu\text{sec}) + \tfrac{1}{2}(4\ \mu\text{sec})(9.1 \text{ ma}) = 73.8 \text{ ma-}\mu\text{sec} \\
A_{\text{II}} &= (9.2 \text{ ma})(16\ \mu\text{sec}) + \tfrac{1}{2}(16\ \mu\text{sec})(4.7 \text{ ma}) = 184.6 \text{ ma-}\mu\text{sec} \\
A_{\text{III}} &= 0
\end{aligned}
$$

The summation of the current areas is 258.4 ma-μsec, and the dynamic average value of i_b is

$$I_{bd} = \frac{258.4 \text{ ma-}\mu\text{sec}}{100\ \mu\text{sec}} = 2.58 \text{ ma}$$

The dynamic Q point coordinates are then $e_b = 173.8$ volts, $i_b = 2.58$ ma. Notice that this point falls on the static load line. It must do so, since we have computed the d-c values that i_b and e_b will have with a signal present. The static load line is the locus of all the d-c operating points that the tube can assume with a fixed E_{bb} supply and d-c load resistance.

What would the static and dynamic load lines for the circuit of Fig. 5-26*a* be?

1. The static load resistance that the tube looks into is made up of R_1 in series with parallel combination of R_2 and R_3. Since R_3 is small in comparison with R_2 and R_1, we shall approximate the load resistance by considering R_1 only. Therefore, as a first step, draw a 10-kilohm load line as shown in Fig. 5-26*b*.

2. With zero signal input, the Q point will lie on the intersection of the load line with the $e_c = 0$ curve at point 1. The plate current is 9.2 ma, and the plate voltage is about 106 volts if read from the curves. Since R_3 is much smaller than R_2, we shall assume all the plate current flows through R_3.

3. At t_0 the tube suddenly cuts off. The inductor current will still be 9.2 ma for an instant. The equivalent circuit at t_0 plus is shown in Fig. 5-26*c*. The initial value of the current source is 9.2 ma. Since all the

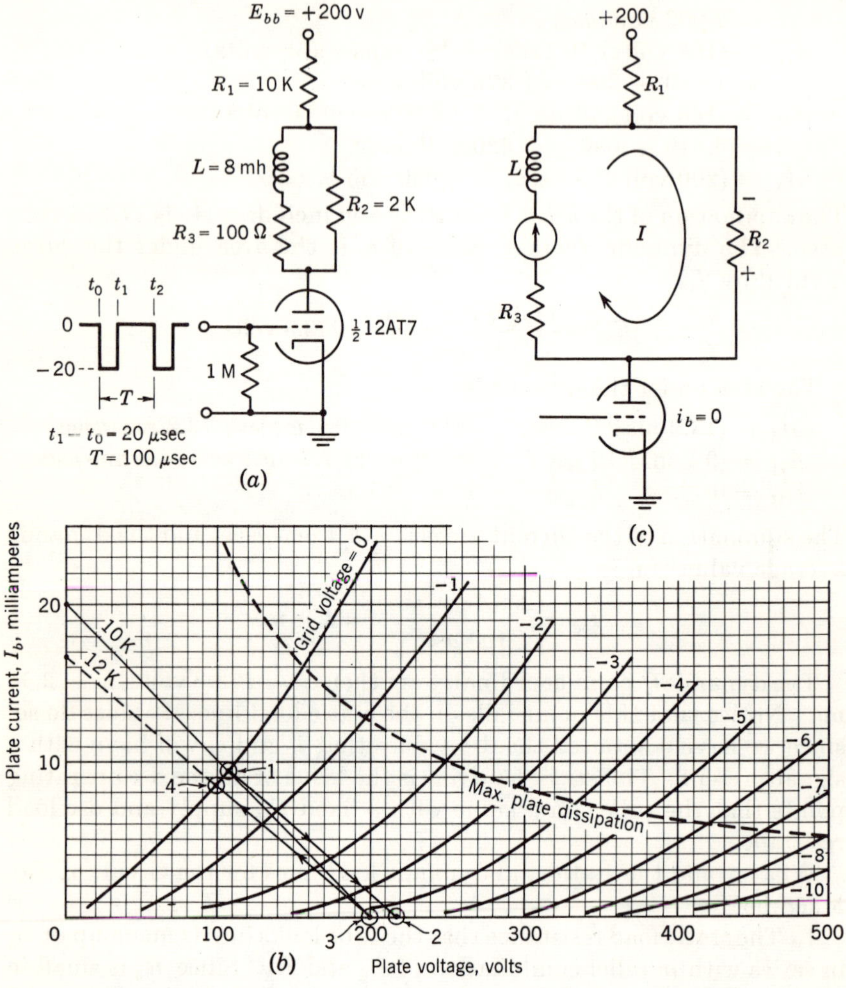

FIG. 5-26. (*a*) The locus of operating points is to be determined with an inductive load. (*b*) A graphic analysis for *a*. (*c*) An equivalent circuit for *a* when the tube is switched off.

9.2 ma must flow through R_2, the voltage across R_2 will jump to 9.2 ma (2 kilohms) = 18.4 volts. This voltage is series-aiding the E_{bb} supply, and therefore e_b jumps to 218.4 volts. The operating point therefore jumps from point 1 to point 2.

If designing a circuit of this nature, where the tube is operated as a switch with an inductive load, the cutoff voltage applied to the grid must be large enough to cut the tube off with the inductive kick added to the E_{bb} supply.

4. The current I will now decay exponentially. The discharge time constant about the loop is $L/(R_2 + R_3) \approx L/R_2 = 4$ μsec. Since the tube is cut off for 20 μsec, the current will have decayed by 100 per cent of its maximum swing, or zero per cent of 9.2 ma will remain. Therefore, I equals zero at t_1 minus. The plate-to-cathode voltage at t_1 minus is then equal to E_{bb} plus the zero voltage drop across R_2. Since the tube is cut off, there is no drop across R_1. At t_1 minus, we then have $e_b = 200$ volts and $i_b = 0$ for point 3.

5. At t_1 the tube suddenly switched on again. At t_1 plus, the inductor current is still zero, and the apparent load resistance that the tube sees is $R_1 + R_2$, or 12 kilohms. A 12-kilohm load line intersects the $e_c = 0$ curve at $i_b = 8.2$ ma, and $e_b = 99$ volts. The tube operating point will therefore jump from point 3 to these coordinates at point 4.

6. As the inductor current builds up, the operating point will shift from point 4 back to point 1. The time constant formed by the inductor and the circuit it looks back into is

$$\tau_C = \frac{L}{R_3 + \dfrac{R_2(R_1 + r_p)}{R_2 + R_1 + r_p}}$$

The value of r_p can be determined by noting the change in i_b due to the change in e_b between points 4 and 1. We shall not analyze this circuit further, since we have determined the path of the dynamic load line, which is our primary concern at this time.

In Fig. 5-27*a* we have an interesting situation. The d-c load that the tube looks back into is simply R_1. The a-c load is R_1 in parallel with R_2, if we assume $X_C \approx 0$ at the signal frequency. What would the dynamic path of operation for the tube look like? By superposition we can decompose Fig. 5-27*a* into Fig. 5-27*b* and *c*. Caution must be used in working with Fig. 5-27*c*, for if it were accepted at face value, it would seem that rectification must occur with $i_p = 0$, when the cathode end of the e_s generator swings positive. We must remember that if E_{bb} is greater than e_s, the diode will always be forward-biased. Assume that the diode has the plate characteristics of Fig. 5-27*d*.

A graphical solution is obtained in the following manner. First draw the static, or d-c, load line for Fig. 5-27*b*. With $E_{bb} = 240$ volts, the 4-kilohm load line intersects the diode curve to give an $I_{b0} = 30$ ma and $E_{b0} = 120$ volts. The next thing to do is to superimpose Fig. 5-27*c* upon Fig. 5-27*b*. To do this, consider the Q point as the origin of the e_p-i_p axis.

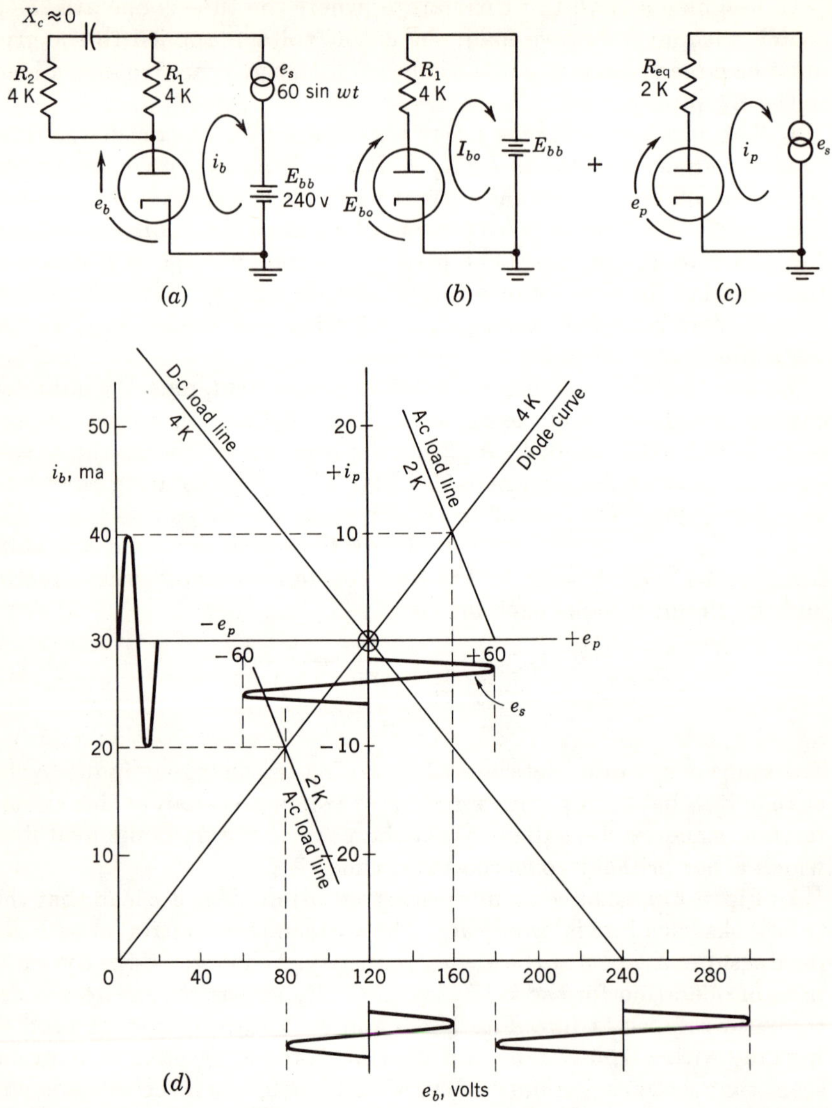

FIG. 5-27. (*a*) The a-c and d-c loads that the tube works into are different. (*b*) The d-c load. (*c*) The a-c load. (*d*) Graphic analysis of *a*.

On this e_p axis the signal voltage e_s is sketched. The peaks of e_p occur at $\pm$ 60 volts as shown, and we shall try to find the tube current at these extremes. If we freeze the e_s source in time at the instant it is +60 volts, this appears to be a plate supply of 60 volts acting through a 2-kilohm load on the tube. To find the corresponding current, draw a 2-kilohm load line, one end of which lies on +60 volts and the other end on 60 volts/2 kilohms = 30 ma above this new origin. Since this 30 ma is added to the 30 ma of I_{b0}, we are at the edge of the chart. Therefore, let us use a halfway point of 30 volts and 15 ma to draw the load line. The intersection of the 2-kilohm a-c load line with the tube curve occurs at $i_p = 10$ ma, $e_p = 40$ volts. A similar load line is constructed for the peak negative signal swing and the point of intersection is at $i_p = -10$ ma, $e_p = -40$ volts. All other intermediate values of e_s will yield operating points lying on the tube curve between these two extremes and indicated by the heavy line in Fig. 5-27*d*.

If the reactance of the capacitor were not negligible, the a-c component of plate current would not be in phase with the plate voltage, and the solid line of Fig. 5-27*d* would open up into an ellipse. The mathematics of determining the operating points along the ellipse is quite tedious and seldom employed. Furthermore, the dynamic operating path will be an ellipse only if the tube characteristics are linear or the signal is small enough to restrict operation along a linear section of the curve.

Figure 5-28*a* is a typical *RC*-coupled amplifier stage. The values of R_L and R_g, however, have been chosen lower than usual in order to avoid crowding everything into the lower part of the tube characteristics. Both C_c and C_K are assumed to have negligible reactance over the operating range of frequencies. Capacitor C_K is usually chosen to have a resistance equal to one-tenth or less of the ohmic value of R_K at the lowest frequency of interest. Similarly, C_c may be chosen so that X_{C_c} is equal to one-tenth of R_g at the lowest frequency of interest. There are other considerations in determining the choice of these capacitors which are beyond the scope of this book. Let us analyze Fig. 5-28*a* and determine the output signal for a 1-volt peak input signal.

The first thing to do is to locate the quiescent operating point. This is done by plotting the total d-c plate load line of 26.17 kilohms $\approx$ 26 kilohms. The bias curve for $R_K = 178$ ohms is then plotted by assuming values of e_c and computing corresponding values of i_b. For example, if $e_c = 0$, $i_b = 0$; if $e_c = -1$ volt, $i_b = 1$ volt/178 ohms $\approx$ 5 ma; if $e_c = -2$ volts, $i_b = 2$ volts/178 ohms $\approx$ 11 ma. The intersection of the total load line with the bias curve is the actual static Q point.

The impedance of C_K, C_c, and the E_{bb} supply is assumed to be negligible at the lowest operating frequency. Therefore, the a-c equivalent circuit seen looking out of the plate and cathode terminals is that of Fig. 5-28*b*. Let the parallel combination of R_L and R_g be referred to as R_{eq}. If we assume values of a-c signal voltage (e_p) across the tube, corresponding values of i_p flowing into R_{eq} can be determined. An a-c, or dynamic, load line can then be drawn through the Q point as the

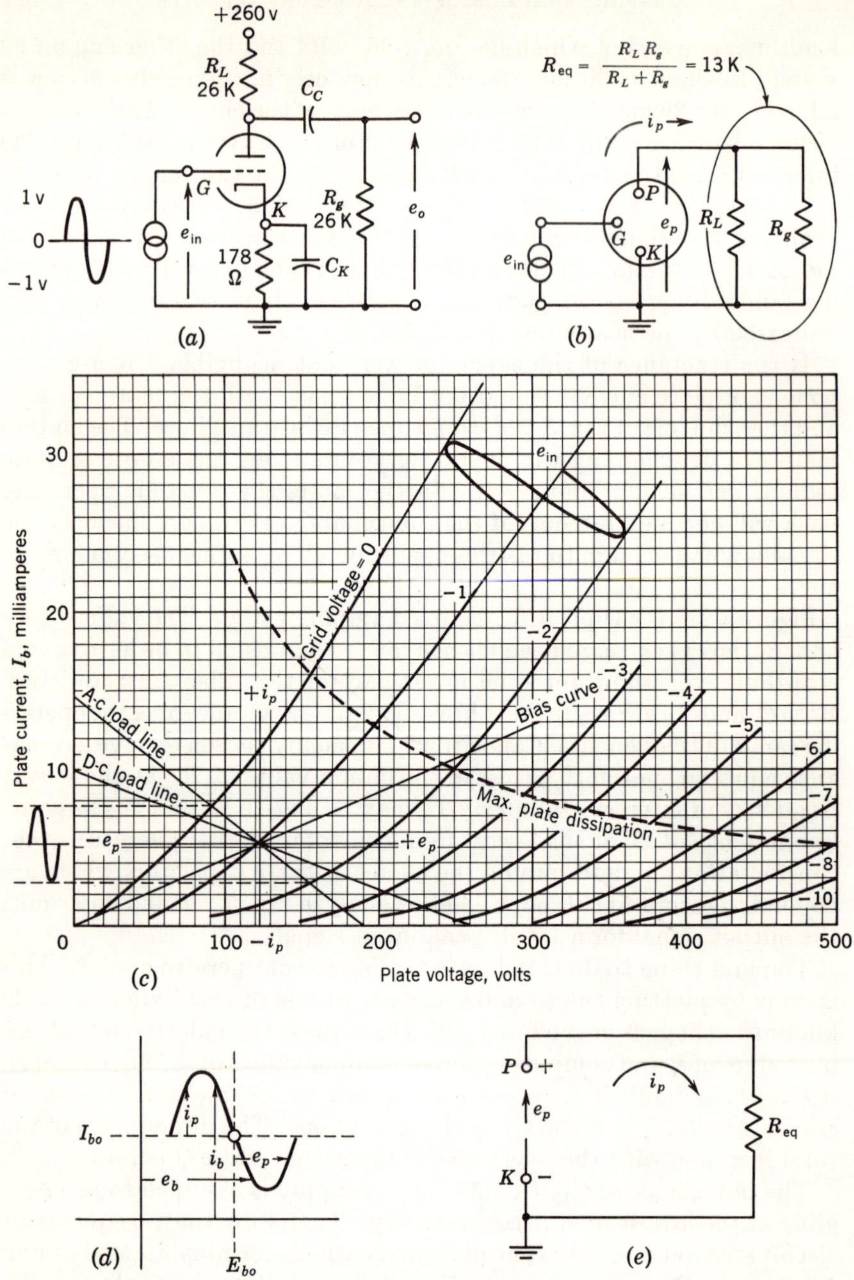

FIG. 5-28. (*a*) An *RC*-coupled amplifier. (*b*) The a-c load that the amplifier works into. (*c*) The a-c load line is necessary to determine the alternating plate current graphically. (*d*) The plate voltage increases as the plate current decreases. (*e*) The plate voltage and plate current are 180° out of phase.

origin about which the signal variations occur. We shall later see that the tube between the P and K terminals behaves as an a-c generator which pumps the alternating current (i_p) through R_{eq}.

The details of drawing the a-c load line are as follows. When there is no a-c signal on the input, e_p must equal zero, and correspondingly i_p also equals zero. Therefore, the static Q point would seem to be a point on the dynamic load line. Using the Q point as an origin, assume an instantaneous a-c signal swing e_p, for example, $+60$ volts with respect to this origin. This means that the plate is momentarily 60 volts positive with respect to the cathode and the internal generator within the tube must really be pumping i_p in the direction opposite to that assumed in Fig. 5-28*b*. The total instantaneous current i_b must then be decreasing. Therefore, when $e_p = +60$, $i_p = 60$ volts/13 kilohms $= 4.5$ ma. With a pair of dividers lay off 60 volts along the e_p axis of Fig. 5-28*c* and then down 4.5 ma along the $-i_p$ direction. Draw the a-c load line through the Q point and this point which must lie on the a-c load line. The reader may be wondering why the 4.5 ma value of i_p was located in the negative direction. Physically we know that the positive e_p that we have assumed must be increasing e_b. As e_b goes up, i_b must go down; and so i_p, which is the varying part of i_b, must be decreasing, as shown in Fig. 5-28*d*. Another way of understanding this point is to write the loop equation for Fig. 5-28*e*. When e_p goes positive, we mean that it does so with respect to the cathode. The positive direction of i_p has been assumed to be that of electron flow from cathode to plate within the tube. We can then write

$$e_p = -i_p R_{eq} \qquad \text{or} \qquad i_p = -\frac{e_p}{R_{eq}}$$

Wherever a grid-to-cathode voltage curve intersects this a-c load line locates the corresponding value of i_b on the i_b-e_b axis or the value of i_p on the i_p-e_p axis about the Q point. Since the input signal in this problem has a 1-volt peak value and the self-bias voltage is -1 volt, the grid will swing from 0 to -2 volts about the bias voltage. The peak positive swing in plate current is from 5.1 to 7.6 ma and the negative peak swing is from 5.1 down to 2.6 ma. The peak-to-peak output current swing is $7.6 - 2.6 = 5.0$ ma. The peak-to-peak output voltage swing is therefore (5.0 ma)(13 kilohms) $= 65$ volts. The voltage gain A_v is then the peak-to-peak output divided by the peak-to-peak input, or $65/2 = 32.5$.

Actually, these results are only approximate, but close enough for practical purposes. We have assumed that the dynamic Q point is the same as the static Q point. The dynamic Q point is located by determining the average plate voltage and current with a signal present. If no distortion or partial rectification occurs, the dynamic and static Q

points will be the same. In this problem, if the negative current swing was greater than the positive swing, a plate d-c meter would read slightly lower with a signal present than without one. With a lower direct current, the d-c tube voltage will increase somewhat, owing to a lower drop in R_L. The true dynamic Q point will then not lie on the static Q point, but instead will be shifted down and to the right. If we knew what this point was, we could redraw the dynamic load line through the dynamic Q point instead of the static Q point and rework the problem for a little more accuracy. Generally, the Q point shift is slight, and therefore the static and dynamic Q points are considered identical. If the a-c load had appreciable reactance at the signal frequency, the locus of e_b-i_b operating points would change from the linear a-c load line of Fig. 5-28*c* to an ellipse.

Another case in which the a-c load for a tube is different from the d-c load is shown in Fig. 5-29*a*. If we assume that the transformer is ideal, the load in the secondary (R_L) will appear across the primary terminals of the ideal transformer as $R_{eq} = R_s(N_p/N_s)^2$. The a-c equivalent circuit is shown in Fig. 5-29*b*. Again assume that C_K is an a-c short circuit.

Since the transformer is ideal, the primary resistance can be considered zero. The d-c load line is then determined by R_K only and will be almost vertical, since R_K is so small. We shall assume that it is vertical. To find the Q point, a bias curve must be drawn on Fig. 5-29*c* by assuming values of e_c, as in previous problems. The Q point lies at $E_{b0} = 300$ volts and $I_{b0} = 8.8$ ma, which is just above $P_d$max. The a-c signal axis is then drawn about the Q point as an origin.

To construct the a-c load line, R_{eq} must be evaluated.

$$R_{eq} = \frac{R_s}{(N_s/N_p)^2} = \frac{10}{(1/30)^2} = 9 \text{ kilohms}$$

In this circuit, the a-c load resistance is greater than the d-c resistance. To sketch the a-c load line, assume an e_p value, say 90 volts negative on the a-c signal axis. The corresponding value of i_p is 10 ma positive, since i_p increases as e_p decreases. This is a point on the a-c load line, and the other point is the Q point. We can now construct a straight line between these two points.

With -3 volts of self-bias on the tube, the input signal swings e_c from -2 to -4 volts. The positive i_p swing is 3.7-ma peak and the negative swing is 2.4-ma peak. The peak-to-peak swing is 6.1 ma. The plate voltage swings from a minimum of 269 volts ($e_{b,\min}$) to a maximum of 322 volts ($e_{b,\max}$). The reason that e_b rises higher than the E_{bb} supply is because of the inductive kick furnished by the transformer. The peak-to-peak signal voltage developed in the load is equal to the peak-to-peak

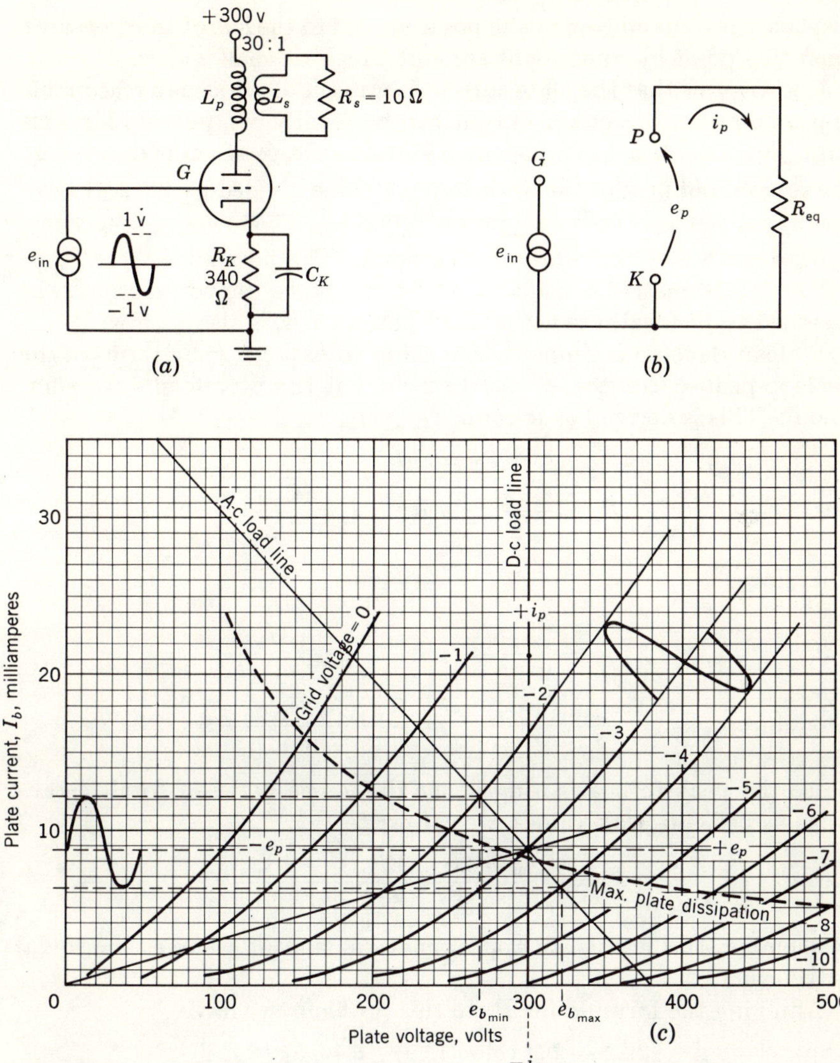

FIG. 5-29. (*a*) A transformer-coupled load. (*b*) The equivalent load that the tube sees will be dependent upon the transformer turns ratio. (*c*) The d-c and a-c load lines are different with a transformer-coupled load.

current swing multiplied by the a-c load resistance, which equals (6.1 ma)(9 kilohms) = 54.9 volts peak to peak. Since the current swings are not symmetrical, the output is distorted and in this case has a d-c component, which will cause a d-c plate milliammeter to read higher when the input signal is present. The average plate voltage with a signal would then decrease. The net effect of both these changes would be to displace the dynamic operating point up and to the left of the quiescent operating point by some slight amount which we shall ignore.

If we assume that the plate current is sinusoidal, which is a reasonable approximation, the power output can be readily computed. The rms value of the plate a-c component is equal to the peak value divided by the square root of 2, or the peak-to-peak value divided by $2\sqrt{2}$. Therefore, $I_p = 6.1\text{ ma}/2\sqrt{2} = 2.16$ ma. The rms signal voltage is $E_p = I_pR_{eq} = 2.16\text{ ma}(9\text{ kilohms}) = 19.4$ volts. The power in R_{eq} is $P_0 = I_pE_p = (2.16\text{ ma})(19.4\text{ volts}) = 41.4$ mw. Since the transformer was assumed to be ideal, the actual load power in R_s is also 42 mw.

We can develop a simple relationship to express P_0 in terms of the peak-to-peak current and voltage swings if the waveforms are sinusoidal. This is derived as follows:

$$P_0 = E_pI_p$$

but
$$E_p = \frac{e_{p,pk}}{\sqrt{2}} \qquad \text{and} \qquad I_p = \frac{i_{p,pk}}{\sqrt{2}}$$

Therefore,
$$P_0 = \frac{e_{p,pk}i_{p,pk}}{\sqrt{2}(\sqrt{2})}$$

but
$$e_{p,pk} = \frac{e_{p,pk-pk}}{2} \qquad \text{and} \qquad i_{p,pk} = \frac{i_{p,pk-pk}}{2}$$

Therefore,
$$\frac{(e_{p,pk-pk})(i_{p,pk-pk})}{2\sqrt{2} \times 2\sqrt{2}} = \frac{(e_{p,pk-pk})(i_{p,pk-pk})}{8}$$

Since the peak-to-peak swing is the difference between total instantaneous peak-to-peak values, we can write

$$P_0 = \frac{(e_{b,\max} - e_{b,\min})(i_{b,\max} - i_{b,\min})}{8}$$

Remember, this is only an approximation, since it assumes sinusoidal waveforms.

Applying the formula for P_0 to this problem, we have

$$P_0 = \frac{(322 - 269\text{ volts})(12.2 - 6.4\text{ ma})}{8} = 38.5\text{ mw}$$

which is close to the previously computed value of 41.4 mw. The discrepancy is primarily due to reading the curves.

5-17. Parallel Operation. Sometimes tubes are connected in parallel as shown in Fig. 5-30a for greater amplification or power output. A simple way of analyzing circuits using this configuration is to generate a plate family of curves for a tube whose characteristics will be equivalent to that of the parallel combination. This is most easily accomplished by assuming an E_b value across V_{1a} and V_{1b}. The same value of

e_c is applied to both tubes. With an e_c and E_b value specified for V_{1a}, I_1 is determined from the curves. If V_{1b} is identical to V_{1a}, I_1 will equal I_2; otherwise, I_1 and I_2 must be individually determined for a given e_c and E_b. The current I_T is equal to $I_1 + I_2$. Therefore, along the assumed

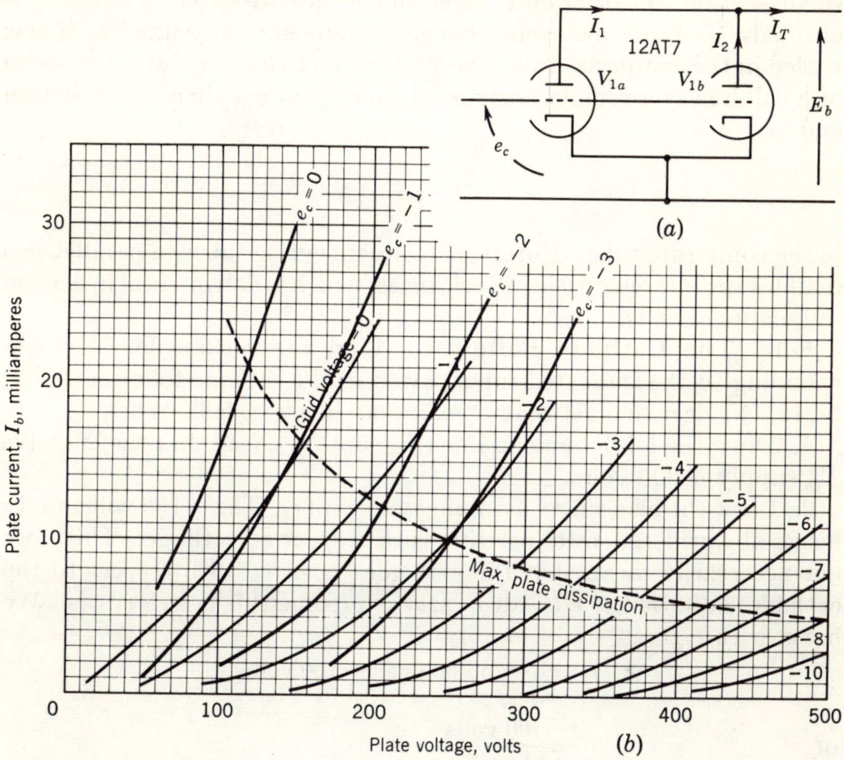

FIG. 5-30. (*a*) Tubes in parallel. (*b*) A composite characteristic may be drawn for parallel tubes.

E_b lay off I_1 for, say, $e_c = 0$ applied to V_{1a}. Next lay off I_2 for V_{1b}. Then with a divider add $I_1 + I_2$ to get an I_T value, which the equivalent tube must pass with the same E_b and e_c applied to the parallel combination. If the tubes are the same, the distance laid off for I_1 need only be doubled. Now, without changing the value of e_c, assume another E_b and determine I_1, I_2, and I_T. Locate this new E_b, I_T point and keep repeating the process while holding e_c fixed at zero. Draw a curve through the E_b, I_T points. This is the equivalent tube characteristic for $e_c = 0$.

A new value of e_c, say $e_c = -1$, can then be chosen and different values of E_b assumed for this e_c. Then I_1, I_2, and I_T are determined, and a new set of points is plotted for the composite characteristic with $e_c = -1$. In Fig. 5-30*b* this was done for V_{1a} and V_{1b} as sections of a 12AT7.

The $P_{d,\max}$ curve of the equivalent tube would naturally be twice as great for each section.

We see that for a fixed e_c, the composite tube passes twice the current if both sections are identical. This means R_p is half as great as for a single section. Furthermore, since the curves are twice as steep, r_p is also halved. Since a current change is doubled at any fixed E_b if e_c is varied in the composite tube, the g_m must be twice as great. The value of μ will however remain unchanged, since $\mu = g_m r_p$ in a single section and

$$\mu = 2g_m \frac{r_p}{2} = g_m r_p$$

in the composite tube. Unfortunately connecting tubes in parallel also doubles the interelectrode capacitances, which is detrimental in certain pulse applications.

5-18. The Cathode Follower. Find the output levels corresponding to the indicated input levels in the cathode follower of Fig. 5-31*a*. Neglect the fact that $P_d$max may be exceeded. When $e_{in} = 0$:

1. Draw the 5-kilohm load line (Fig. 5-31*b*). Note that the effective E_{bb} supply is 300 volts.

2. Draw the bias curve, assuming values of e_c. The -100 volt supply biases the grid 100 volts positive with respect to cathode. This does not necessarily mean that the grid goes positive with respect to the cathode, because the $i_b R_K$ drop may overcome this positive bias. We have, with $e_{in} = 0$,

$$e_c = -i_b(5\text{ kilohms}) + 100\text{ volts}$$

or

$$i_b = \frac{100\text{ volts} - e_c}{5\text{ kilohms}}$$

If $e_c = 0$, $\qquad i_b = 20\text{ ma}$

If $e_c = -1$, $\qquad i_b = \dfrac{101\text{ volts}}{5\text{ kilohms}} = 20.2\text{ ma}$

Connect this pair of e_c, i_b points with a line.

3. Note that with $e_{in} = 0$ the quiescent value of plate current is 20 ma. The output voltage is therefore

$$e_o = -100 + 20\text{ ma }(5\text{ kilohms}) = 0\text{ volts}$$

When $e_{in} = +10$ volts:

1. The general expression for e_c in the circuit of Fig. 5-31*a* is

$$e_c = -i_b R_K + 100 + e_{in}$$

or $$i_b = \frac{100 + e_{in} - e_c}{R_K} = \frac{100 + 10 - e_c}{5 \text{ kilohms}}$$

If $e_c = 0$, $$i_b = \frac{110 \text{ volts}}{5 \text{ kilohms}} = 22 \text{ ma}$$

If $e_c = -1$, $$i_b = \frac{111 \text{ volts}}{5 \text{ kilohms}} = 22.2 \text{ ma}$$

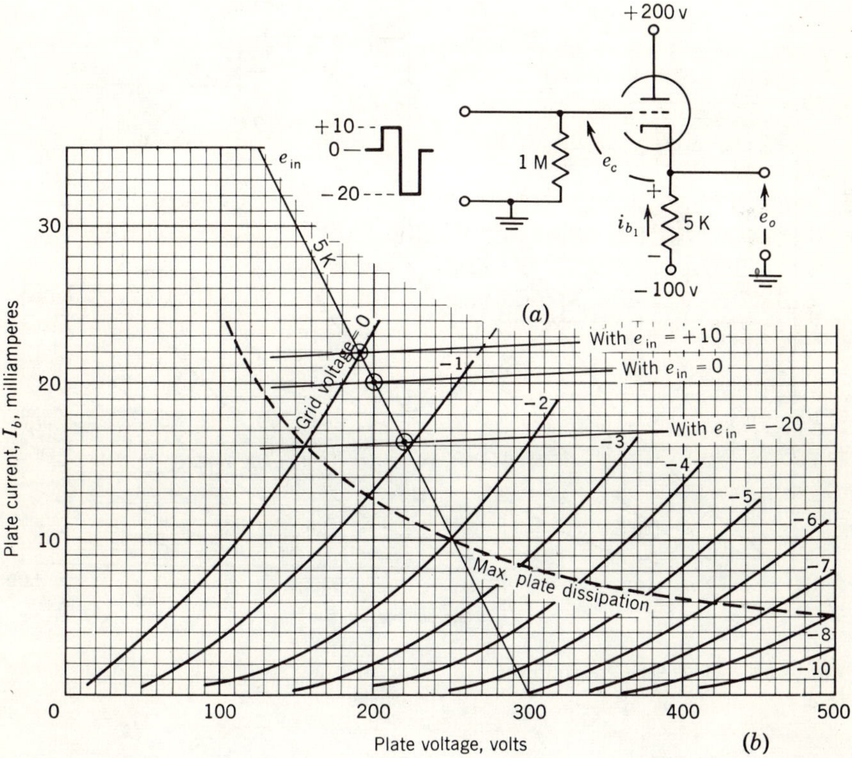

FIG. 5-31. (*a*) A cathode follower. (*b*) Graphic analysis of the cathode follower in *a*.

Both of these points fall on the right-hand side of the load line (Fig. 5-31*b*), which indicates that the grid-to-cathode voltage has gone slightly positive. The $e_c = 0$, $i_b = 22$-ma point is virtually on the load line, however, and so we assume that with $e_{in} = +10$ the value of i_b is 22 ma.

Therefore, $$e_o \approx -100 + 22 \text{ ma}(5 \text{ kilohms}) = +10 \text{ volts}$$

If the grid-to-ground voltage were driven to more than +10 volts, the grid would conduct and the grid would look like the plate of a diode to the cathode. Since the drop across the internal resistance of a conducting diode is small, we could, as a first approximation, say that the grid is clamped to the cathode. Therefore, if e_{in} should rise to, say, +20 volts, e_o would also be about +20 volts. The circuit (which happens to be a cathode follower) is now loading down the source. The

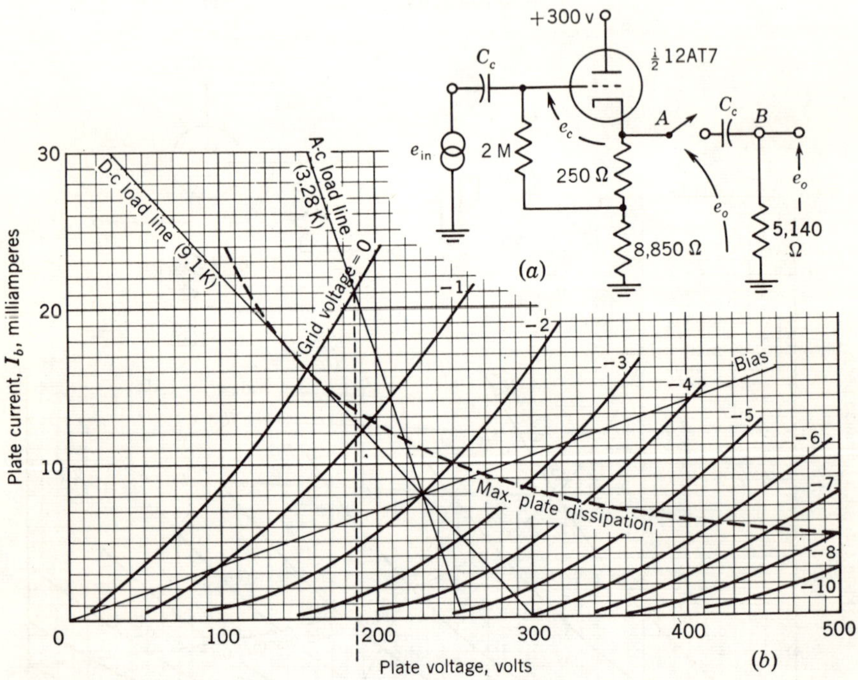

FIG. 5-32. (*a*) A cathode follower. (*b*) Graphic analysis of the cathode follower in *a*.

cathode current is 120 volts/5 kilohms = 24 ma. The grid-to-cathode voltage is approximately zero, however, and with $e_c = 0$, $i_b = 22$ ma on the load line. The grid current is then 2 ma, the difference between the cathode current and the plate current. Remember that this is approximate, since the grid will actually rise a fraction or more of a volt positive with respect to the cathode and will cause i_b, the plate current, to increase. With $e_{in} = -20$,

$$e_c = -i_b R_K + 100 + e_{in}$$

where $e_{in} = -20$, or

$$i_b = \frac{100 - 20 - e_c}{5 \text{ kilohms}}$$

If $e_c = 0$, $\qquad i_b = 16$ ma

If $e_c = -2$, $\qquad i_b = \dfrac{82 \text{ volts}}{5 \text{ kilohms}} = 16.4$ ma

Connecting these points with a straight line, we see that $i_b = 16.1$ ma. The corresponding output voltage is

$$e_o = -100 \text{ volts} + 16.1 \text{ ma } (5 \text{ kilohms}) = -19.5 \text{ volts}$$

Notice how closely the output voltage follows the input in both sign and magnitude. In a previous graphical analysis of a cathode follower, we showed that the output impedance was quite low—on the order of 500 ohms. In the cathode follower we have a unique means of generating various voltages for use in establishing clipping or clamping levels. For example, suppose that in modifying some piece of equipment we find that a +10-volt bias supply is needed. Rather than redesign or add a separate power supply, it might be feasible to simply put in a cathode follower and feed the grid from a +10-volt point on a divider across the E_{bb} supply. The cathode or output voltage will closely equal the 10 volts applied to the grid but at a much lower impedance level than would be obtained with the usual resistive divider.

Determine the input-vs.-output voltage characteristics for the cathode-follower circuit of Fig. 5-32*a* at point *A* with the switch open and then at point *B* with the switch closed. Assume that the capacitors have zero reactance.

The first thing we shall do is construct the d-c and a-c load lines for the switch open and closed, respectively.

1. Plot the d-c load line for a total load resistance of 250 + 8,850 = 9,100 ohms. The e_b intercept of this load line lies on $e_b = 300$ volts. The i_b intercept lies on $i_b = 300$ volts/9.1 kilohms = 33 ma.

2. To determine the Q point assume values of e_c and compute corresponding values of i_b from $i_b = e_c/250$ ohms:

e_c	0	. . .	−2	. . .	−4
i_b, ma . .	0	. . .	8	. . .	16

Plot these points and draw the bias curve as shown in Fig. 5-32*b*. The Q point is at the intersection of the d-c load line and the bias curve. The

Q point coordinates are $I_{b0} = 8$ ma and $E_{b0} = 230$ volts. The quiescent grid bias may be read from the curves as -2 volts or can be computed from $e_{c0} = -I_{b0}(250 \text{ ohms}) = -2$ volts. We are now in a position to construct the a-c load line (see Fig. 5-32*b*).

3. The a-c load resistance is 9,100 ohms in parallel with 5,140 ohms, which is an equivalent resistance of 3,280 ohms. The a-c load line will then have a slope of $-1/3{,}280$ mhos and pass through the Q point. To plot the a-c load line, imagine an i_p, e_p axis with its origin at the Q point. If we assume a decrease in e_p of, say, 30 volts, the alternating plate current must have increased by 30 volts/3.28 kilohms, or 9.15 ma, above I_{b0}. Therefore, locate a point 30 volts to the left of the Q point and 9.1 ma above it. Draw a line from this point through the Q point for the a-c load line as shown in Fig. 5-32*b*.

To determine the input-output characteristics at point A with the switch open, do the following:

1. Assume a published value of e_c.
2. From the curves note the value of i_b at the intersection of this assumed value of e_c and the d-c load line.
3. For each of the i_b values calculate e_o from $e_o = i_b(9.1 \text{ kilohms})$.
4. Determine the e_{in} required to develop this e_o from $e_{in} = e_o + e_c$ (e_c carries its own sign).

Tabulating steps 1 to 4 for various values of e_c, we have

e_c	0	-1	-2	-3	-4	-5	-6
i_b	16	12	8	5	3	1.3	0
e_o	145	109	72.7	45.5	27.3	11.8	0
e_{in}	145	108	70.7	42.5	23.3	6.8	-6

The output-vs.-input characteristic at point A with the switch open is plotted in Fig. 5-33*b*. The e_{in} of Fig. 5-33*b* is not the actual a-c input voltage but also includes the d-c bias component. That is, with the generator out of the circuit, e_o equals 72.7 volts because of I_{b0}. Therefore, 72.7 volts is really the axis about which the a-c component of e_{in} varies. So far as the tube is concerned, the input signal is derived from a circuit equivalent to Fig. 5-33*a*. Notice in Fig. 5-33*b* that e_{in} may swing from -6 to $+145$ volts with negligible distortion. If the input swings above 145 volts, grid current will start to flow, since $e_c = 0$ with this peak input. If the 2-meg grid resistor of Fig. 5-32*a* were not returned to a tap on the cathode resistor but instead went to ground, a larger self-bias voltage would cause the quiescent point to shift much closer to cutoff and the zero axis for the a-c input signal in Fig. 5-33*b* would be far to the left. While this means that a larger positive swing in e_{in} can be handled, the negative swing would quickly cut the tube off and cause clipping.

In some specific applications this might be desirable but generally such distortion is unwanted.

The voltage gain at point A with the switch open can be readily computed. With a peak-to-peak input swing from -6 to $+145$ volts or

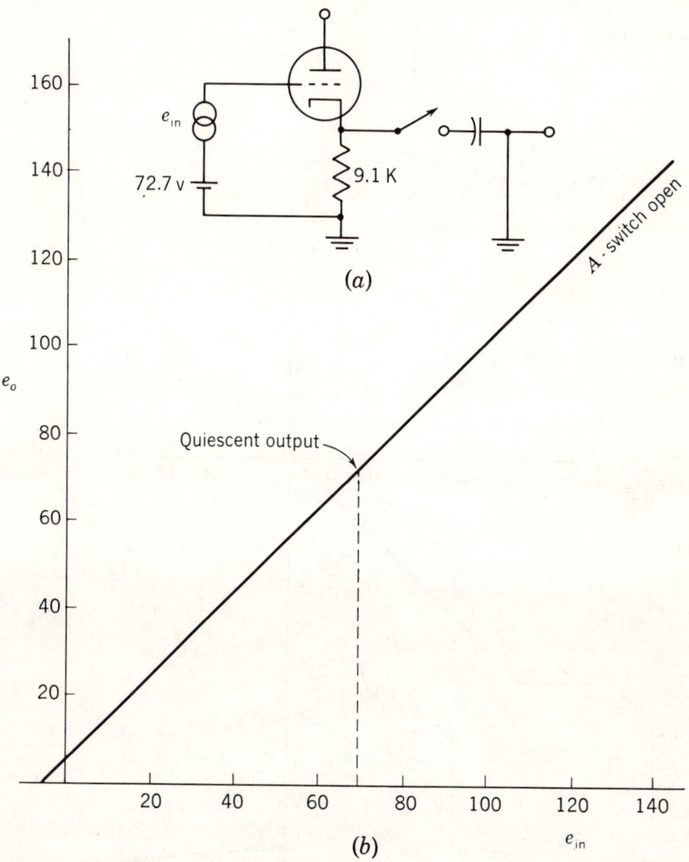

FIG. 5-33. (a) Circuit used for determining the d-c output characteristics. (b) The d-c output of the cathode follower varies linearly with the input.

151 volts, the output swings from 0 to 145 volts. The gain is therefore $^{145}/_{151} = 0.96$, which is typical for a cathode follower with a moderately high value of cathode resistance.

Now let us determine the a-c input-output characteristic, that is, e_o at point B versus e_{in}, with the switch closed:

1. Imagine or sketch in an e_p, i_p axis with its origin at the Q point. If the a-c input signal should cause e_c to shift from the quiescent value of -2 volts to -1 volt, the instantaneous value of i_p must be 6.1 ma, since

this is where $e_c = -1$ volt intersects the a-c load line. This 6.1 ma must flow through the equivalent a-c load of 3.28 kilohms and therefore develop an instantaneous e_o value of (6.1 ma)(3.28 kilohms) = 20 volts.

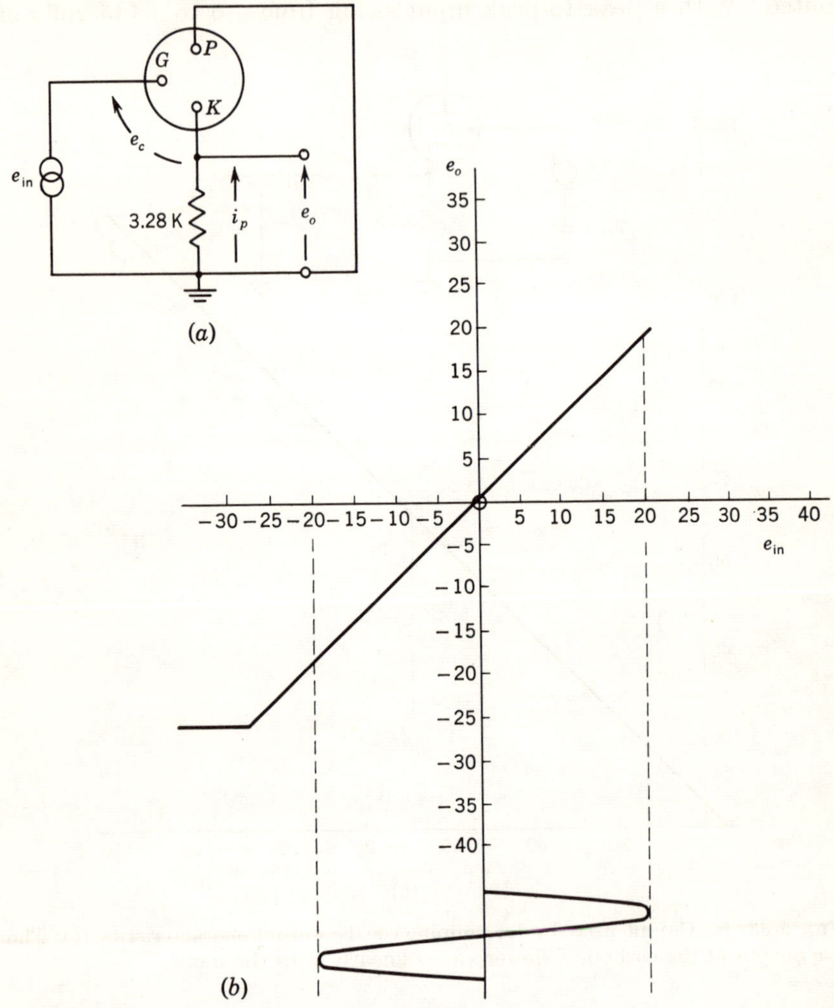

FIG. 5-34. (*a*) The equivalent a-c load that the cathode follower works into. (*b*) The a-c output is also very linear for the cathode follower.

The a-c circuit the tube looks into is shown in Fig. 5-34*a*. The corresponding value of e_{in} is then determined from $e_{in} = e_o + e_c$ and is (20 + 1) volts for this example. Note that we do not use the minus sign with e_c, since the e_c value we are using now is really Δe_c, the positive 1-volt increase from $e_c = -2$ to $e_c = -1$.

2. Repeat the procedure of step 1, assuming other values of shift in e_c from the quiescent value, and tabulate:

Shift in e_c . .	0	+1	+2	−1	−2	−3	−4
i_p	0	6.1	12.9	−3.8	−6.1	−7.7	−8
e_o	0	20	42.3	−12.5	−20	−25.2	−26.2
e_{in}	0	21	44.3	−13.5	−22	−28.3	−30.2

The results of this tabulation are plotted in Fig. 5-34*b*. Notice that the output is virtually undistorted if the a-c input swing is limited between +21 and −27.5 volts. Below −27.5 volts the tube is cut off, and above 44.3 volts grid current will flow.

The a-c voltage gain can be determined from Fig. 5-34*b* by assuming an a-c input signal of, say, 40 volts peak-to-peak and noting the corresponding output, which is about 37 peak-to-peak volts. The gain is then $^{37}\!/_{40} = 0.93$. The gain is lower in this case than at point *A* with the switch open because the a-c load is smaller than the d-c load.

Another way to determine the gain readily is as construct the vertical projections to the a-c load line as shown in Fig. 5-32*b*. Note that for a 4-volt peak-to-peak e_c swing the peak-to-peak load swing is $250 - 187 = 63$ volts. The required value of e_{in} to produce this output swing across the load is $e_{in} = 63 + 4$, or 67 peak-to-peak volts. The gain is therefore $^{63}\!/_{67} = 0.94$, which is in agreement with our previous result.

A general expression for the small-signal gain of the cathode follower can be developed. This formula is

$$A_v = \frac{\mu R_K}{r_p + R(\mu + 1)}$$

where $R_K = R_{eq}$, or 3.28 kilohms

$\mu = 54$

$r_p = 11$ kilohms

for a 12AT7 at the *Q* point in this problem. Substituting

$$A_v = \frac{54(3.28 \text{ kilohms})}{11 \text{ kilohms} + 3.28 \text{ kilohms}(55)} = 0.92$$

which again is in close agreement. The formula does not in any way allow for clipping or other forms of distortion and must be used with caution if large input signals are considered. For example, an input signal of 1,000 volts might be used and acceptance of this formula at face value would indicate that a 920-volt output was available, which is impossible with a plate supply of 300 volts and this tube.

5-19. The Grounded-grid Amplifier. Find the output voltage response of the grounded-grid amplifier circuit of Fig. 5-35*a* to the pulse train shown:

1. Plot the $R_K + R_L = 10$-kilohm load line.
2. Plot the R_K bias line with $e_{in} = 0$ for determining e_o during the

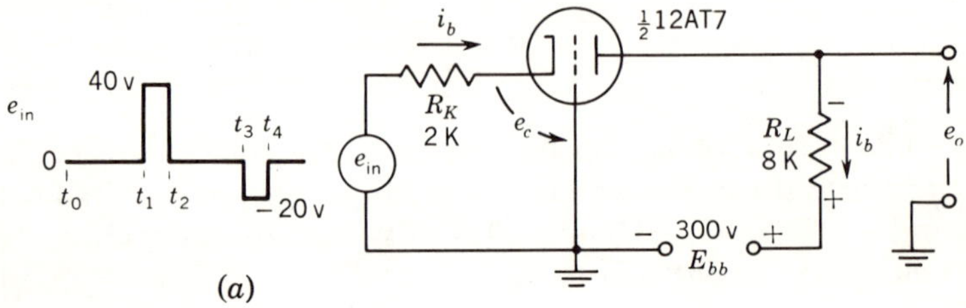

(*a*)

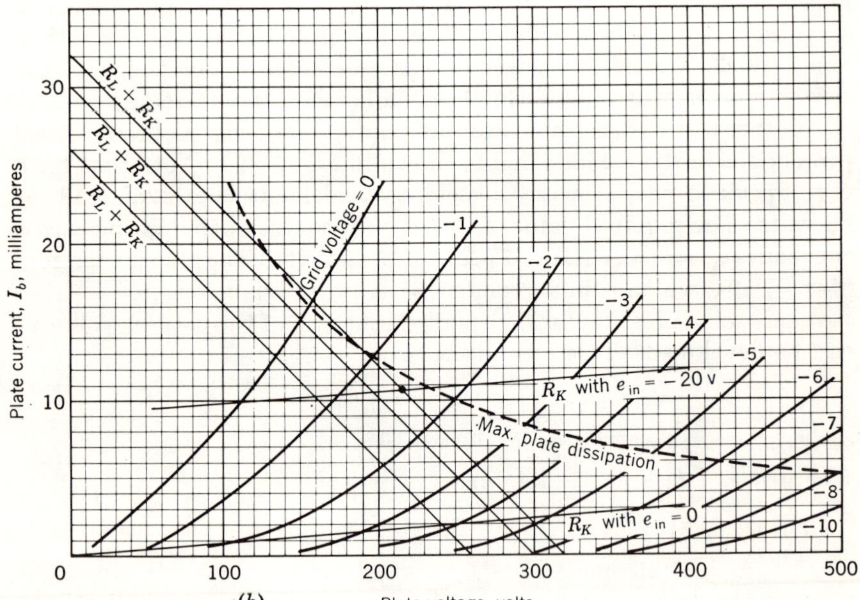

(*b*)

FIG. 5-35. (*a*) A grounded-grid amplifier. (*b*) Graphic analysis of a grounded-grid amplifier.

interval t_0 to t_1 as shown in Fig. 5-35*b*. Assume values of e_c and calculate corresponding values of i_b:

$$e_c = -i_b R_K + e_{in}$$

$$i_b = \frac{e_{in} - e_c}{R_K}$$

For example, if $e_c = -2$, $i_b = \dfrac{0 - (-2 \text{ volts})}{2 \text{ kilohms}} = 1$ ma

e_c	0	−2	−4
i_b, ma ...	0	1	2

3. The intersection of the R_K bias line with $e_{in} = 0$ occurs at $i_b = I_{b0} = 2.1$ ma. The drop across the 8-kilohm load is $I_{b0}R_L = 16.8$ volts.

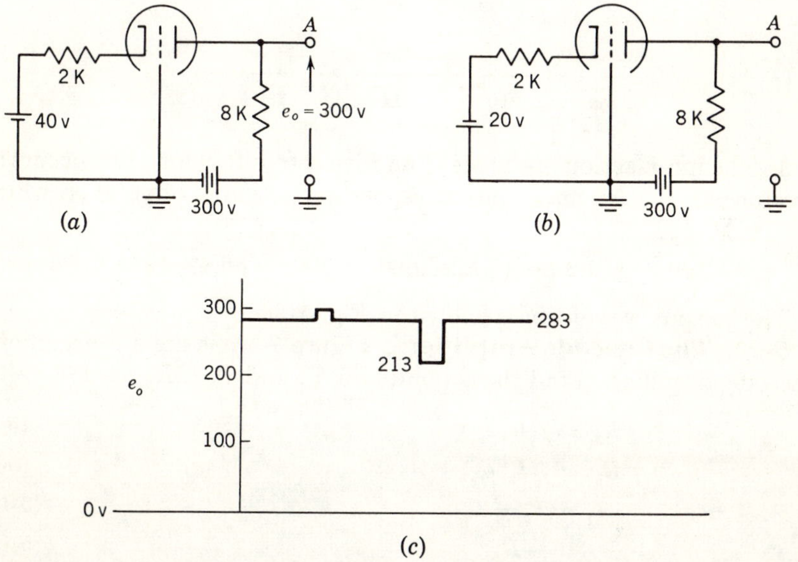

FIG. 5-36. (a) Equivalent circuit when the input is 40 volts. (b) Equivalent circuit when the input is minus 20 volts. (c) The output voltage waveform.

Since e_o is measured with respect to ground, we can determine e_o by starting at ground, going through the 300-volt battery, which brings us up 300 volts and then down (from + to −) 16.8 volts through the 8-kilohm load resistor to point A. The voltage of point A with respect to ground is therefore 283.2 volts with $e_{in} = 0$. At t_1, e_{in} jumps to 40 volts with respect to ground. The circuit is effectively that of Fig. 5-36a.

1. Relocate the $R_L + R_K$ load line because the effective E_{bb} supply between plate and cathode is (300 − 40) volts = 260 volts.

2. The tube is plainly cut off, since the input signal is seen to drive the grid 40 volts negative with respect to the cathode. With $i_b = 0$, there is no drop across R_L and therefore the voltage at point A is 300 volts. At t_2 the input drops back to zero and the output returns to 283.2 volts. At t_3 the input jumps 20 volts negative with respect to ground. The equivalent circuit is that shown in Fig. 5-36b.

1. Relocate the $R_L + R_K$ load line because the effective E_{bb} supply is now 320 volts.

2. Draw a bias curve assuming values of e_c and computing corresponding values of i_b:

$$i_b = \frac{e_{in} - e_c}{R_K}$$

For example, if $e_c = -4$, $i_b = \dfrac{20 - (-4) \text{ volts}}{2 \text{ kilohms}} = 12$ ma:

e_c	0	−2	−4	−6
i_b, ma ..	10	11	12	13

3. The intersection of the bias line with the shifted load line occurs at a value of $i_b = 10.8$ ma. The corresponding potential of point A, which is e_o, is

$$e_o = 300 - (10.8 \text{ ma})(8 \text{ kilohms}) = 300 - 86.4 = 213.6 \text{ volts}$$

The output waveform is shown in Fig. 5-36*c*.

5-20. The Cascode Amplifier. Figure 5-37*a* is the d-c circuit of a cascode amplifier. Find the Q points for V_1 and V_2 if $E_{g_2} = 178$ volts.

$$E_{b_1} = f(I_1, e_{c_1}) \tag{5-8}$$
$$E_{b_2} = f(I_2, e_{c_2}) \tag{5-9}$$
$$I_1 = I_2 = I_x \tag{5-10}$$
$$e_{c_1} = -E_K \tag{5-11}$$
$$e_{c_2} = -E_{b_1} - E_K + E_{g_2} \tag{5-12}$$
$$E_{b_1} = E_{bb} - I_x(20 \text{ kilohms}) - E_{b_2} - E_K \tag{5-13}$$
$$E_{b_2} = E_{bb} - I_x(20 \text{ kilohms}) - E_K - E_{b_1} \tag{5-14}$$
$$E_K = I_x R_K \tag{5-15}$$

Equations (5-8) to (5-15) can be written for the circuit of Fig. 5-37*a*. The solution, based upon these equations, is as follows:

1. Assume published values of e_{c_1}.
2. Compute I_x from Eqs. (5-11) and (5-15).
3. Using Eq. (5-8), read off E_{b_1} for assumed e_{c_1}, and I_{b_1}.
4. Using Eqs. (5-11) and (5-12) and the results of step 3, determine corresponding e_{c_2}. For example,

$$e_{c_2} = -E_{b_1} + e_{c_1} + E_{g_2}$$

when $e_{c_1} = -4$

$$e_{c_2} = -343 + (-4) + 178 = -169$$

In looking over the results of this calculation, we see that with $e_{c_1} = -3$, e_{c_2} is biased well beyond cutoff. With $e_{c_1} = -1$, e_{c_2} is positive. If we assume no positive grid current, then the correct value of e_{c_1} must lie well between -1 and -3.

Unfortunately, there are no published values for $e_c = -1.5$ or other in-between values. We can either guess at the location of these curves or construct the transfer characteristic (grid family) as follows. Choose a plate voltage, for example $e_b = 50$ volts and plot i_b versus e_c by noting the e_c, i_b values along the vertical line $e_b = 50$ volts. Repeat for other values of e_b until the grid family is plotted as shown in Fig. 5-37*b*. For greater accuracy we can now replot odd values of grid voltage on the original plate characteristic of Fig. 5-37*c*. For example, in Fig. 5-37*c* we see that with $e_c = -1.5$ we have $i_b = 2$ ma and $e_b = 100$ volts. With e_c still at -1.5 volts but $e_b = 150$, $i_b = 5$ ma. This procedure is then repeated for other values of e_c until fractional values of e_c are sketched into the plate family, as shown in Fig. 5-37*c*.

With e_{c_1} assumed at -1.5 and -2.5, the corresponding values of e_{c_2} again are unrealistic. Let us try $e_{c_1} = -1.75$ and -2.25. Inspection of the resulting values of e_{c_2} indicates that e_{c_1} lies between -1.75 and -2.25 volts.

We must go on splitting hairs until we have assumed a value of e_{c_1} which will result in a value of I_x such that, with the e_{c_2} determined by this I_x, the value of E_{b_2} from Eq. (5-9), when substituted in (5-14), will actually yield the E_{bb} applied. In other words, when we finally assume the correct e_{c_1}, the corresponding I_x will be such that the drop across R_K, V_1, V_2, and R_L will equal E_{bb}.

Let us try $e_{c_1} = -1.9$ and -2.1 volts and again tabulate the results. The location of these e_{c_1} values is determined by visually interpolating:

e_{c_1}	0	−1	−1.5	−1.75	−1.9	−2
I_x	0	2	3	3	3.8	4
E_{b_1}	0	80	126	150	170	180
e_{c_2}	...	97	50.5	26.3	6	−4

e_{c_1}	−2.1	−2.25	−2.5	−3	−4	−5
I_x	4.2	4.5	5	6	8	10
E_{b_1}	186	196	224	266	343	420
e_{c_2}	−10	−20.3	−48.5	−91	−169	

Since we have assumed that V_2 does not draw grid current, e_{c_1} must be more negative than -1.9 volts. Therefore, I_x must at least be in excess of 3.8 ma. On the other hand, neither tube can be cut off if any plate current is flowing and develop a self-bias voltage. This means that e_{c_1}

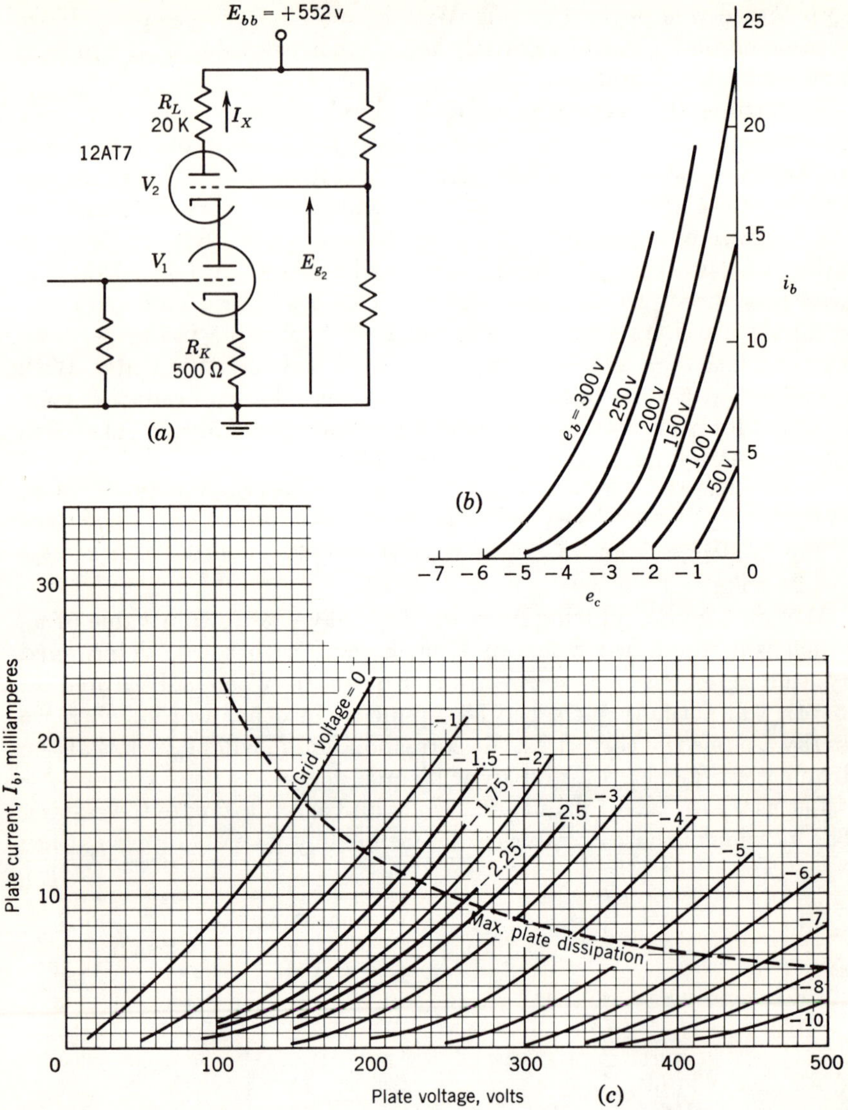

FIG. 5-37. (a) A cascode amplifier. (b) Transfer curves used in analyzing the cascode amplifier. (c) Locating additional values of e_c from the transfer curves of b.

must be more positive than −2.1 volts and I_x must be less than 4.2 ma. We can therefore conclude that 3.8 ma $< I_x <$ 4.2 ma. The exact value assumed in making up this problem was 2 ma. The difficulty in this problem is that e_{c_2} is the difference between two big numbers which can easily result in a number greater than the cutoff voltage for a high-μ tube.

5-21. The Cathode-coupled Amplifier. The circuit of Fig. 5-38*a* is that of a cathode-coupled amplifier. At this time, we shall confine our efforts to determining the quiescent operating points of V_1 and V_2. There are several ways of evaluating the Q points in Fig. 5-38*a*, and the following discussion is primarily intended to acquaint the reader with the type of thinking involved.

1. Assume a value of E_K, since E_K is measured at the junction of three parallel branches. For each E_K value an I_K value can be calculated from $I_K = E_K/330$ ohms:

E_K	0	1	2	3	4	5	6
I_K	0	3	6.1	9.1	12.1	15.1	18.2

2. Plainly, $e_{c_1} = -E_K + E_{g_1}$ or just $-E_K$, since E_{g_1} is zero. We also see that $E_{b_1} = E_{bb} - E_K$. Therefore, for any assumed E_K we can determine the corresponding e_{c_1} and E_{b_1}. Furthermore, for any value of e_{c_1} and E_{b_1} a corresponding I_{b_1} can be determined from the tube curves. We can now construct Table 5-3.

TABLE 5-3

E_K	0	1	2	3	4	5	6
e_{c_1}	0	−1	−2	−3	−4	−5	−6
I_K	0	3	6.1	9.1	12.1	15.1	18.2
E_{b_1}	300	299	298	297	296	295	294
I_{b_1}	...	...	15.9	8.5	4.1	1.4	0

Using the e_{c_1}, E_{b_1} or e_{c_1}, I_{b_1} values of Table 5-3, we can plot the locus of possible operating points of V_1 for any assumed value of E_K (see Fig. 5-38*b*).

3. For every E_K value in Table 5-3 we have a corresponding I_K and I_{b_1}. The corresponding values of I_{b_2} can readily be determined from $I_{b_2} = I_K - I_{b_1}$. This gives us Table 5-4.

TABLE 5-4

E_K	0	1	2	3	4	5	6
e_{c_1}	0	−1	−2	−3	−4	−5	−6
I_K	0	3	6.1	9.1	12.1	15.1	18.2
E_{b_1}	300	299	298	297	296	295	294
I_{b_1}	...	...	15.9	8.5	4.4	1.4	0
I_{b_2}	...	...	−8.9	−0.6	7.7	13.7	18.2

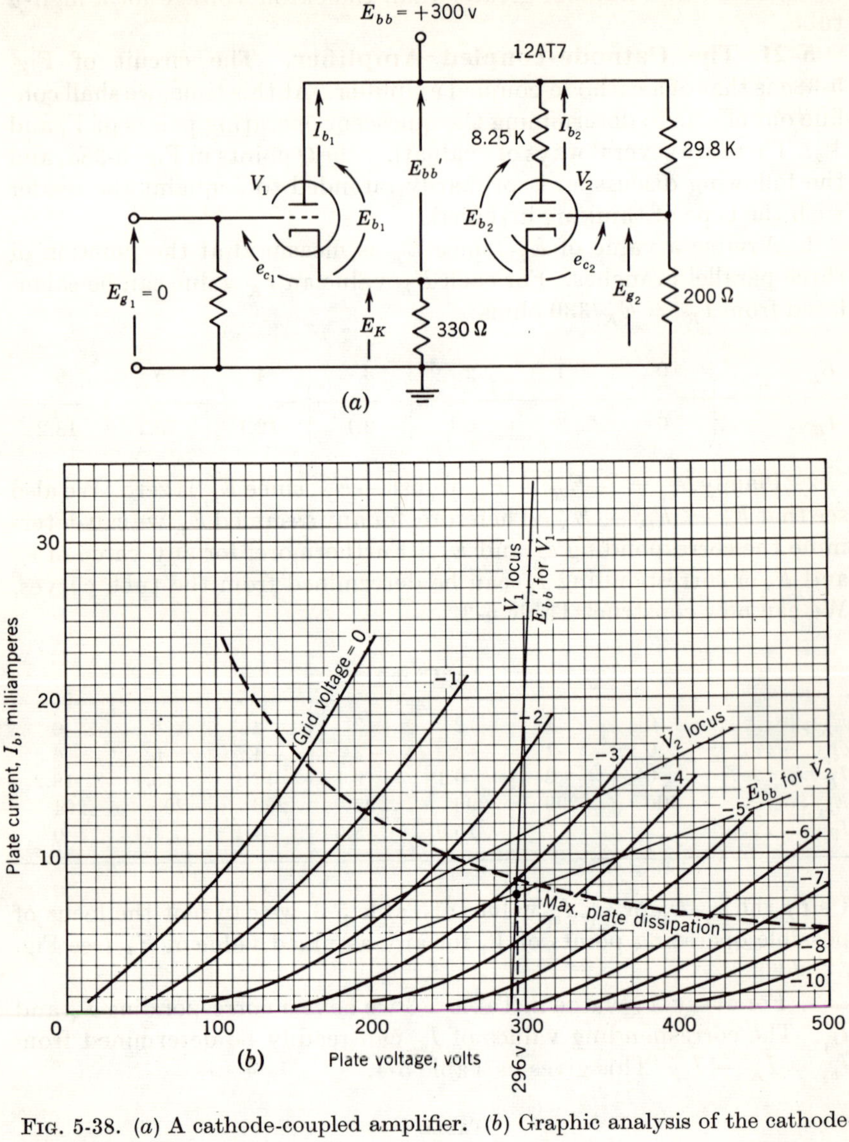

FIG. 5-38. (*a*) A cathode-coupled amplifier. (*b*) Graphic analysis of the cathode-coupled amplifier.

4. The divider in the grid circuit of V_2 can be thevenized to yield a $V_{oc} = E_{g_2} = 2$ volts. This is assuming that V_2 does not draw grid current. With this assumption there is no need to compute Z_{th}. Plainly, $e_{c_2} = -E_K + E_{g^2} = -E_K + 2$ volts. Therefore, for each assumed

value of E_K, we can compute the corresponding e_{c_2}. Furthermore, we have from Table 5-4 values of I_{b_2} corresponding to each E_K value. Knowing an I_{b_2} and e_{c_2} enables us to plot the locus of operating points for V_2 and also read corresponding E_{b_2} values from the curves, as in Table 5-5.

TABLE 5-5

E_K	0	1	2	3	4	5	6
I_{b_2}	...	...	−8.9	−0.6	7.7	13.7	18.2
e_{c_2}	+2	+1	0	−1	−2	−3	−4
E_{b_2}	...	...	...	...	228	347	

5. For every value of E_{b_2} that corresponds to an assumed E_K, we can compute E'_{bb} for V_2, since $E'_{bb} = E_{b_2} + I_{b_2}$ (8.25 kilohms). When $E_{b_2} =$ 228 volts, $I_{b_2} = 7.7$ ma, and therefore $E'_{bb} = 295.5$ volts as an example. Similarly, with $E_{b_2} = 347$ and $I_{b_2} = 13.7$ ma, we find $E'_{bb} = 465$ volts. A curve is then drawn between these E'_{bb} values for V_2 as shown in Fig. 5-38*b*. Now E'_{bb} is common for both V_1 and V_2. There is only one value of E'_{bb} that will satisfy both tubes, and that value of E'_{bb} is the point where the E'_{bb} locus for V_2 intersects the E'_{bb} locus for V_1 at 296 volts. For V_1 the E'_{bb} locus is the same as the V_1 locus, since $E_{b_1} = E'_{bb}$. We also have $E_K = E_{bb} - E'_{bb} = 4$ volts. Since $E_K = -e_{c_1} = 4$ volts, the intersection of the $e_{c_1} = -4$-volt curve and the locus of possible operating points for V_1 is the Q point for V_1. This is $I_{10} = 4.1$ ma. If the quiescent value of E_{K0} is 4 volts and $R_K = 330$ ohms, I_{K0} must equal 4 volts/330 ohms = 12.1 ma. We then know I_{20} from $I_{20} = I_{K0} - I_{10} = 12.1 - 4.1 = 8$ ma. All other quantities can readily be computed.

Analyze the modified cathode-coupled amplifier of Fig. 5-39*a* for the Q points of V_1 and V_2. About how many volts would have to be put on the grid of V_1 to cut off V_2? The solution for the Q points is obtained in the following manner:

1. Assume values of E_K.
2. This fixes e_{c_1}, since $e_{c_1} = -E_K$.
3. E_{b_1} is also fixed from $E_{b_1} = 200 - E_K$.
4. With a given e_{c_1} and E_{b_1}, a corresponding I_1 can be read from the tube curves.
5. Any assumed E_K also fixes I_K, since

$$I_K = \frac{E_K}{R_{K_1} + R_{K_2}}$$

6. A corresponding I_2 is computed from $I_2 = I_K - I_1$.
7. We also have $e_{c_2} = -I_K R_{K_1}$.

8. A locus of possible operating points for V_2 is plotted by connecting all the I_2, e_{c_2} points. This also represents possible E_{b_2} values.

9. The possible E_{b_2} values must also satisfy the equation $E_{b_2} = 240 - E_K - I_2R_L$. Therefore, plot this equation using the assumed E_K values and corresponding I_2 values from step 6.

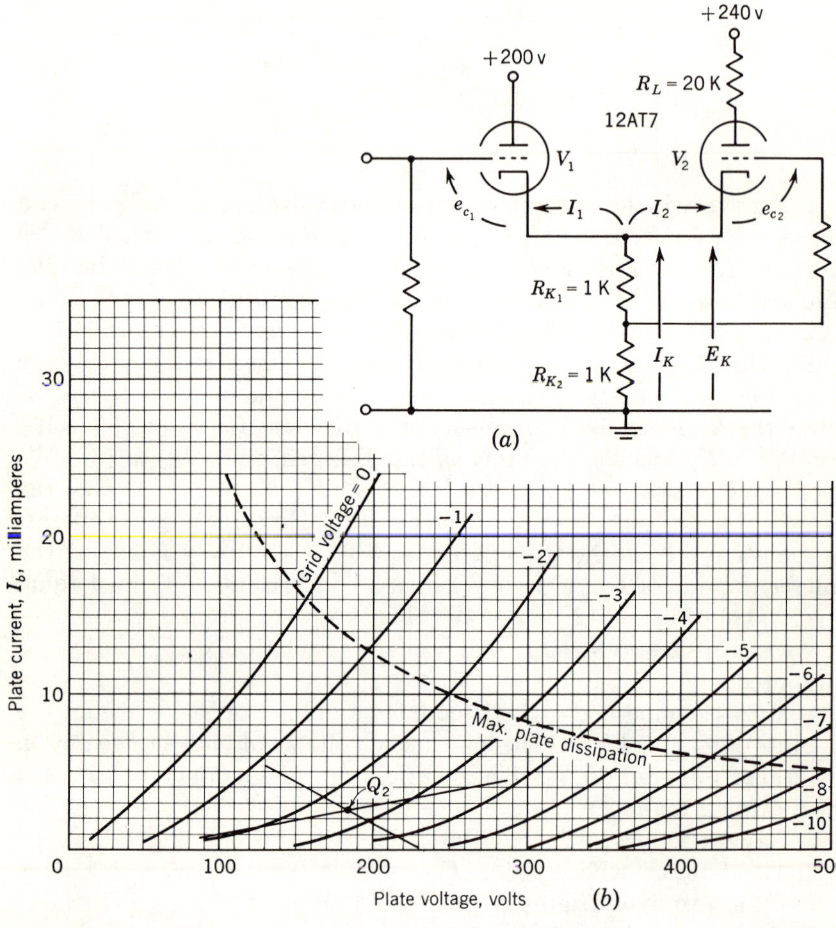

FIG. 5-39. (a) The bias for V_2 is adjusted by tapping the cathode resistor. (b) Graphic analysis of a.

Table 5-6 compiles the results of steps 1 to 9. The graphic constructions are shown in Fig. 5-39b.

For the intersection of the lines generated by steps 8 and 9, we read off the value of E_{b_2} which satisfies both equations. The corresponding current is $I_2 = I_{20} = 2.6$ ma. From the curves we see that the bias on

V_2 is about $e_{c_2} = -2.6$ volts. But e_{c_2} is the drop across R_{K_1}. Therefore, $I_K = e_{c_2}/R_{K_1}$ or 2.6 volts/1 kilohm $= 2.6$ ma. Since $I_1 = I_K - I_2$ or 2.6 ma $-$ 2.6 ma, we see that $I_1 = 0$ and V_1 is cut off. It must be cut off, since $e_{c_1} = -I_K(R_{K_1} + R_{K_2})$ or 2.6 ma(2 kilohms) $= -5.2$ volts, which is just enough to cut the cathode-follower section off with 200 volts on the plate. Actually, the plate voltage is even less, owing to the slight drop across R_{K_1} and R_{K_2}.

TABLE 5-6

1. E_K	0	1	2	3	4	5	6
2. e_{c_1}	0	−1	−2	−3	−4	−5	−6
3. E_{b_1}	200	199	198	197	196	195	194
4. I_1......	23.1	12.9	5.3	1.8	0.4	0	0
5. I_K	0	0.5	1	1.5	2	2.5	3
6. I_2	−23.1	...	...	...	1.6	2.5	3
7. e_{c_2}	0	−0.5	−1	−1.5	−2	−2.5	−3
8. E_{b_2}	...	...	...	...	131	180	222
9. E_{b_2}	...	...	...	...	204	185	174

To approximate the required input to V_1 to cut V_2 off, we shall first assume that the effective plate supply seen by V_2 is $E_{bb} = 240$ volts, instead of 240 volts minus the slight drop across the cathode resistors. From the curves it takes about -5.5 volts of e_{c_2} to cut V_2 off with a 240-volt plate supply. This means that I_K must equal 5.5 volts/1 kilohm $=$ 5.5 ma. Since V_2 is cut off, this means that $I_1 = I_K = 5.5$ ma. With $E_K = (5.5 \text{ ma})(2 \text{ kilohms}) = 11$ volts, E_{b_1} equals $200 - 11 = 189$ volts. With 189 volts for E_{b_1} and 5.5 ma for I_1, the bias on V_1 must be $e_{c_1} \approx -1.3$ volts as estimated from the curves. The required input is then $E_{in} = e_{c_1} + E_K = -1.3 + 11 = 9.7$ volts.

Occasionally one encounters cathode-coupled amplifiers with a load in each plate circuit, as shown in Fig. 5-40*a*. Determine the Q point of V_1 and V_2. This circuit, as well as some others in this section, is not necessarily a desirable one but instead has been specifically designed to serve as an exercise in locating Q points.

We can write the following equations for Fig. 5-40*a*:

$$E_{b_1} = E_{bb_1} - E_K - I_1 R_{L_1} \tag{5-16}$$

$$E_{b_2} = E_{bb_2} - E_K - I_2 R_{L_2} \tag{5-17}$$

$$E_K = (I_1 + I_2)R_K \tag{5-18}$$

$$e_{c_1} = -E_K \tag{5-11}$$

$$e_{c_2} = -E_K \tag{5-19}$$

$$E_{b_1} = f(I_1, e_{c_1}) \tag{5-8}$$

$$E_{b_2} = f(I_2, e_{c_2}) \tag{5-9}$$

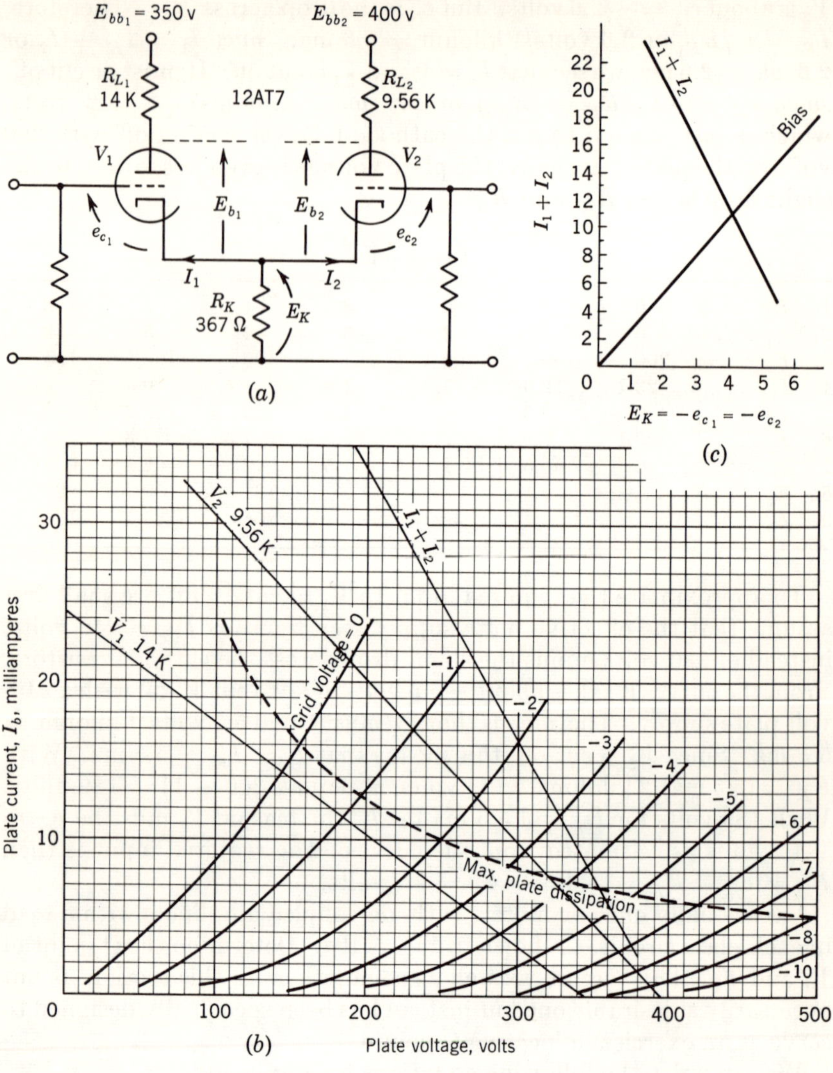

FIG. 5-40. (*a*) Another cathode-coupled amplifier which may also be driven from two separate inputs. (*b* and *c*) Graphic analysis of *a*.

Equations (5-8) and (5-9) represent the tube curves and merely state that the plate voltage across a tube is a function of the plate current and grid voltage.

1. We could start out by assuming an E_K value. This fixes e_{c_1}, e_{c_2}, and $I_1 + I_2$ in Eqs. (5-11), (5-19), and (5-18) respectively.
2. With this E_K draw the load line for Eq. (5-16).

3. Similarly, draw the load line for Eq. (5-17). E_K must be between zero and about -10.5 volts or both tubes will be cut off. The load line constructions for Eqs. (5-16) and (5-17) will hardly differ whether we assume $E_K = 0$ or -10.5 volts. Therefore, let us assume an E_K of 5 volts just for these two load-line constructions.

4. The intersection for the load line for V_1 with the e_{c_1} curves represents possible I_1 values for different assumed E_K values, since $e_{c_1} = -E_K$.

5. Similarly, the intersection of the load line for V_2 with the e_{c_2} curves represents possible I_2 values.

6. A curve of possible $I_1 + I_2$ values is plotted as shown in Fig. 5-40*b* and then transferred to Fig. 5-40*c*.

7. A bias curve using Eq. (5-18) is also plotted on Fig. 5-40*c*. The intersection of the bias curve with the $I_1 + I_2$ curve from steps 3 and 4 represents a point where the value of $I_1 + I_2$ satisfies Eq. (5-18) as well as the others. Therefore, the quiescent value of E_K is about 4.1 volts and 4.1 volts of bias exists on both tubes. If we imagine a bias curve for $e_c = -4.1$ volts, between the $e_c = -4$ and -5 volt curves, the V_2 load line is intersected at $I_2 = I_{20} = 6.8$ ma. The V_1 load line is intersected at about $I_1 = I_{10} = 3.9$ ma.

5-22. A Balanced Amplifier. Figure 5-41*a* illustrates a balanced amplifier. If the grids are excited by equal voltages, but 180° out of phase, and the output is taken between the two plates, the circuit is called a push-pull amplifier. Determine the Q point for each tube in this symmetrical circuit.

The load line of Fig. 5-41*b* is virtually identical for $E_K = 0$ and $E_K = 5$ volts:

1. $E_K = -e_c$ ...	0	1	2	3	4	5	6
2. I_K	0	1	2	3	4	5	6
3. $I_1 = I_2$	2.1	1.6	1	0.4	0.2	0	0
4. $I_1 + I_2$	4.1	2.6	2	0.8	0.4	0	

Inspection of the above data shows that when E_K equals 2 volts, the sum of $I_1 + I_2$ in step 4 gives the same I_K value as in step 2. The quiescent current for each tube is then 1 ma.

5-23. A Clipper. The circuit of Fig. 5-42*a* is sometimes used to convert sine waves to square waves and may be classified as a clipping circuit. Let us analyze the response of the circuit to the input waveform. With E_{in} at zero, the static current and voltage of the tube can be determined from the load-line intersection with the $e_c = 0$ grid-voltage characteristic, as shown in Fig. 5-42*b*. We see that $I_{b0} = 6.1$ ma and $E_{b0} = 78$ volts.

As E_{in} starts swinging positive with respect to ground, the grid is driven slightly positive and the flow of grid current commences. The grid looks like the plate of a diode to the cathode, which for most tubes has about 500 to 1,000 ohms of resistance when conducting. Because

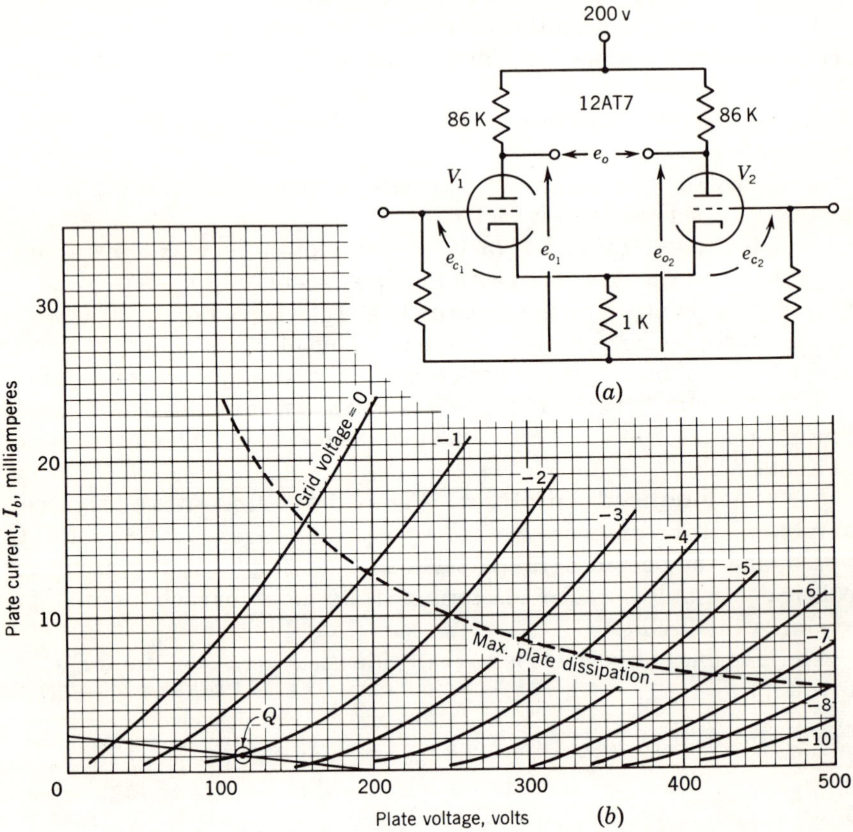

FIG. 5-41. (*a*) A symmetrical circuit which may be used as a push-pull amplifier. (*b*) Graphic analysis of *a*.

the 1-meg resistor is in series with this small diode resistance, virtually all of the positive input signal is dropped across R_g. Therefore, e_c is approximately zero during this interval, as shown in Fig. 5-42*c*. At best the grid current cannot exceed 10 volts/1 meg $= 10\ \mu a$. If 10 μa flows through a 1-kilohm diode resistance, the diode voltage is only 0.01 volt. When the input swings negative, the grid follows the input, since there is no drop across R_g any more.

From the curves we see that when $e_c = -5$ volts, which is the same as e_{in} for the negative swing, the tube is completely cut off. To sketch

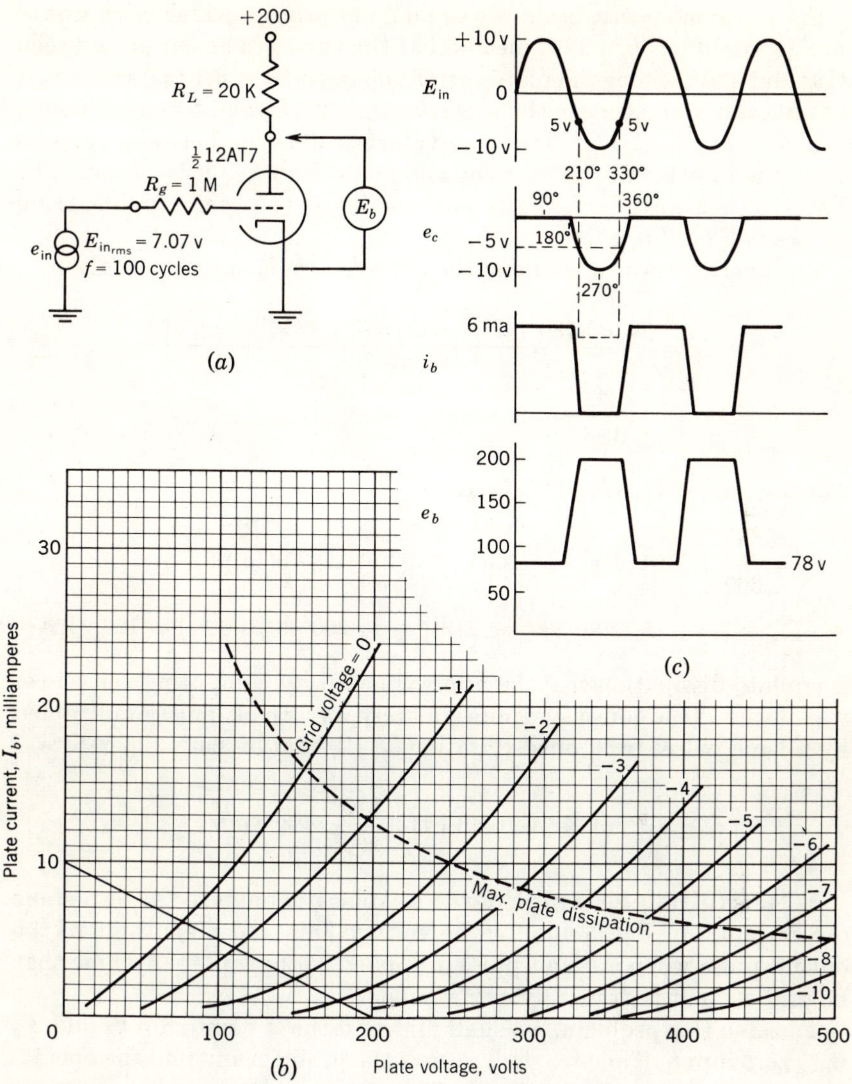

FIG. 5-42. (*a*) Approximately what will the d-c voltmeter E_b read? What is the P_d of the tube? (*b*) Graphic analysis of *a*. (*c*) Waveforms for *a*.

the e_c waveform with a little detail in it, let us determine the time in degrees that it takes e_c to get to −5 volts. The general expression for e_{in} is

$$e_{in} = 10 \sin \omega t = 10 \sin \theta$$

If we let $e_{in} = 5$, we have

$$\theta = \sin^{-1} 5/10 = \sin^{-1} 0.5$$

From trigonometric tables we could find the angle which has a sine of 0.5; it would be 30°. This means that the tube will be cut off between 210° and 330°. Since the plate current changes from 6.1 to 0 ma in just 30°, we can approximate the i_b waveform with straight-line segments, as shown in Fig. 5-42*c*. The e_b waveform will swing between 78 volts when the tube is fully conducting and 200 volts when it is cut off. The plate-voltage waveform shows how the input has been amplified and squared off or clipped.

The average, or d-c, value of the i_b waveform is approximately

$$I_b = \frac{(6\text{ ma})(180°) + \frac{1}{2}(6\text{ ma})(30°) + \frac{1}{2}(6\text{ ma})(30°)}{360°}$$

$$= \frac{1{,}260}{360} = 3.5\text{ ma}$$

The d-c value of e_b is approximately

$$E_b = \frac{1}{360°}[(78)(180°) + \tfrac{1}{2}(200 - 78)(30°) + (78)(30°)(2)$$
$$+ (200)(330° - 210°) + \tfrac{1}{2}(200 - 78)(30°)] = 126\text{ volts}$$

The plate dissipation may be approximated by computing the power waveform. This consists, approximately, of 78-volt pulses multiplied by 6.1-ma pulses for a pulse duration of 210° out of 360°. Therefore,

$$P_d = (78\text{ volts})(6.1\text{ ma})\frac{210°}{360°} = 280\text{ mw}$$

5-24. Multistage Amplifier. Graphically analyze the two-stage d-c amplifier of Fig. 5-43*a* for the voltage gain and magnitude of the output pulse when a 2-volt pulse is applied to the input. Assume that no grid current is drawn by either tube.

To solve this problem, we shall first find the Q points for V_1 and V_2 with zero input. Then we shall rework the problem and find the operating points with the input signal. The output voltage and the voltage gain (A_v) can then be determined. A step-by-step procedure would be as follows:

1. Find the zero signal Q point of V_1. To do this, we must plot the plate load line and the cathode load line, or bias curve, as in Fig. 5-43*c*. The intersection of the bias curve with the load line is the Q point and $I_{b0} = 2.1$ ma. The 2.88- and 2.95-meg resistors are so large compared with 8 kilohms that we can assume that the -300 volt supply is isolated from the plate of V_1. The voltage between plate and ground of V_1 with

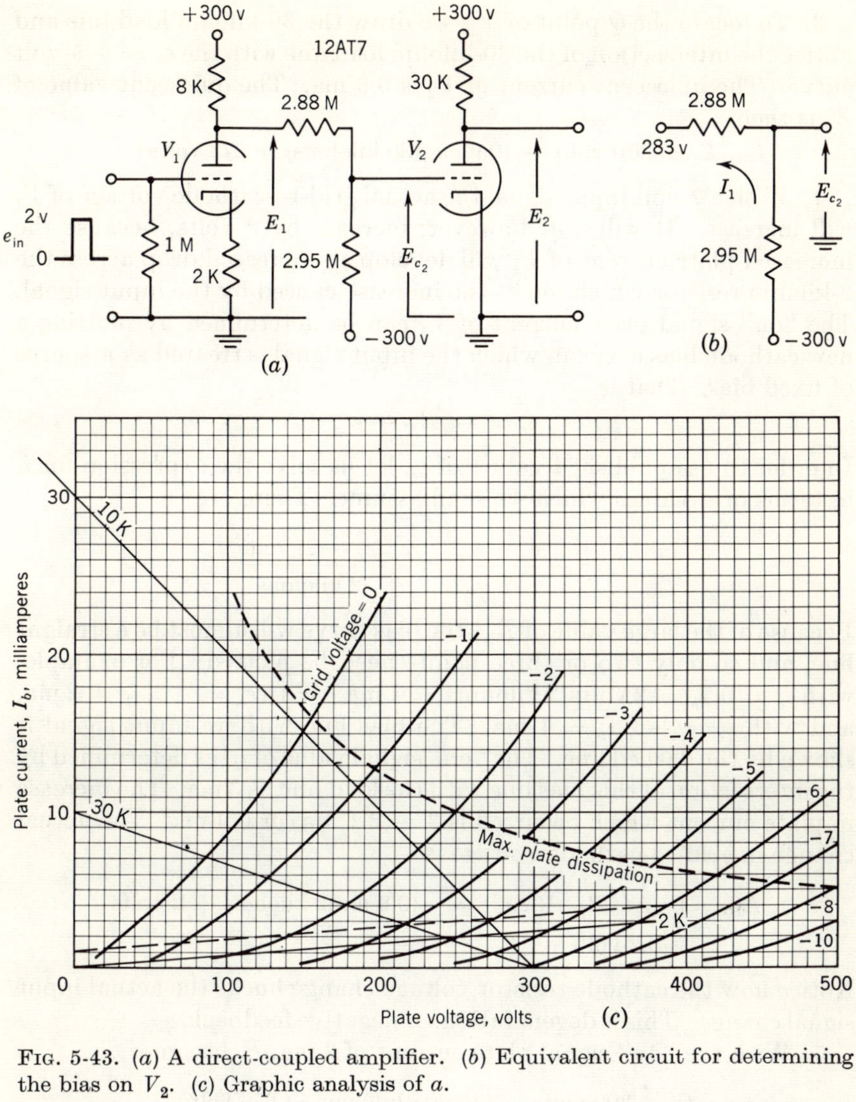

FIG. 5-43. (*a*) A direct-coupled amplifier. (*b*) Equivalent circuit for determining the bias on V_2. (*c*) Graphic analysis of *a*.

zero signal is then

$$E_1 = 300 \text{ volts} - (2.1 \text{ ma})(8 \text{ kilohms}) = 283 \text{ volts}$$

2. The voltage between grid and cathode of V_2 is readily determined by considering the voltage divider of Fig. 5-43*b*. The divider current is

$$I_1 = \frac{[283 - (-300)] \text{ volts}}{(2.88 + 2.95) \text{ meg}} = 100\ \mu\text{a}$$

$$E_{c_2} = -300 \text{ volts} + (100\ \mu\text{a})(2.95 \text{ meg}) = -5 \text{ volts}$$

3. To locate the Q point of V_2, we draw the 30-kilohm load line and notice the intersection of the 30-kilohm load line with the $e_c = -5$-volt curve. The quiescent current of V_2 is 0.9 ma. The quiescent value of E_2 is then

$$E_2 = 300 \text{ volts} - (0.9 \text{ ma})(30 \text{ kilohms}) = 273 \text{ volts}$$

4. With a 2-volt input signal the actual grid-to-cathode voltage of V_1 will increase. It will not, however, increase by 2 volts, because the increased plate current of V_1 will develop an increased drop across the 2-kilohm resistor which bucks the increase caused by the input signal. The "on" signal plate current of V_1 can be determined by plotting a new cathode bias curve in which the input signal is treated as a source of fixed bias. That is,

$$e_c = -i_b R_K + e_{in}$$

In order to use published values of e_c, let us solve the expression for i_b in terms of e_c values which we shall assume. Therefore

$$i_b = -\frac{e_c - e_{in}}{R_K} = -\frac{(e_c - 2) \text{ volts}}{2 \text{ kilohms}}$$

Because of the large value of R_K, the bias curve will almost be a straight line, and so only two or three points need be plotted. For example, with $e_c = 0$, $i_b = 2$ volts/2 kilohms $= 1$ ma. With $e_c = -2$, $i_b = 2$ ma, and with $e_c = -6$, $i_b = 4$ ma. The bias line with an input signal is shown by the dashed line. The "on" signal value of i_b, as determined by the intersection of this bias line with the load line, is 3 ma. The increase in plate current which occurred is $3 - 2.1$ ma, or 0.9 ma. The actual change in grid-to-cathode voltage is

$$\begin{aligned}\Delta e_c &= (-\Delta i_b)R_K + \Delta e_{in} = -(0.9 \text{ ma})2 \text{ kilohms} + 2 \text{ volts} \\ &= +0.2 \text{ volt}\end{aligned}$$

Notice how the cathode resistor voltage change bucks the actual input signal change. This is degeneration, or negative feedback.

5. With the "on" signal plate current of 3 ma, E_1 becomes

$$E_1 = 300 \text{ volts} - (3 \text{ ma})8 \text{ kilohms} = 276 \text{ volts}$$

6. The new value of E_{c_2} as determined in step 2 is

$$I_1 = \frac{276 - (-300) \text{ volts}}{2.88 + 2.95 \text{ meg}} = 98.8\ \mu\text{a}$$

$$E_{c_2} = -300 + (98.8\ \mu\text{a})(2.95 \text{ meg}) = -8 \text{ volts}$$

7. Now locate the "on" signal Q point for V_2 with $E_{c_2} = -8$ volts. From the 30-kilohm load line plot of V_2, we see that with -8 volts V_2

is cut off. Therefore E_2 is 300 volts. The output pulse has changed from a zero signal down level of 273 volts to an "on" signal up level of 300 volts. The voltage gain (A_v) is the ratio of the output level change to the input level change and is

$$A_v = \frac{\Delta E_2}{\Delta e_{in}} = \frac{300 - 273 \text{ volts}}{2 \text{ volts}} = \frac{27}{2} = 13.5$$

Problem. Find e_{c_1}, e_{b_1}, E_{g_2}, e_{c_2}, e_{b_2}, E_{g_3}, E_K, e_{c_3}, e_{b_3}, and I_{dc} in Fig. 5-44*a*. This is not meant to be a practical circuit but is rather intended as an exercise in load line techniques.

Solution. With zero input level, thevenize, looking out of the plate of V_1 to determine load line and effective E_{bb} supply. Assume that e_{c_2} is negative so that no grid current flows in V_2. Later we can thevenize the circuit that V_2 grid looks into and see if e_{c_2} is really negative. If it is not negative, assume an effective diode resistance of 800 ohms between V_2 grid and cathode.

The open-circuit divider current in Fig. 5-44*b* is 400 volts/40 kilohms = 10 ma. Therefore, V_{oc} = 300 volts − 10 ma(5 kilohms) = 250 volts.

$$Z_{th} = \frac{(5)(35)}{5 + 35} = 4.38 \text{ kilohms}$$

We can now draw a load line for V_1 which intersects the e_b axis at 250 volts and the i_b axis at 250 volts/4.38 kilohms, or 57 ma. Since the current intercept lies off the chart, draw the load line through the point 125 volts and 28.5 ma. With $e_{c_1} = 0$, the quiescent value of i_{b_1} is I_{b_1} = 18.1 ma. The $P_{d,\max}$ value is slightly exceeded at this Q point. The value of e_{b_1} is e_{b_1} = 250 volts − (18.1 ma)(4.38 kilohms) = 170.5 volts. To find E_{g_2}, assuming e_{c_2} is negative, solve the simple divider problem of Fig. 5-44*c*:

$$E_{g_2} = -100 + \frac{270.5(15 \text{ kilohms})}{35 \text{ kilohms}} = -100 + 116 = +16 \text{ volts}$$

$$e_{c_2} = -38.2 + 16 = -22.2 \text{ volts}$$

$$e_{b_2} = 300 - 38.2 = 261.8 \text{ volts}$$

Since the cathode of V_2 is returned to +38.2 volts, e_{c_2} will equal −22.2 volts, which cuts V_2 off. Our initial assumption that V_2 draws no grid current is then valid. The load line for V_2 will pass through E_{bb} = 300 − 38.2 = 261.8 volts and a current intercept of 261.8 volts/30 kilohms = 8.7 ma, as shown in Fig. 5-44*d*. With V_2 cut off, E_{g_3} = 300 volts. To determine the Q point of V_3 with zero signal find the Thévenin's equivalent circuit that the cathode of V_3 looks into, as shown in Fig. 5-45. The total load line for V_3 will go through an e_b intercept of e_b = 500 − 120 = 380 volts and an i_b intercept of 380 volts/20 = 19 ma.

To find the actual quiescent point for V_3, we shall have to construct a bias curve by assuming values of e_c and computing the corresponding values of i_b. We start at the cathode of V_3 and work our way around to the grid. Noting all voltages that try to drive the grid negative with a minus sign and all those that try to drive the grid positive with a plus sign, we have from Fig. 5-45,

$$e_c = -i_b R_K - 120 \text{ volts} + 300 \text{ volts} = -i_b R_K + 180 \text{ volts}$$

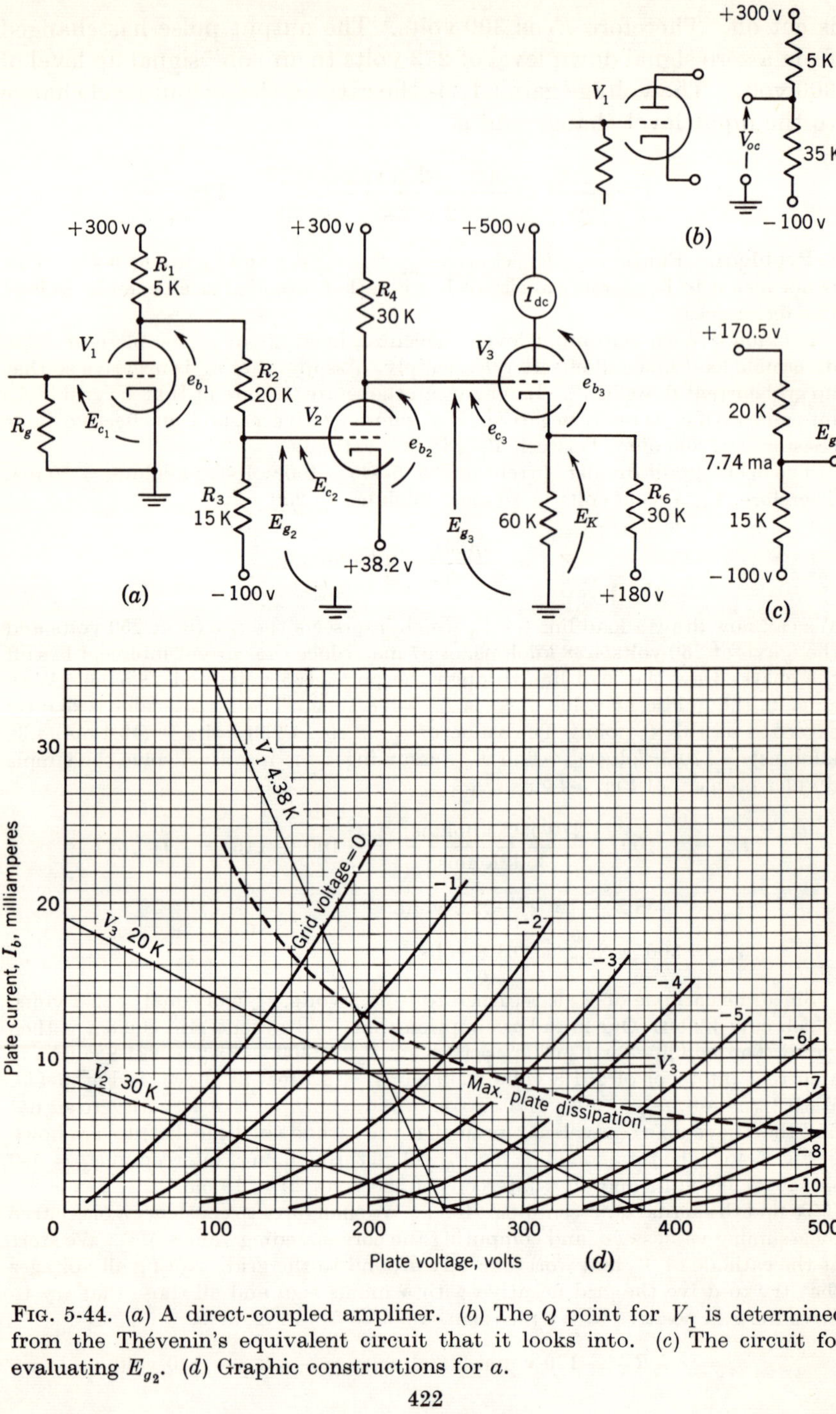

Fig. 5-44. (*a*) A direct-coupled amplifier. (*b*) The Q point for V_1 is determined from the Thévenin's equivalent circuit that it looks into. (*c*) The circuit for evaluating E_{g_2}. (*d*) Graphic constructions for *a*.

or
$$-i_b R_K = e_c - 180 \text{ volts}$$
$$i_b = \frac{180 - e_c}{20 \text{ kilohms}}$$

If we assume $e_c = 0$,
$$i_b = \frac{180 \text{ volts}}{20 \text{ kilohms}} = 9 \text{ ma}$$

If we assume $e_c = -4$,
$$i_b = \frac{184 \text{ volts}}{20 \text{ kilohms}} = 9.2 \text{ ma}$$

The intersection of the bias curve with the load line occurs at a value of $i_b \approx 9.05$ ma. This is the quiescent current of V_3. Therefore, $E_K = +120 + 9.05$ ma (20 kilohms) $= 120 + 181.2 = 301.2$ volts, $e_{b_3} = 500 - 301.2 \approx 198.8$ volts, and $e_{c_3} \approx -301.2 + 300 = -1.2$ volts.

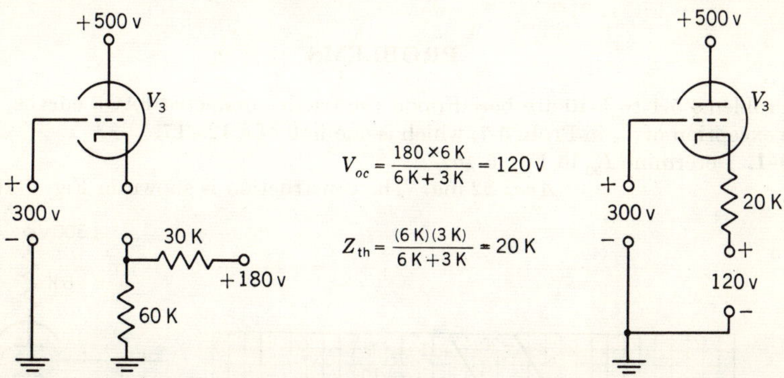

FIG. 5-45. Equivalent circuits used in locating the Q point of V_3.

Summing up the solutions, we have

$$\begin{aligned} e_{c_1} &= 0 & E_{g_3} &= 300 \text{ volts} \\ e_{b_1} &= 170.5 \text{ volts} & E_K &= 301.2 \text{ volts} \\ E_{g_2} &= 16 \text{ volts} & e_{c_3} &= -1.2 \text{ volts} \\ e_{b_2} &= 261.8 \text{ volts} & e_{b_3} &= 198.8 \text{ volts} \\ e_{c_2} &= -22.2 \text{ volts} & I_{dc} &= 9.06 \text{ ma} \end{aligned}$$

SUMMARY

In this chapter we have endeavored to develop techniques suitable for the graphic analysis of vacuum-tube circuits. While a variety of circuits have been considered, the primary intent is to give the reader a feeling for the type of thinking involved. The reader should try to associate each operation as a step in the graphic solution of the circuit equations. Once the Q points are determined, the results can be checked by substituting the I_{b0} values and seeing if the various voltage drops total the E_{bb} supply.

If required, the small-signal parameters, μ, g_m, and r_p, can be evaluated at the operating point. We have discussed the significance of these parameters, their definitions and interrelationship:

$$\mu = \frac{de_b}{de_c}\bigg/ i_b \qquad r_p = \frac{de_b}{di_b}\bigg/ e_c$$

$$g_m = \frac{di_b}{de_c}\bigg/ e_b \qquad \mu = g_m r_p$$

While there is little more that can be said in summarizing this chapter, its importance is in no way diminished. The reader should work every problem, since significant points are occasionally highlighted within a problem that may not have been emphasized in the body of the chapter.

PROBLEMS

Problems 5-1 to 5-10 are based upon the triode connected 6AQ5 curves, with the exception of V_1 in Prob. 5-7, which is one-half of a 12AT7.

5-1. Determine I_{b0} in Fig. 5-46*a*.

Ans. 22 ma. The construction is shown in Fig. 5-46*b*.

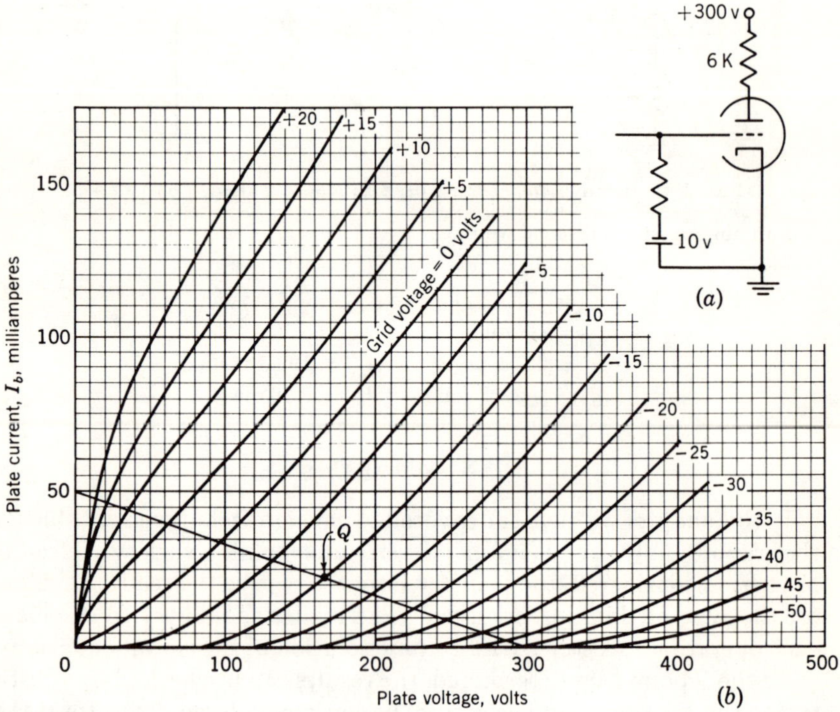

FIG. 5-46. Problem 5-1.

5-2. Determine I_b d-c in Fig. 5-47*a*.

Ans. 13.5 ma. The construction is shown in Fig. 5-47*b*.

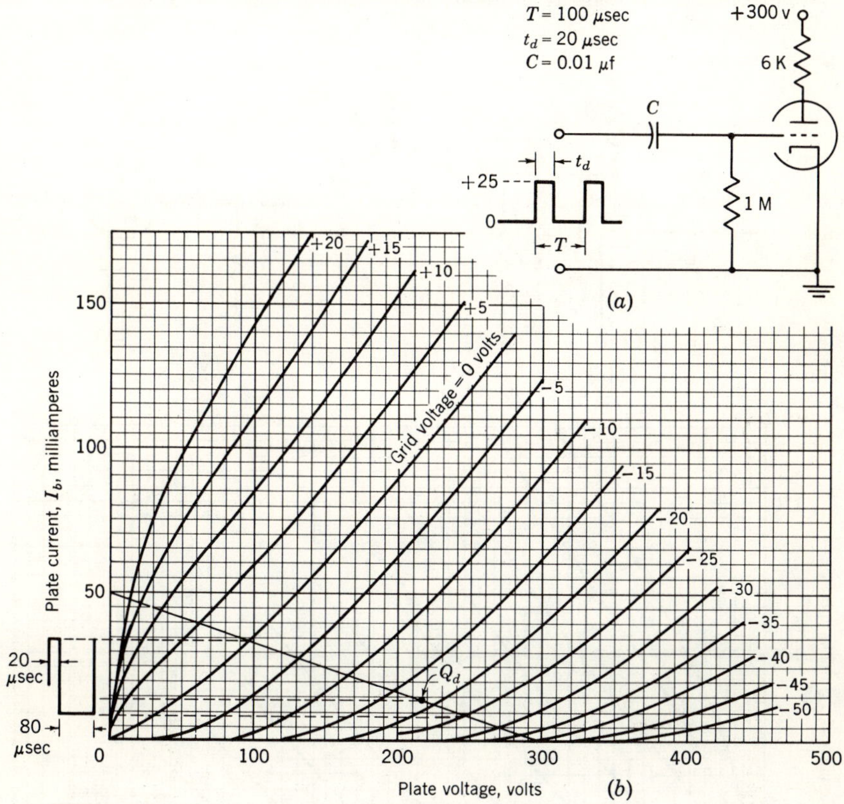

FIG. 5-47. Problem 5-2.

5-3. Determine I_{b0} in Fig. 5-48*a*.

Ans. 0 ma. The circuit of Fig. 5-48*a* can be simplified to that of Fig. 5-48*b*. The construction for Fig. 5-48*b* is shown in Fig. 5-48*c*, which shows that the tube is cut off.

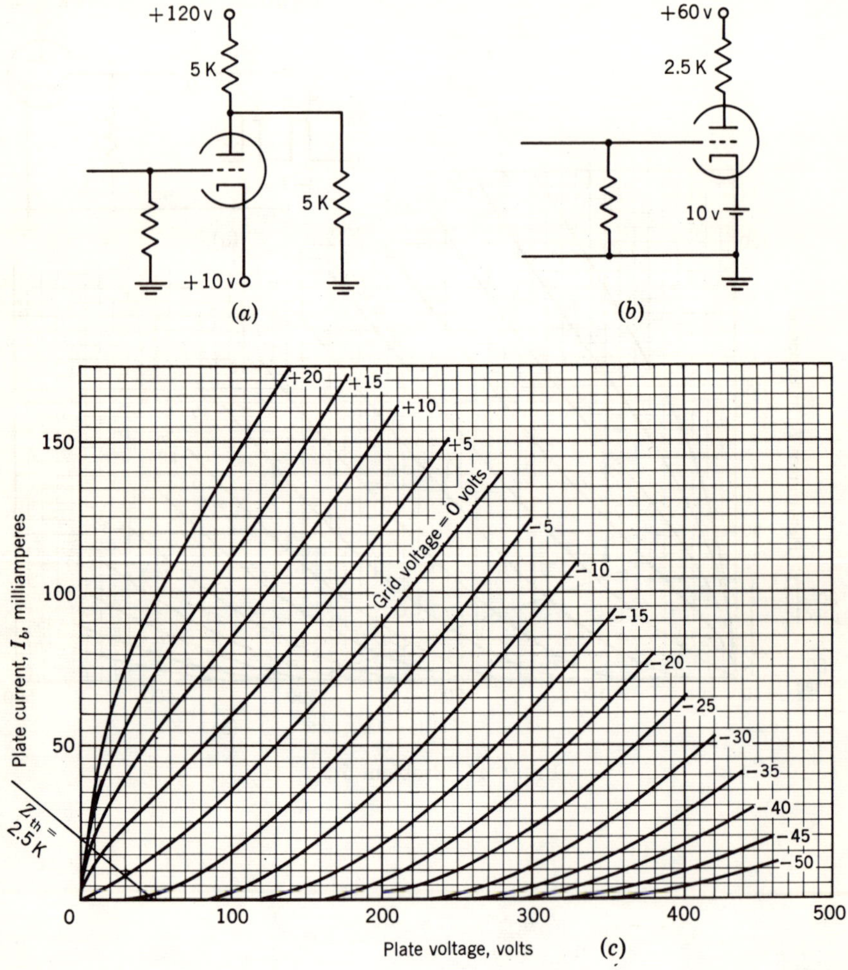

FIG. 5-48. Problem 5-3.

5-4. Determine I_{b0} in Fig. 5-49a.

Ans. 17.5 ma. The construction is shown in Fig. 5-49b.

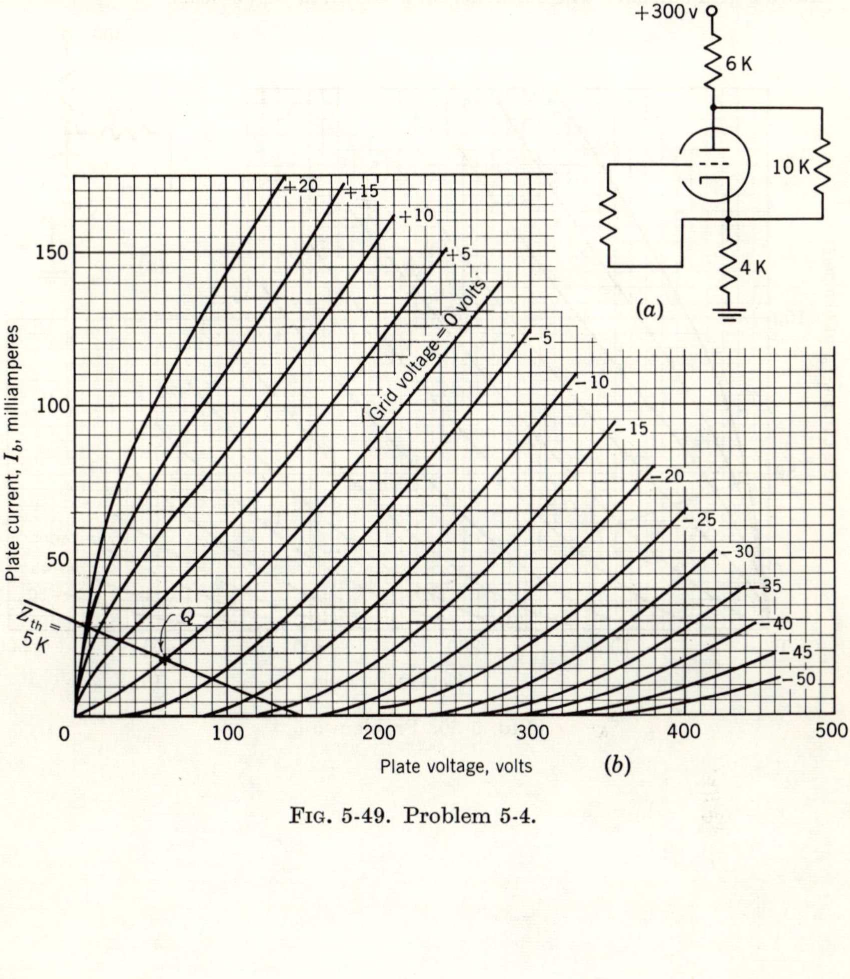

FIG. 5-49. Problem 5-4.

5-5. Determine I_{b0} in Fig. 5-50*a*.

Ans. 34 ma. HINT: Assume 500-ohm grid-to-cathode resistance if grid is drawing grid current. The construction is shown in Fig. 5-50*b*.

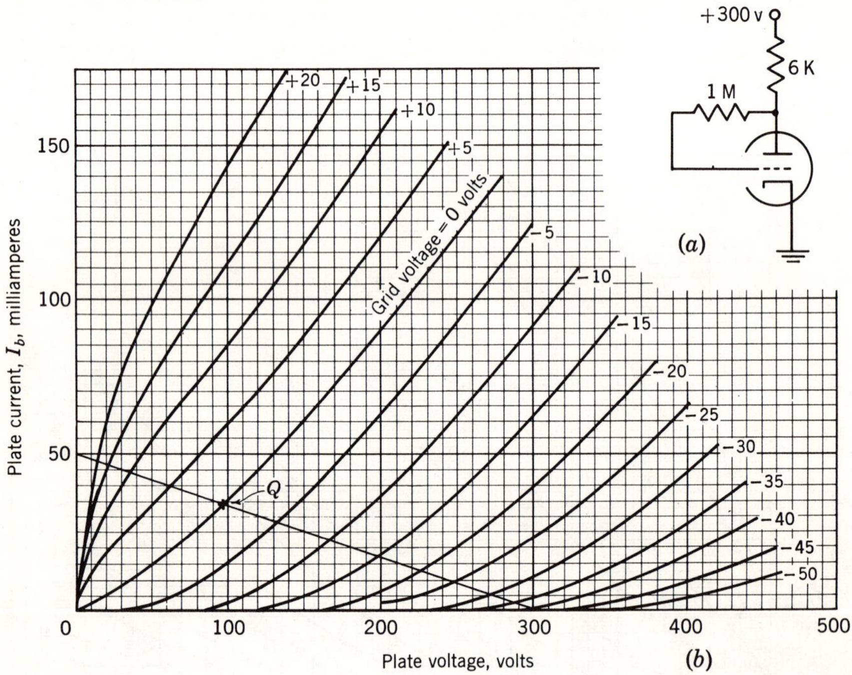

FIG. 5-50. Problem 5-5.

5-6. Determine I_{b0} in Fig. 5-51*a*.

Ans. 13 ma. The construction is shown in Fig. 5-51*b*.

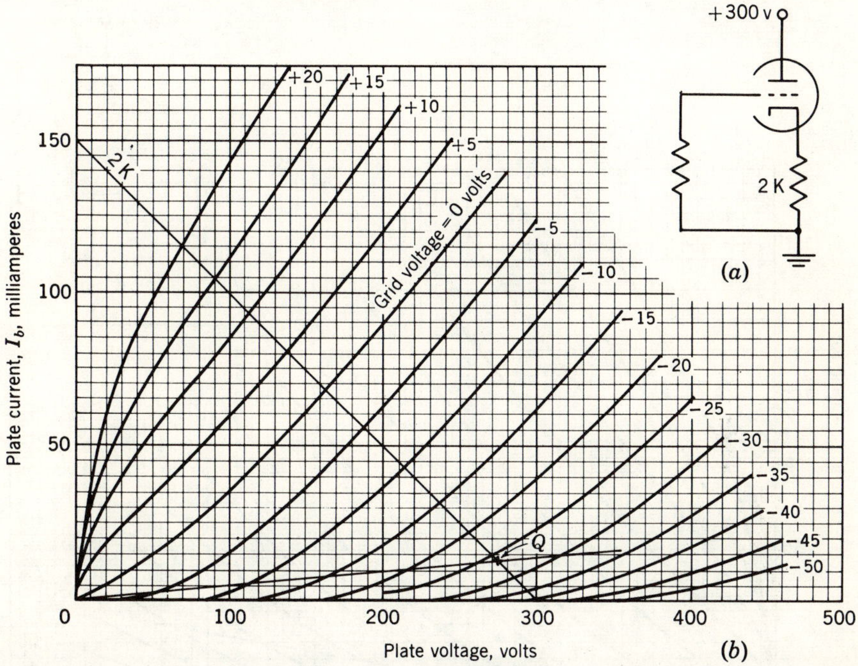

FIG. 5-51. Problem 5-6.

5-7. Determine I_{b0} in Fig. 5-52*a*.

Ans. 13 ma. The construction is shown in Fig. 5-52*b*.

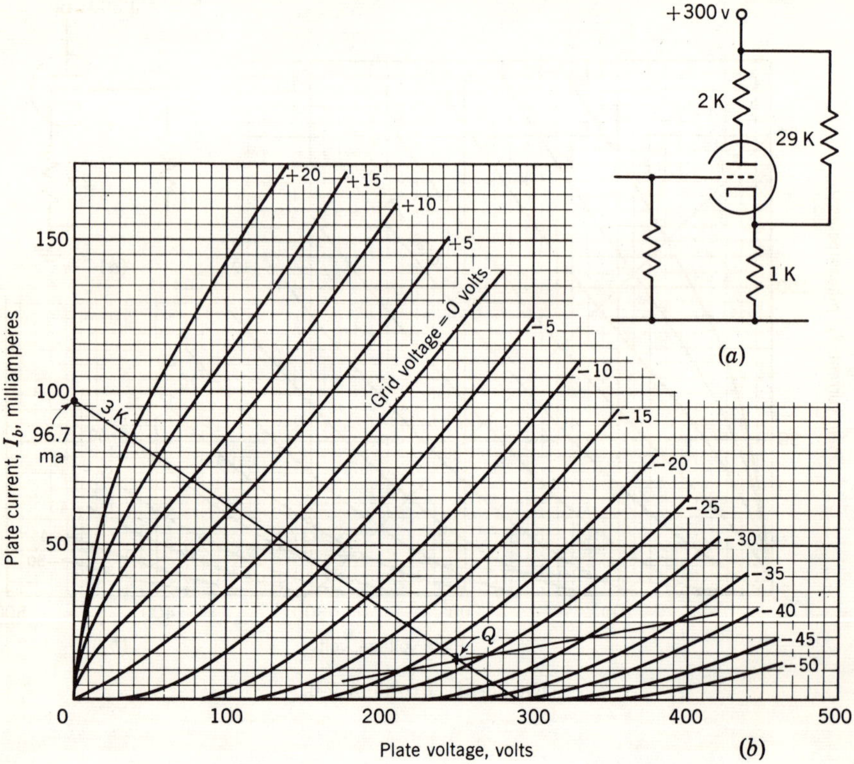

FIG. 5-52. Problem 5-7.

5-8. (*a*) Determine I_{b0} in Fig. 5-53*a* if $E_{in} = 20$ volts. *Ans.* 25 ma.

(*b*) What is the maximum allowable input if the grid-to-cathode voltage is not to exceed zero volts? *Ans.* +87.5 volts.

(*c*) What input voltage is required for zero output volts?

Ans. −14.6 volts. The construction is shown in Fig. 5-53*b*.

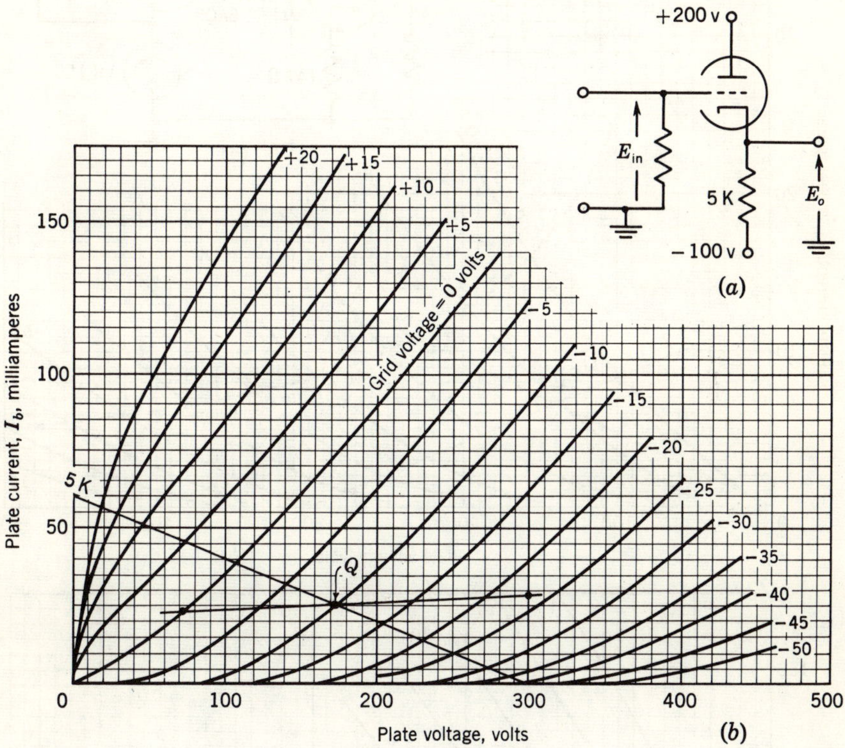

FIG. 5-53. Problem 5-8.

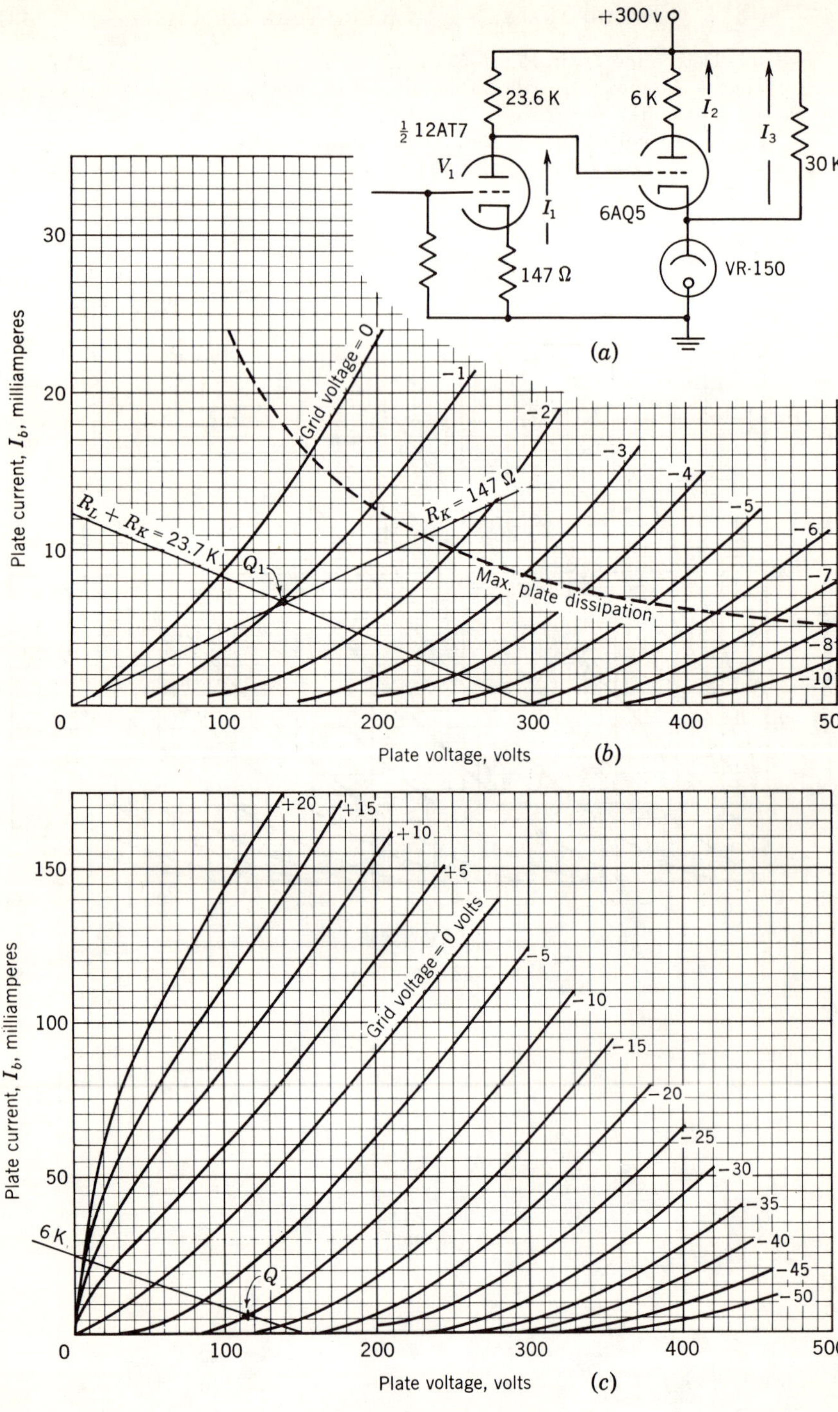

FIG. 5-54. Problem 5-9.

5-9. Determine I_1, I_2, and I_3 in Fig. 5-54*a*.

Ans. $I_1 = 6.8$ ma, $I_2 = 16$ ma, $I_3 = 6$ ma. The constructions for V_1 and V_2 are shown in Fig. 5-54*b* and *c*, respectively.

5-10. (*a*) Determine I_{b0} in Fig. 5-55*a*. *Ans.* 25 ma.

(*b*) What value of E_{in} is required for a grid-to-cathode voltage of zero? *Ans.* 90 volts.

(*c*) What value of E_{in} is required for E_o to equal zero?

Ans. -14.5 volts. The construction is shown in Fig. 5-55*b*. To evaluate part *b*, note that 38 ma of I_b are required (along the load line) for $e_c = 0$. To evaluate part *c*, note that 20 ma must flow through R_K to bring E_o to zero. The required e_c (along the load line) to develop 20 ma through R_K is about -14.5 volts. Therefore, $E_{in} = E_o + e_c = 0 + (-14.5) = -14.5$ volts.

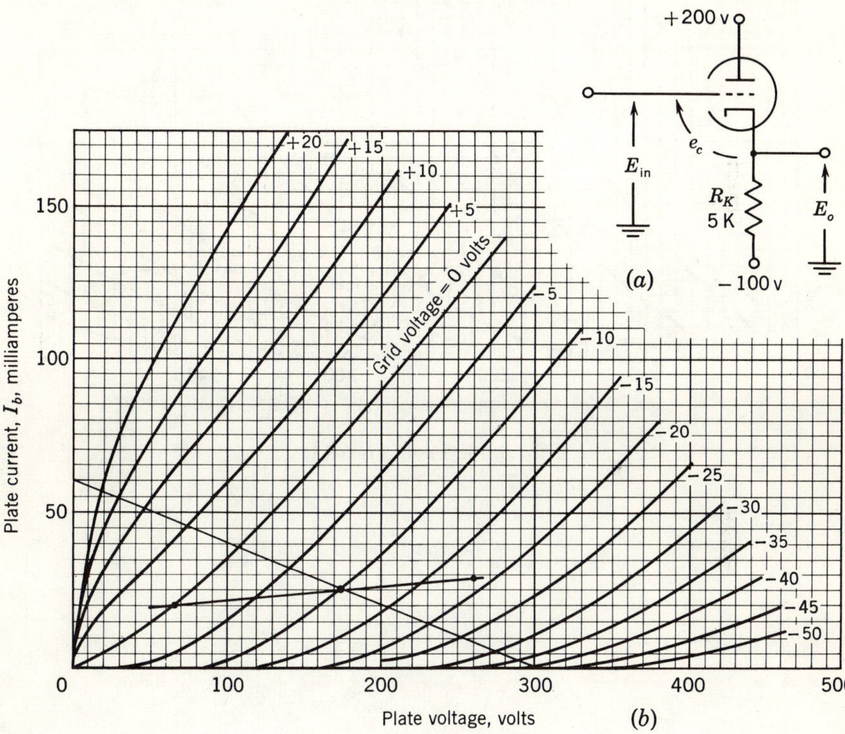

FIG. 5-55. Problem 5-10.

5-11. Determine I_{b0} in Fig. 5-56*a* with the switch in position *a* and then in position *b*.

Ans. (*a*) 6.4 ma; (*b*) 15 ma. The construction is shown in Fig. 5-56*b*.

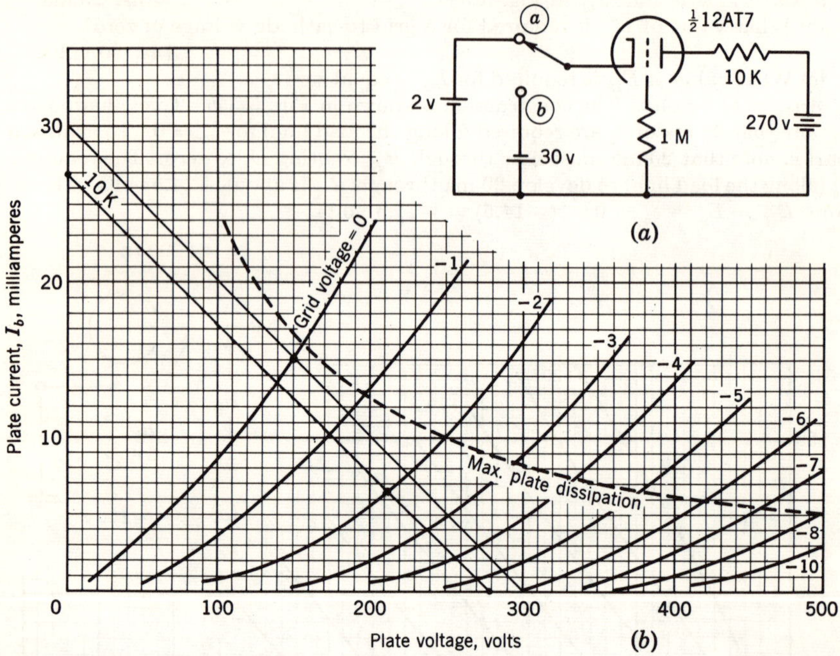

FIG. 5-56. Problem 5-11.

5-12. Determine I_{b0} in Fig. 5-57*a*. *Ans.* 6.3 ma (Fig. 5-57*b*).

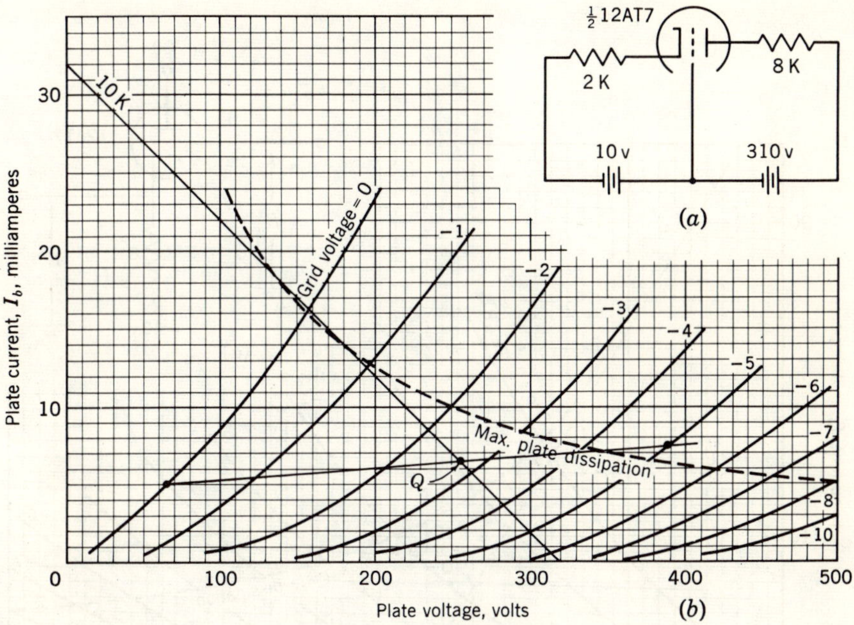

Fig. 5-57. Problem 5-12.

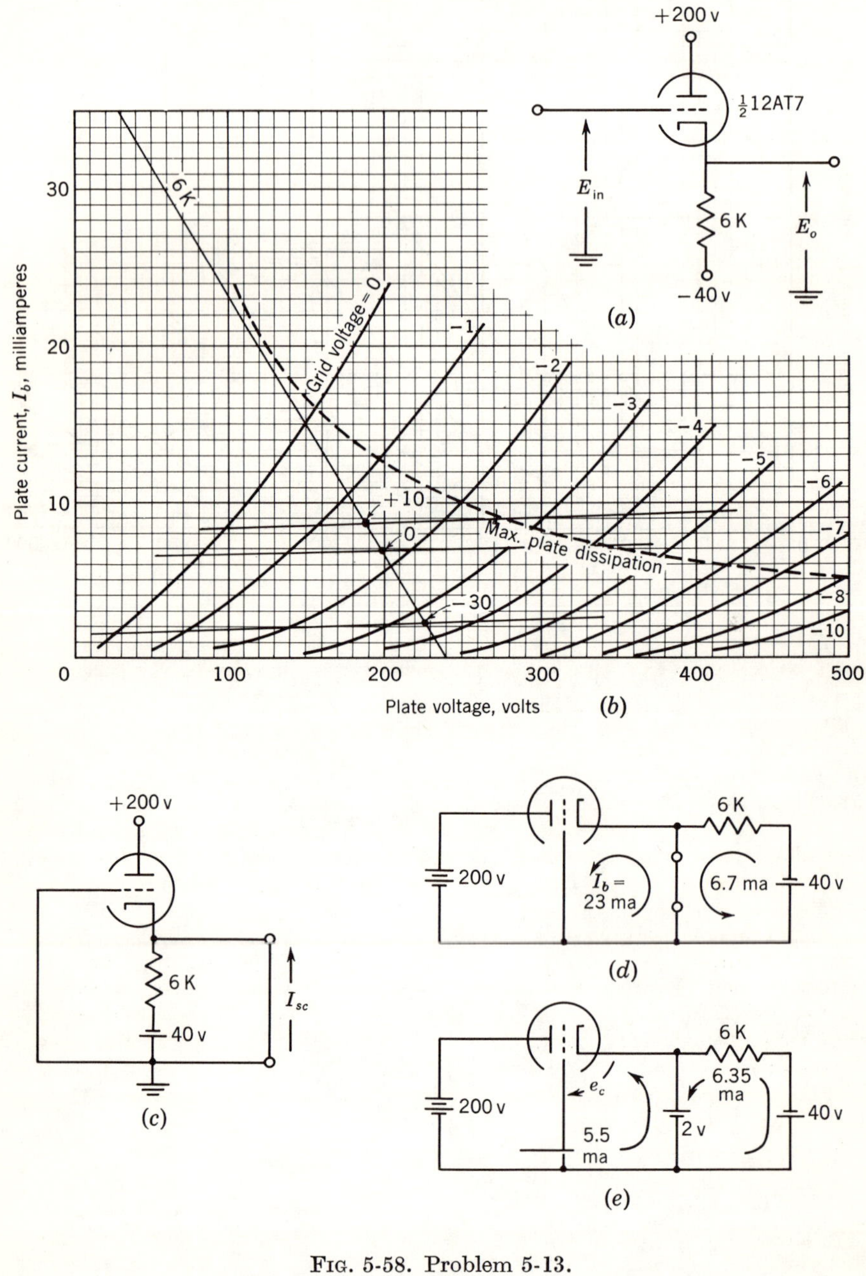

FIG. 5-58. Problem 5-13.

5-13. If the input signal in Fig. 5-58*a* shifts between +10 and −30 volts, determine (*a*) E_o for +10 volts input; (*b*) E_o for −30 volts input; (*c*) the peak-to-peak output swing; (*d*) the voltage gain, A_v; and (*e*) the impedance seen looking into the output terminals. Constructions are shown in Fig. 5-58*b*.

Solution

$$e_c = -I_b(6 \text{ kilohms}) + 40 \text{ volts} + E_{in}$$

$$I_b = -\frac{e_c - 40 - E_{in}}{6 \text{ kilohms}}$$

Assuming e_c values with $E_{in} = +10$,

e_c	0	−5
I_b	8.3	9.2

Ans. (*a*) $I_{b0} = 8.8$ ma, $E_o = 12.8$ volts.

Assuming e_c values with $E_{in} = -30$,

e_c	0	−5
I_b	1.7	2.5

Ans. (*b*) $I_{b0} = 2.2$ ma, $E_o = -26.8$ volts.

$$A_v = \frac{\Delta E_o}{\Delta E_{in}} = \frac{12.8 - (-26.8)}{10 - (-30)} = \frac{39.6}{40} = 0.99$$

To determine the output impedance Z_o we can determine the open-circuit value of E_o with E_{in} held at some convenient value—say, zero volts, since this is between +10 and −30. We can then short-circuit the output terminals and determine I_{sc}. As shown in Chap. 1, $Z_{th} = V_{oc}/I_{sc}$ and Z_{th} is Z_o in this case.

If the bias curve for $E_{in} = 0$ is constructed, it intersects the load line at 6.95 ma. Therefore, the open circuit (unloaded) value of E_o with $E_{in} = 0$ is $E_{o_{oc}} = -40 +$ (6.95 ma)(6 kilohms) = 1.8 volts.

When the output terminals are shorted (cathode tied directly to ground), the circuit appears as shown in Fig. 5-58*c*, which is redrawn as Fig. 5-58*d*. The actual value of I_{sc} is $I_{sc} = 23 - 6.7$ ma $= 16.3$ ma. Therefore, $Z_{th} = 1.8$ volts/16.3 ma = 110 ohms.

An alternative approach to evaluating Z_o would be to hang a fictitious generator across the output terminals and determine what current flows out of the generator into the circuit. For example, assume the generator has zero internal impedance and an instantaneous output of zero volts as in Fig. 5-58*d*. We then see that with zero volts the generator current is 16.3 ma. Now assume that the generator has an instantaneous output of 2 volts as in Fig. 5-58*e*. With $e_c = -2$ volts, $I_b = 5.5$ ma and the net generator current is −0.85 ma. The output impedance of the circuit, which is the input impedance seen by this generator, is

$$Z_o = \frac{\Delta E_o}{\Delta I_o} = \frac{2 - 0 \text{ volts}}{16.3 - (0.85) \text{ ma}} = 117 \text{ ohms}$$

which agrees closely with the previously calculated value of Z_o. Both approaches determine the dynamic output impedance.

5-14. Determine E_o and the load current in Fig. 5-59, which is identical with Fig. 5-58*a*, except that a 6-kilohm load has been put across the output terminals.

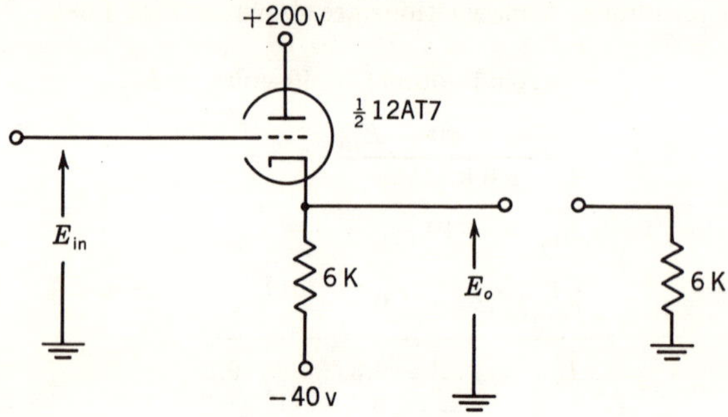

FIG. 5-59. Problem 5-14.

5-15. If a half section of a 12AT7 has only 1 kilohm of resistance in the cathode, with a 300-volt B supply, would the tube burn out?

Ans. No, P_d is only 1.2 watts. The construction is shown in Fig. 5-60.

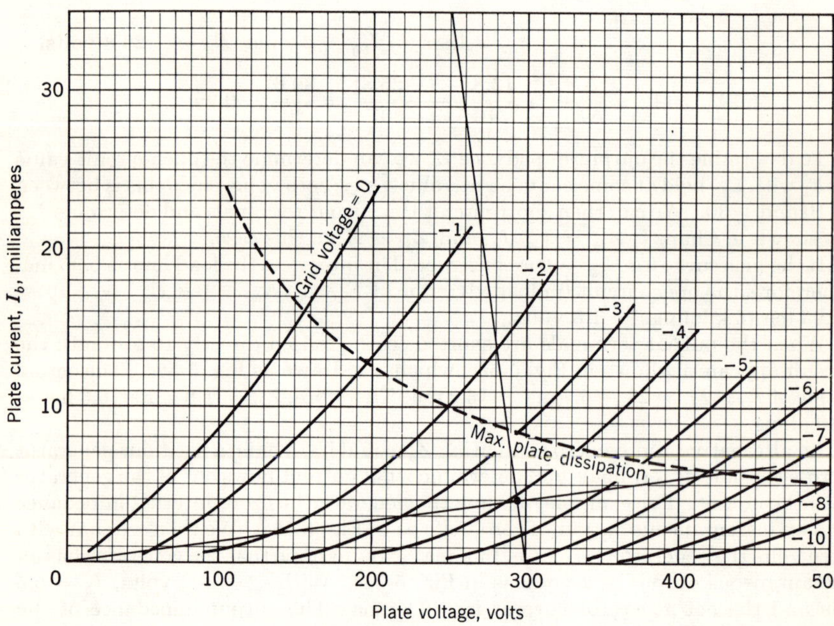

FIG. 5-60. Problem 5-15.

5-16. Design a self-biased amplifier, using ½12AT7. The quiescent operating point of the tube should be at $e_c = -2$ volts, $I_{b0} = 10$ ma. The E_{bb} supply is 300 volts. The circuit should look like Fig. 5-61*a*. Find R_L and R_K.

Solution. First locate the specified Q point on the curves. Then draw a load line for $R_L + R_K$ passing through this point and $E_{bb} = 300$ volts, as shown in Fig. 5-61. Now if $e_c = -2$ volts, $E_K = 2$ volts. Since 10 ma must flow through R_K, we have $R_K = 2$ volts/10 ma $= 200$ ohms. The slope of the $R_L + R_K$ load line indirectly tells us what $R_L + R_K$ is equal to, since the horizontal distance

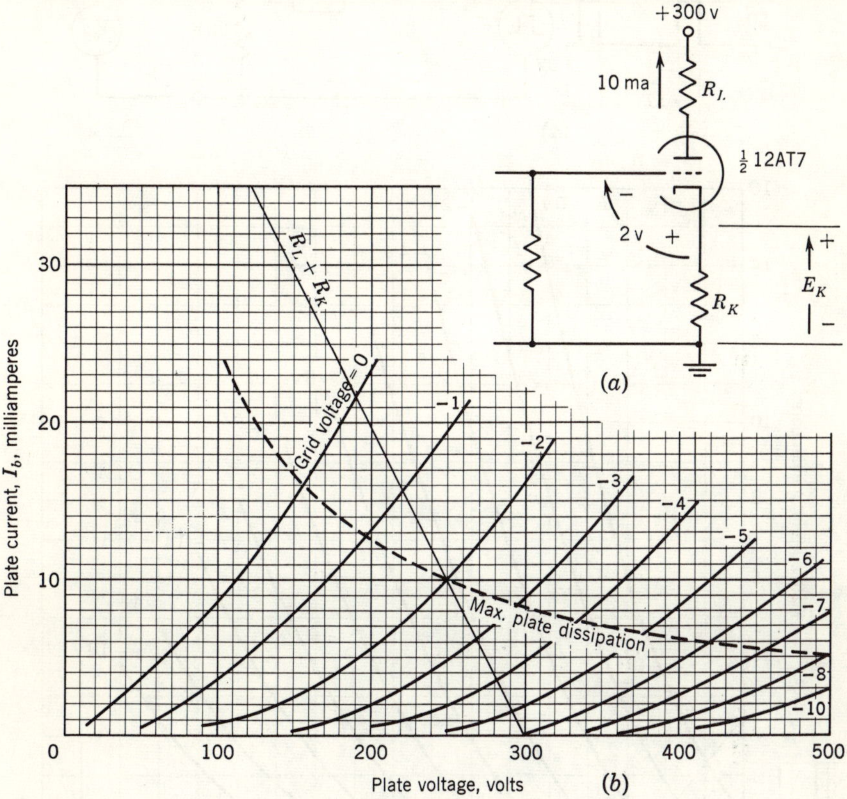

FIG. 5-61. Problem 5-16.

from the vertical line $E_{bb} = 300$ to any point on the load line is the voltage across $R_L + R_K$. We see that, in causing the current through $R_L + R_K$ to increase from zero to 10 ma, the drop across $R_L + R_K$ changed from 300 to 250, or 50 volts. Therefore, $R_L + R_K = 50$ volts/10 ma $= 5$ kilohms. Since we have already determined that $R_K = 200$ ohms, R_L must equal 4,800 ohms. The construction is shown in Fig. 5-61*b*.

5-17. What will the $I_{\text{d-c}}$ and $V_{\text{d-c}}$ meters read when the circuit is driven by the pulse waveform shown in Fig. 5-62*a*? The first thing to do is construct a 20-kilohm load line with the input signal at its peak value of 50 volts. The construction is shown in Fig. 5-62*b*. From the curves we see that the peak current is 1.3 ma. The current waveform is shown in Fig. 5-62*c*.

$$I_{\text{d-c}} = 1.3 \text{ ma} \times \frac{20\ \mu\text{sec}}{100\ \mu\text{sec}} = 0.26 \text{ ma}$$

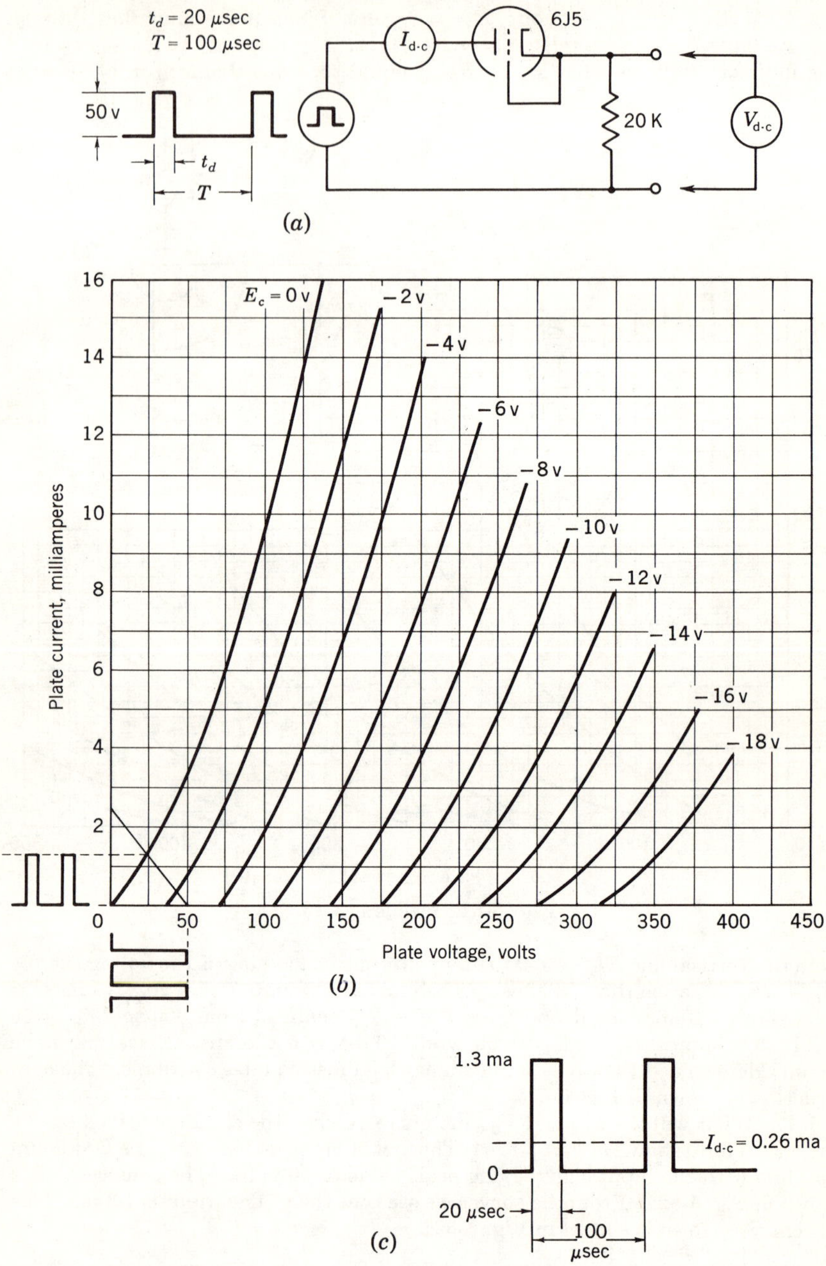

FIG. 5-62. Problem 5-17.

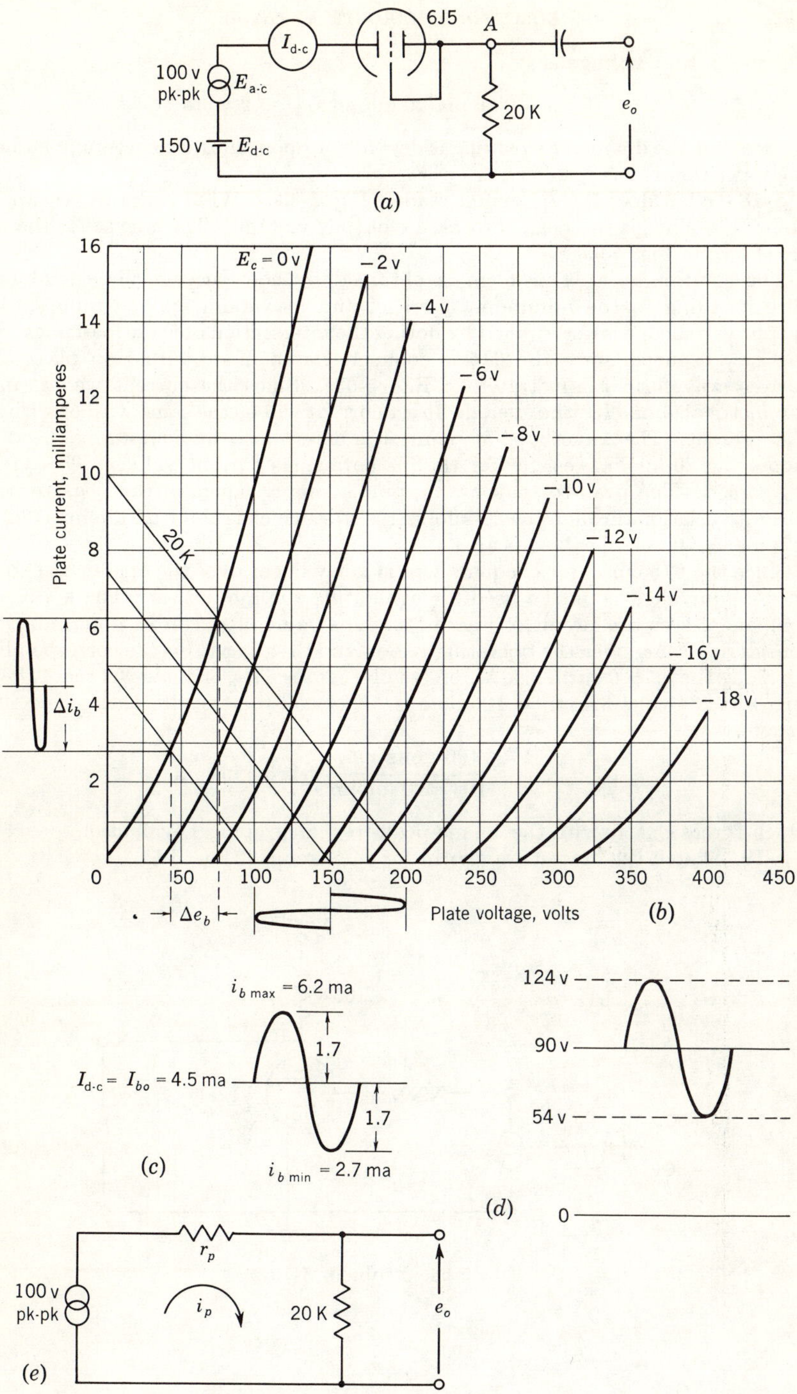

FIG. 5-63. Problem 5-18.

The d-c output voltage is

$$V_{\text{d-c}} = 0.26 \text{ ma}(20 \text{ kilohms}) = 5.2 \text{ volts}$$

Notice that the d-c meter's reading is dependent upon the duty cycle only and not the repetition rate.

5-18. What does the $I_{\text{d-c}}$ meter read in Fig. 5-63*a*? What is the waveshape at point *A*? What is the peak-to-peak a-c output voltage? The curves for the 6J5 are shown in Fig. 5-63*b*.

The solution to this problem is obtained by constructing three load lines corresponding to the minimum, average, and maximum applied inputs. The output waveshape is determined by noting the intersection of these load lines with the $e_c = 0$ curve, since the 6J5 is connected as a diode with zero bias. The current waveform is illustrated in Fig. 5-63*c*. Since the current waveform is symmetrical about I_{b0}, the average direct current will be the same with or without $E_{\text{a-c}}$ present. The waveshape at point *A* is determined by computing the drop across the 20-kilohm resistor for the instantaneous current values. This yields Fig. 5-63*d*. The peak-to-peak a-c output is simply equal to the peak-to-peak current variation through the 20-kilohm resistor, which is about 20 kilohms (6.2 − 2.7) ma = 70 volts peak-to-peak.

Since the 6J5 curve for e_c equals zero is fairly linear over the operating region, an a-c equivalent could be used for evaluating the a-c output. The a-c signal component sees the circuit of Fig. 5-63*e*. The value of r_p can be determined by considering a Δe_b over the operating region and dividing it by the corresponding Δi_b along the $e_c = 0$ curve. A Δe_b of 34 volts causes a Δi_b of about 3.4 ma. Therefore, $r_p = 34 \text{ volts}/3.4 \text{ ma} \approx 10$ kilohms. The peak-to-peak alternating current is therefore

$$i_{pk\text{-}pk} = \frac{100 \text{ volts } pk\text{-}pk}{10 + 20 \text{ kilohms}} = 3.3 \text{ ma}$$

which agrees closely with the 3.4 ma peak-to-peak graphical evaluation.

5-19. Sketch the output waveform for the circuit of Fig. 5-64.

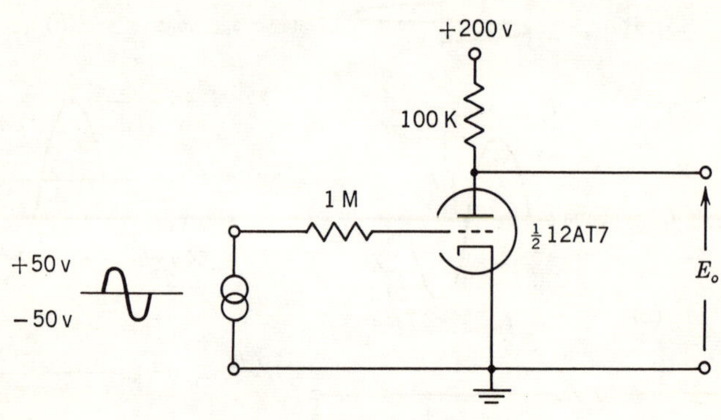

FIG. 5-64. Problem 5-19.

5-20. Sketch the output waveform for the circuit of Fig. 5-65.

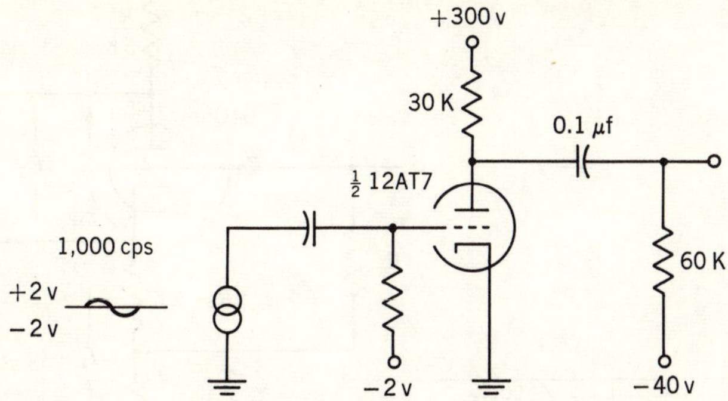

FIG. 5-65. Problem 5-20.

5-21. Determine I_{b0} in Fig. 5-66*a* for the triode-connected 6AQ5.

Solution. There are several ways of solving this problem, one of which is to write the basic equations:

$$E_b = E_{bb} - I_1(R_1 + R_2 + R_3) \qquad (5\text{-}20)$$

$$e_c = -I_1R_2 \qquad (5\text{-}21)$$

$$I_1 = I_b + I_4 \qquad (5\text{-}22)$$

$$I_4 = \frac{E_b}{R_4} \qquad (5\text{-}23)$$

We cannot readily plot (5-20) on the tube curves, because I_1 is not the same as I_b. We can, however, wrestle (5-20) into a function of I_b by substituting (5-23) into (5-22) and the result into (5-20). That is,

$$I_1 = I_b + \frac{E_b}{R_4} \qquad (5\text{-}22a)$$

$$\begin{aligned} E_b &= E_{bb} - \left(I_b + \frac{E_b}{R_4}\right)(R_1 + R_2 + R_3) \\ &= \frac{E_{bb}}{1 + [(R_1 + R_2 + R_3)/R_4]} - I_b\frac{R_1 + R_2 + R_3}{1 + [(R_1 + R_2 + R_3)/R_4]} \\ &= \frac{360 \text{ volts}}{1 + (9 \text{ kilohms}/30 \text{ kilohms})} - I_b\frac{9 \text{ kilohms}}{1 + (9 \text{ kilohms}/30 \text{ kilohms})} \\ &= 277 \text{ volts} - (6.92 \text{ kilohms})I_b \end{aligned} \qquad (5\text{-}20a)$$

The final form of (5-20*a*) can be plotted as a straight line on the tube curves. The construction is shown in Fig. 5-66*b*. This is exactly the same result obtained by thevenizing out of the tube terminals, since $V_{oc} = 277$ volts and $Z_{th} = 6.92$ kilohms.

Substituting (5-22*a*) into (5-21) gives

$$e_c = -R_2\left(I_b + \frac{E_b}{R_4}\right) \qquad (5\text{-}21a)$$

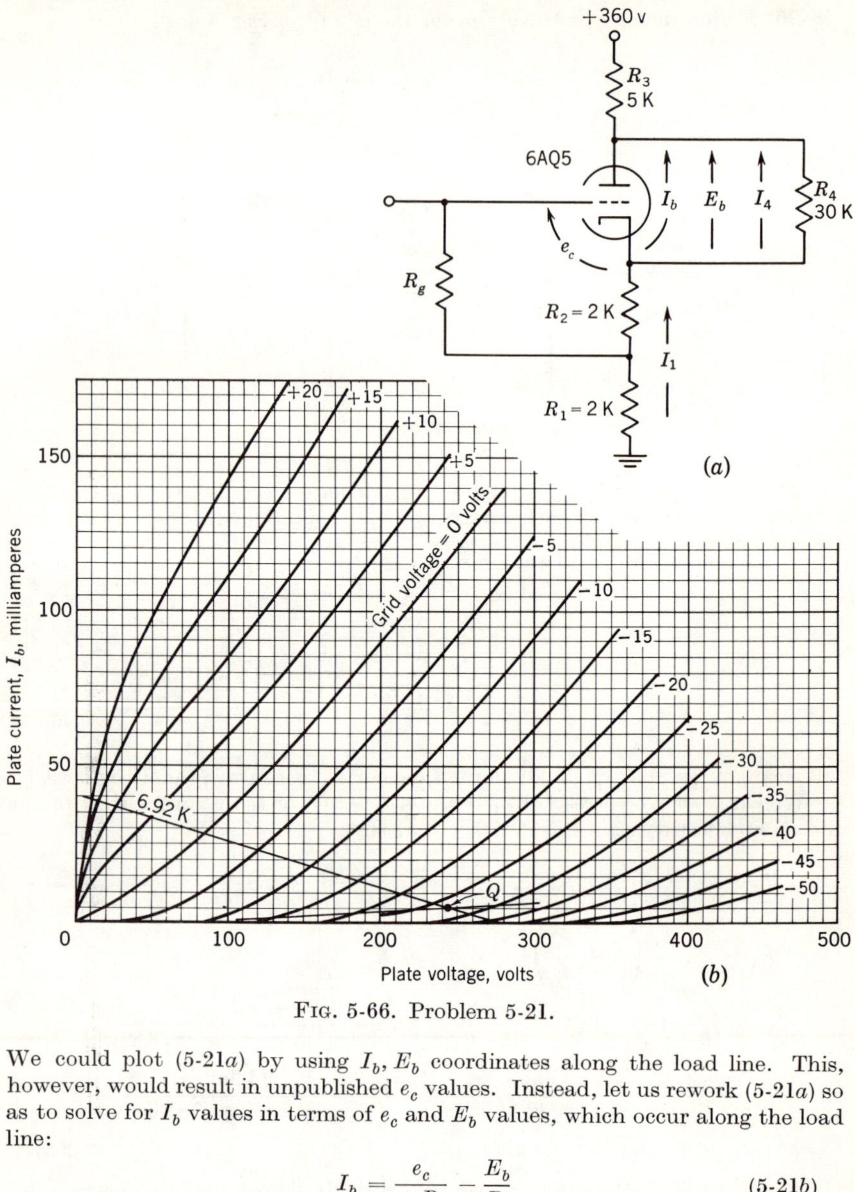

FIG. 5-66. Problem 5-21.

We could plot (5-21*a*) by using I_b, E_b coordinates along the load line. This, however, would result in unpublished e_c values. Instead, let us rework (5-21*a*) so as to solve for I_b values in terms of e_c and E_b values, which occur along the load line:

$$I_b = \frac{e_c}{-R_2} - \frac{E_b}{R_4} \tag{5-21b}$$

e_c	0	−20	−10	−15	−30
E_b	86	212	155	185	260
I_b	...	2.9	−0.2	1.3	6.3

If we plot the bias curve of (5-21b), it intersects the load line at $I_{b0} = 5$ ma and $E_{b0} = 242$ volts. We can check the result as follows:

If $E_{b0} = 242$ volts, $I_4 = 242$ volts/30 kilohms $= 8.1$ ma. Therefore

$$I_1 = I_{b0} + I_4 = 5 + 8.1 \text{ ma} = 13.1 \text{ ma}$$

$$\begin{aligned} E_{bb} &= I_1(R_1 + R_2 + R_3) + E_{b0} \\ &= 13.1 \text{ ma}(9 \text{ kilohms}) + 242 = 359.9 \text{ volts} \end{aligned}$$

which closely checks the actual E_{bb} of 360 volts.

5-22. Determine I_{b0} in Fig. 5-67a.

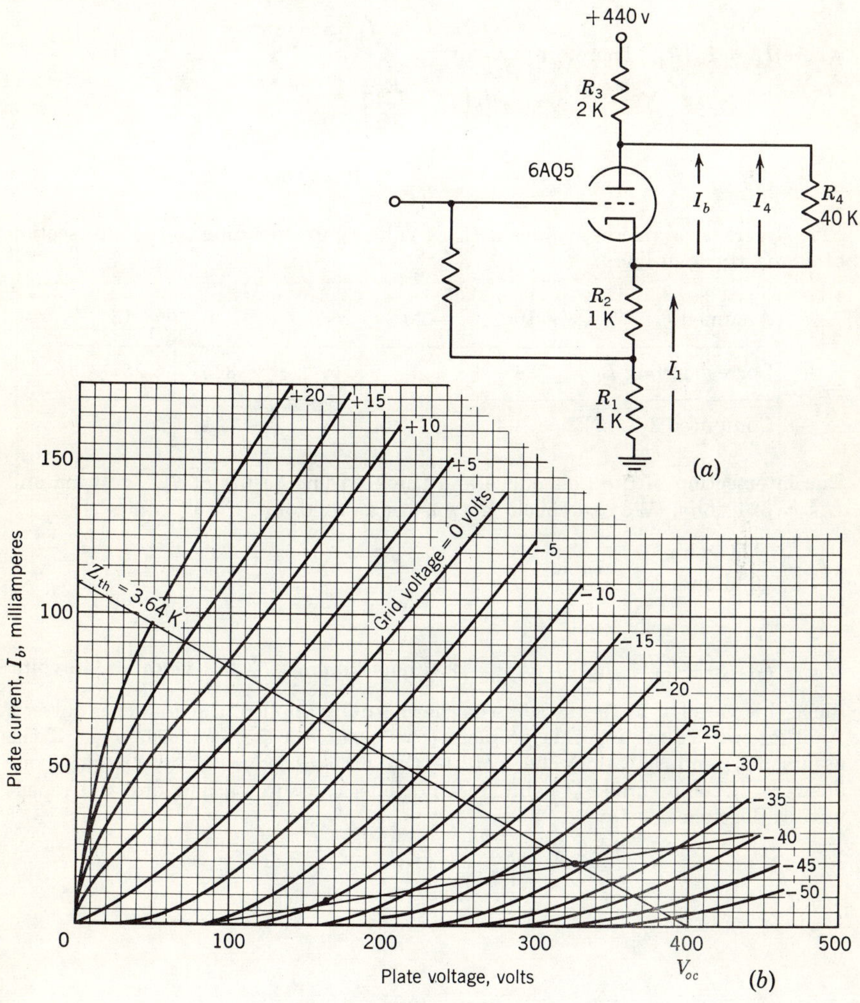

FIG. 5-67. Problem 5-22.

To construct a load line, thevenize looking out of the plate and cathode terminals.

$$V_{oc} = 440 \text{ volts}\left(\frac{40 \text{ kilohms}}{44 \text{ kilohms}}\right) = 400 \text{ volts}$$

$$Z_{th} = \frac{(40)(4)}{44} = 3.64 \text{ kilohms}$$

$$I_{sc} = \frac{400 \text{ volts}}{3.64 \text{ kilohms}} = 110 \text{ ma}$$

Plot the load line as shown in Fig. 5-67*b*.

$$e_c = -I_1R_2$$

$$I_1 = I_b + I_4$$

where $I_4 = E_b/R_4$. Therefore

$$e_c = -R_2\left(I_b + \frac{E_b}{R_4}\right)$$

$$I_b = \frac{e_c}{-R_2} - \frac{E_b}{R_4} \qquad \text{(bias curve)}$$

Plot I_b versus assumed e_c values with E_b values corresponding to the intersection of e_c and the load line:

Assumed e_c	−10	−20	−30	−40	−15
Corresponding E_b	224	...	...	365	254
Computed I_b	4.4	...	...	31	8.6

The intersection of the bias curve and the load line occurs at $I_{b0} = 21$ ma and $E_{b0} = 324$ volts. We can check this answer as follows:

$$I_4 = \frac{324 \text{ volts}}{40 \text{ kilohms}} = 8.1 \text{ ma}$$

$$I_1 = I_{b0} + I_4 = 29.1 \text{ ma}$$

$$E_{bb} = I_1(R_1 + R_2 + R_3) + E_{b0} = 29.1 \text{ ma}(4 \text{ kilohms}) + 324 \text{ volts} = 440 \text{ volts}$$

Since this value of E_{bb} is the same as that actually applied, the answer is correct. Another way of solving this problem is to assume a Q point. Then see if the assumed I_{b0} and E_{b0} values check out by the same procedure we used to check our answer. If the assumed Q point does not check, assume a new one and repeat until it does check.

BIBLIOGRAPHY

"Alternating-current Circuits," 3d ed., by Russell M. Kerchner and George F. Corcoran, John Wiley & Sons, Inc., New York, 1951.

"Alternating-current Circuits," by K. Y. Tang, International Textbook Company, Scranton, Pa., 1940.

"Analysis of Electric Circuits," by William H. Middendorf, John Wiley & Sons, Inc., 1956.

"Basic Electric Circuit Theory," by Walter W. Lewis and Clarence F. Goodhart, The Ronald Press Company, New York, 1958.

"Basic Mathematics for Electronics," 2d ed., by Nelson M. Cooke, McGraw-Hill Book Company, Inc., New York, 1960.

"Basic Mathematics for Radio and Electronics," 3d ed., by F. M. Colebrook and J. W. Mead, Philosophical Library, Inc., New York, 1957.

"Calculus for Electronics," by A. E. Richmond, McGraw-Hill Book Company, Inc., New York, 1958.

"Calculus for the Practical Man," by J. E. Thompson, D. Van Nostrand Company, Inc., Princeton, N.J., 1931.

"Communication Engineering," 3d ed., by W. L. Everitt and G. E. Anner, McGraw-Hill Book Company, Inc., New York, 1956.

"Electrical Engineering Circuits," by Hugh Hildreth Skilling, John Wiley & Sons, Inc., New York, 1957.

"Electrical Network Calculations," by W. E. Richardson, D. Van Nostrand Company, Inc., Princeton, N.J., 1946.

"Electronic Transformers and Circuits," 2d ed., by Reuben Lee, John Wiley & Sons, Inc., New York, 1955.

"Engineering Electronics," by John D. Ryder, McGraw-Hill Book Company, Inc., New York, 1957.

"Essentials of Radio," by Morris Slurzberg and William Osterheld, McGraw-Hill Book Company, Inc., New York, 1948.

"General Network Analysis," by Wilbur R. LePage and Samuel Seeley, McGraw-Hill Book Company, Inc., New York, 1952.

"Introduction to Circuit Analysis," by Abner R. Knight and Gilbert H. Fett, Harper & Brothers, New York, 1943.

"Introduction to Electrical Engineering," 3d ed., by George V. Mueller, McGraw-Hill Book Company, Inc., New York, 1957.

"Mathematical and Physical Principles of Engineering Analysis," by Walter C. Johnson, McGraw-Hill Book Company, Inc., New York, 1944.

"Mathematics for Communication Engineers," by S. J. Cotton, The Macmillan Company, New York, 1959.

"Mathematics for Electronics with Applications," by Henry M. Nodelman and Frederick W. Smith, Jr., McGraw-Hill Book Company, Inc., New York, 1956.

"Mathematics for Radio Engineers," by Leonard Mautner, Pitman Publishing Corporation, New York, 1947.

"Networks, Lines and Fields," by John D. Ryder, Prentice-Hall, Inc., Englewood Cliffs, N.J., 1949.

"Principles of Electric and Magnetic Circuits," 2d ed., by Warren B. Boast, Harper & Brothers, New York, 1957.

"Principles of Radar," 3d ed., by J. Francis Reintjes and Godfrey T. Coate, McGraw-Hill Book Company, Inc., New York, 1952.

"Pulse Techniques," by Sidney Moskowitz and Joseph Racker, Prentice-Hall, Inc., Englewood Cliffs, N.J., 1951.

"Recurrent Electrical Transients," by L. W. Von Tersch and A. W. Swago, Prentice-Hall, Inc., Englewood Cliffs, N.J., 1953.

"Servomechanisms and Regulating System Design," by Harold Chestnut and Robert W. Mayer, John Wiley & Sons, Inc., New York, 1951.

"Servomechanism Practice," by William R. Ahrendt, McGraw-Hill Book Company, Inc., New York, 1954.

"The Algebra of Electronics," by Chester H. Page, D. Van Nostrand Company, Inc., Princeton, N.J., 1958.

"Transformation Calculus and Electrical Transients," by Stanford Goldman, Prentice-Hall, Inc., Englewood Cliffs, N.J., 1949.

"Transient Electric Currents," 2d ed., by Hugh H. Skilling, McGraw-Hill Book Company, Inc., New York, 1952.

"Transients in Electrical Circuits," by Gladwyn V. Lago and Donald L. Waidelich, The Ronald Press Company, New York, 1958.

INDEX